Basic Transport Phenomena in Biomedical Engineering

Third Edition

Basic Transport Phenomena in Biomedical Engineering

Third Edition

Ronald L. Fournier

The University of Toledo
Ohio, USA

CRC Press
Taylor & Francis Group
Boca Raton   London   New York

CRC Press is an imprint of the
Taylor & Francis Group, an **informa** business

CRC Press
Taylor & Francis Group
6000 Broken Sound Parkway NW, Suite 300
Boca Raton, FL 33487-2742

Printed in the United States of America on acid-free paper
Version Date: 20110427

International Standard Book Number: 978-1-4398-2670-6 (Paperback)

Library of Congress Cataloging-in-Publication Data

Fournier, Ronald L.
    Basic transport phenomena in biomedical engineering / by Ronald L. Fournier. -- 3rd ed.
       p. ; cm.
    Includes bibliographical references and index.
    ISBN 978-1-4398-2670-6 (hardcover : alk. paper)
    1. Biological transport. 2. Biomedical engineering. 3. Biotechnology. I. Title.
    [DNLM: 1. Biological Transport. 2. Biomedical Engineering. 3. Blood--metabolism. 4. Body Fluids. 5. Cell Membrane--metabolism. QU 120]

R857.B52F68 2012
571.6′4--dc22

2011011806

**Visit the Taylor & Francis Web site at
http://www.taylorandfrancis.com**

**and the CRC Press Web site at
http://www.crcpress.com**

In fond memory of my good friend and former student,
James P. Byers, PhD (1961–2007)

Contents

Preface

The challenge in presenting a text on biomedical engineering is the fact that it is a very broad field that encompasses a variety of different engineering disciplines and life sciences. There are many recent edited works and advanced textbooks that provide specialized treatments and summaries of current research in various areas of biomedical engineering and transport phenomena. These works tend to be written more for the biomedical researcher and the advanced student than for the beginning student. Thus, there is a need for an entry-level book that introduces some of the basic concepts. This book is designed to meet this goal. This work brings together fundamental engineering and life science principles to provide a focused coverage of key momentum and mass transport concepts in biomedical engineering.

In accomplishing this task within a finite volume, the coverage was limited to those areas that emphasize chemical and physical transport processes with applications toward the development of artificial organs, bioartificial organs, controlled drug delivery systems, and tissue engineering. With this focus, the book first provides a basic review of units and dimensions, some tips for solving engineering problems, and a discussion on material balances. This is followed by a review of thermodynamic concepts with an emphasis on the properties of solutions. The remaining chapters focus on such topics as the body fluids, osmosis and membrane filtration, the physical and flow properties of blood, solute transport, oxygen transport, and pharmacokinetic analysis. This is followed by application of these principles to extracorporeal devices and the relatively new areas of tissue engineering and bioartificial organs. Throughout the book, considerable emphasis is placed on developing a quantitative understanding of the underlying physical, chemical, and biological phenomena. Therefore, mathematical models are developed using the conceptually simple "shell balance" or compartmental approaches.

Numerous examples are presented based on these mathematical models, and they are compared in many cases with actual experimental data from the research literature. The student is encouraged to rework these examples using the mathematical software package of their choice. Working through the examples and the end of chapter problems using mathematical software packages will develop skill and expertise in engineering problem solving, while at the same time making the mathematics less formidable. These mathematical software packages also provide students with the opportunity to explore various aspects of the solution on their own, or apply these techniques as starting points for the solution to their own problems. In this way, the student should be able to gain confidence in the development of mathematical models for relatively simple systems and then be able to apply these concepts to a wide variety of problems of even greater complexity.

I would like to take this opportunity to thank my student Whitney Young for the many hours she spent editing this edition and for the solutions she provided to many of the end of chapter problems. Also, many thanks to Denise Turk and Wiona Porath for their assistance in many tasks related to the preparation of this text. I would also like to thank Michael King, University of Rochester; Guillermo Aguilar, University of California, Riverside; Harold Knickle, University of Rhode Island; George Pins, Worcester Polytechnic Institute; H. Wally Wu, Texas A&M University; and Michael Deem, Rice University; for their constructive criticism of an earlier edition of this book.

It is hoped that this book is timely and useful to engineers and researchers in the biomedical community, as well as to students in chemical engineering, mechanical engineering, biotechnology, bioengineering, and the life sciences.

Ronald L. Fournier

Author

Ronald L. Fournier is a professor in the Department of Bioengineering at The University of Toledo in Ohio. He was also the founding chair of this department. During his twenty-six years at Toledo, he has taught a variety of chemical engineering and bioengineering subjects, including courses in biochemical engineering, biomedical engineering transport phenomena, design, biomechanics, and artificial organs. His research interests and scholarly publications are in the areas of bioartificial organs, tissue engineering, novel bioreactors, photodynamic therapy, and pharmacokinetics.

Notation

a	Molecular radius
a_f	Radius of a macromolecule
A	Helmholtz free energy defined as $U - TS$
A	Area
A	Total amount of a drug in the body
A_E	Total amount of a drug in the extracellular fluid
A_P	Total amount of a drug in the plasma
A_R	Total amount of a drug everywhere else in the body
A_P	Total cross-sectional area of pores in a membrane
A_S	Surface area
A_{xs}	Cross-sectional area of a reactor
$\text{AUC}^{0 \to \infty}$	Area under the concentration time curve
$\bar{B}, \bar{C}, \bar{D}$ and B, C, D	Virial coefficients
c	Weight concentration
C	Concentration
C	Total concentration of drug in the apparent distribution volume
C_{50}	Concentration where drug response is 50% of maximum response
C_b	Concentration in blood
C_b'	Concentration of oxygen bound to hemoglobin
C_d	Concentration in exchange fluid
C_g	Concentration in gas
C_G	Concentration of glucose
C_{CAPD}	Concentration of a solute in exchange fluid
C_E	Total concentration of drug in extracellular fluid
C_{EB}	Concentration of bound drug in extracellular fluid
C_L	Ligand concentration
$C_{L \cdot M}$	Ligand-macromolecule concentration
C_{max}	Maximum concentration
C_M	Macromolecule concentration
C_P, C_{plasma}	Concentration in plasma
C_{tissue}	Concentration in tissue
C_R	Total concentration of drug everywhere else in the body
C_{SS}	Steady state concentration
C_U	Unbound concentration of a drug
C'	Concentration of oxy-hemoglobin
C_{SAT}'	Concentration of oxygen-saturated hemoglobin
$\bar{C}$	Concentration in the tissue
CL	Clearance
CL_{CAPD}	Clearance of continuous ambulatory peritoneal dialysis
CL_D	Clearance of dialysis
CL_{plasma}	Plasma clearance
CL_{renal}	Clearance due to the kidneys
C_P	Heat capacity at constant pressure
C_{total}	Total concentration of a drug, bound and unbound
C_V	Heat capacity at constant volume
ΔC_{LM}	Log mean concentration difference
d	Diameter
D	Diameter
D	Diffusivity
D	Dilution rate

D	Dose of a drug
D_0	Diffusivity in interstitial fluid
Da	Dahmkohler number
D_B	Dialysance
D_C	Capillary diameter
D_{cell}	Diffusivity in the cell
D_e	Effective diffusivity
$D_{effective}$	Effective diffusivity in blood
D_H	Hydraulic diameter
D_m	Membrane pore diffusivity
D_{plasma}	Diffusivity in plasma
D_T	Diffusivity in tissue
E	Enzyme concentration
E	Extraction factor
E	Extraction ratio
E	Solute extraction factor
E_K	Kinetic energy
E_P	Gravitational potential energy
f	Fraction of drug absorbable
f	Fraction of macromolecule bound to ligand
f	Friction factor
f	Fugacity
f_i	Fugacity of pure component i
$\hat{f}_i$	Fugacity of component i in a mixture
f_R	Cumulative fraction of the drug released
f_U	Fraction of unbound drug
f_{UT}	Fraction of unbound drug in the tissue
F	Faraday's constant
F	Feed flow rate
F	Force
g	Acceleration of gravity
g	Membrane conductance
G	Gibbs free energy defined as $H - TS$
$\bar{G}_i$	Partial molar Gibbs free energy species i
G_B	Glucose concentration in plasma
G^E	Excess Gibbs free energy
$\bar{G}_i^E$	Partial molar excess Gibbs free energy
GFR	Glomerular filtration rate
G_I	Glucose concentration in the implant chamber
h	Height, thickness, or capillary rise height
$h_{friction}$	Frictional effects affecting the flow of a fluid
H	Enthalpy defined as $U + PV$
H	Hematocrit
H_C	Hematocrit of core
H_F	Hematocrit of feed
H_T	Hematocrit of the tube
H	Henry's constant
ΔH^m	Enthalpy of fusion
ΔH^{VAP}	Heat of vaporization
i	Refers to the current flow of a particular component
I	Current
I_B	Plasma insulin concentration
I_I	Insulin concentration in the implant chamber
I_{IF}	Interstitial fluid insulin concentration
I_0	Drug infusion rate

j_S	Diffusion flux
J_S	Solute diffusion rate
J_V	Filtration rate of solute relative to the solvent
k_a	Absorption rate constant
k_b	Mass transfer coefficient for blood
k_{cat}	Enzyme rate constant
k_e	Mass transfer coefficient for exchange fluid
k_g	Mass transfer coefficient for a gas
k_i	Elimination rate constant
k_m	Mass transfer coefficient
$\bar{k}_m$	Length averaged mass transfer coefficient
$k_{metabolic}$	First order rate constant for metabolic drug consumption
k_{renal}	Renal elimination rate constant
k_{te}	Total elimination rate constant
K	Partition coefficient
K_a	Drug–protein affinity constant
$K_{fitting}$	Fitting friction factor
K_i	Distribution coefficient for component i
K_m	Constant in Michaelis–Menten model
$K_{O/W}$	Octanol-water partition coefficient
K_0	Overall mass transfer coefficient
K_S	Monod constant
L	Latent heat of a phase change
L	Length
L	Liquid flow rate
L_P	Hydraulic conductance of a membrane
m	Slope of the oxygen–hemoglobin dissociation curve
$\dot{m}$	Mass flow rate
M	Molar property value
M^E	Excess property value
M_i	Molar property value of component i
M_i	Mass transfer rate of species i across a dialysis membrane
$\bar{M}_i$	Partial molar property value of component i
M_{urine}	Mass of drug in the urine
MW	Molecular weight
n	Constant in Hill equation
n	Number of moles
n_i	Number of moles of component i
N	Number of equilibrium stages
N_A	Avogadro's number
N_{fiber}	Number of fibers in a hollow fiber unit
N_S	Solute diffusion rate across a permeable membrane
N_T	Number of transfer units
N_V	Volumetric gas transport rate
P	Absolute pressure
pCO_2	Partial pressure of carbon dioxide
pH	$-\log_{10}$ (hydrogen ion concentration)
pK	$-\log_{10}$ (dissociation equilibrium constant)
pO_2	Partial pressure of oxygen
$\langle pO_2 \rangle$	Average partial pressure of oxygen
P	Power
P, P'	Product concentration
P_{50}	Constant in the Hill equation
P_c	Permeability of the capillary wall to oxygen
P_C	Hydrodynamic pressure in the capillary

P_C	Critical pressure
Pe	Peclet number
P_i	Partial pressure of component i
P_{IF}	Hydrodynamic pressure of interstitial fluid
P_m	Membrane permeability
P^{SAT}	Saturation or vapor pressure
P_{SC}	Permeability of the stratus corneum
ΔP	Effective pressure drop
q	Filtration flux
q_b	Tissue blood perfusion rate
Q	Blood flow rate
Q	Filtration rate
Q	Heat
Q	Volumetric flow rate of a fluid
Q_b	Volumetric flow rate of blood
Q_d	Volumetric flow rate of exchange fluid
Q_f	Filtration volumetric flow rate
Q_g	Volumetric flow rate of a gas
$Q_{capillary}$	Volumetric flow rate in a capillary
r	Ratio of ion concentration inside to that outside
r	Radius of a bubble or droplet
r_{anoxic}	Radius defining anoxic region in the Krogh tissue cylinder
r_C	Capillary radius
r_{islet}	Islet insulin release rate
r_S	Rate of substrate consumption in an enzyme reaction
$r_{metabolic}$	Metabolic rate of drug consumption
r_T	Krogh tissue cylinder radius
R	Universal gas constant
R	Radius
$R(\bar{C})$	Volumetric reaction rate
R_0	Zero-order volumetric reaction rate
Re	Reynolds number
Re_x	Local Reynolds number at x
$R_{E/I}$	Ratio of the amount of drug-binding protein in extracellular fluid to that in plasma
R_{max}	Saturation drug response at high drug concentration
s	Parameter in the Casson equation
s	Capillary surface area per volume of tissue
S	Entropy
S	Substrate concentration
S	Surface area of a membrane
S_a	Sieving coefficient
S_c	Capillary surface area
S_0	Sieving coefficient at high filtration rates
Sc	Schmidt number
Sh	Sherwood number
SSE	Sum of the square of the errors
t	Temperature in °C or °F
t	Time
$t_{1/2}$	Half-life
t_m	Membrane thickness
T	Absolute temperature in K or R
T_C	Critical temperature
T_f	Freezing temperature of a mixture
T_m	Normal melting temperature
U	Internal energy

$\bar{U}$	Reduced average velocity
U_B	Bosanquet velocity
v	Capillary volume fraction in tissue
v_i	Molar volume
v_x	Velocity in the x-direction
v_y	Velocity in the y-direction
v_z	Velocity in the z-direction
V	Molar, specific volume, or total volume
V	Velocity, usually of a plate or bulk solution
V	Voltage
$V_{apparent}$	Apparent volume of distribution
$V_{average}$	Average velocity
V_{Bg}	Glucose distribution volume
V_{Bi}	Insulin distribution volume
V_C	Critical volume
V_{CAPD}	Volume of exchange fluid in continuous ambulatory peritoneal dialysis
V_d	Apparent dialysate fluid velocity
V_{device}	Device volume
V_E	Extracellular fluid volume
V_{IF}	Interstitial fluid distribution volume
V_{max}	Constant in Michaelis–Menten model
V_P, V_{plasma}	Plasma volume
V_R	Fluid volume of all other body fluids
V_{tissue}	Tissue volume
V_T	Total gas volume
V_T	Total volume of tissue
W	Width
W	Work
W_b	Membrane area per unit length on blood side
W_g	Membrane area per unit length on gas side
W_p	Work of a pump
W_S	Shaft work as in a flow process
x_i	Mole fraction of component i, usually the liquid phase
X	Cell density
y_i	Mole fraction of component i, usually the vapor or gas phase
Y	Fraction of hemoglobin that is saturated
$Y_{X/S}$, $Y_{P/S}$, $Y_{P'/S}$	Yield coefficients
z	Ion charge
z	Local position
z	Ratio of blood flow rate to exchange fluid flow rate
Z	Compressibility factor
Z	Elevation relative to a datum level
$\bar{Z}_i$	Partial molar compressibility factor for species i

Superscripts

c	Core
E	Excess property value
G	Refers to the gas
I, II, ... π	Phase I, phase II, phase π
L	Refers to the liquid
p	Plasma
(P)	Refers to a planar surface
R	Denotes a residual thermodynamic property
S	Refers to the solid
SAT	Refers to the saturated state

STP	Refers to standard temperature and pressure
T	Refers to tissue
V	Refers to the vapor
∞	Denotes the value at infinite dilution
∞	Denotes the value after a very long time

Subscripts

0	Denotes initial value
0	Denotes interstitial fluid value
A	Denotes arterial value
AVG, average	Denotes average value
b	Refers to the value in a bulk solution
b	Refers to the value in blood
bs	Refers to the value at the surface of a membrane or particle
BP	Refers to the boiling point
c	Core
C	Capillary
CAPD	Refers to continuous ambulatory peritoneal dialysis
f	Filtrate side of the membrane
g	Refers to gas
IF	Interstitial fluid
L	Local value at $x = L$
m	Denotes membrane value
max, maximum	Denotes a maximum of the value
min, minimum	Denotes a minimum of the value
p	Plasma
plasma	Refers to plasma
S	Shaft work
tube	Refers to a tube or other conduit
T	Denotes tissue
urine	Refers to the urine
V	Denotes venous value
W	Refers to the wall
x	Local value at position x

Greek letters

α	Kinetic energy correction term
ρ	Mass density
δ	Marginal or plasma layer thickness
δ	Thickness of the velocity boundary layer
δ_C	Concentration boundary layer thickness
δ_{cell}	Thickness of a cellular layer
δ_{device}	Device thickness
δ_i	Solubility parameter for species i
δ_{SC}	Thickness of stratus corneum
μ	Specific growth rate
μ	Viscosity
$\mu_{apparent}$	Apparent viscosity
μ_i	Chemical potential of component i
ν	Kinematic viscosity, μ/ρ
τ	Residence time
τ	Shear stress
τ	Tortuosity of a membrane
τ_{max}	Time when concentration peaks
τ_{rz}	Shear stress, viscous transport of z momentum in the r-direction

τ_w	Wall shear stress
τ_y	Yield stress
τ_{yx}	Shear stress, viscous transport of x momentum in the y-direction
γ	Heat capacity ratio defined as C_P/C_V
γ	Surface energy, surface tension
$\dot{\gamma}$	Shear rate
$\dot{\gamma}_w$	Shear rate at the wall
γ_i	Activity coefficient of component i
ϕ	Cell volume fraction
ϕ	Thiele modulus
ϕ_i	Fugacity coefficient of pure component i
$\hat{\phi}_i$	Fugacity coefficient of component i in a mixture
Φ_1	Volume fraction of the solvent
π	Osmotic pressure
η	Effectiveness factor
η	Overall efficiency of an equilibrium stage
η	Efficiency of a pump or a turbine
Δ	Ratio of concentration and velocity boundary thicknesses
λ	Ratio of the molecular radius to the pore radius
λ	Eigen value
Γ	Gamma function
$\Gamma_{metabolic}$	Metabolic oxygen consumption rate
ω_r	Hindered diffusion parameter
σ	Reflection coefficient
ε	Tissue void volume fraction
ε	Void volume
ε_j	Extent of reaction j
ε_{SC}	Void volume fraction of stratus corneum
ν_{ij}	Stoichiometric coefficient of species i in reaction j

Additional notes

$-$	An overbar denotes a partial molar property
$\wedge$	Circumflex denotes the value of a property in a mixture
Δ	Difference operator, final state minus initial state

1 Introduction

Before we can begin our study of biomedical engineering transport phenomena, we first must spend time reviewing some basic concepts that are essential for understanding the material in this book. You may have come across some of this material in other courses, such as in chemistry, physics, and perhaps thermodynamics. Reviewing these concepts once again is very important since they form the basis of our approach to analyzing and solving biomedical engineering problems.

1.1 REVIEW OF UNITS AND DIMENSIONS

1.1.1 UNITS

Careful attention must be given to *units* and *dimensions* when solving engineering problems. Otherwise, serious errors can occur in your calculations.

Units are how we describe the size or amount of a dimension. For example, a second is a common unit that is used for the dimension of time. In this book, we primarily use the *International System of Units*, which is also known by its abbreviation SI, for *Système International*. Other systems of units are also used, such as the English and American engineering systems and the centimeter-gram-second (cgs) system. We will come across some of these non-SI units as well in our study. The units of these other systems may be related to the SI units by appropriate conversion factors. Table 1.1 provides a convenient summary of common conversion factors that relate these other units to the SI system.

It is important to remember that in engineering calculations you must always attach units to the numbers that arise in your calculations, unless they are already unitless. Furthermore, within a calculation, it is important to use a consistent system of units, and in this book we recommend that you work with the SI system. In the event that a number has a non-SI unit, you will first need to convert those units into SI units using the conversion factors found in Table 1.1. Also, treat the units associated with a number as algebraic symbols. Then, as long as the units are the same, you can perform operations such as addition, subtraction, multiplication, and division on the like units, thereby combining and in some cases even cancelling them out.

1.1.2 FUNDAMENTAL DIMENSIONS

The measurement of the physical properties we will be interested in is derived from the *fundamental dimensions* of length, mass, time, temperature, and mole. Table 1.2 summarizes the basic SI units for these fundamental dimensions. The SI unit for *length* is meter (m) and that for *time* is second (sec).

1.1.2.1 Mass and Weight

The *mass* of an object refers to the total amount of material that is in the object. The mass is a property of matter and is the same no matter where the object is located. For example, the mass of an object is the same on Earth, on Neptune, or if it is just floating along somewhere in space. In SI units, we measure the mass of an object in kilograms (kg). Remember that the mass of an object is different from the weight of an object. *Weight* is the force exerted by gravity on the object. Since weight is a force, we can use Newton's second law (F = ma) to relate weight (F) and mass (m).

TABLE 1.1
Conversion Factors

Dimension	Conversion Factors	
Length	1 m = 100 cm	
	1 m = 3.28084 ft	
	1 m = 39.37 in	
Volume	$1 \text{ m}^3 = 10^6 \text{ cm}^3$	1 liter (L) = 1000 cm^3
	$1 \text{ m}^3 = 35.3147 \text{ ft}^3$	1 gallon (US) = 3.7853 L
	$1 \text{ m}^3 = 61{,}023.38 \text{ in}^3$	1 gallon (US) = 0.1337 ft^3
Mass	1 kg = 1000 g	
	1 kg = 2.20462 lb_m	
Force	1 N = 1 kg m sec^{-2}	1 dyne = 1 g cm sec^{-2}
	1 N = 100,000 dyne	
	1 N = 0.22481 lb_f	
Pressure	1 bar = 100,000 kg m^{-1} sec^{-2}	1 Pa = 1 N m^{-2}
	1 bar = 100,000 N m^{-2}	100,000 Pa = 1 bar
	1 bar = 10^6 dyn cm^{-2}	1 atm = 760 mmHg
	1 bar = 0.986923 atm	1 atm = 14.7 psi
	1 bar = 14.5038 psi	1 atm = 101,325 Pa
	1 bar = 750.061 mmHg or torr	
Density	1 g cm^{-3} = 1000 kg m^{-3}	
	1 g cm^{-3} = 62.4278 lb_m ft^{-3}	
Energy	1 J = 1 kg m^2 sec^{-2}	
	1 J = 1 N m	
	1 J = 1 m^3 Pa	
	1 J = 10^{-5} m^3 bar	
	1 J = 10 cm^3 bar	
	1 J = 9.86923 cm^3 atm	
	1 J = 10^7 dyn cm	
	1 J = 10^7 erg	
	1 J = 0.239 cal	
	1 J = 5.12197×10^{-3} ft^3 psia	
	1 J = 0.7376 ft lb_f	
	1 J = 9.47831×10^{-4} Btu	
Power	1 kW = 10^3 W	
	1 kW = 1000 kg m^2 sec^{-3}	
	1 kW = 1000 J sec^{-1}	
	1 kW = 239.01 cal sec^{-1}	
	1 kW = 737.562 ft lb_f sec^{-1}	
	1 kW = 0.947831 Btu sec^{-1}	
	1 kW = 1.34102 hp	
Viscosity	1 P = 1 g cm^{-1} sec^{-1}	1 P = 0.1 Pa sec
	1 P = 100 cP	1 cP = 0.001 Pa sec
	1 P = 1 dyn sec cm^{-2}	
	1 P = 0.1 N sec m^{-2}	

TABLE 1.2
SI Units for the Fundamental Dimensions

Fundamental Dimension	Unit	Abbreviation
length	meter	m
mass	kilogram	kg
time	second	sec or s
temperature	Kelvin	K
mole	mole	mol

Example 1.1

Calculate the weight in SI units of an average person on Earth.

Solution

On Earth, the acceleration due to gravity (g) is 9.8 m sec^{-2}. The average mass of a human is 75 kg, so on Earth the weight of the average human is 75 kg $\times$ 9.8 m sec^{-2} = 735 kg m sec^{-2}. Next, we will see that 1 newton (N) is a special SI unit defined as 1 kg m sec^{-2}, so an average human also has a weight of 735 N.

In the United States with its American engineering system of units, a very confusing situation can occur when working with mass and force (weight). The following discussion should be sufficient for you to see the advantages of working in the SI system of units. However, the American engineering system of units is still widely used, and it is important that you understand the following in order to properly handle these units in the event that they arise in your calculations.

What is commonly referred to as a pound is really a *pound force* (weight) in the American engineering system of units. The pound force is abbreviated as lb$_f$, where subscript f must be present to indicate that this is a pound force. Mass is also expressed as pounds, but in reality what is meant is the *pound mass* (lb$_m$) and, once again, subscript m must be present to indicate that this is a pound mass. Using Newton's second law, we can develop the following relationship between the pound force and the pound mass:

$$(\text{force}) \; \text{lb}_f = (\text{mass}) \; \text{lb}_m \times (\text{acceleration of gravity}) \; \text{ft sec}^{-2} \times 1/g_c. \tag{1.1}$$

The g_c term in Equation 1.1 is a conversion factor needed to make the units in this equation work out properly, that is to convert lb$_m$ to lb$_f$. Note that unlike the acceleration of gravity, the value of g_c does not depend on location and is a constant. In the American engineering system of units, the values of force (pound force or lb$_f$) and mass (pound mass or lb$_m$) are defined in such a way as to be the same at sea level and at 45° latitude. The acceleration of gravity (g) in the American system of units at sea level and 45° latitude is 32.174 ft sec^{-2}. The value of g_c in Equation 1.1 is therefore 32.174 ft lb$_m$ sec^{-2} lb$_f^{-1}$. It is important to recognize though that the acceleration of gravity depends on location. A pound force and a pound mass have the same value only at sea level, where the acceleration of gravity is 32.174 ft sec^{-2} or whenever the ratio of the local acceleration of gravity to g_c is equal to one. For calculations done on the surface of the earth, it is usually just assumed that a pound mass (lb$_m$) equals a pound force (lb$_f$), since the acceleration of gravity really does not vary that much. So, if an object with a mass of 10 lb$_m$ is on Earth, then for practical purposes its weight is 10 lb$_f$. However, if this object is taken to the moon, where the acceleration of gravity is 5.309 ft sec^{-2}, the mass is still 10 lb$_m$; however, by Equation 1.1 its weight is quite a bit lower, as shown in the following calculation:

$$10 \; \text{lb}_m \times 5.309 \, \frac{\text{ft}}{\text{sec}^2} \times \frac{\text{sec}^2 \; \text{lb}_f}{32.174 \; \text{ft lb}_m} = 1.65 \; \text{lb}_f.$$

1.1.2.2 Temperature

Temperature is defined as an *absolute* temperature instead of a relative temperature like the commonly used Fahrenheit and Celsius scales. The Fahrenheit scale is based on water freezing at 32°F and boiling at 212°F, whereas the Celsius scale sets these at 0°C and 100°C, respectively. The following equations may be used to relate a Celsius temperature to a Fahrenheit temperature and vice versa:

$$t°F = 1.8 \ t°C + 32 \quad \text{or} \quad t°C = \tfrac{5}{9}(t°F - 32). \tag{1.2}$$

The SI unit for the measurement of absolute temperature is Kelvin (K). The absolute temperature scale sets the absolute zero point, or 0 Kelvin (0 K), as the lowest possible temperature at which matter can exist. The temperature unit of the absolute Kelvin temperature scale is the same as that of the Celsius scale, so 1 K = 1°C. The following equation may be used to relate the Celsius temperature (t) scale to the absolute Kelvin temperature (T) scale. In Celsius units, the absolute zero point is therefore –273.15°C:

$$t°C = T\,K - 273.15. \tag{1.3}$$

The absolute temperature can also be defined by the same unit of temperature measurement as the Fahrenheit scale. This is referred to as the *absolute Rankine* (R) *temperature* scale and 1 R = 1°F. Note that 0 K is the same as 0 R, since the absolute Kelvin and the absolute Rankine temperature scales are based on the same reference point of absolute zero. Equation 1.4 may be used to relate the Fahrenheit temperature scale (t) to the absolute Rankine temperature (T) scale. In Fahrenheit units, the absolute zero point is –459.67°F:

$$t°F = T\,R - 459.67. \tag{1.4}$$

Using Equations 1.2 through 1.4, one can also show that the relationship between the Rankine and Kelvin temperatures is given by

$$T\,R = 1.8\,T\,K. \tag{1.5}$$

1.1.2.3 Mole

The *mole* is used to describe the amount of a substance that contains the same number of atoms or molecules as there are atoms in 0.012 kg (or 12 g) of carbon-12, that is 6.023×10^{23} atoms, which is also called *Avogadro's number*. This is also called the *gram mole*, which can be written as *gmole*, *mole*, or just *mol*. In the American engineering system, the *pound mole* (lbmole or lbmol) is used and is defined as $6.023 \times 10^{23} \times 454$ atoms. Therefore, 1 pound mole is equal to 454 moles.

The *molecular weight* is defined as the mass per mole of a given substance. For example, the molecular weight of glucose is 0.180 kg mol^{-1}, whereas the molecular weight of water is 0.018 kg mol^{-1}. A mole of glucose or a mole of water both contain the Avogadro number of molecules. Even though one mole of glucose and water contains the same number of molecules, the masses of each are quite different because the molecules of glucose are much larger than those of water. That is, one mole of glucose has a mass of 0.180 kg and one mole of water has a mass of 0.018 kg.

1.1.3 Derived Dimensional Quantities

From these fundamental dimensions, we can derive the units for a variety of other dimensional quantities of interest. Some of these quantities will occur so often that it is useful to list them as shown in Table 1.3. In addition, some of these quantities, like force and energy, have their own special SI units, whereas others, like mass density, are based on the fundamental dimensions listed in Table 1.2. Many times, we will use special symbols to denote a quantity. For example, the Greek symbol ρ is used to denote the mass density and μ is often used for the viscosity of a substance. The viscosity, discussed in Chapter 4, is a physical property of a fluid and is a measure of its resistance to flow. A listing of the symbols used in this book to denote a variety of quantities is given in the Notation section that follows the Table of Contents.

Oftentimes, a particular unit is expressed as a multiple or a decimal fraction. For example, 1/1000th of a meter is known as a millimeter, where the prefix "milli-" means to multiply the base unit of meter by the factor of 10^{-3}. The prefix "kilo-" means to multiply the base unit by 1000. So, a kilogram is the same as 1000 g. Table 1.4 summarizes a variety of prefixes that are commonly used to scale a base unit.

TABLE 1.3
SI Units for Other Common Dimensional Quantities

Quantity	Unit	Abbreviation	Fundamental Units	Derived Units
Force	Newton	N	$kg\ m\ sec^{-2}$	$J\ m^{-1}$
Energy, work, and heat	Joule	J	$kg\ m^2\ sec^{-2}$	$N\ m$
Pressure and stress	Pascal	Pa	$kg\ m^{-1}\ sec^{-2}$	$N\ m^{-2}$
Power	Watt	W	$kg\ m^2\ sec^{-3}$	$J\ sec^{-1}$
Volume	–	–	m^3	–
Mass density	–	–	$kg\ m^{-3}$	–
Molar concentration	–	–	$mol\ m^{-3}$	–
Specific volume	–	–	$m^3\ kg^{-1}$	–
Viscosity	–	–	$kg\ m^{-1}\ sec^{-1}$	$Pa\ sec$
Diffusivity	–	–	$m^2\ sec^{-1}$	–
Permeability and mass transfer coefficient	–	–	$m\ sec^{-1}$	–

TABLE 1.4
Common Prefixes for SI Units

Prefix	Multiplication Factor	Symbol
femto	10^{-15}	f
pico	10^{-12}	p
nano	10^{-9}	n
micro	10^{-6}	μ
milli	10^{-3}	m
centi	10^{-2}	c
deci	10^{-1}	d
deka	10	da
hecto	10^2	h
kilo	10^3	k
mega	10^6	M
giga	10^9	G
tera	10^{12}	T

1.1.3.1 Pressure

Pressure is defined as a force per unit area. Pressure is also expressed on either a relative or an absolute scale, and its unit of measure in the SI system is the Pascal (Pa), which is equal to a newton per square meter. How the pressure is measured will affect whether or not it is a relative or an absolute pressure. *Absolute pressure* is based on reference to a perfect vacuum. A *perfect vacuum* on the absolute pressure scale defines the zero point or 0 Pa. The zero point for a relative pressure scale is usually the pressure of the air or atmosphere where the measurement is taken. This local pressure is called either the *atmospheric pressure* or the *barometric pressure* and is measured with a *barometer*, which is a device for measuring atmospheric pressure. It is important to remember that the pressure of the surrounding air or atmosphere is not a constant but depends on location, elevation, and other factors related to the weather. That is why it is always important to log in your record book the atmospheric pressure in the laboratory when doing experiments that involve pressure measurements.

Most pressure measurement devices, or pressure gauges, measure the pressure relative to the surrounding atmospheric pressure or barometric pressure. These so-called *gauge pressures* are also relative pressures. A good example of a gauge pressure is the device used to measure the inflation pressure of your tires. If the tire pressure gauge has a reading of 250 kPa, then that means that the pressure in the tire is 250 kPa *higher* than the surrounding atmospheric pressure. A gauge pressure of 0 Pa means that the pressure is the same as the local atmospheric pressure, or you have a flat tire! A negative gauge pressure (i.e., suction or partial vacuum) means that the pressure is that amount *below* the atmospheric pressure. If the pressure measurement device measures the pressure relative to a perfect vacuum, then that pressure reading is referred to as an absolute pressure reading. The relationship between relative or gauge pressure and absolute pressure is given by:

$$(\text{gauge pressure}) + (\text{atmospheric pressure}) = (\text{absolute pressure}). \tag{1.6}$$

The *standard atmosphere* is equivalent to the absolute pressure exerted at the bottom of a column of mercury that is 760 mm in height at a temperature of 0°C. This value is nearly the same as the atmospheric pressure that one may find on a typical day at sea level. The standard atmosphere in a variety of units is summarized in Table 1.5. Using Equation 1.6, and the pressure conversion factors shown in Table 1.5, we can say that your tire pressure is 250 kPa gauge or 351.325 kPa absolute.

1.1.3.2 Volume

Volume refers to the amount of space that an object occupies and depends on the mass of the object. The *mass density* is the mass of an object divided by the volume of space that it occupies. *Specific volume* is the volume of an object divided by its mass. Specific volume is, therefore, just the inverse of the mass density. For solids and liquids, the volume of an object of a given mass is not strongly

TABLE 1.5
Standard Atmosphere in a Variety of Units

1.000 atmosphere (atm)

14.696 pounds force per square inch (psi)

760 millimeters of mercury (mmHg)

33.91 feet of water (ft H_2O) = 10.34 meters of water (m H_2O)

29.92 inches of mercury (in Hg)

101,325 Pascals (Pa)

TABLE 1.6
Gas Constant (*R*)

8.314 m³ Pa mol⁻¹ K⁻¹

8.314 J mol⁻¹ K⁻¹

82.06 cm³ atm mol⁻¹ K⁻¹

0.0821 L atm mol⁻¹ K⁻¹

0.7302 ft³ atm lbmol⁻¹ R⁻¹

10.73 ft³ psi absolute lbmol⁻¹ R⁻¹

---

dependent on the temperature or the pressure. For solids and liquids, the volume is then the mass multiplied by the specific volume or the mass divided by the density of the object. For gases, the volume of a gas is also strongly dependent on both the temperature and the pressure. For gases, we need to use an equation of state to properly relate the volume of the gas to the temperature, pressure, and moles (mass divided by molecular weight).

1.1.3.3 Equations of State

An *equation of state* is a mathematical relationship that relates the pressure, volume, and temperature of a gas, liquid, or solid. For our purposes, the ideal gas law is an equation of state that will work just fine for the types of problems involving gases that we will consider here. Recall that an ideal gas has mass; however, the gas molecules themselves have no volume and these molecules do not interact with one another. The ideal gas law works well for gases like hydrogen, oxygen, and air at pressures around atmospheric. Recall that the ideal gas law is given by the following relationship:

$$PV = n R T, \qquad (1.7)$$

where P denotes the absolute pressure, V the volume, T the absolute temperature, and n the number of moles. R is called the *universal gas constant,* and a suitable value must be used to make the units in Equation 1.7 work out properly. Table 1.6 summarizes commonly used values of the gas constant in a variety of different units.

1.2 DIMENSIONAL EQUATION

A *dimensional equation* is an equation that contains both numbers and their associated units. The dimensional equation usually arises from a specific formula that is being used to solve a problem or when a particular number has nonstandard units associated with it. In the latter case, a series of conversion factors is used to put the number in some other system of units, e.g., the SI system of units. To arrive at the final answer, which is usually a number with its associated units, the dimensional equation is solved by performing the arithmetical operations on the numbers and, as discussed earlier, algebraic operations on the various units.

Example 1.2 illustrates the use of the dimensional equation to convert a quantity in nonstandard units to SI units. Example 1.3 illustrates the use of the dimensional equation to determine the volume of a gas at a given temperature and pressure. When solving engineering problems, make sure to use the dimensional equation approach illustrated in these examples. In this way, you can easily handle the conversion of the units associated with the quantities involved in your calculation and arrive at the correct answer. Remember to refer to Tables 1.1 and 1.3 through 1.6 where appropriate.

Example 1.2

A particular quantity was reported to have a value of 1.5 cm³ h⁻¹ m⁻² mmHg⁻¹. Recall that mmHg is a measurement of pressure, where 760 mm of mercury (Hg) is equal to a pressure of 1 atmosphere (1 atm). Convert this dimensional quantity to SI units of meter squared second per kilogram.

Solution

The dimensional equation for the conversion of this quantity to SI units is

$$1.5 \frac{cm^3}{hm^2 \; mmHg} \times \frac{1 \, h}{3600 \, sec} \times \frac{760 \, mmHg}{1 \, atm} \times \frac{1 \, atm}{101,325 \, Pa} \times \frac{1 \, Pa}{1 \, N \, m^{-2}} \times \frac{1 \, N}{kg \, m \, sec^{-2}}$$

$$\times \frac{1 \, m^3}{(100 \, cm)^3} = 3.125 \times 10^{-12} \frac{m^2 \, sec}{kg}.$$

Example 1.3

Calculate the volume occupied by 100 lb$_m$ of oxygen (molecular weight equals 0.032 kg mol⁻¹) at a pressure of 40 ft of water and a temperature of 20°C. Express the volume in cubic meters.

Solution

The dimensional equation for the calculation of the volume of oxygen is based on Equation 1.7 rearranged to solve for the volume, i.e., $V = nRT/P$:

$$V = \left(100 \; lb_m \times \frac{1 \, kg}{2.2046 \, lb_m} \times \frac{1 \, mol}{0.032 \, kg} \right) \times 8.314 \; \frac{m^3 \, Pa}{mole \, K} \times (273.15 + 20)K$$

$$\times \frac{1}{\left(40 \; ft \; of \; water \times \dfrac{1 \, atm}{33.91 \, ft \; of \; water} \times \dfrac{101,325 \, Pa}{1 \, atm} \right)} = 28.90 \; m^3.$$

1.3 TIPS FOR SOLVING ENGINEERING PROBLEMS

In this section, we outline a basic strategy that can be used to solve the type of engineering problems considered in this book and those you are likely to encounter as a biomedical engineer. This basic strategy is useful since it eliminates approaching the solution to each new problem or system as if it is a unique situation. The numerous examples throughout the book as well as the problems at the end of each chapter will provide you with the opportunity to apply the strategy outlined here and, at the same time, develop your skills in using the engineering analysis techniques discussed in this book. One of the unique and challenging aspects of engineering is that the number of problems that you may come across during the course of your career is limitless. However, it is important to understand that the underlying engineering principles and the solution strategy are always the same.

The engineering problems we will consider are derived from some type of process. A *process* is defined as any combination of physical operations and chemical reactions that act on or change the substances involved in the process. A process also occurs within what is also more generally called the *system*. Everything else that is affected by the system or interacts with the system is known as the *surroundings*. A biological example of a process or system is the kidney. Blood enters the kidney (system) and through a variety of physical and chemical operations, the essential nutrients in the blood are recovered and the blood is purified of waste products. These waste products are stored as urine in the bladder. Several times a day the urine that is stored in the bladder (another system) is eliminated through the process of urination. The urine eliminated from the bladder is then added to the surroundings. The engineering analysis of all problems can therefore be considered within the context of the system and its surroundings.

There are several important steps that you should follow before attempting to make any engineering calculations on a system and its surroundings. For ease of reference, these steps are also summarized in Table 1.7. One of the first things that you should do when solving a particular problem is to draw a sketch that defines your system. Drawing a sketch of the system for a given problem will allow you to identify and focus your attention on the key features of the problem. The sketch does not have to be elaborate; you can use boxes to represent the system and lines with arrows can be used to indicate the flow of streams that enter and leave the system. Figure 1.1 shows a simple sketch for the kidney example discussed earlier. If you know them or can easily calculate them, make sure you also write down the flow rates of the flowing streams and the amount or concentration of each substance in each of the flowing streams. On your sketch, you will also write down the available data that you already have concerning the problem. Examples of additional data that may be important in solving your problem include temperatures, pressures, experimental data, and even conversion factors. Next, you should obtain any additional information or data needed to solve the problem. You should also write down any chemical reactions that need to be considered and make sure that these are also balanced.

TABLE 1.7
Steps for Solving Engineering Problems

1. Draw a sketch that defines your system.
2. Indicate with lines and arrows the flow of streams that enter and leave the system.
3. Indicate the flow rates and the amounts or concentrations of each of the substances in each of the flowing streams. For those that are unknown, see if you can easily calculate them using material balance equations.
4. Write down any chemical reactions that need to be considered and make sure that these are also balanced.
5. Write down the available data that you already have concerning the problem.
6. Obtain any additional information or data needed to solve the problem.
7. Identify the specific engineering principles and formulas that govern the physical and chemical operations that occur within your system.
8. Solve the problem using the above information.

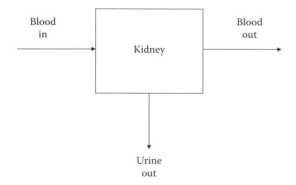

FIGURE 1.1   Engineering sketch of the kidney.

Obtaining the answer to the problem illustrated by your sketch of the system will require the use of specific engineering principles and formulas that govern the physical and chemical operations that occur within your process. Formulating the solution and solving engineering problems is a skill that one needs to develop and refine in order to be successful in engineering. That is the goal of this book; to provide you with the mathematical techniques and skills for solving a variety of biomedical engineering problems that involve chemical reactions and transport processes such as fluid flow and mass transfer. Building on the skills learned in this book, you will then be able to design biomedical equipment and devices.

1.4 CONSERVATION OF MASS

One of the most fundamental relationship that should always be considered when performing an engineering analysis on a system is the *conservation of mass*. This is also called a *material balance*. In its simplest terms, conservation of mass is the statement of fact that the sum of the masses of all substances that enter the system must equal the sum of the masses of all the substances that leave the system. However, there can be changes in the individual masses or amounts of various chemical entities because of chemical reactions that may also be occurring within the system. Remember that mass is always conserved and moles are not conserved if chemical reactions occur in the system. The only exception to conservation of mass is for nuclear reactions where energy is converted to mass and vice versa. Nuclear reactions are not included in the systems that we will be considering.

1.4.1 LAW OF CONSERVATION

Equation 1.8 provides a generalized statement of the *law of conservation* for our system, which we will find extremely useful for the solution of a variety of problems:

$$\text{Accumulation} = \text{In} - \text{Out} + \text{Generation} - \text{Consumption.} \tag{1.8}$$

This equation may be used to account for changes in quantities such as mass, moles, energy, momentum, and even money.

The *Accumulation* term accounts for the change with time of the quantity of interest within the system volume. The accumulation term in this equation is a time derivative of the quantity of interest and defines what is called an *unsteady problem*. The accumulation term can be either positive or negative. If the accumulation term is zero, then the quantity of interest within the volume of the system is not changing with time, and we refer to this situation as a *steady-state problem*. The *In* term accounts for the entry of the quantity of interest into the system by all routes. For example, as carried in by the flow of fluid, by diffusion, or in the case of energy, by heat conduction. The *Out* term is similar to the *In* term, but accounts for the loss of the quantity of interest from the system. The *Generation* term accounts for the production of the quantity of interest within the system volume, whereas the *Consumption* term represents the loss of the quantity of interest within the system. The generation and consumption of the quantity of interest occurs through chemical reactions and the metabolic processes that occur within cells.

The following *non-engineering* example illustrates a practical use for Equation 1.8, which is calculating the periodic payment on a loan. Remember this example since you can use the results to calculate payments for items such as automobiles, your house, and your credit card balance.

Example 1.4

You have just graduated from college as a bioengineer and have accepted your first job with a medical device company. You decide to buy yourself a nice sports car and in order to buy it you will need to borrow from the bank a total of $65,000. The bank is offering you a 5-year loan at an annual interest rate of 6% compounded continuously.* What is your monthly payment? What is the total amount that you will pay back to the bank? How much interest will you pay?

Solution

Let A represent the amount of money at any time that you owe the bank and R will represent the annual payment that you make to the bank on the loan. The interest rate is represented by i. The amount of interest that the bank charges you for the loan is equal to the interest rate multiplied by the amount at any time that you owe them. Using Equation 1.8, we can then write the following equation that expresses how the amount you owe the bank changes with time

$$\frac{dA}{dt} = iA - R.$$

Note that the amount you owe the bank *Accumulates* or, in this case since this is a loan, *Deaccumulates* according to the term dA/dt. The amount you owe is also *Generated* by the interest term represented by iA and the amount you owe is decreased (*Out*) by the amount of your annual payment represented by R. Since R will be greater than iA, then dA/dt will be < 0 and the amount that you owe the bank will decrease with time.

This is a first order differential equation and represents what is also called an initial value problem. In order to solve this equation, we need to specify an initial condition (IC), which is the amount you owe the bank at time equal to zero, which is when you take out the loan. The initial condition can be written as

$$\text{IC}: \ t = 0 \quad A = A_0 = \$65,000.$$

Several methods can be used to solve this differential equation, including integrating factors and Laplace transforms. Table 4.5 summarizes the Laplace transforms for a variety of functions and one of the end of chapter problems asks you to resolve this example using Laplace transforms. However, an easy way to solve the previous equation (since R is a constant) is to let A be the sum of the homogeneous solution (i.e., when $R = 0$) represented by A_h and a constant C_1 yet to be determined. So, we let $A = A_h + C_1$. We then substitute this equation into the differential equation above and obtain:

$$\frac{dA_h}{dt} = iA_h \quad \text{and} \quad -iC_1 = -R.$$

We immediately see that $C_1 = R/i$ and the homogeneous differential equation can be readily integrated to give $A_h = C_2 e^{it}$, where C_2 is another constant that we evaluate from the initial condition. Our solution for the amount that we owe the bank at any time is then given by

$$A(t) = C_2 e^{it} + \frac{R}{i}.$$

Now, we use the initial condition that at $t = 0$, $A = A_0$ and we find that the constant $C_2 = A_0 - R/i$. Our result for the amount that you owe the bank at any time after the loan inception is then given by:

$$A(t) = A_0 e^{it} + \left(\frac{R}{i}\right)(1 - e^{it}).$$

---

* Note that most loans are compounded on some basis where interest is calculated, e.g., yearly, quarterly, monthly, or even daily, in the limit becoming continuous compounding. For practical purposes, the basis of the compounding is not a significant factor in estimating a loan payment.

Now, what we want to find is the annual payment needed to pay off the loan after T years. Then, since $A(T) = 0$, we can solve the previous equation for R and obtain the following equation for the annual loan payment

$$R = \frac{i\,A_0}{1 - e^{-it}}.$$

We can now insert the numbers to calculate the loan payment amount

$$R = \frac{\dfrac{0.06}{\text{yr}} \times \$65{,}000}{1 - e^{-(0.06/\text{yr}) \times 5\ \text{yr}}} = \$15{,}047.35\,\frac{1}{\text{yr}} = \$1{,}253.95\,\frac{1}{\text{mo}}.$$

So, we see that the monthly payment for the sports car is about \$1254. The total amount paid to the bank over the 5-year period of the loan is \$15,047.35 $\text{yr}^{-1} \times 5$ yr = \$75,236.77, and the total amount of interest paid to the bank for the loan is \$75,236.77 − \$65,000 = \$10,236.77.

1.4.2 Chemical Reactions

In the case of chemical reactions or cellular metabolism occurring within the system, you must also take into account the reaction stoichiometry as given by your balanced chemical equations for those substances that are involved in the chemical reactions. In addition, it is important to remember that for biological reactions, there can be an increase in the number of cells and your balanced chemical reaction must also consider the formation of biomass from the substances that take part in the reaction, as shown in the following example.

Example 1.5

Consider the anaerobic fermentation of glucose to ethanol by yeast. Glucose ($C_6H_{12}O_6$) is converted into yeast, ethanol (C_2H_5OH), the by-product glycerol ($C_3H_8O_3$), carbon dioxide, and water. An empirical chemical formula for yeast can be taken as $CH_{1.74}N_{0.2}O_{0.45}$ (Shuler and Kargi 2001). Assuming ammonia (NH_3) is the nitrogen source, we can write the following empirical chemical equation to describe the fermentation

$$C_6H_{12}O_6 + a\ NH_3 \rightarrow b\ CH_{1.74}N_{0.2}O_{0.45}(\text{yeast}) + c\ C_2H_5OH + d\ C_3H_8O_3 + e\ CO_2 + f\ H_2O.$$

Suppose we found that 0.12 mol of glycerol was formed for each mole of ethanol produced and 0.08 mol of water was formed for each mole of glycerol. Determine the *stoichiometric coefficients*, which are the letters a, b, c, d, e, and f in front of the chemical formulas in the previous equation. Remember that the stoichiometric coefficients need to be determined so that the total number of carbon, hydrogen, oxygen, and nitrogen atoms is the same on each side of the equation.

Solution

Balancing the previous equation by inspection is not an easy thing to do. Hence, we use the more formal approach as outlined below, where we write a balance for each element, i.e., carbon, hydrogen, oxygen, and nitrogen

Carbon balance : $6 = b + 2c + 3d + e$

Hydrogen balance : $12 + 3a = 1.74b + 6c + 8d + 2f$

Oxygen balance: $6 = 0.45b + c + 3d + 2e + f$

Nitrogen balance: $a = 0.2b$

Other constraints: $d = 0.12c$ and $f = 0.08d$.

These relationships for each element can then be arranged into matrix form as follows. We see that the solution for a, b, c, d, e, and f involves the solution of six algebraic equations for these six unknowns

$$
\begin{bmatrix}
0 & 1 & 2 & 3 & 1 & 0 \\
-3 & 1.74 & 6 & 8 & 0 & 2 \\
0 & 0.45 & 1 & 3 & 2 & 1 \\
1 & -0.2 & 0 & 0 & 0 & 0 \\
0 & 0 & -0.12 & 1 & 0 & 0 \\
0 & 0 & 0 & 0.08 & 0 & -1
\end{bmatrix}
\begin{bmatrix}
a \\ b \\ c \\ d \\ e \\ f
\end{bmatrix}
=
\begin{bmatrix}
6 \\ 12 \\ 6 \\ 0 \\ 0 \\ 0
\end{bmatrix}.
$$

Solution of this matrix gives $a = 0.048$, $b = 0.239$, $c = 1.68$, $d = 0.202$, $e = 1.796$, and $f = 0.016$. The balanced chemical equation for the anaerobic fermentation of glucose by yeast can then be written as

$$
C_6H_{12}O_6 + 0.048\ NH_3 \rightarrow 0.239\ CH_{1.74}N_{0.2}O_{0.45}\ (\text{yeast}) + 1.68\ C_2H_5OH + 0.202\ C_3H_8O_3
$$

$$
+ 1.796\ CO_2 + 0.016\ H_2O.
$$

Example 1.6

Now, determine the biomass yield coefficient ($Y_{X/S}$) and the product yield coefficients ($Y_{P/S}$, $Y_{P'/S}$), where X, S, P, and P' denote biomass, glucose, ethanol, and glycerol, respectively. Use the balanced chemical equation from the previous example, which takes into account the formation of biomass. Compare these results to the theoretical maximum, which is based on the conversion of all the glucose to just ethanol and carbon dioxide, i.e., $C_6H_{12}O_6 \rightarrow 2\ C_2H_5OH + 2\ CO_2$.

Solution

The yield coefficients based on the balanced empirical equation obtained in the previous example are defined as

$$
Y_{X/S} = \frac{b}{1} = \frac{0.239\ \text{mol yeast}}{1\ \text{mol glucose}} \times \frac{23.74\ \text{gmol}^{-1}}{180\ \text{gmol}^{-1}} = \frac{0.032\ \text{g yeast}}{1\ \text{g glucose}},
$$

$$
Y_{P/S} = \frac{c}{1} = \frac{1.68\ \text{mol ethanol}}{1\ \text{mol glucose}} \times \frac{46\ \text{gmol}^{-1}}{180\ \text{gmol}^{-1}} = \frac{0.4293\ \text{g ethanol}}{1\ \text{g glucose}},
$$

$$
Y_{P'/S} = \frac{d}{1} = \frac{0.202\ \text{mol glycerol}}{1\ \text{mol glucose}} \times \frac{92\ \text{gmol}^{-1}}{180\ \text{gmol}^{-1}} = \frac{0.1032\ \text{g glycerol}}{1\ \text{g glucose}}.
$$

For comparison, the theoretical yield of ethanol is given as

$$
Y_{P/S\ \text{theoretical}} = \frac{2\ \text{mol ethanol}}{1\ \text{mol glucose}} \times \frac{46\ \text{gmol}^{-1}}{180\ \text{gmol}^{-1}} = \frac{0.511\ \text{g ethanol}}{1\ \text{g glucose}}.
$$

1.4.3 MATERIAL BALANCES

In applying the law of mass conservation to your system, you can write a material balance using Equation 1.8 for each component. If you have N components, then this will provide a total of N equations. You can also write a total material balance giving a total of $N + 1$ equations. However, these $N + 1$ equations are not independent since we can also obtain the total material balance equation by just summing the N component material balances. Remember that you need as many equations as you have unknowns in order to completely describe your system. If you have more unknowns than equations relating these unknowns, then you have that many *degrees of freedom*, that is

$$\text{Degrees of Freedom} = \text{Number of Unknowns} - \text{Number of Equations}, \tag{1.9}$$

and you must specify the values of some of these unknowns. In solving engineering problems or designing equipment and devices, it is quite common to have more unknowns than equations that relate these unknowns. These degrees of freedom are also referred to as design variables or quantities that you specify as part of the design of the system.

The following two examples illustrate the application of material balance principles.

Example 1.7

Drug B has a solubility of 2.5 g B 100^{-1} g^{-1} water at 80°C, and at 4°C the solubility of B is only 0.36 g B 100^{-1} g^{-1} of water. If a saturated drug solution at 80°C has 1000 g of the drug in it, determine how much water is also present with the drug. If this solution is then cooled to 4°C, how much of the drug will be precipitated from the solution?

Solution

As a basis for the calculation, we start with the fact that there is 1000 g of the drug in the initial solution. Since the drug has a solubility of 2.5 g 100^{-1} g^{-1} of water at 80°C, we can calculate the amount of water present in this saturated solution as follows:

$$\frac{100 \text{ g water}}{2.5 \text{ g B}} \times 1000 \text{ g B} = 40{,}000 \text{ g water}.$$

So, the initial saturated solution consists of 1000 g of drug B and 40,000 g of water. This solution is then cooled to 4°C and the decrease in drug solubility results in the precipitation of B out of the solution. Here, we see that the 40,000 g of water is conserved within the cooled solution; however, the mass of drug B in the solution at 4°C and in the precipitate is unknown. A total mass balance on this process says that $F = S + P$, where F is the mass of the initial solution, S is the mass of the saturated solution after cooling to 4°C, and P is the mass of drug B in the solid precipitate. Therefore, 41,000 g $= S + P$. Here, we still have two unknowns, S and P, so we need another equation. However, a component material balance on drug B can also be written to find the mass of the drug precipitate, P, as shown below. The term on the left-hand side is the amount of drug B in the original solution, the first term on the right provides the amount of drug B that can exist in the saturated solution at 4°C, and P is the mass of the drug precipitate:

$$1000 \text{ g B} = \frac{0.36 \text{ g B}}{100 \text{ g water}} \times 40{,}000 \text{ g water} + P,$$

$$P = 856 \text{ g of drug B in the precipitate}.$$

With P now known, the overall material balance can be used to find that the mass of the cooled solution is

$$S = F - P = 41{,}000 \text{ g} - 856 \text{ g} = 40{,}144 \text{ g}.$$

The percentage recovery in the precipitate of drug B from the initial saturated solution at 80°C by cooling to a saturated solution at 4°C is given by

$$\% \text{ drug B recovery} = \frac{856 \text{ g drug B}}{1000 \text{ g drug B}} \times 100 = 85.6\,\%.$$

Example 1.8

A continuous membrane filtration process is being used to separate components A and B into a product stream that is mostly A and a waste stream that is primarily B. A and B enter the process together at a flow rate (F) of 100 kg h^{-1}. The composition of this feed stream is 40% A and 60% B by mass. The product stream (P) being removed from the process has a composition of 90% A and 10% B, whereas the waste stream (W) leaving the process has a composition of 2% A and 98% B. Calculate the mass flow rates of the product and waste streams.

Solution

First we can write an overall material balance that says, $F = 100$ kg h$^{-1} = P + W$. With two unknowns P and W, there is a need for another equation. So, we can write a component material balance on A that says, 0.40×100 kg h$^{-1} = 0.90 \times P + 0.02 \times W$. Since $W = 100$ kg h$^{-1} - P$ from the overall material balance, the component material balance can be rewritten in terms of just P, or 40 kg h$^{-1} = 0.90\,P + 2$ kg h$^{-1} - 0.02\,P$. Solving for the product flow rate we obtain $P = 43.18$ kg h^{-1}. From the overall material balance we then find that $W = 100$ kg h$^{-1} - 43.18$ kg h$^{-1} = 56.82$ kg h^{-1}.

The following examples illustrate both unsteady and steady-state material balance calculations in a bioreactor where a chemical reaction also occurs.

Example 1.9

Consider the production of ethanol by the fermentation of glucose by yeast in a batch fermentor. Let X represent the yeast cell density (g L^{-1}) in the bioreactor. We then define μ (1 h^{-1}) as the specific growth rate of the yeast in the fermentor and by definition $\mu = (1/X)(dX/dt)$. The *Monod equation* is an empirical expression that describes how the specific growth rate of cells depends on a limiting substrate, which in this case is glucose,

$$\mu = \frac{\mu_{max} S}{K_S + S},$$

where S (g L^{-1}) represents the concentration of glucose in the fermentor, K_S is a constant, and μ_{max} is the maximum specific growth rate, i.e., when $S \gg K_S$. Assuming that the glucose fermentation stoichiometry is given by the empirical chemical equation obtained in Example 1.5, calculate as a function of time the concentration in the fermentor of the yeast cells (X), glucose (S), ethanol (P), and glycerol (P'). Assume that the initial glucose concentration in the fermentor is 100 g L^{-1} and that the initial yeast cell density after inoculation of the fermentor is 0.2 g L^{-1}. The Monod constants are $K_S = 1.5$ g L^{-1} and $\mu_{max} = 0.075$ h^{-1}.

Solution

After the growth medium in the fermentor is inoculated with yeast, there is a short period of time called the lag phase (t_0) where the yeast cells are adapting to their new environment. The yeast then enters a phase where they grow very rapidly, according to the Monod equation shown earlier. A material balance on the yeast cells in the fermentor may then be written using Equation 1.8 as

$$\text{Yeast} \quad V\frac{dX}{dt} = \mu\, X\, V = \frac{\mu_{max}\, S}{K_S + S} X\, V.$$

Note, there is no yeast flowing into or out of the fermentor; hence the accumulation of yeast ($V(dX/dt)$) within the fermentor is due only to the production of yeast as described by the Monod equation $((\mu_{max}S/K_S + S)\, XV)$. We can also write material balance equations for glucose, ethanol, and glycerol as shown below. Note that the definition of the yield coefficients from Example 1.6 allows us to relate the changes in the glucose (S), ethanol (P), and glycerol (P') concentrations to the change in the yeast concentration (X)

$$\text{Glucose} \quad V\frac{dS}{dt} = -\frac{V}{Y_{X/S}}\frac{dX}{dt},$$

$$\text{Ethanol} \quad V\frac{dP}{dt} = -Y_{P/S}V\frac{dS}{dt} = \frac{V\, Y_{P/S}}{Y_{X/S}}\frac{dX}{dt},$$

$$\text{Glycerol} \quad V\frac{dP'}{dt} = -Y_{P'/S}V\frac{dS}{dt} = \frac{V\, Y_{P'/S}}{Y_{X/S}}\frac{dX}{dt}.$$

Since the volume of the fermentor (V) is constant, we can cancel out the volumes and rewrite these equations as shown below

$$\text{Yeast} \quad \frac{dX}{dt} = \mu\, X = \frac{\mu_{max}\, S}{K_S + S} X,$$

$$\text{Glucose} \quad \frac{dS}{dt} = -\frac{1}{Y_{X/S}}\frac{dX}{dt},$$

$$\text{Ethanol} \quad \frac{dP}{dt} = \frac{Y_{P/S}}{Y_{X/S}}\frac{dX}{dt},$$

$$\text{Glycerol} \quad \frac{dP'}{dt} = \frac{Y_{P'/S}}{Y_{X/S}}\frac{dX}{dt}.$$

Now we have four differential equations that can be solved to provide X, S, P, and P' as a function of time. The differential equations for glucose, ethanol, and glycerol all depend on X. If we multiply both sides of these equations by dt, we obtain

$$dS = -\frac{1}{Y_{X/S}}dX,$$

$$dP = \frac{Y_{P/S}}{Y_{X/S}}dX,$$

$$dP' = \frac{Y_{P'/S}}{Y_{X/S}}dX.$$

These equations may be integrated from an initial condition where $t = t_0$ and $S = S_0$, $P = P_0$, $P = P_0'$, and $X = X_0$, to some later time in the fermentation (t) to give

$$S(t) = S_0 - \frac{1}{Y_{X/S}}(X(t) - X_0),$$

$$P(t) = P_0 + \frac{Y_{P/S}}{Y_{X/S}}(X(t) - X_0),$$

$$P'(t) = P_0' + \frac{Y_{P'/S}}{Y_{X/S}}(X(t) - X_0).$$

So now we have replaced three differential equations for glucose, ethanol, and glycerol, with three algebraic equations whose solution depends only on $X(t)$. Once we solve the remaining differential equation shown below for $X(t)$, we can then get the glucose, ethanol, and glycerol concentrations by the previous three equations. Note that the maximum yeast cell concentration occurs when $S = 0$; hence $X_{max} = Y_{X/S}S_0 + X_0$

$$\frac{dX}{dt} = \mu X = \frac{\mu_{max}S}{K_S + S}X.$$

Note that the right-hand side of this differential equation depends on both S and X. However, we can use the algebraic relationship we obtained previously to relate S to X, that is

$$S = S_0 - \frac{1}{Y_{X/S}}(X - X_0).$$

Hence, our differential equation for X becomes

$$\frac{dX}{dt} = \frac{\mu_{max}\left[S_0 - \frac{1}{Y_{X/S}}(X - X_0)\right]}{K_S + \left[S_0 - \frac{1}{Y_{X/S}}(X - X_0)\right]}X.$$

The initial condition needed for the solution of this equation is that at $t = t_0$, $X = X_0$, where t_0 represents the end of the lag period and the beginning of rapid cell growth. All the terms involving X may be collected on the left-hand side of the previous equation as

$$\int_{X_0}^{X} \frac{\left\{K_S + \left[S_0 - \frac{1}{Y_{X/S}}(X - X_0)\right]\right\}}{\left[S_0 - \frac{1}{Y_{X/S}}(X - X_0)\right]X}dX = \mu_{max}\int_{t_0}^{t}dt = \mu_{max}(t - t_0).$$

The integral on the left-hand side can then be integrated to provide the following analytical solution

$$\mu_{max}(t - t_0) = \left(\frac{K_S Y_{X/S}}{Y_{X/S}S_0 + X_0} + 1\right)\ln\left(\frac{X}{X_0}\right) + \left(\frac{K_S Y_{X/S}}{Y_{X/S}S_0 + X_0}\right)\ln\left(\frac{Y_{X/S}S_0}{Y_{X/S}S_0 - (X - X_0)}\right).$$

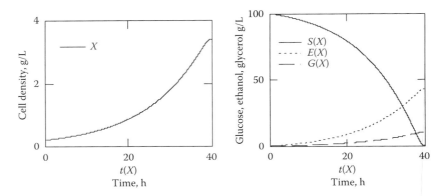

FIGURE 1.2   Results from the simulation of a batch fermentor.

This equation may then be solved for the time (t) needed to achieve a certain cell density $X(t)$ in the fermentor assuming the cell concentration, X_0, at the end of the lag phase is 0.2 g L^{-1}. Using the values of the parameters from the problem statement for this example, we can solve this equation, as well as the algebraic equations, for the concentrations of glucose, ethanol, and glycerol. The graphs in Figure 1.2 show the solution, assuming that the lag time is negligible.

Example 1.10

Suppose the fermentation considered in the previous example takes place at steady state in a continuous stirred-tank fermentor (CSTF). Assume that the feed stream to the fermentor contains no yeast cells (i.e., sterile), ethanol, or glycerol. In addition, the glucose concentration in the feed stream is 100 g L^{-1}. The fermentor has a volume of 2000 L and the feed flow rate to the fermentor (F) is 100 L h^{-1}. Determine the exiting concentrations of the yeast cells, glucose, ethanol, and glycerol.

Solution
Once again, we use Equation 1.8 and write a material balance for yeast, glucose, ethanol, and glycerol as

$$\text{Yeast} \quad V\frac{dX}{dt} = F\,X_{in} - F\,X_{out} + \frac{V\,\mu_{max}\,S}{K_S + S}X,$$

$$\text{Glucose} \quad V\frac{dS}{dt} = F\,S_{in} - F\,S_{out} - \frac{V}{Y_{X/S}}\frac{\mu_{max}S}{K_S + S}X,$$

$$\text{Ethanol} \quad V\frac{dP}{dt} = F\,P_{in} - F\,P_{out} + \frac{V\,Y_{P/S}}{Y_{X/S}}\frac{\mu_{max}\,S}{K_S + S}X,$$

$$\text{Glycerol} \quad V\frac{dP'}{dt} = F\,P'_{in} - F\,P'_{out} + \frac{V\,Y_{P'/S}}{Y_{X/S}}\frac{\mu_{max}\,S}{K_S + S}X,$$

where X, S, P, and P' represent the concentrations of cells, glucose, ethanol, and glycerol, respectively, within the fermentor. Since the fermentor is considered to be well mixed, the exiting concentrations, i.e., X_{out}, S_{out}, P_{out}, and P'_{out}, are the same as X, S, P, and P'. Since the fermentor operates at steady state, we can also eliminate the time derivatives on the left-hand side of the previous equations. With X_{in}, P_{in}, and P'_{in} also equal to zero, the above equations may then be written as

$$\text{Yeast}\quad FX = \frac{V\mu_{max}S}{K_S + S}X,$$

$$\text{Glucose}\quad FS = FS_{in} - \frac{V}{Y_{X/S}}\frac{\mu_{max}S}{K_S + S}X,$$

$$\text{Ethanol}\quad FP = \frac{VY_{P/S}}{Y_{X/S}}\frac{\mu_{max}S}{K_S + S}X,$$

$$\text{Glycerol}\quad FP' = \frac{VY_{P'/S}}{Y_{X/S}}\frac{\mu_{max}S}{K_S + S}X.$$

The dilution rate (D) is defined as F/V and has units of time^{-1}. For this problem:

$$D = \frac{100\ \text{L h}^{-1}}{2000\ \text{L}} = 0.05\ \text{h}^{-1},$$

and from the yeast equation we can solve for the value of S as given by

$$S = \frac{DK_S}{\mu_{max} - D} = \frac{0.05\ \text{h}^{-1} \times 1.5\ \text{g L}^{-1}}{0.075\ \text{h}^{-1} - 0.05\ \text{h}^{-1}} = 3\ \text{g L}^{-1},$$

giving a glucose conversion of 97%. Note that this equation is only valid provided that $\mu_{max} > D$. If this condition is not met, then the growth rate of the cells is not sufficient to keep up with the cells leaving the outlet stream. This then leads to washout of the cells from the fermentor and the fermentation process ends.

From the glucose material balance equation, we can then solve for the exiting yeast concentration as

$$X = \frac{F(S_{in} - S)}{\dfrac{V}{Y_{X/S}}\dfrac{\mu_{max}S}{K_S + S}} = \frac{100\ \text{L h}^{-1}}{2000\ \text{L}}\frac{(100-3)\ \text{g L}^{-1}}{0.075\ \text{h}^{-1}}\frac{}{0.032}\frac{3\ \text{g L}^{-1}}{1.5\ \text{g L}^{-1} + 3\ \text{g L}^{-1}} = 3.104\ \text{g L}^{-1}.$$

Similarly, we can rearrange the material balance equations for ethanol and glycerol and obtain their exiting concentrations as

$$P = \frac{V}{F}\frac{Y_{P/S}}{Y_{X/S}}\frac{\mu_{max}S}{K_S + S}X = \frac{2000\ \text{L} \times 0.4293}{100\ \text{L h}^{-1} \times 0.032}\frac{0.075\ \text{h}^{-1} \times 3\ \text{g L}^{-1}}{1.5\ \text{g L}^{-1} + 3\ \text{g L}^{-1}} \times 3.104\ \text{g L}^{-1}.$$

$$P = 41.64\ \text{g L}^{-1}.$$

$$P' = \frac{V}{F}\frac{Y_{P'/S}}{Y_{X/S}}\frac{\mu_{max}S}{K_S + S}X = \frac{2000\ \text{L} \times 0.1032}{100\ \text{L h}^{-1} \times 0.032}\frac{0.075\ \text{h}^{-1} \times 3\ \text{g L}^{-1}}{1.5\ \text{g L}^{-1} + 3\ \text{g L}^{-1}} \times 3.104\ \text{g L}^{-1}.$$

$$P' = 10.01\ \text{g L}^{-1}.$$

TABLE 1.8
Comparison of Results Obtained for the Batch and Continuous Fermentors

	Batch	CSTF
Glucose conversion	97%	97%
Residence time (h)	38.58	20
Productivity (g ethanol L^{-1} h^{-1})	1.08	2.082
Yeast (g L^{-1})	3.304	3.104
Glucose (g L^{-1})	3.0	3.0
Ethanol (g L^{-1})	41.64	41.64
Glycerol (g L^{-1})	10.01	10.01

Table 1.8 compares the results obtained in Example 1.9 for the batch fermentor to those obtained in this example for a CSTF. The batch reactor results shown in the table are based on the time needed to obtain the same 97% conversion of the glucose as in this example. For the batch fermentor this is 38.58 h. This is also known as the *residence time* for the batch fermentor. In the CSTF, the residence time is equal to the volume of the fermentor divided by the flow rate or

$$\text{Residence time} \quad \tau = \frac{V}{F} = \frac{1}{D} = \frac{2000 \text{ L}}{100 \text{ L h}^{-1}} = 20 \text{ h}.$$

Productivity is a measure of the amount of product produced per unit time per unit volume of fermentor. For the batch fermentor, productivity can be shown to be equal to the final product concentration divided by the batch residence time. For the CSTF, productivity is equal to $(PF/V) = PD$, assuming that there is no product in the feed to the fermentor. Note that in this case the CSTF is nearly twice as productive as the batch fermentor.

PROBLEMS

1. $R = 8.314$ J mol^{-1} K^{-1}, find the equivalent in calories per mole per Kelvin.
2. Convert 98.6°F to degree Celsius and then Kelvin.
3. During cell migration, fibroblasts can generate traction forces of approximately 2×10^4 μdyn. What is the equivalent force in kilonewtons?
4. Endothelial cells in random motion were recorded to move at the *lightning speed* of 27 μm h^{-1}, what would their speed be in miles per second?
5. A process requires 48 MW of power to convert A to B. What is the power needed in centigrams per square centimeter per cubic hour? What about in kilojoules per minute?
6. Which is moving faster; a plane moving at 400 mi h^{-1} or a molecule moving at 6.25×10^{15} nm min^{-1}? Show the answer in units of millimeters per second.
7. Resolve Example 1.4 using the Laplace transform technique.
8. Rework Example 1.5 assuming there are 0.21 mole of glycerol formed for each mole of ethanol and 0.13 mole of water formed for each mole of glycerol.
9. Draw a sketch of Example 1.9 and label all the components added to the batch fermentor. Show for a glucose conversion of 97% that the material balance on all the components that take part in the fermentation is satisfied.
10. Draw a sketch of Example 1.10 and label all the streams entering and exiting the fermentor. Show for a glucose conversion of 97% that the material balance on all the components that take part in the fermentation is satisfied.
11. Rework Example 1.9 assuming the specific growth rate (μ_{max}) of the yeast cells is 0.15 h^{-1}.

12. Rework Example 1.10 assuming the specific growth rate (μ_{max}) of the yeast cells is 0.15 h^{-1}.

13. A hollow fiber membrane separator with a nominal molecular weight cutoff of 100,000 is fed a solution of proteins at the rate of 250 mL min^{-1}. The composition of the protein solution is protein A (4 g L^{-1}, MW = 20,000), protein B (7 g L^{-1}, 150,000), and protein C (6 g L^{-1}, MW = 300,000). The filtrate flow rate is found to be 116.4 mL min^{-1} and the flow rate of the retentate is 133.5 mL min^{-1}. The concentration of protein A in the retentate is found to be 5.84 g L^{-1}. What is the concentration of protein A in the filtrate? Also, what are the concentrations of proteins B and C in the filtrate and in the retentate?

14. Explore the productivity of the fermentor discussed in Example 1.10 as a function of the residence time $\tau = V/F$. Is there an optimal residence time that maximizes the ethanol productivity in the fermentor?

15. An artificial kidney is a device that removes water and wastes from blood. In one such device, the hollow fiber hemodialyzer, blood flows from an artery through the insides of a bundle of cellulose acetate fibers. Dialyzing fluid, which consists of water and various dissolved salts, flows on the outside of the fibers. Water and wastes—principally urea, creatinine, uric acid, and phosphate ions—pass through the fiber walls into the dialyzing fluid, and the purified blood is returned to a vein. At some time during the dialysis of a patient in kidney failure, the arterial and venous blood conditions are

	Arterial Blood—In	Venous Blood—Out
Flow rate (mL min^{-1})	200	195
Urea concentration (mg mL^{-1})	1.90	1.75

a. Calculate the rates at which urea and water are being removed from the blood.

b. If the dialyzing fluid enters at the rate of 1500 mL min^{-1} and the exiting dialyzing solution (dialysate) leaves at about the same rate, calculate the concentration of urea in the dialysate.

c. Suppose we want to reduce the patient's urea level from an initial value of 2.7 mg mL^{-1} to a final value of 1.1 mg mL^{-1}. If the total blood volume is 5 L and the average rate of urea removal is that calculated in part (a), how long must the patient be dialyzed? (Neglect the loss in total blood volume due to the removal of water in the dialyzer.)

2 A Review of Thermodynamic Concepts

Thermodynamics (power from heat) as a science began its development during the nineteenth century and was first used to understand the operation of work-producing devices such as steam engines. In broadest terms, thermodynamics is concerned with the relationships between different types of energy. Here, we are mainly interested in a specialized area of thermodynamics related to solutions. But before we can understand solution thermodynamics, we first must review some basic concepts of thermodynamics that will then lead us to the relationships we need for understanding the thermodynamics of solutions. Our goal here is to understand the mathematical basis of the field and apply the most useful results to our study of biomedical engineering transport phenomena.

2.1 FIRST LAW OF THERMODYNAMICS

There are three general laws of thermodynamics and these will be stated here in forms that do not consider the effects of nuclear reactions. The *first law* is a statement of energy conservation and must be applied to both the system and the surroundings. In order to describe how the system and surroundings may exchange energy, we first need to define the type of system that is being considered. The most basic type of system is called an *isolated system*. In an isolated system, there is no exchange of mass or energy between the system and its surroundings. Therefore, for an isolated system its energy and mass is constant. We define a *closed system* as a system that can exchange energy with its surroundings but not mass. Hence, the mass of a closed system does not change. The *open system* is defined as one that can exchange both mass and energy with its surroundings. We will discuss open systems in greater detail during our discussion of solution thermodynamics.

2.1.1 CLOSED SYSTEMS

Energy exchange between a closed system and its surroundings is in the form of *heat* (Q) and *work* (W). Careful attention must be given to the signs carried by Q and W. The sign convention that is used is that Q and W are positive for transfer of energy from the surroundings to the system. Hence, if heat is added to the system, or work done on the system, then Q and W are positive.

The total energy of a closed system consists of a summation of its *internal energy* (U) and external energies, known as *potential energy* (E_P) and *kinetic energy* (E_K). The potential energy depends on the position of the system in a gravitational field and the kinetic energy is a result of system motion. The internal energy of the system is energy possessed by the molecules that make up the chemical substances within the system. These molecular energies include the kinetic energy of translation, rotation, vibration, and intermolecular forces.

With respect to the system, we are usually not really interested in the absolute values of these energies, but only in their changes. Hence, we can state that the change in the total energy of the system is equal to $\Delta U + \Delta E_P + \Delta E_K$, where the Greek symbol delta (Δ) signifies the final state minus the initial state.

For a closed system, the change in the total energy of the system ($\Delta U + \Delta E_P + \Delta E_K$) must be equal to the energy transferred as heat and work with the surroundings. Equation 2.1 is therefore a statement of the *first law of thermodynamics* for a closed system:

$$\Delta U + \Delta E_P + \Delta E_K = Q + W. \tag{2.1}$$

If there are no changes in the kinetic and potential energies of the system, then we simply have that, for a closed system:

$$\Delta U = Q + W. \tag{2.2}$$

Most of the time, ΔE_P and ΔE_K are much smaller than ΔU, Q, and W, and we can just use Equation 2.2. The exceptions are those cases where large changes in system position or velocity are expected or desired. For an isolated system we have $\Delta U = 0$, since both Q and W are zero.

2.1.2 STEADY FLOW PROCESSES

We can also apply the first law of thermodynamics to the steady flow process shown in Figure 2.1. We now consider the energy changes that occur within a unit mass of material that enters at plane 1 and leaves the process at plane 2. We take our system to be this unit mass of material. The total energy of this unit mass of fluid can change as a result of changes in its internal energy, potential energy, and kinetic energy. Equation 2.1 still applies to this situation as well; however, the kinetic energy term represents the change in kinetic energy of the unit mass of flowing fluid and the potential energy term represents the change in potential energy of the unit mass of flowing fluid due to changes in its elevation within its gravitational field. Also, it is important to remember that the work term W also includes the work done on or by the unit mass of fluid as it flows through the process. This work done on or by the unit mass of fluid is called the *shaft work* (W_S).

Examples of shaft work include the work effect associated with reciprocating engines, pumps, turbines, and compressors. Therefore, in Equation 2.1, W represents the shaft work as well as the work done on or by the unit mass of fluid as it enters and exits the process. Recall that work is defined as the displacement of an external force, that is $W = -\int_{x_1}^{x_2} F_{external} dx$. Here, the force would be the pressure at the entrance of the process (P_1) multiplied by the cross-sectional area (A_1) normal to the flow of the unit mass of fluid. The displacement of the unit mass of fluid would equal the

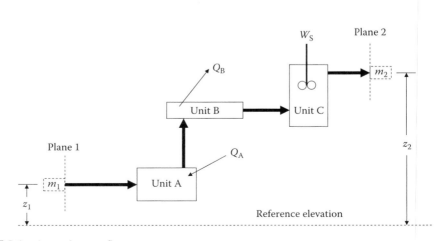

FIGURE 2.1   A steady state flow process.

specific volume of the fluid (V_1) divided by the cross-sectional area (A_1). So, the work performed by the surroundings to push the unit mass of fluid into the process at the entrance is given by

$$W_1 = P_1 A_1 \times \frac{V_1}{A_1} = P_1 V_1. \tag{2.3}$$

Note that W_1 is positive since work is done by the surroundings on the unit mass of fluid. At the exit of the process, the work is done by the system on the surroundings and W_2 is given by the following expression:

$$W_2 = -P_2 V_2. \tag{2.4}$$

For our system taken as a unit mass of fluid, we can now rewrite Equation 2.1 and obtain the following equation between planes 1 and 2 as shown in Figure 2.1:

$$\Delta U + \Delta E_\mathrm{P} + \Delta E_\mathrm{K} = Q + W_\mathrm{S} + P_1 V_1 - P_2 V_2. \tag{2.5}$$

This equation can also be rearranged and written as

$$\Delta(U + PV) + \Delta E_\mathrm{P} + \Delta E_\mathrm{K} = Q + W_\mathrm{S}. \tag{2.6}$$

The combination of the internal energy (U) and the quantity (PV) is defined as the *enthalpy* (H), and we can write Equation 2.6 as

$$\Delta H + \Delta E_\mathrm{P} + \Delta E_\mathrm{K} = Q + W_\mathrm{S}. \tag{2.7}$$

In many cases, as mentioned earlier, the changes in the potential and kinetic energy of the fluid are negligible in comparison to the other terms, so Equation 2.7 simplifies to

$$\Delta H = Q + W_\mathrm{S}. \tag{2.8}$$

Equations 2.7 and 2.8 are expressions of the first law of thermodynamics for steady flow processes, as shown in Figure 2.1. Recall that for a closed nonflow system the first law is given by Equations 2.1 and 2.2.

As written, Equations 2.2 and 2.8 apply for finite changes in the energy quantities ΔU and ΔH. We will find it useful later to work with differential changes of these energy quantities, and Equations 2.2 and 2.8 may then be written in their differential form as

$$dU = dQ + dW \quad \text{and} \quad dH = dQ + dW_\mathrm{S}. \tag{2.9}$$

2.2   SECOND LAW OF THERMODYNAMICS

The *second law of thermodynamics* states that there is a property of matter called the *entropy* (S) and that for any process, the sum of the entropy changes of the system and the surroundings (i.e., ΔS_Total) is always greater than or equal to zero. Therefore, entropy is not conserved like energy. No process is feasible if the total change in entropy is less than zero. This statement of the second law of thermodynamics is given by

$$\Delta S_{\text{Total}} = \Delta S_{\text{System}} + \Delta S_{\text{Surroundings}} \geq 0. \tag{2.10}$$

For an isolated system, there is no exchange of mass or energy between the system and the surroundings, so $\Delta S_{\text{Surroundings}} = 0$ and ΔS_{System} (isolated) ≥ 0.

At the molecular level, entropy is a measure of disorder. The more disorder, the higher the entropy. It is important to recognize that a system can become more organized ($\Delta S_{\text{System}} < 0$); however, there must be an even greater increase in the disorder of the surroundings such that, in an overall sense, Equation 2.10 is satisfied. The *third law of thermodynamics* sets a lower limit on the entropy. The third law states that at a temperature of absolute zero, the entropy for a perfectly ordered crystal of a given substance is zero.

2.2.1 REVERSIBLE PROCESSES

Processes for which $\Delta S_{\text{Total}} > 0$ are known as *irreversible processes*, whereas processes for which $\Delta S_{\text{Total}} = 0$ are known as *reversible processes*. A reversible process is one that can be reversed by an infinitesimal change in the variable that controls the process; in other words, the system is only infinitesimally removed from its equilibrium state.

In thermodynamics, an equilibrium state is defined as one in which there is no tendency for change on a macroscopic scale. For example, for a mechanically reversible process, imagine a frictionless piston that confines a gas within a cylinder closed at one end. At equilibrium, the pressure of the gas within the cylinder exactly balances the weight of the piston and the pressure of the surroundings. A reversible expansion of this gas requires that the pressure within the cylinder be increased by an infinitesimal amount. Similarly, a reversible compression requires that the pressure of the surroundings be increased by an infinitesimal amount. The driving force for the reversible expansion or compression of the gas within the cylinder is the infinitesimal difference in pressure between the confined gas and the surroundings.

Similarly, thermal equilibrium requires that an object and its surroundings be at the same temperature. Reversible heat transfer occurs when the object and its surroundings differ in temperature by an infinitesimal amount. The driving force for reversible transfer of heat being the infinitesimal difference in temperature between the object and its surroundings.

A reversible process is an idealization since no real process is reversible. However, it is a useful concept because it does allow us to set limits on a real process and perform useful calculations. For example, for a mechanically reversible process like the piston and cylinder example previously discussed, the work term in Equations 2.1 and 2.2 can be evaluated as follows. Recall that work is defined as the displacement of an external force, that is $W = -\int_{x_1}^{x_2} F_{\text{external}} dx$. Then, for the reversible work, we can replace the external force by the product of the gas pressure, P, and the cross-sectional area of the cylinder, A, which is the force generated by the gas pressure within the cylinder. Since the external and internal forces are only infinitesimally different, then $F_{\text{external}} = PA$. The differential displacement of the piston (i.e., dx) by the expanding or contracting gas within the cylinder would be equivalent to the differential change in volume of the gas divided by the cross-sectional area of the cylinder, i.e., $dx = dV/A$. Therefore, Equation 2.11 results for the mechanically reversible work of the gas contained within our piston cylinder system:

$$W_{\text{Reversible}} = -\int_{V_1}^{V_2} P \, dV, \tag{2.11}$$

or in differential form, we can say that $dW_{\text{Reversible}} = -P \, dV$. Provided we know how the pressure P varies with the change in volume, we can evaluate the integral in Equation 2.11 and obtain the reversible work, $W_{\text{Reversible}}$.

Example 2.1

Consider the reversible expansion or compression of one mole of an ideal gas at constant temperature (isothermal) from an initial volume V_1 to a final volume V_2. Find an expression for the heat and work effects of this isothermal process.

Solution

For an isothermal process involving an ideal gas, there is no change in the internal energy since the temperature is constant, so $\Delta U = 0$, and from Equation 2.2 we have that $Q = -W_{\text{Reversible}}$. For an ideal gas, we can also use the ideal gas law with $P = RT/V$ and substitute this into Equation 2.11. On integration, we obtain the following equations for the heat and reversible work per mole of gas:

$$Q = -W_{\text{Reversible}} = \int_{V_1}^{V_2} \frac{RT}{V} \, dV = RT \, \ln\left(\frac{V_2}{V_1}\right) = RT \, \ln\left(\frac{P_1}{P_2}\right).$$

2.3 PROPERTIES

Through the previous discussion of the three laws of thermodynamics, we have introduced three new properties of matter. These are internal energy (U), enthalpy (H), and entropy (S). Along with pressure (P), volume (V), and temperature (T), these are all known as properties or state variables. State variables only depend on the state of the system and not on how one arrived at that state. Q and W (or W_S) are not properties since, according to Equations 2.2 and 2.8, one can envision different processes or paths involving the transfer of Q and W (or W_S), all giving the same change in the internal energy (ΔU), enthalpy (ΔH), or entropy (ΔS). Therefore, Q and W (or W_S) are path-dependent quantities and depend on how the process is actually carried out. For example, for a mechanically reversible process, Equation 2.11 allows calculation of the work, provided one knows how P changes with V.

For a pure substance, only two of these properties (i.e., T, P, V, U, and S) need to be specified in order to completely define the thermodynamic state of the substance. Our equation of state also provides an additional relationship between P, V, and T. Usually, we fix the state of a pure component by specifying the temperature and pressure. So, from the equation of state, we can calculate the volume. Other properties, such as internal energy, enthalpy, and entropy, are also known since they too will depend only on the temperature and pressure, as shown in the following discussion.

2.3.1 HEAT CAPACITY

Calculating the changes in the internal energy, enthalpy, and entropy is facilitated by the definition of heat capacities. The *heat capacity* relates the change in temperature of an object to the amount of heat that was added to it. However, heat (Q) is not a property but is dependent on how the process of heat transfer was carried out. So, it is convenient to define heat capacity in such a manner that it too is a property. Therefore, we have two types of heat capacity, one defined at constant volume (C_V) and the other defined at constant pressure (C_P) as given by the following equations:

$$C_V \equiv \left(\frac{\partial U}{\partial T}\right)_V \quad \text{heat capacity at constant volume.} \tag{2.12}$$

$$C_P \equiv \left(\frac{\partial H}{\partial T}\right)_P \quad \text{heat capacity at constant pressure.} \tag{2.13}$$

Since they only depend on properties, C_P and C_V are also properties of a substance and their values for specific substances may be found or estimated as described in such reference books as Poling et al. (2001). For monatomic gases, C_P is approximately $(5/2)R$ and for diatomic gases like nitrogen, oxygen, and air, C_P is approximately $(7/2)R$. We will also show in Example 2.2 that for an ideal gas, $C_P = C_V + R$.

Now for a constant volume process, or for any process where the final volume is the same as the initial volume, Equation 2.12, when integrated, tells us that the change in internal energy is given by

$$\Delta U = \int_{T_1}^{T_2} C_V \, dT. \tag{2.14}$$

If the process occurs reversibly and at constant volume, i.e., the volume never changes, then from Equation 2.11, $W_{\text{Reversible}} = 0$ and $Q = \Delta U$. So, for a constant volume process, the heat transferred is equal to the change in the internal energy.

In a similar manner, for the constant pressure process, or a process in which the initial and final pressures are the same, Equation 2.13 can be integrated to obtain the change in enthalpy:

$$\Delta H = \int_{T_1}^{T_2} C_P \, dT. \tag{2.15}$$

If the process occurs reversibly and at constant pressure, i.e., the pressure never changes, then from Equations 2.2 and 2.11 we have that $W_{\text{Reversible}} = -P\Delta V$ and $\Delta U = Q - P\Delta V$ or $Q = \Delta H$. Hence, for a constant pressure process, the heat transferred is equal to the change in enthalpy.

2.3.2 Calculating the Change in Entropy

The entropy change of a closed system during a reversible process has been shown to be given by

$$\Delta S = \int \frac{dQ_{\text{Reversible}}}{T}. \tag{2.16}$$

In differential form, we can write this as $dQ_{\text{Reversible}} = T \, dS$.

2.3.2.1 Entropy Change of an Ideal Gas

We can use this result to calculate the entropy change for a substance. First, we consider an ideal gas and assume that we have one mole of gas and write the first law of thermodynamics for a reversible process as $dU = dQ_{\text{Reversible}} - P\,dV = T\,dS - P\,dV$. Next, we use the definition of enthalpy ($H = U + PV$) to arrive at the fact that $dH = dU + P\,dV + V\,dP$. We now solve this relationship for dU and use the result shown previously for dU to obtain the following expression for dS:

$$dS = \frac{dH}{T} - \frac{V}{T} \, dP. \tag{2.17}$$

For one mole of an ideal gas, we know that $dH = C_P\, dT$ and from the ideal gas law $V/T = R/P$, so Equation 2.17 becomes

$$dS = C_P \frac{dT}{T} - R\frac{dP}{P}.$$

(2.18)

This equation can then be integrated from an initial state (T_1, P_1) to a final state (T_2, P_2) as given by

$$\Delta S = \int_{T_1}^{T_2} C_P \frac{dT}{T} - R\ln\left(\frac{P_2}{P_1}\right).$$

(2.19)

Note, all that is needed to calculate the entropy change of an ideal gas is the heat capacity, C_p. Note that at constant P, $\Delta S = \int_{T_1}^{T_2} C_P(dT/T)$.

For a constant volume process, we can use the fact that from the ideal gas law, $(dP/P) = (RdT/PV) = (dT/T)$ and Example 2.2 shown below we have that for an ideal gas, $C_P = C_V + R$, so when these two relationships are used in Equation 2.18 and the result is integrated, we obtain $\Delta S = \int_{T_1}^{T_2} C_V(dT/T)$ for the constant volume process.

It is important to note that although we started the derivation of Equations 2.18 and 2.19 assuming a reversible process, the resulting equations only contain properties and are therefore independent of how the process was actually carried out between the initial state of (T_1, P_1) and the final state of (T_2, P_2). Therefore, Equations 2.18 and 2.19 can be used to calculate the entropy change of an ideal gas regardless of whether or not the process is reversible.

Example 2.2

Calculate the final temperature and the work produced for the reversible adiabatic expansion of 10 moles of an ideal gas from 10 atm and 500 K to a final pressure of 1 atm. Assume that $C_V = (5/2)R$ and that the heat capacity ratio $\gamma = C_P/C_V$ is also a constant and equal to 1.4.

Solution

Since the process is reversible, we can state that $\Delta S_{Total} = \Delta S_{System} + \Delta S_{Surroundings} = 0$. An adiabatic process means that there is no heat transfer (i.e., $Q = 0$) between the system (i.e., the gas) and its surroundings. Therefore, $\Delta S_{Surroundings} = 0$ and we then have that $\Delta S_{System} = 0$. A process for which $\Delta S_{System} = 0$ is also called an *isentropic process*. Next, we can use Equation 2.19 and set that equal to zero and obtain.

$$C_P \ln\left(\frac{T_2}{T_1}\right) = R\ln\left(\frac{P_2}{P_1}\right).$$

Next, we solve this equation for the final temperature T_2, which is given by

$$\frac{T_2}{T_1} = \left(\frac{P_2}{P_1}\right)^{R/C_P} = \left(\frac{P_2}{P_1}\right)^{(\gamma-1)/\gamma}.$$

Note that we have used the fact that since $H = U + PV$, then for an ideal gas $H = U + RT$ and $dH = dU + R\, dT$. Since $dH = C_P\, dT$ and $dU = C_V\, dT$, we have for an ideal gas that $C_P = C_V + R$. The heat capacity ratio is defined by $\gamma = C_P/C_V$. We can now use the previous equation to calculate the final temperature of the gas following the reversible expansion:

$$T_2 = 500 \text{ K}\left(\frac{1 \text{ atm}}{10 \text{ atm}}\right)^{(1.4-1)/1.4} = 258.98 \text{ K}.$$

Next, we can calculate the amount of work that is done by the 10 moles of gas. For a closed system, we have from the first law of thermodynamics that $n \Delta U = Q + W$, where n is the number of moles of gas. Since the process is adiabatic, then $Q = 0$ and $W = n \Delta U = n C_V (T_2 - T_1)$. Now $C_V = (5/2)R$ and using the value of $R = 8.314$ J mol^{-1} K^{-1} from Table 1.6 gives a value of $C_V = 20.78$ J mol^{-1} K^{-1}. So the work effect of this expansion can now be calculated as

$$W = 10 \text{ mol} \times 20.78 \text{ J mol}^{-1} \text{K}^{-1} \times (258.98 \text{ K} - 500 \text{ K})$$

$$= -50,084 \text{ J} = -50.084 \text{ kJ}.$$

In this example, the system is doing the work so the sign is negative. The change in the internal energy is also the same as the work in this case, so the energy used to perform the work by the system is at the expense of the internal energy of the gas. That is why the final temperature of the gas is much lower than the initial temperature of the gas.

2.3.3 Gibbs and Helmholtz Free Energy

In addition to the internal energy (U), entropy (S), and enthalpy ($H = U + PV$), there are two additional properties that can be derived from the primary properties of P, V, T, U, and S. These are the *Gibbs free energy* (G) and the *Helmholtz free energy* (A).

2.3.3.1 Gibbs Free Energy

The Gibbs free energy (G) is also a property and is defined in terms of the enthalpy, temperature, and entropy as $G = H - TS$ or $G = U + PV - TS$. The Gibbs free energy change (ΔG) at constant temperature can be calculated from the change in enthalpy and entropy as $\Delta G = \Delta H - T \Delta S$. The Gibbs free energy can be shown to equal the maximum amount of useful work that can be obtained from a reversible process at constant T and P. This can be shown as follows.

We have at constant temperature and pressure that $\Delta G = \Delta H - T \Delta S = \Delta U + P \Delta V - T \Delta S$. But for a reversible process at constant temperature, we also know that $\Delta U = Q_{\text{Reversible}} + W_{\text{Reversible}} = T \Delta S + W_{\text{Reversible}}$. Therefore, $\Delta G = W_{\text{Reversible}} + P \Delta V$. At constant pressure, $- P \Delta V$ represents the work (W_{PV}) obtained as a result of any volume changes that occur in the system. So, we define the useful work over and above any PV work as $W_{\text{Useful}} = W_{\text{Reversible}} - W_{\text{PV}} = \Delta G$.

The Gibbs free energy is also useful for determining whether or not a given process will occur at constant temperature and pressure. For example, from the second law of thermodynamics, we know that for an isolated system $\Delta S = (Q_{\text{Reversible}}/T) \geq 0$. From the first law of thermodynamics, we know that for a reversible process at constant pressure $Q_{\text{Reversible}} = \Delta U + P\Delta V = \Delta H$. Therefore, we have that $\Delta S \geq \Delta H/T$ or $\Delta H - T \Delta S \leq 0$. But $\Delta H - T \Delta S$ is the same as ΔG at constant temperature, so $\Delta G \leq 0$. We thus obtain the following criterion for determining the feasibility of a given process at constant temperature and pressure:

$$\text{Spontaneous process} \quad \Delta G < 0,$$

$$\text{Equilibrium} \quad \Delta G = 0, \tag{2.20}$$

$$\text{No spontaneous process} \quad \Delta G > 0.$$

2.3.3.2 Helmholtz Free Energy

The Helmholtz free energy (A) is defined in terms of the internal energy, temperature, and entropy as $A = U - TS$. The Helmholtz free energy is useful for determining whether or not a given process will occur at constant temperature and volume. The criterion for the feasibility of a process at constant temperature and volume is

$$\text{Spontaneous process } \Delta A < 0,$$

$$\text{Equilibrium } \Delta A = 0,$$

$$\text{No spontaneous process } \Delta A > 0.$$

(2.21)

2.4   FUNDAMENTAL PROPERTY RELATIONS

With this background on the first and second laws of thermodynamics, we can now develop what are known as the fundamental property relations for one mole of a single fluid phase of constant composition. From Equation 2.9 in differential form, we have that $dU = dQ + dW$. For a reversible process (cf. Equations 2.11 and 2.16), we also have that $dQ_{\text{Reversible}} = TdS$ and $dW_{\text{Reversible}} = -PdV$. Hence:

$$dU = TdS - PdV.$$

(2.22)

As discussed before, even though this equation was derived for the special case of a reversible process, since it only contains properties of the system, it is valid for all processes, reversible or not.

Now, since $H = U + PV$, we can differentiate this to obtain $dH = dU + PdV + VdP$. We can then substitute for dU from Equation 2.22 to obtain $dH = TdS - PdV + PdV + VdP$. So, we get that $dH = TdS + VdP$. Next, we have that $G = H - TS$, so $dG = dH - TdS - SdT$. Using the relationship we just obtained for dH, we then get that $dG = TdS + VdP - TdS - SdT$ and obtain the result $dG = -SdT + VdP$. Finally, since $A = U - TS$, we then have $dA = dU - TdS - SdT$. Using Equation 2.22, we get $dA = TdS - PdV - TdS - SdT$ and we obtain $dA = -PdV - SdT$.

So, for one mole of a single fluid phase of constant composition, we can state our fundamental property relationships as

$$dU = TdS - PdV,$$

(2.23)

$$dH = TdS + VdP,$$

(2.24)

$$dG = -SdT + VdP,$$

(2.25)

$$dA = -PdV - SdT.$$

(2.26)

These equations also provide functional relationships for U, H, G, and A. Therefore, we have that $U = U(S,V)$, $H = H(S,P)$, $G = G(T,P)$, and $A = A(V,T)$.

2.4.1   EXACT DIFFERENTIALS

The fundamental property relations shown earlier are also exact differentials. Recall from calculus that the total differential of a function $F(x,y)$ is given by

$$dF = \left(\frac{\partial F}{\partial x}\right)_y dx + \left(\frac{dF}{dy}\right)_x dy = Mdx + Ndy,$$

(2.27)

with $M = (\partial F/\partial x)_y$ and $N = (\partial F/\partial y)_x$. We can also differentiate M and N with respect to y and x and obtain

$$\left(\frac{\partial M}{\partial y}\right)_x = \frac{\partial^2 F}{\partial y \partial x} \quad \text{and} \quad \left(\frac{\partial N}{\partial x}\right)_y = \frac{\partial^2 F}{\partial x \partial y}. \tag{2.28}$$

Because the order of differentiation with respect to x and y in Equations 2.27 and 2.28 is not important, we obtain

$$\left(\frac{\partial M}{\partial y}\right)_x = \left(\frac{\partial N}{\partial x}\right)_y, \tag{2.29}$$

which is the criterion for exactness of the total differential given by Equation 2.27. We can now use Equation 2.29 on our fundamental property relations given in Equations 2.23 through 2.26 to obtain a set of relationships between our properties. These relationships are known as the *Maxwell equations* and several of these, derived from Equations 2.23 through 2.26, are

$$\begin{aligned}
\left(\frac{\partial T}{\partial V}\right)_S &= -\left(\frac{\partial P}{\partial S}\right)_V, \\[2mm]
\left(\frac{\partial T}{\partial P}\right)_S &= \left(\frac{\partial V}{\partial S}\right)_P, \\[2mm]
\left(\frac{\partial V}{\partial T}\right)_P &= -\left(\frac{\partial S}{\partial P}\right)_T, \\[2mm]
\left(\frac{\partial P}{\partial T}\right)_V &= \left(\frac{\partial S}{\partial V}\right)_T.
\end{aligned} \tag{2.30}$$

The Maxwell equations and the fundamental property relations are useful for deriving thermodynamic relationships between the various properties. The following example illustrates this.

Example 2.3

Obtain expressions for the T and P dependence of the enthalpy and the entropy, i.e., $H(T,P)$ and $S(T,P)$.

Solution

The approach is to use the fundamental property relations and the Maxwell equations to obtain final expressions that only depend on P, V, T, and C_P and heat capacity. P, V, and T can then be related by an equation of state, the simplest being the ideal gas law for which $PV = RT$, or experimental PVT data. With this as our strategy, we first take the total differential of H and S and obtain

$$dH = \left(\frac{\partial H}{\partial T}\right)_P dT + \left(\frac{\partial H}{\partial P}\right)_T dP \quad \text{and} \quad dS = \left(\frac{\partial S}{\partial T}\right)_P dT + \left(\frac{\partial S}{\partial P}\right)_T dP.$$

Now, since $(\partial H/\partial T)_P = C_P$, from Equation 2.24 we can also get that $(\partial H/\partial T)_P = T(\partial S/\partial T)_P$ and $(\partial S/\partial T)_P = (1/T)(\partial H/\partial T)_P = C_P/T$. From Equation 2.24, we can also write that $(\partial H/\partial P)_T = T(\partial S/\partial P)_T + V$.

Using the third Maxwell equation in Equation 2.30, replace $(\partial S/\partial P)_T$ with $-(\partial V/\partial T)_P$ and we then have that $(\partial H/\partial P)_T = V - T(\partial V/\partial T)_P$. Our expressions for dH and dS can then be written as

$$dH = C_P\, dT + \left[V - T\left(\frac{\partial V}{\partial T}\right)_P\right] dP \quad \text{and} \quad dS = C_P \frac{dT}{T} - \left(\frac{\partial V}{\partial T}\right)_P dP.$$

These equations can be integrated to obtain the change in enthalpy and entropy between two states (T_1, P_1) and (T_2, P_2):

$$\Delta H = \int_{T_1}^{T_2} C_P\, dT + \int_{P_1}^{P_2}\left[V - T\left(\frac{\partial V}{\partial T}\right)_P\right] dP \quad \text{and} \quad \Delta S = \int_{T_1}^{T_2} C_P \frac{dT}{T} - \int_{P_1}^{P_2}\left(\frac{\partial V}{\partial T}\right)_P dP.$$

These equations provide the temperature and pressure dependence for the enthalpy and entropy for one mole of a single phase fluid of constant composition. The heat capacity and experimental *PVT* data, or an equation of state, are needed to solve these equations. For the special case of an ideal gas, we can use the fact that $PV = RT$ and that for an ideal gas $(\partial V/\partial T)_P = R/P$. On substituting this expression into the previous equations, we obtain the following results for an ideal gas:

$$\Delta H = \int_{T_1}^{T_2} C_P\, dT \quad \text{and} \quad \Delta S = \int C_P \frac{dT}{T} - R\ln\left(\frac{P_2}{P_1}\right).$$

Note that we obtain the result that, for an ideal gas, the enthalpy and for that matter the internal energy only depend on temperature. However, the entropy of an ideal gas depends on both the temperature and pressure. Also note that this result for ΔS for the special case of the ideal gas is the same result we obtained earlier as Equation 2.19.

2.5 SINGLE PHASE OPEN SYSTEMS

Now suppose that we have a single phase open system containing N chemical substances. For an open system, there can be an exchange of matter between the system and the surroundings. Therefore, the composition or the number of moles of each substance present in the system can change. Let n represent the total moles of these N chemical substances that are present. So, for a total of n moles and any property M, we can state that $nM = nM (T, P, n_1, n_2, n_3,\ldots, n_N)$ and n_i represent the moles of each substance that is present in the system. This tells us that the property M depends on the temperature, pressure, and the number of moles of each species present in the solution. The total differential of nM is then given by

$$d(nM) = \left[\frac{\partial(nM)}{\partial T}\right]_{P,n} dT + \left[\frac{\partial(nM)}{\partial P}\right]_{T,n} dP + \sum_i \left[\frac{\partial(nM)}{\partial n_i}\right]_{T,P,n_j} dn_i. \tag{2.31}$$

In the last term of Equation 2.31, the summation is over all substances that are present and the partial derivative within the bracketed term in the summation is taken with respect to substance i at constant temperature, pressure, and the moles of all other substances held constant.

2.5.1 PARTIAL MOLAR PROPERTIES

In solution thermodynamics, the partial derivative of a property (M) with respect to the number of moles of a given substance (n_i) is called a *partial molar property* $(\bar{M}_i)$ and, in general, is defined by

$$\bar{M}_i = \left[\frac{\partial(nM)}{\partial n_i}\right]_{T,P,n_j}. \tag{2.32}$$

Of particular interest in subsequent discussions is the partial molar Gibbs free energy ($\bar{G}_i$), which is also known as the *chemical potential* (μ_i) of component i in the mixture. The chemical potential is therefore defined as

$$\mu_i = \bar{G}_i = \left[\frac{\partial(nG)}{\partial n_i} \right]_{T, P, n_j}. \tag{2.33}$$

We can use Equation 2.32 and rewrite Equation 2.31 as

$$d(nM) = \left[\frac{\partial(nM)}{\partial T} \right]_{P, n} dT + \left[\frac{\partial(nM)}{\partial P} \right]_{T, n} dP + \sum_i \bar{M}_i \, dn_i. \tag{2.34}$$

Next, we can define the mole fraction of substance i^* as $x_i = n_i/n$, so $n_i = x_i n$ and $\sum_i x_i = 1$. We also have that $dn_i = x_i \, dn + n \, dx_i$ and $d(nM) = ndM + Mdn$. Using these relations, and recognizing that n in the first two bracketed terms of Equation 2.34 is constant, Equation 2.34 can be rearranged into the following form:

$$\left[dM - \left(\frac{\partial M}{\partial T} \right)_{P, x} dT - \left(\frac{\partial M}{\partial P} \right)_{T, x} dP - \sum_i \bar{M}_i dx_i \right] n + \left[M - \sum_i x_i \bar{M}_i \right] dn = 0. \tag{2.35}$$

Since n and dn are arbitrary, the only way that Equation 2.35 can be satisfied is for both bracketed terms to equal zero. Hence, we obtain the following two equations:

$$dM = \left(\frac{\partial M}{\partial T} \right)_{P, x} dT + \left(\frac{\partial M}{\partial P} \right)_{T, x} dP + \sum_i \bar{M}_i dx_i, \tag{2.36}$$

$$M = \sum_i x_i \bar{M}_i \quad \text{or} \quad nM = \sum_i n_i \bar{M}_i. \tag{2.37}$$

Now Equation 2.36 is nothing more than Equation 2.34 written on a mole fraction basis. Equation 2.37, however, shows how the mixture property M depends on the composition of the solution. Note that the mixture property is a mole fraction or mole weighted average of the component partial molar property $\bar{M}_i$. The partial molar property for component i represents the property value for that component as it exists in the solution, which can be quite different from the pure component value of that property, M_i.

From Equation 2.37, we can also write that $dM = \sum_i x_i d\bar{M}_i + \sum_i \bar{M}_i dx_i$. Using this relationship along with Equation 2.36 results in

$$\left(\frac{\partial M}{\partial T} \right)_{P, x} dT + \left(\frac{\partial M}{\partial P} \right)_{T, x} dP - \sum_i x_i d\bar{M}_i = 0. \tag{2.38}$$

Equation 2.38 places another restriction on the property changes for a single phase solution. This equation is known as the *Gibbs–Duhem equation* and can be used to test the thermodynamic

---

* Note that most of the time x is used to denote liquid phase mole fractions and y is used for vapor or gas phase mole fractions.

consistency of experimental mixture property data, since this equation must be satisfied. For the special case of constant temperature and pressure, we have that

$$\sum_i x_i d\bar{M}_i = 0. \tag{2.39}$$

2.5.1.1 Binary Systems

To illustrate the application of the previous equations, let us apply them to a binary solution at constant temperature and pressure. For a binary solution, we can write from Equation 2.37 that

$$M = x_1 \bar{M}_1 + x_2 \bar{M}_2 \quad \text{and} \quad dM = x_1 d\bar{M}_1 + \bar{M}_1 dx_1 + x_2 d\bar{M}_2 + \bar{M}_2 dx_2. \tag{2.40}$$

The Gibbs–Duhem equation for a binary system can be written as

$$x_1 d\bar{M}_1 + x_2 d\bar{M}_2 = 0. \tag{2.41}$$

Since $x_1 + x_2 = 1$ and $dx_1 = -dx_2$, we can combine Equations 2.40 and 2.41 and obtain

$$dM = \bar{M}_1 dx_1 - \bar{M}_2 dx_1 \quad \text{or} \quad \frac{dM}{dx_1} = \bar{M}_1 - \bar{M}_2. \tag{2.42}$$

Now we can solve for $\bar{M}_1$ and $\bar{M}_2$ from Equation 2.40 and substitute these results into Equation 2.42 to obtain the following expressions for $\bar{M}_1$ and $\bar{M}_2$:

$$\bar{M}_1 = M + (1 - x_1)\frac{dM}{dx_1} \quad \text{and} \quad \bar{M}_2 = M - x_1\frac{dM}{dx_1}. \tag{2.43}$$

These equations are very important since they allow for the calculation at constant temperature and pressure of the partial molar properties $\bar{M}_1$ and $\bar{M}_2$ from the composition dependence of the mixture property M. Note that as $x_2 \to 0$, then component 2 is becoming infinitely dilute and from Equation 2.43, $\bar{M}_2 \to \bar{M}_2^\infty \to M_1 - (dM/dx_1)|_{x_1 \to 1}$ where also $\bar{M}_1 \to M_1$ or the pure component property value M_1 and the superscript ∞ on $\bar{M}_2$ denotes that component 2 is infinitely dilute. Likewise, as $x_1 \to 0$, then component 1 is becoming infinitely dilute and $\bar{M}_1 \to \bar{M}_1^\infty \to M_2 + (dM/dx_1)|_{x_2 \to 1}$. Figure 2.2 shows the composition dependence of these property values at constant T and P. Note that $(dM/dx_1)|_{x_1 \to 1}$ and $(dM/dx_1)|_{x_2 \to 1}$ are the respective tangents of the M curve as $x_2 \to 0$ and $x_1 \to 0$.

2.5.1.2 Property Changes of Mixing

The difference between the mixture property value (M) and the mole fraction weighted sum of the pure component property values ($\sum_i y_i M_i$) evaluated at the same T and P is known as the *property change of mixing* (ΔM^{mix}) and is given by

$$\Delta M^{\text{mix}} = \sum_i y_i \bar{M}_i - \sum_i y_i M_i = M - \sum_i y_i M_i. \tag{2.44}$$

2.5.1.3 Ideal Gas

The above-mentioned relationships are valid for describing the solution behavior of real solutions whether they are solids, liquids, or gases. Now let us see what these relationships tell us for the special case of an ideal gas. Recall that the ideal gas model assumes that the molecules have negligible

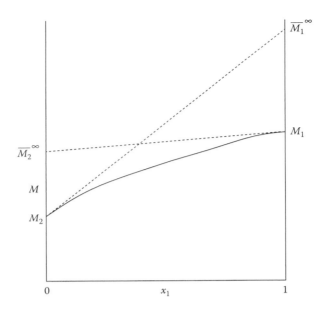

FIGURE 2.2   Partial molar properties in a binary system at constant T and P.

volume and that the molecules do not interact with one another. In other words, each species in the ideal gas acts as if no other species are present and all of the gas molecules can move independently within the whole volume (V_T) of the container.

Consider an ideal gas consisting of a total of n moles and N components at a given temperature T and in a total volume V_T. We know from the ideal gas law that $P = nRT/V$. For the same value of the temperature, each of the n_i moles of component i considered separately will also occupy the same total volume V_T. Therefore, each component i exerts a contribution to the total pressure P called the *partial pressure* (P_i), which is given by $P_i = n_i RT/V_T$. The ratio of P_i to P is given by $(P_i/P) = (n_i/n) = y_i$, which defines the gas phase mole fraction of component i in the mixture, that is y_i. Hence, we obtain the fact that the *partial pressure* of component i in an ideal gas mixture is

$$P_i = y_i P \quad \text{with} \quad y_i = \frac{n_i}{n} = \frac{P_i}{P}. \tag{2.45}$$

Since $\Sigma_i y_i = 1$, then $\Sigma_i P_i = P$, or the sum of the partial pressures of all components equals the total pressure. In addition, we can deduce that for an ideal gas the partial molar volume for component i is $\bar{V}_i^{\text{ideal gas}} = V_i^{\text{ideal gas}} = RT / P$. Since $V_T = \Sigma_i n_i V_i^{\text{ideal gas}}$, then we have that $V_T = \Sigma_i n_i V_i^{\text{ideal gas}} = \Sigma_i n_i (RT/P) = (nRT/P)$, and we see that this definition of the ideal gas partial molar volume satisfies the ideal gas law for the mixture. Therefore, for an ideal gas the partial molar volume is the same for all components and is equal to RT/P.

With the exception of the volume, for any other property, the partial molar value of the property for component i in an ideal gas mixture at T and P is the same as the corresponding pure component molar property at the mixture temperature and at a pressure that is equal to that component's partial pressure in the mixture, that is

$$\bar{M}_i^{\text{ideal gas}}(T,P) = M_i^{\text{ideal gas}}(T,P_i). \tag{2.46}$$

In Example 2.3, we showed that for an ideal gas the enthalpy and the internal energy do not depend on the pressure. Furthermore, since there are no molecular interactions in an ideal gas, the enthalpy or internal energy of component i in the mixture for a given T and P is the same as its pure component value at T and P_i. Therefore, for an ideal gas, we can state that $\bar{H}_i\,(T,P) = H_i^{\text{ideal gas}}\,(T,P_i)$ and $\bar{U}_i\,(T,P) = U_i^{\text{ideal gas}}\,(T,P_i)$, where $H_i^{\text{ideal gas}}$ and $U_i^{\text{ideal gas}}$ represent the pure component ideal gas values of the enthalpy and internal energy at the mixture T and at component is partial pressure, P_i. Hence from Equation 2.37, for an ideal gas mixture we have that

$$H^{\text{ideal gas}} = \sum_i y_i H_i^{\text{ideal gas}} \quad \text{and} \quad U^{\text{ideal gas}} = \sum_i y_i U_i^{\text{ideal gas}}, \tag{2.47}$$

and this can be generalized using Equation 2.46 for any ideal gas property M as

$$M^{\text{ideal gas}}(T,P) = \sum_i y_i M_i^{\text{ideal gas}}(T,P_i). \tag{2.48}$$

We also saw in Example 2.3 that the entropy of an ideal gas depends on the pressure and the temperature. Consider pure component i at T and P_i. We can ask what is the entropy change for component i when it is placed in an ideal gas mixture at T and P? Since T is constant and the pressure changes from P_i to P, we can use the result for the entropy change of an ideal gas that was obtained in Example 2.3, that is

$$\Delta S_i^{\text{ideal gas}} = S_i^{\text{ideal gas}}(T,P) - S_i^{\text{ideal gas}}(T,P_i) = -R\ln\left(\frac{P}{P_i}\right) = -R\ln\left(\frac{P}{y_i\,P}\right) = R\ln\,y_i, \tag{2.49}$$

and after rearranging we obtain the result that

$$S_i^{\text{ideal gas}}(T,P_i) = S_i^{\text{ideal gas}}(T,P) - R\ln\,y_i. \tag{2.50}$$

Using Equation 2.48, we then obtain the following expression for the entropy of an ideal gas mixture:

$$S^{\text{ideal gas}} = \sum_i y_i S_i^{\text{ideal gas}}(T,P) - R\sum_i y_i \ln\,y_i. \tag{2.51}$$

Example 2.4

Determine an expression for the enthalpy, internal energy, and entropy change that occurs when N pure components at T and P are mixed to form one mole of a solution at the same T and P. Assume that the pure components and the resulting mixture are ideal gases.

Solution
Equation 2.44 defines the property change of mixing as

$$\Delta M^{\text{mix}} = M - \sum_i y_i M_i.$$

Comparing this equation with Equation 2.47 shows that the enthalpy and internal energy change of mixing is zero for the formation of an ideal gas mixture from its pure components. However, using Equation 2.51, the entropy change of mixing for this case is given by

$$\Delta S^{\text{mix}} = S^{\text{ideal gas}} - \sum_i y_i S_i^{\text{ideal gas}}(T,P) = -R\sum_i y_i \ln\,y_i.$$

The $-R\Sigma_i y_i \ln y_i$ term is always positive, indicating that the process of mixing results in an increase in the entropy of the system. Since the enthalpy of mixing is also zero, there is no change in the entropy of the surroundings, so the total entropy change is also positive. Therefore, mixing these pure ideal gas components is an irreversible process.

2.5.1.4 Gibbs Free Energy of an Ideal Gas Mixture

With the entropy of an ideal gas mixture given by Equation 2.51, we can now calculate the Gibbs free energy of an ideal gas mixture, that is $G^{\text{ideal gas}} = H^{\text{ideal gas}} - T S^{\text{ideal gas}}$. Using the above-mentioned relationships that were derived for $H^{\text{ideal gas}}$ and $S^{\text{ideal gas}}$, we can obtain the following expression for the Gibbs free energy of an ideal gas mixture:

$$G^{\text{ideal gas}} = \sum_i y_i H_i^{\text{ideal gas}} - T\left(\sum_i y_i S_i^{\text{ideal gas}} - R\sum_i y_i \ln y_i\right), \tag{2.52}$$

which can also be written as follows, since $G_i^{\text{ideal gas}} = H_i^{\text{ideal gas}} - T S_i^{\text{ideal gas}}$:

$$G^{\text{ideal gas}} = \sum_i y_i G_i^{\text{ideal gas}} + RT\sum_i y_i \ln y_i. \tag{2.53}$$

Comparing Equation 2.53 with Equations 2.46 and 2.48, we see that the partial molar Gibbs free energy, or the chemical potential, of component i in an ideal gas mixture is then given by

$$\mu_i^{\text{ideal gas}} = \overline{G}_i^{\text{ideal gas}} = G_i^{\text{ideal gas}} + RT\ln y_i. \tag{2.54}$$

We also know from our fundamental property relations (see Equation 2.25) for pure component i at constant T that $dG_i^{\text{ideal gas}} = V_i^{\text{ideal gas}} dP = (RT/P)dP = RT\, d\ln(P)$. This result can be integrated from an arbitrary pressure P_0 to pressure P to give the following result for the pure component Gibbs free energy of an ideal gas:

$$G_i^{\text{ideal gas}} = (G_i^{0,\,\text{ideal gas}} - RT\ln P_0) + RT\ln P = G_i^{0,\text{ideal gas}} + RT\ln\frac{P}{P_0} = \mu_i^{0,\,\text{ideal gas}} + RT\ln\frac{P}{P_0}, \tag{2.55}$$

where $G_i^{0,\text{ideal gas}}$ or $\mu_i^{0,\text{ideal gas}}$ depends only on the temperature and is the Gibbs free energy or chemical potential per mole of pure component i at a pressure equal to P_0. Combining the previous result with Equation 2.54 provides the following alternative expression for the chemical potential of component i in an ideal gas mixture at T and P relative to a reference pressure P_0:

$$\mu_i^{\text{ideal gas}} = \mu_i^{0,\text{ideal gas}} + RT\ln\left(\frac{y_i P}{P_0}\right). \tag{2.56}$$

2.5.2 Pure Component Fugacity

For a pure component (i.e., $y_i = 1$) in the ideal gas state, Equation 2.56 tells us that the pure component Gibbs free energy of an ideal gas is given by the following expression: $G_i^{\text{ideal gas}} = (\mu_i^{0,\text{ideal gas}} - RT\ln P_0) + RT\ln P$. This result can be generalized to the real gas by defining a new property called the *pure component fugacity* (f_i) to replace the pressure P. The fugacity may be thought of as a "corrected" pressure that provides the value of the pure component Gibbs free energy of a real gas at pressure P. Therefore, the following expression defines the pure component fugacity for a real gas:

$$G_i = \left(\mu_i^{0,\text{ideal gas}} - RT \ln P_0\right) + RT \ln f_i. \tag{2.57}$$

Now, if we subtract Equation 2.55 from Equation 2.57, we obtain the following result:

$$G_i - G_i^{\text{ideal gas}} = RT \ln \frac{f_i}{P} = RT \ln \phi_i. \tag{2.58}$$

The pure component fugacity coefficient, ϕ_i, is defined as the ratio of the pure component fugacity, f_i, to the pressure, P. Therefore, $\phi_i = f_i/P$. Clearly, we see that for an ideal gas the pure component fugacity, f_i, is equal to the pressure, or $f_i^{\text{ideal gas}} = P$ and $\phi_i^{\text{ideal gas}} = 1$.

The difference in the Gibbs free energy in Equation 2.58, i.e., $G_i - G_i^{\text{ideal gas}}$, defines what is also called a *residual property* or in this case the residual Gibbs free energy, $G_i^{\text{R}} = G_i - G_i^{\text{ideal gas}}$, where both G_i and $G_i^{\text{ideal gas}}$ are evaluated at the same T and P. This leads to the following general definition of a residual property as the difference between the actual value of a property and its value in the ideal gas state:

$$M^{\text{R}} = M - M^{\text{ideal gas}}. \tag{2.59}$$

Equation 2.59 applies to any of our properties, i.e., U, H, S, G, A, and V.

2.5.2.1 Calculating the Pure Component Fugacity

As shown in the following development, the pure component fugacity (f_i) can be calculated from experimental *PVT* data or an appropriate equation of state. First, it is convenient to define the *compressibility factor* (Z), which describes the deviation of a real gas from the ideal gas state. The compressibility factor is a dimensionless quantity and is defined by

$$Z = \frac{P\,V}{R\,T}. \tag{2.60}$$

Note that for an ideal gas $Z = 1$. A variety of equations of state have been developed for calculating the compressibility factor for real gases and liquids at high pressures (Poling et al. 2001).

We can also use the following mathematical relationship for $d(G/RT)$, and expanding this differential we obtain

$$d\left(\frac{G}{RT}\right) = \frac{RT\,dG - G\,R\,dT}{(RT)^2} = \frac{1}{RT}dG - \frac{G}{R\,T^2}dT, \tag{2.61}$$

and then substituting for dG from Equation 2.25 and recognizing that $G = H - TS$, we have that

$$d\left(\frac{G}{RT}\right) = \frac{V}{RT}dP - \frac{H}{R\,T^2}dT. \tag{2.62}$$

From this equation, we can also write the following additional relationships at constant pressure and constant temperature:

$$\frac{H}{R\,T} = -T\left[\frac{\partial\left(\dfrac{G}{RT}\right)}{\partial\,T}\right]_P \quad \text{and} \quad \frac{V}{RT} = \left[\frac{\partial\left(\dfrac{G}{RT}\right)}{\partial\,P}\right]_T. \tag{2.63}$$

These relationships are also valid for the residual properties G^R, H^R, and V^R by simply adding a superscript to the property.

Now the residual volume, V^R, is defined as $V - V^{\text{ideal gas}}$. So, we can write the residual volume as follows using the definition given earlier for the compressibility factor:

$$V^R = \frac{RT}{P}(Z-1). \tag{2.64}$$

Using these relationships, we can now obtain the following expressions for the residual properties. First, we can take the second expression in Equation 2.63 at constant temperature and write it for the residual property. Then, we integrate at constant temperature from zero pressure (G^R is equal to zero since this is an ideal gas state) to any pressure P:

$$\frac{G^R}{RT} = \int_0^P \frac{V^R}{RT}dP = \int_0^P (Z-1)\frac{dP}{P} \quad \text{(constant } T\text{).} \tag{2.65}$$

To obtain the residual enthalpy, we can use the first expression in Equation 2.63. Next, we evaluate $[\partial(G^R/RT)/\partial T]_P$ using Equation 2.65 and then obtain

$$\frac{H^R}{RT} = -T\int_0^P \left(\frac{\partial Z}{\partial T}\right)_P \frac{dP}{P} \quad \text{(constant } T\text{).} \tag{2.66}$$

Since $TS^R = H^R - G^R$, we can also write that $(S^R/R) = (H^R/RT)-(G^R/RT)$ and we can use these relationships for H^R and G^R to obtain

$$\frac{S^R}{R} = -T\int_0^P \left(\frac{\partial Z}{\partial T}\right)_P \frac{dP}{P} - \int_0^P (Z-1)\frac{dP}{P} \quad \text{(constant } T\text{).} \tag{2.67}$$

Equations 2.65 through 2.67 are very important relationships since they allow us to calculate the real property value M from the residual property value M^R and the ideal gas property value $M^{\text{ideal gas}}$. Provided we have experimental PVT data, or an appropriate equation of state, we can use these equations to calculate the values of the Gibbs free energy, the enthalpy, and entropy for real gases and even liquids at high pressures. In addition, we can also use Equation 2.65 to calculate the fugacity of a pure component at any T and P. Comparing Equation 2.65 with Equation 2.58, we see that Equation 2.65 is also equal to the natural logarithm of the fugacity coefficient (ϕ_i) or f_i/P, that is

$$\ln \phi_i = \ln\left(\frac{f_i}{P}\right) = \int_0^P (Z-1)\frac{dP}{P}. \tag{2.68}$$

Therefore, Equation 2.68 allows us to determine the fugacity of a pure component at any T and P from knowledge of its PVT behavior.

Example 2.5

Obtain an expression for the fugacity coefficient using the pressure explicit form of the virial equation of state truncated after the second coefficient. If the second virial coefficient ($\bar{B}$) of a particular gas is -0.01 atm^{-1}, calculate the compressibility factor and the fugacity of this gas at a temperature of 500 K and a pressure of 10 atm.

Solution

The simplest equation of state for a real gas is the virial equation of state given by the following volume and pressure explicit forms:

$$Z = 1 + \frac{B}{V} + \frac{C}{V^2} + \frac{D}{V^3} + \cdots$$

$$Z = 1 + \bar{B} P + \bar{C} P^2 + \bar{D} P^3 + \cdots,$$

where B, C, D and $\bar{B}$, $\bar{C}$, $\bar{D}$ are known as the second, third, and fourth virial coefficients, respectively, and are only a function of temperature. If we truncate the above-mentioned series after the second coefficient, then we have for the pressure explicit form that $Z = 1 + \bar{B} P$ or $Z - 1 = \bar{B} P$. Using Equation 2.68, we then obtain

$$\ln \phi_i = \bar{B} \int_0^P dP = \bar{B} P.$$

Hence, the value of $Z = 1 - 0.01$ atm$^{-1} \times 10$ atm $= 0.90$, and the fugacity coefficient and fugacity is calculated as

$$\ln \phi_i = -0.01 \text{ atm}^{-1} \times 10 \text{ atm} = -0.10$$

$$\phi_i = 0.905 \text{ and since } \frac{f_i}{P} = \phi_i, \text{ then } f_i = 0.905 \times 10 \text{ atm} = 9.05 \text{ atm,}$$

and we see that the fugacity of this gas at these conditions is 9.05 atm compared to 10 atm if the gas was an ideal gas.

2.5.3   FUGACITY OF A COMPONENT IN A MIXTURE

For a component in a mixture of real gases or a liquid solution, we can generalize Equation 2.57 for an ideal gas mixture to provide the fugacity of component i as it exists in the real mixture:

$$\mu_i = (\mu_i^{0,\text{ideal gas}} - RT \ln P_0) + RT \ln \hat{f}_i, \tag{2.69}$$

where $\hat{f}_i$ is defined as the fugacity of component i as it exists in the mixture. We can also solve Equation 2.57 for the value of $(\mu_i^{0,\text{ideal gas}} - RT \ln P_0)$ in terms of the pure component values of G_i and f_i to obtain

$$\mu_i = \bar{G}_i = G_i + RT \ln\left(\frac{\hat{f}_i}{f_i}\right). \tag{2.70}$$

Equation 2.70 shows the relationship between the chemical potential of component i as it exists in the mixture in terms of its pure component Gibbs free energy and fugacity (G_i and f_i), and its mixture fugacity ($\hat{f}_i$).

Now, if we subtract Equation 2.56 from Equation 2.69 for the same T and P, we obtain

$$\mu_i - \mu_i^{\text{ideal gas}} = \bar{G}_i^R = RT \ln\left(\frac{\hat{f}_i}{y_i P}\right) = RT \ln \hat{\phi}_i, \tag{2.71}$$

where $\hat{\phi}_i = \hat{f}_i / y_i P$ is defined as the component fugacity coefficient. For component i in an ideal gas $\hat{\phi}_i = 1$ since $\mu_i = \mu_i^{\text{ideal gas}}$ we have

$$\hat{f}_i = \hat{f}_i^{\text{ideal gas}} = y_i P. \tag{2.72}$$

We can calculate the values of $\hat{\phi}_i$ provided we have mixture PVT data or an equation of state that describes the mixture (Poling et al. 2001). To see how this is done, we can rewrite Equation 2.65 for n moles of our mixture as

$$\frac{nG^{\text{R}}}{RT} = \int_0^P (nZ - n) \frac{dP}{P} \quad \text{(constant } T\text{).} \tag{2.73}$$

Then, from Equation 2.71 we have that $\ln \hat{\phi}_i = \bar{G}_i^{\text{R}} / RT = \left[\partial (nG^R / RT) / \partial n_i \right]_{\text{T,P},n_j}$, which allows us to perform this operation on Equation 2.73, giving the following result:

$$\ln \hat{\phi}_i = \int_0^P \left[\frac{\partial (nZ - n)}{\partial n_i} \right]_{\text{T, P, } n_j} \frac{dP}{P} = \int_0^P (\bar{Z}_i - 1) \frac{dP}{P}, \tag{2.74}$$

with the partial molar compressibility factor $\bar{Z}_i = \left[\partial (nZ) / \partial n_i \right]_{\text{T,P},n_j}$ and $(\partial n / \partial n_i) = 1$. Provided we have mixture PVT data or a mixture equation of state, we can use Equation 2.74 to calculate the component fugacity coefficient (Poling et al. 2001).

2.5.4 IDEAL SOLUTION

The ideal solution model includes ideal gas mixtures as well as liquids and solids. The ideal solution model is useful for describing mixtures of substances whose molecules do not differ much in their size and chemical nature. For example, a liquid solution of ethanol and propanol would be expected to form an ideal solution, whereas a solution of ethanol and water would be expected to form a nonideal or real solution. The ideal solution model therefore serves as a useful basis of comparison to real solution behavior.

Recall that for an ideal gas the chemical potential of component i was given by Equation 2.54. For an ideal solution, we generalize this result by replacing the Gibbs free energy of component i in the ideal gas state, $G_i^{\text{ideal gas}}$, with G_i, the pure component Gibbs free energy at the same T, P, and physical state (i.e., solid, liquid, or gas) as the mixture. Therefore, an ideal solution is defined by Equation 2.75 for the partial molar Gibbs free energy or chemical potential of component i:

$$\mu_i^{\text{ideal solution}} = \bar{G}_i^{\text{ideal solution}} = G_i + RT \ln x_i, \tag{2.75}$$

where x_i is defined as the mole fraction of component i in the mixture. We can also write Equation 2.70 for an ideal solution as $\mu_i^{\text{ideal solution}} = G_i + RT \ln \left(\hat{f}_i^{\text{ideal solution}} / f_i \right)$ and combining this result with Equation 2.75, we obtain the important result that

$$\hat{f}_i^{\text{ideal solution}} = x_i f_i. \tag{2.76}$$

Equation 2.76 is also known as the *Lewis–Randall rule* and says that the fugacity of a component in an ideal solution is proportional to its mole fraction. Furthermore, the proportionality constant is the pure component fugacity evaluated at the same T and P as the solution being considered. Since $\hat{\phi}_i = \left(\hat{f}_i / x_i P \right)$, we then see, using Equation 2.76, that for an ideal solution, $\hat{\phi}_i^{\text{ideal solution}} = f_i / P = \phi_i$.

Example 2.6

Show that with the (i.e., Equation 2.75) mentioned definition of the partial molar Gibbs free energy of component i in an ideal solution earlier, the enthalpy of mixing, the internal energy of mixing, and the volume of mixing are equal to zero for an ideal solution.

Solution

Based on Equation 2.31, we can write the mixture Gibbs free energy as

$$d(nG) = \left[\frac{\partial(nG)}{\partial T}\right]_{P,n} dT + \left[\frac{\partial(nG)}{\partial P}\right]_{T,n} dP + \sum_i \left[\frac{\partial(nG)}{\partial n_i}\right]_{T,P,n_j} dn_i.$$

For a closed system containing n moles, we can also write from Equation 2.25 that $d(nG) = -(nS)\,dT + (nV)\,dP$. Comparing this result to the first two terms of the previous equation for constant n, we see that the first two partial derivatives can be replaced by $-(nS)$ and (nV), respectively. In addition, we know that $[\partial(nG)/\partial n_i]_{T,P,n_j} = \bar{G}_i = \mu_i$, the chemical potential of component i. Therefore, the previous equation may be written as

$$d(nG) = -(nS)\,dT + (nV)\,dP + \sum_i \mu_i\,dn_i.$$

We can now use the criterion of exactness (Equation 2.29) on this equation and obtain the following relationships:

$$\left(\frac{\partial \mu_i}{\partial T}\right)_{P,n} = -\left(\frac{\partial(nS)}{\partial n_i}\right)_{P,T,n_j} = -\bar{S} \quad \text{and} \quad \left(\frac{\partial \mu_i}{\partial P}\right)_{T,n} = \left(\frac{\partial(nV)}{\partial n_i}\right)_{P,T,n_j} = \bar{V}_i.$$

Using the above-mentioned relationships along with Equation 2.75, we then have for an ideal solution:

$$\left(\frac{\partial \mu_i^{\text{ideal solution}}}{\partial T}\right)_{P,n} = -\bar{S}_i^{\text{ideal solution}} = \left(\frac{\partial G_i}{\partial T}\right)_P + R\ln x_i,$$

$$\left(\frac{\partial \mu_i^{\text{ideal solution}}}{\partial P}\right)_{T,n} = \bar{V}_i^{\text{ideal solution}} = \left(\frac{\partial G_i}{\partial P}\right)_T.$$

Next, we can use our fundamental property relations for component i (Equation 2.25) to show that $(\partial G_i/\partial T)_P = -S_i$ and $(\partial G_i/\partial P)_T = V_i$. Hence, we obtain the following equations for the partial molar entropy and partial molar volume for an ideal solution:

$$\bar{S}_i^{\text{ideal solution}} = S_i - R\ln x_i \quad \text{and} \quad \bar{V}_i^{\text{ideal solution}} = V_i.$$

By Equation 2.44, we can easily see that the volume change of mixing for an ideal solution is zero and that $V^{\text{ideal solution}} = \sum_i x_i V_i$. For the enthalpy change of mixing, we can use the fact that $\bar{H}_i^{\text{ideal solution}} = \bar{G}_i^{\text{ideal solution}} + T\bar{S}_i^{\text{ideal solution}}$ and using the above-mentioned relationships we obtain $\bar{H}_i^{\text{ideal solution}} = G_i + TS_i = H_i$. Therefore, the enthalpy change of mixing for an ideal solution is also zero and $H^{\text{ideal solution}} = \sum_i x_i H_i$. In a similar fashion, we can use the fact that $\bar{U}_i^{\text{ideal solution}} = \bar{H}_i^{\text{ideal solution}} - P\bar{V}_i^{\text{ideal solution}} = H_i - PV_i = U_i$ and that the internal energy change of mixing is also zero.

2.6   PHASE EQUILIBRIUM

With this background on solution thermodynamics, we can now address the criterion for equilibrium between two phases (I and II) contained within a closed system. Equilibrium between the two phases also implies that the two phases are at the same temperature. Also, the two phases must have the same pressure, unless they are separated by a semipermeable rigid barrier or membrane. We will consider the situation where the pressures may be different later in this chapter when we discuss osmotic equilibrium.

Each phase is also an open system and is free to exchange mass with the other phase. Since we are considering the closed system containing the two phases to be at constant T and P, we can use the result from Example 2.6 that $d(nG) = -(nS)dT + (nV)dP + \sum_i \mu_i\, dn_i$ and write this equation for each phase as

$$d(nG)^{\mathrm{I}} = -(nS)^{\mathrm{I}}\, dT + (nV)^{\mathrm{I}}\, dP + \sum_i \mu_i^{\mathrm{I}} dn_i^{\mathrm{I}},$$

$$d(nG)^{\mathrm{II}} = -(nS)^{\mathrm{II}}\, dT + (nV)^{\mathrm{II}}\, dP + \sum_i \mu_i^{\mathrm{II}} dn_i^{\mathrm{II}}. \tag{2.77}$$

Using these equations, then for the closed system we can also write that

$$d(nG) = -(nS)dT + (nV)dP = d(nG)^{\mathrm{I}} + d(nG)^{\mathrm{II}}$$

$$= -\left[(nS)^{\mathrm{I}} + (nS)^{\mathrm{II}}\right]dT + \left[(nV)^{\mathrm{I}} + (nV)^{\mathrm{II}}\right]dP + \sum_i \mu_i^{\mathrm{I}} dn_i^{\mathrm{I}} + \sum_i \mu_i^{\mathrm{II}}\, dn_i^{\mathrm{II}}. \tag{2.78}$$

Since $(nS)^{\mathrm{I}} + (nS)^{\mathrm{II}} = (nS)$ and $(nV)^{\mathrm{I}} + (nV)^{\mathrm{II}} = (nV)$, Equation 2.78 simplifies to

$$\sum_i \mu_i^{\mathrm{I}}\, dn_i^{\mathrm{I}} + \sum_i \mu_i^{\mathrm{II}}\, dn_i^{\mathrm{II}} = 0, \tag{2.79}$$

where dn_i^{I} and dn_i^{II} represent the differential changes in the moles of component i as a result of mass transfer between phases I and II. Clearly, the law of mass conservation requires that $dn_i^{\mathrm{I}} = -dn_i^{\mathrm{II}}$. With this result, Equation 2.79 may then be written as

$$\sum_i (\mu_i^{\mathrm{I}} - \mu_i^{\mathrm{II}})dn_i^{\mathrm{I}} = 0. \tag{2.80}$$

The values of dn_i^{I} and dn_i^{II} are arbitrary, so the only way that Equation 2.80 can be satisfied is for each term in parentheses (i.e., $\mu_i^{\mathrm{I}} - \mu_i^{\mathrm{II}}$) to be equal to zero. So, our criterion for phase equilibrium in a two-phase system can be expressed by

$$\mu_i^{\mathrm{I}} = \mu_i^{\mathrm{II}}, \tag{2.81}$$

which simply states that the chemical potential for each component i is the same in phase I and phase II at the given T and P.

This result can be easily generalized to provide the criterion for phase equilibrium in π phases at the same T and P as

$$\mu_i^{\mathrm{I}} = \mu_i^{\mathrm{II}} = \cdots = \mu_i^{\pi}. \tag{2.82}$$

If we only have a pure component in each phase, Equations 2.81 or 2.82 still apply and since $\mu_i = \bar{G}_i = G_i$, then for a pure component, phase equilibrium requires that

$$G_i^{\mathrm{I}} = G_i^{\mathrm{II}} = \cdots = G_i^{\pi} \quad \text{(pure component).} \tag{2.83}$$

Using Equation 2.82 along with $\mu_i = \bar{G}_i + RT \ln(\hat{f}_i / f_i)$ from Equation 2.70, we can write an alternative expression for the criterion for phase equilibrium in multicomponent systems in terms of the component mixture fugacities as

$$\hat{f}_i^{\mathrm{I}} = \hat{f}_i^{\mathrm{II}} = \cdots = \hat{f}_i^{\pi}. \tag{2.84}$$

2.6.1 PURE COMPONENT PHASE EQUILIBRIUM

Figure 2.3 shows a pressure–volume or PV curve for a pure component. Shown in the figure are several dashed lines of constant temperature or isotherms. The area within the dome-shaped region defines conditions, where two phases, e.g., vapor and liquid, are at equilibrium for a given T and P. At the top of the dome is the *critical point* (C), where the liquid and vapor phases become identical and thus have the same physical properties. The isotherm that passes through the critical point is known as the critical isotherm, T_C. The critical temperature, pressure, and volume (T_C, P_C, and V_C) are also pure component physical properties and play an important role in the estimation of a variety of physical properties (Poling et al. 2001).

Below this critical isotherm, we see that the other isotherms (i.e., subcritical isotherms) have three separate segments. Within the dome, we see that the isotherm is horizontal and represents the constant temperature and pressure phase change between the *saturated liquid* and *saturated vapor* states. The left most point on this horizontal segment denotes the saturated liquid and the right most point on the horizontal segment denotes the saturated vapor. Saturated means that for the pure component the vapor and liquid phases are at equilibrium at $T = T^{\mathrm{Sat}}$ and $P = P^{\mathrm{Sat}}$, where P^{Sat} is called the *saturation pressure* or the vapor pressure of the pure component. P^{Sat} for a pure component only depends on the saturation temperature, i.e., $P^{\mathrm{Sat}} = P^{\mathrm{Sat}}(T^{\mathrm{Sat}})$. The normal boiling point (T_{BP}) of a pure component occurs when the value of P^{Sat} is the same as the atmospheric pressure or $P^{\mathrm{Sat}} = 1$ atm $= P^{\mathrm{Sat}}(T^{\mathrm{Sat}}) = P^{\mathrm{Sat}}(T_{\mathrm{BP}})$. Points along the curve ABC define saturated liquids and points along the curve CDE define saturated vapor. Points along the horizontal isotherm within the dome represent saturated mixtures of liquid and vapor. The actual molar volume of the vapor/liquid mixture will

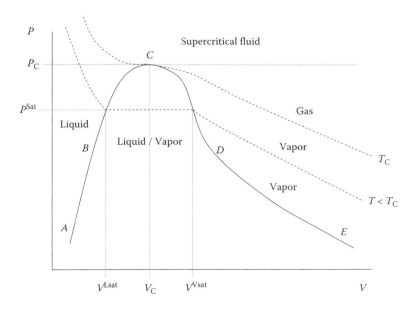

FIGURE 2.3 *PV* diagram for a pure component.

depend on the relative amounts of each phase present, i.e., $V = (1-x) \, V^{\text{Lsat}} + x \, V^{\text{Vsat}}$, where x is the vapor fraction of the vapor/liquid mixture.

The region to the left of the curve ABC denotes the liquid state, and the isotherms here rise very quickly because liquid volume changes are very small as the pressure is increased and, for the most part, liquids at modest pressures are therefore considered to be incompressible. The region to the right of curve CDE denotes the vapor state provided for $T < T_C$ and $P < P_C$, and a gas for $T > T_C$ and $P < P_C$. The conditions where $T > T_C$ and $P > P_C$ are known as the supercritical fluid region.

Now let us consider the vapor (V) and liquid (L) equilibrium of a pure component i at T and $P = P^{\text{Sat}}(T)$. From Equation 2.83, we can write that $G^V = G^L$ and it follows that $dG^V = dG^L$. Next, we can use Equation 2.25 and express the previous result as

$$dG^V = -S^V \, dT + V^V \, dP^{\text{Sat}} = -S^L \, dT + V^L \, dP^{\text{Sat}} = dG^L. \qquad (2.85)$$

This equation may be rearranged as

$$\frac{dP^{\text{Sat}}}{dT} = \frac{S^V - S^L}{V^V - V^L} = \frac{\Delta S^{\text{LV}}}{\Delta V^{\text{LV}}}. \qquad (2.86)$$

Now ΔS^{LV} and ΔV^{LV} represent the changes in the entropy and volume of the pure component when it is transferred from the liquid phase to the vapor phase at the equilibrium temperature (T) and pressure (P^{Sat}). If we integrate Equation 2.24 for the change in phase from liquid to vapor at constant T and P, we obtain $\Delta H^{\text{LV}} = T\Delta S^{\text{LV}}$ or $\Delta S^{\text{LV}} = \Delta H^{\text{LV}}/T$. ΔH^{LV} is also known as the *heat of vaporization* (ΔH^{Vap}) and represents the difference in enthalpy between the saturated vapor and the saturated liquid. These results can be substituted into Equation 2.86 to give

$$\frac{dP^{\text{Sat}}}{dT} = \frac{\Delta H^{\text{Vap}}}{T \, \Delta V^{\text{LV}}}. \qquad (2.87)$$

This equation is also known as the *Clausius–Clapeyron* equation and is an equilibrium relationship between the saturated liquid and the saturated vapor. For a given change in the temperature (dT), Equation 2.87 provides the change in pressure (dP^{Sat}) required to maintain equilibrium between the saturated liquid and its saturated vapor. For solid–liquid or solid–vapor equilibrium, one may substitute the *enthalpy of melting* or the *enthalpy of sublimation* for the enthalpy of vaporization in Equation 2.87.

For liquid–vapor and solid–vapor phase changes at saturation pressures near atmospheric, one can neglect the molar volume of the solid or liquid phase in comparison to the volume of the vapor and assume that the vapor phase behaves as an ideal gas, i.e., $\Delta V^{\text{LV}} = V^V = RT/P^{\text{Sat}}$. Equation 2.87 may then be written as

$$\frac{d \ln P^{\text{Sat}}}{d \dfrac{1}{T}} = -\frac{L}{R}, \qquad (2.88)$$

where L now represents either ΔH^{LV} (heat of vaporization) or ΔH^{SV} (heat of sublimation). Equation 2.88 provides a relationship between the saturation pressure and the temperature in terms of the enthalpy change associated with the phase change. Over narrow ranges of temperature, L is pretty much constant, and Equation 2.88 can be integrated from an arbitrary temperature (T_0) to provide an approximate relationship between the saturation pressure and the temperature:

$$\ln P^{Sat} = \left(\ln P_0^{Sat} + \frac{L}{RT_0} \right) - \left(\frac{L}{R} \right) \frac{1}{T} = A - \frac{B}{T}, \tag{2.89}$$

where A and B represent constants that can be fitted to pure component vapor pressure data. Equation 2.89 shows that over a narrow range of temperatures there is a linear relationship between the natural logarithm of the saturation pressure and the inverse of the temperature. Additional empirical equations for determining the saturation or vapor pressure of a given component as a function of temperature can be found in Poling et al. (2001).

Since we have a pure component i, Equation 2.84 can be written in terms of the pure component fugacities as

$$f_i^L = f_i^V = f_i^{Sat}, \tag{2.90}$$

where f_i^{Sat} denotes the fugacity of either the saturated liquid or the saturated vapor.

Since $f_i^{Sat} = \phi_i^{Sat} P_i^{Sat}$, we then have that

$$f_i^L = f_i^V = f_i^{Sat} = \phi_i^{Sat} P_i^{Sat}, \tag{2.91}$$

where ϕ_i^{Sat} would be given by Equation 2.68 with $P = P^{Sat}$ using an appropriate set of PVT data or an equation of state to perform the integration. Therefore

$$f_i^L = f_i^V = f_i^{Sat} = P_i^{Sat} \exp \left(\int_0^{P_i^{Sat}} (Z-1) \frac{dP}{P} \right). \tag{2.92}$$

2.6.1.1 Fugacity of a Pure Component as a Compressed Liquid

For a given temperature T, if the pressure P is greater than the value of $P^{Sat}(T)$ for a pure component, then the liquid is considered to be subcooled or a compressed liquid and is not in equilibrium with its vapor. Calculation of the fugacity of a pure component i as a compressed liquid starts with our fundamental property relation (Equation 2.25), i.e., $dG_i = -S_i dT + V_i dP$. For constant temperature, we can write this as $dG_i = V_i dP$ and integrate this equation from P_i^{Sat} to P as

$$G_i - G_i^{Sat} = \int_{P_i^{Sat}}^{P} V_i \, dP. \tag{2.93}$$

Using Equation 2.57, we can rewrite Equation 2.93 as

$$G_i - G_i^{Sat} = RT \ln \frac{f_i}{f_i^{Sat}} = \int_{P_i^{Sat}}^{P} V_i \, dP. \tag{2.94}$$

As mentioned earlier, liquid molar volumes do not depend that strongly on P, so Equation 2.94 can be written as follows, after setting the subcooled molar liquid volume (V_i) equal to the saturated liquid volume (V_i^{Sat}) at the same T:

$$f_i = f_i^{Sat} \exp \left(\frac{V_i^{Sat} (P - P_i^{Sat})}{RT} \right) = \phi_i^{Sat} P_i^{Sat} \exp \left(\frac{V_i^{Sat} (P - P_i^{Sat})}{RT} \right). \tag{2.95}$$

Comparing Equation 2.95 with Equation 2.91, we see that the exponential term is a correction of the saturated fugacity to account for the fact that the pressure of the subcooled liquid is greater than its saturation pressure for a given T. The exponential term is also known as the *Poynting factor*.

2.6.2 Excess Properties

Recall that we previously defined a residual property as the difference between the real property value and its corresponding value as an ideal gas ($M^R = M - M^{\text{ideal gas}}$). We showed in Section 2.5.2.1 that the residual properties can be determined from *PVT* data or an equation of state. We also found that the definition of the residual properties provided a convenient method for calculating the pure component fugacity coefficient and the fugacity coefficient of a component in a mixture, i.e., Equations 2.68 and 2.74. Accordingly, residual properties and fugacity coefficients are usually used for describing the behavior of real gases, since the residual properties and the fugacity coefficients express the deviation of the real gas from ideal gas behavior.

For liquids, it is usually easier to compare real liquid solution behavior to an ideal solution. An excess property (M^E) is then defined as the difference between the property value in a real solution and the value it would have in an ideal solution at the same T, P, and composition. Therefore, we have that

$$M^E = M - M^{\text{ideal solution}}. \tag{2.96}$$

In terms of partial molar properties, Equation 2.96 becomes

$$\bar{M}_i^E = \bar{M}_i - \bar{M}_i^{\text{ideal solution}}. \tag{2.97}$$

For phase equilibrium calculations, we are primarily interested in the Gibbs free energy, so the partial molar excess Gibbs free energy is given by

$$\bar{G}_i^E = \bar{G}_i - \bar{G}_i^{\text{ideal solution}}. \tag{2.98}$$

Using Equations 2.70 and 2.75 for $\bar{G}_i$ and $\bar{G}_i^{\text{ideal solution}}$, we obtain, using Equation 2.98:

$$\bar{G}_i^E = \bar{G}_i - \bar{G}_i^{\text{ideal solution}} = RT \ln \frac{\hat{f}_i}{x_i f_i} = RT \ln \gamma_i. \tag{2.99}$$

The term ($\hat{f}_i / x_i f_i$) is dimensionless and is also known as the *activity coefficient*, γ_i. In terms of the activity coefficient, we can then express the fugacity of component i as it exists in the solution as $\hat{f}_i = \gamma_i x_i f_i$. Note that for an ideal solution, all γ_i equal unity and $\hat{f}_i^{\text{ideal solution}} = x_i f_i$. We can then write the partial molar excess Gibbs free energy in terms of the activity coefficient as

$$\bar{G}_i^E = \left[\frac{\partial (n G^E)}{\partial n_i} \right]_{T, P, n_j} = RT \ln \gamma_i. \tag{2.100}$$

Using Equation 2.37 and the Gibbs–Duhem relationship (Equation 2.39) at constant T and P, we also have that

$$G^E = RT \sum_i x_i \ln \gamma_i, \tag{2.101}$$

$$\sum_i x_i \, d \ln \gamma_i = 0. \tag{2.102}$$

Activity coefficients for a component in a mixture are obtained from experimental data, and the results are correlated with a model of how the excess Gibbs free energy depends on the composition of all the components within the mixture, i.e., $G^E/RT = f(x_1, x_2, \ldots, x_N)$ at constant T, since for the most part we can ignore the effect of pressure on the activity coefficients.

Equation 2.102 serves as a thermodynamic consistency check of experimentally determined activity coefficients or empirical equations that are used to calculate activity coefficients. For a binary system at constant T and P, Equation 2.102 can also be written as

$$x_1 \left(\frac{\partial \ln \gamma_1}{\partial x_1} \right)_{T,P} + x_2 \left(\frac{\partial \ln \gamma_2}{\partial x_1} \right)_{T,P} = 0. \tag{2.103}$$

For a binary system, we can also write Equation 2.101 as

$$\frac{G^E}{RT} = x_1 \ln \gamma_1 + x_2 \ln \gamma_2. \tag{2.104}$$

Equation 2.104 can then be differentiated with respect to x_1 at constant T and P to give

$$\frac{d\left(\dfrac{G^E}{RT} \right)}{dx_1} = x_1 \frac{\partial \ln \gamma_1}{\partial x_1} + \ln \gamma_1 + x_2 \frac{\partial \ln \gamma_2}{\partial x_1} + \ln \gamma_2 \frac{dx_2}{dx_1}. \tag{2.105}$$

Since $dx_2/dx_1 = -1$ and using the Gibbs–Duhem equation from Equation 2.103, we then have that

$$\frac{d\left(\dfrac{G^E}{RT} \right)}{dx_1} = \ln \left(\frac{\gamma_1}{\gamma_2} \right). \tag{2.106}$$

This equation can then be integrated over x_1 as shown in the following equation:

$$\int_0^1 \frac{d\left(\dfrac{G^E}{RT} \right)}{dx_1} dx_1 = \left(\frac{G^E}{RT} \right)\bigg|_{x_1=1} - \left(\frac{G^E}{RT} \right)\bigg|_{x_1=0} = \int_0^1 \ln \left(\frac{\gamma_1}{\gamma_2} \right) dx_1. \tag{2.107}$$

Now the excess Gibbs free energy at $x_1 = 0$ and $x_1 = 1$ by definition is zero; hence we obtain the following as a condition on the activity coefficients:

$$\int_0^1 \ln \left(\frac{\gamma_1}{\gamma_2} \right) dx_1 = 0. \tag{2.108}$$

Equation 2.108 is also called the *area test* of activity coefficients. It provides a convenient method for testing the thermodynamic consistency of activity coefficient data. One can simply plot the values of $\ln(\gamma_1/\gamma_2)$ vs. x_1 and if the area under the resulting curve is equal to zero, then the test for thermodynamic consistency is satisfied.

A variety of models of varying complexity for describing the excess Gibbs free energy of liquid solutions have been developed (Poling et al. 2001). Most of the modern activity coefficient models easily handle multicomponent solutions. In addition, the so-called *UNIFAC model* allows the activity coefficients to be predicted from the molecular structure of the species that are in the solution. The following example illustrates the calculation of an expression for the activity coefficients for a binary system from a model of the excess Gibbs free energy.

Example 2.7

The simplest model to describe the excess Gibbs free energy of a binary liquid solution is given by the following power series:

$$\frac{G^E}{x_1 x_2 RT} = A_0 + A_1(x_1 - x_2) + A_2(x_1 - x_2)^2 + \cdots$$

Retaining only the lead constant A_0, find expressions for γ_1 and γ_2.

Solution

We therefore have that $(G^E/RT) = A_0 x_1 x_2 = A_0(n_1/n)(n_2/n)$, where n_1 and n_2 are the moles of components 1 and 2, respectively, and $n = n_1 + n_2$. Multiplying both sides of this expression by n and using Equation 2.100, we have that

$$\ln \gamma_1 = \left[\frac{\partial\left(\frac{nG^E}{RT}\right)}{\partial n_1} \right]_{T, P, n_2} = \frac{(n_1 + n_2)A_0 n_2 - A_0 n_1 n_2}{(n_1 + n_2)^2} = A_0 \frac{n_2}{n} - A_0 \frac{n_1 n_2}{n^2}$$

$$= A_0(x_2 - x_1 x_2) = A_0 x_2(1 - x_1) = A_0 x_2^2.$$

So, we find that the activity coefficient for component 1 based on this model for G^E is given by $\ln \gamma_1 = A_0 x_2^2$. In a similar manner, we can show that the activity coefficient for component 2 is given by $\ln \gamma_2 = A_0 x_1^2$. Expressions for the activity coefficients based on more complex models of G^E for both binary and multicomponent systems may be found in Poling et al. (2001).

Example 2.8

From Example 2.7, determine the expressions for the activity coefficients of components 1 and 2 at infinite dilution.

Solution

At infinite dilution for component 1, we have that $x_1 = 0$ and $x_2 = 1$. Therefore, from the results in Example 2.7, we have that $\ln \gamma_1 = A_0$ and $\ln \gamma_2 = 0$ or $\gamma_2 = 1$. At infinite dilution for component 2, we have that $x_1 = 1$ and $x_2 = 0$. Hence, $\ln \gamma_1 = 0$ or $\gamma_1 = 1$ and $\ln \gamma_2 = A_0$. In this example, for either component, the natural logarithm of the activity coefficient at infinite dilution is equal to A_0.

Example 2.9

Show that the activity coefficient expressions found in Example 2.7 satisfy the Gibbs–Duhem relationship, i.e., Equation 2.103.

Solution

For a binary system, Equation 2.103 states that

$$x_1 \left(\frac{\partial \ln \gamma_1}{\partial x_1} \right)_{T,P} + x_2 \left(\frac{\partial \ln \gamma_2}{\partial x_1} \right)_{T,P} = 0.$$

Substituting in the fact that $\ln \gamma_1 = A_0 x_2^2$ and $\ln \gamma_2 = A_0 x_1^2$, we then have that

$$2 A_0 x_1 (1 - x_1)(-1) + 2 A_0 x_1 (1 - x_1) = 0$$

or

$$0 = 0,$$

which shows that the expression used in Example 2.7 for the excess Gibbs free energy satisfies the Gibbs–Duhem consistency test.

2.6.3 Applications of Equilibrium Thermodynamics

With the development of the above thermodynamic equilibrium relationships, we can now address specific topics in solution thermodynamics, such as the solubility of a solid in a liquid solvent, freezing point depression, solid–gas equilibrium, gas solubility, osmotic pressure, the distribution of a solute between two liquid phases, vapor–liquid equilibrium, flammability limits, the thermodynamics of surfaces, equilibrium dialysis, and chemical equilibrium in an ideal solution. The following discussion will form the foundation for our understanding in later chapters of solute transport in biological systems. Solute transport occurs across the interface between phases and at the interface we assume that the solute is in phase equilibrium.

2.6.3.1 Solubility of a Solid in a Liquid Solvent

We will consider a binary system where the solvent is denoted by subscript 1 and the solute by subscript 2. We assume that the solvent has negligible solubility in the solid, hence the solid solute will exist as a pure phase. We can then use Equations 2.84 and 2.99 as follows to express the phase equilibrium between the solid and liquid phases:

$$f_2^S = \gamma_2 \, x_2 \, f_2^L, \tag{2.109}$$

where f_2^S refers to the fugacity of the pure solid solute, x_2 represents the equilibrium solubility of the solid in the solvent phase expressed as a mole fraction, and f_2^L represents the fugacity of pure liquid solute at the equilibrium temperature and pressure of the solution. Since this temperature must be less than the melting temperature of the solid solute, it follows that the solute in the solution exists as a subcooled liquid. Since the solute does not ordinarily exist as a liquid at these conditions, we will have to estimate the value of f_2^L as shown in the following discussion.

With these assumptions, we can rearrange Equation 2.109 to solve for the solubility (x_2) as follows:

$$x_2 = \frac{f_2^S}{\gamma_2 \, f_2^L}. \tag{2.110}$$

Equation 2.110 shows that the solubility is directly proportional to the ratio of the pure component fugacities of the solid and its subcooled liquid and inversely proportional to the value of the activity coefficient of the solid in the solvent solution.

To determine an expression for the solid solubility, we first must calculate the ratio of these fugacities for the pure solid. This can be accomplished by recognizing the fact that the *triple point* (T_{tp}) for a pure substance (see Figure 2.4) defines that unique equilibrium temperature and pressure where all three phases (solid, liquid, and vapor) coexist in equilibrium. In addition, as shown in Figure 2.4, the melting temperature (T_m) of a pure substance does not depend strongly on the pressure and is therefore very close to the triple point temperature. Now from Equation 2.57, we can write that at constant temperature $dG = RT\, d \ln f$. Combining this result with Equation 2.62, we have that

$$d\left(\frac{G}{RT}\right) = d \ln f = -\frac{H}{RT^2} dT + \frac{V}{RT} dP. \tag{2.111}$$

This equation may then be written for the solid and subcooled liquid phases, with the difference between the solid and subcooled liquid fugacities given by

$$d \ln \frac{f_2^S}{f_2^L} = \frac{H^L - H^S}{RT^2} dT - \frac{V^L - V^S}{RT} dP. \tag{2.112}$$

The term ($H^L - H^S$) is the *enthalpy of fusion* and is dependent on the temperature. Using the enthalpy of fusion at the triple point temperature (T_{tp}) as a reference, we can write that $H^L - H^S$ at some other temperature T is given by

$$H^L - H^S = \left(H^L\big|_{tp} + \int_{T_{tp}}^{T} C_{PL}\, dT\right) - \left(H^S\big|_{tp} + \int_{T_{tp}}^{T} C_{PS}\, dT\right), \tag{2.113}$$

which can also be written as follows, assuming that the heat capacities do not vary much with the temperature between T_{tp} and T:

$$H^L - H^S = \Delta H_{tp} + \Delta C_P (T - T_{tp}), \tag{2.114}$$

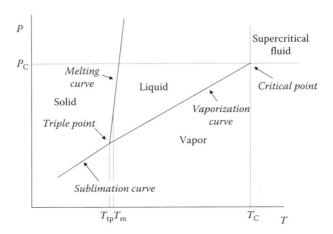

FIGURE 2.4 *PT* diagram for a pure component.

where ΔH_{tp} is the enthalpy of fusion of the solid at the triple point temperature and ΔC_P represents the difference in the heat capacities, $\Delta C_P = C_{PL} - C_{PS}$.

We can now substitute Equation 2.114 into Equation 2.112 and integrate between (T_{tp}, P_{tp}) and the conditions of interest (i.e., T, P). In performing this integration, we also make use of the fact that, for the most part, the difference in the specific volumes ($\Delta V = V^L - V^S$) is independent of the pressure. In addition, we recognize that at the triple point $f_2^L = f_2^S$:

$$\ln \frac{f_2^S}{f_2^L} = \frac{\Delta H_{tp}}{R}\left(\frac{1}{T_{tp}} - \frac{1}{T}\right) - \frac{\Delta C_P}{R}\left(\ln\frac{T_{tp}}{T} - \frac{T_{tp}}{T} + 1\right) - \frac{\Delta V}{RT}(P - P_{tp}). \qquad (2.115)$$

We can now substitute Equation 2.115 into Equation 2.110 and obtain the following general equation for the solubility of a solid in a liquid solvent at T and P:

$$x_2 = \frac{1}{\gamma_2}\exp\left[\frac{\Delta H_{tp}}{R}\left(\frac{1}{T_{tp}} - \frac{1}{T}\right) - \frac{\Delta C_P}{R}\left(\ln\frac{T_{tp}}{T} - \frac{T_{tp}}{T} + 1\right) - \frac{\Delta V}{RT}(P - P_{tp})\right]. \qquad (2.116)$$

We can now make some reasonable approximations that simplify the use of Equation 2.116. In most cases, the pressure correction in the last term of Equation 2.116 is negligible. In addition, the heat capacity difference also provides only a minor contribution and can be ignored. In addition, as mentioned earlier, the triple point temperature (T_{tp}) is close to the melting point temperature (T_m) (cf. Figure 2.4) of the solid at atmospheric pressure, and we can then replace the enthalpy of fusion at T_{tp} with the enthalpy of fusion at the normal melting temperature, i.e., ΔH^m. For a variety of organic molecules, the ratio of the enthalpy of fusion (ΔH^m, calories per gram mole) to the normal atmospheric melting point (T_m, Kelvin) is approximately in the range from 9 to 11. With these simplifications, Equation 2.116 may be written as

$$x_2 = \frac{1}{\gamma_2}\exp\left[\frac{\Delta H^m}{R}\left(\frac{1}{T_m} - \frac{1}{T}\right)\right]. \qquad (2.117)$$

If the solid solute and the solvent are chemically similar, then we would expect them to form an ideal solution. For an ideal solution, we can set γ_2 in Equation 2.117 equal to unity. For an ideal solution with $\gamma_2 = 1$, the solubility of the solid solute can be predicted only from the enthalpy of fusion and the melting temperature. Note that the ideal solubility is based only on the pure component properties of the solute and is the same regardless of the solvent. For nonideal solutions, one must use an appropriate solution model for the activity coefficient (γ_2) (Poling et al. 2001) in order to calculate the solubility. In the limit of negligible solubility in the solvent, one may use the infinite dilution activity coefficient (γ_2).

In the case of nonpolar solutes and solvents, it is worth mentioning at this point that the *Scatchard–Hildebrand equation* allows estimation of the solute activity coefficient from the pure component properties of the solute and the solvent. Using the Scatchard–Hildebrand equation, the solute activity coefficient is given by

$$\ln \gamma_2 = \frac{V_2^L(\delta_1 - \delta_2)^2 \Phi_1^2}{RT}, \qquad (2.118)$$

where V_2^L represents the molar volume of the solute as a subcooled liquid that, in practice, is usually taken to be the same as the molar volume of the solute as a liquid at the melting temperature, δs are the *solubility parameters* for the solute and the solvent, and Φ_1 is the volume fraction of the solvent defined by

$$\Phi_1 = \frac{x_1 V_1^L}{x_1 V_1^L + x_2 V_2^L}. \tag{2.119}$$

The square of the solubility parameter (δ_i^2) is defined as the ratio of the change in the internal energy for complete vaporization to the molar liquid volume. The internal energy change of vaporization is the same as the enthalpy of vaporization minus RT assuming an ideal gas. The solubility parameter for component i is then given by

$$\delta_i = \left(\frac{\Delta H_i^{Vap} - RT}{V_i^L} \right)^{1/2}. \tag{2.120}$$

One usually ignores the temperature dependence of ΔH_i^{Vap} and V_i^L and simply uses the values at the normal melting point for the solute and at 25°C for the solvent. For a solid solute, the heat of vaporization would also be equal to the difference between the heat of sublimation and the heat of fusion, i.e., $\Delta H^{Vap} = \Delta H^{Sub} - \Delta H^m$. If an expression for the vapor pressure of the solid is known, e.g., Equation 2.89 and similar, then the heat of sublimation can be calculated using the Clausius–Clapeyron equation, i.e., Equation 2.88. The more chemically similar the solute and solvent are, the closer the values of their respective solubility parameters, and from Equation 2.118 we see that the resulting solution then approaches ideality. Hence, comparing the solubility parameters provides a quick method for determining chemical similarity and the degree of nonideality between the solute and the solvent.

Example 2.10

A drug has a molecular weight of 230 and a melting temperature of 155°C. Estimate the solubility of this drug in benzene and in n-hexane at 25°C, assuming they form an ideal solution. Also, determine the solubility based on the Scatchard–Hildebrand equation. The following data are also provided:

Heat of fusion of the drug	4300 cal mol⁻¹
Density of the drug	1.04 g cm⁻³
Vapor pressure of the solid drug	$\ln P^{Sat} (mmHg) = 27.3 - \dfrac{8926}{T(K)}$
Molar volume of benzene	89.4 cm³ mol⁻¹
Solubility parameter for benzene	9.2 (cal cm⁻³)¹ᐟ²
Molar volume of n-hexane	131.6 cm³ mol⁻¹
Solubility parameter for n-hexane	7.3 (cal cm⁻³)¹ᐟ²

Solution

To calculate the ideal solubility of the drug, we use Equation 2.117 with $\gamma_2 = 1$.

$$x_2 = \exp\left[\frac{4300 \text{ cal gmol}^{-1}}{1.987 \text{ cal gmol}^{-1}\text{K}^{-1}} \left(\frac{1}{273.15+155} - \frac{1}{298.15} \right) \text{K}^{-1} \right] = 0.110$$

This is the same value whether the solvent is benzene or n-hexane. To calculate the solubility based on the Scatchard–Hildebrand equation, we need to estimate the solubility parameter for the drug. First, the heat of sublimation can be found as follows from the vapor pressure of the solid drug using Equation 2.88:

$$\Delta H^{Sub} = -R\frac{d\ln P^{Sat}}{d\frac{1}{T}} = 1.987 \text{ cal gmol}^{-1}\text{K}^{-1} \times 8926 \text{ K} = 17736.9 \text{ cal mol}^{-1}.$$

Then the solubility parameter for the drug is found from Equation 2.120, assuming that the molar volume of the drug as a liquid and as a solid is similar:

$$\delta_2 = \left(\frac{17736.9 \text{ cal mol}^{-1} - 4300 \text{ cal mol}^{-1} - 1.987 \text{ cal mol}^{-1}\text{K}^{-1} 298.15 \text{ K}}{\frac{1}{1.04} \text{ cm}^3\text{g}^{-1} \times 230 \text{ g mol}^{-1}}\right)^{1/2} = 7.62 \left(\text{cal mol}^{-1}\right)^{1/2}.$$

Next, we can combine Equations 2.117 through 2.119 to give

$$x_2 = \frac{\exp\left[\frac{\Delta H^m}{R}\left(\frac{1}{T_m} - \frac{1}{T}\right)\right]}{\exp\left\{\frac{V_2^L\left(\delta_1 - \delta_2\right)^2}{RT}\left[\frac{\left(1 - x_2\right)V_1^L}{\left(1 - x_2\right)V_1^L + x_2 V_2^L}\right]^2\right\}}.$$

This equation is an implicit algebraic equation in the solubility x_2. Numerical techniques need to be used to solve for the value of x_2. One approach that can be used is *direct iteration*. In this technique, one assumes a value for x_2, substitutes this value on the right-hand side of the previous equation, and calculates a new value of x_2 called $\hat{x}_2$. One then lets $x_2 = \hat{x}_2$, and this new value of x_2 is then substituted into the right-hand side of the previous equation and the process is repeated until there is no longer any significant difference in the values of x_2 and $\hat{x}_2$ and convergence is then obtained. Although direct iteration is simple to implement, there is no guarantee of convergence to the solution. In addition, numerous iterations are sometimes required. The second approach is based on *Newton's method*. Newton's method has more reliable convergence and requires fewer iterations. To implement Newton's method, one first forms the following difference equation:

$$f(x_2) = x_2 - \frac{\exp\left[\frac{\Delta H^m}{R}\left(\frac{1}{T_m} - \frac{1}{T}\right)\right]}{\exp\left\{\frac{V_2^L\left(\delta_1 - \delta_2\right)^2}{RT}\left[\frac{\left(1 - x_2\right)V_1^L}{\left(1 - x_2\right)V_1^L + x_2 V_2^L}\right]^2\right\}}.$$

The goal is to find the value of x_2 that makes $f(x_2) = 0$. This is accomplished through the following iterative equation, which provides an update for the value of x_2 based on the evaluation of $f(x_2)$ and $df(x_2)/dx_2$ at the previous value of x_2:

$$\hat{x}_2 = x_2 - \frac{f\left(x_2\right)}{\dfrac{df\left(x_2\right)}{dx_2}}.$$

One then assumes $x_2 = \hat{x}_2$ and the process is repeated until convergence. In terms of choosing an initial starting value of x_2 for either approach, one could use the ideal solubility or simply assume that $x_2 = 0$. After substituting the values of the known quantities into the earlier equation for $f(x_2)$ and performing the iterative calculations, it is found that the solubility based on the

Scatchard–Hildebrand model for the solute activity coefficient is 0.054 when the solvent is benzene and 0.107 when the solvent is n-hexane. The drug and n-hexane solubility parameters are nearly identical and the solubility of the drug is nearly the same as that found assuming an ideal solution. On the other hand, for benzene as the solvent, the drug solubility is considerably lower than the value based on an ideal solution.

2.6.3.2 Depression of the Freezing Point of a Solvent by a Solute

Consider a small amount of a solute dissolved in a solvent. As the temperature of this solution is decreased, a temperature (T_f) is reached where the pure solvent just starts to separate out of the solution as a solid phase. The freezing temperature of this mixture is T_f. This temperature is lower than the freezing point of the pure solvent (T_m). One is usually interested in determining the freezing point depression, which is then defined as $\Delta T = T_m - T_f$. This problem is very similar to the one we just addressed concerning the solubility of a solute in a solvent. However, now the focus is on the solvent, which is component 1.

To solve this problem, we assume that equilibrium exists between the first amount of solvent that freezes and the rest of the solution. We also assume that the solvent that freezes exists as pure solvent, and we assume that the minute amount of solid solvent formed has no effect on the composition of the solution. We can then write an equilibrium equation like Equation 2.109 but now for the solvent:

$$f_1^S = \gamma_1 x_1 f_1^L, \tag{2.121}$$

which can be rearranged to give

$$\ln \gamma_1 x_1 = \ln \frac{f_1^S}{f_1^L}. \tag{2.122}$$

Following a similar development we used for the solute in the previous section, we can write that

$$d \ln \frac{f_1^S}{f_1^L} = \frac{H^L - H^S}{RT^2} dT - \frac{V^L - V^S}{RT} dP, \tag{2.123}$$

and we also have that

$$H^L - H^S = \Delta H^m + \Delta C_P (T_f - T_m), \tag{2.124}$$

where for the solvent we use the normal melting temperature of the pure solvent as the reference temperature, and T_f is the mixture freezing temperature. We also assumed that the heat capacities do not vary much as the temperature changes from T_m to T_f. We can now substitute Equation 2.124 into Equation 2.123 and integrate between (T_m, P_m) and the conditions of interest, i.e., (T_f, P_f). In performing this integration, we also make use of the fact that, for the most part, the difference in the specific volumes ($\Delta V = V^L - V^S$) is independent of the pressure. In addition, we recognize that at the melting point of the pure solvent, $f_1^L = f_1^S$:

$$\ln \frac{f_1^S}{f_1^L} = \frac{\Delta H^m}{R} \left(\frac{1}{T_m} - \frac{1}{T_f} \right) - \frac{\Delta C_P}{R} \left(\ln \frac{T_m}{T_f} - \frac{T_m}{T_f} + 1 \right) - \frac{\Delta V}{RT} (P_f - P_m). \tag{2.125}$$

Combining Equation 2.125 with Equation 2.122, we obtain

$$\ln \gamma_1 x_1 = \frac{\Delta H^m}{R} \left(\frac{1}{T_m} - \frac{1}{T_f} \right) - \frac{\Delta C_P}{R} \left(\ln \frac{T_m}{T_f} - \frac{T_m}{T_f} + 1 \right) - \frac{\Delta V}{RT} (P_f - P_m). \tag{2.126}$$

Once again, we can ignore the effect of the pressure term and since $T_f \approx T_m$, we obtain

$$\ln \gamma_1 x_1 = \frac{\Delta H^m}{R} \left(\frac{1}{T_m} - \frac{1}{T_f} \right) = -\frac{\Delta H^m}{RT_m^2} (T_m - T_f). \tag{2.127}$$

The freezing point depression is then given by

$$\Delta T = T_m - T_f = -\frac{RT_m^2}{\Delta H^m} \ln \gamma_1 x_1. \tag{2.128}$$

Note that if the freezing depression is measured, then Equation 2.128 provides a convenient means for determining the activity coefficient of the solvent for a solution of given composition.

For ideal solutions, we have that $\gamma_1 = 1$. For solutions that are very dilute in the solute, then $\gamma_1 \approx 1$ and $\ln x_1 = \ln(1-x_2) \approx -x_2$. So, for dilute solutions, we can write that

$$\Delta T = T_m - T_f = \frac{RT_m^2}{\Delta H^m} x_2, \tag{2.129}$$

where x_2 is the mole fraction of the solute in the mixture.

Example 2.11

Estimate the freezing point depression of a benzene solution containing the drug considered in Example 2.10. Assume that the drug concentration is 0.04 g cm^{-3}. The normal melting point for benzene is 278.7 K.

Solution

We expect the mole fraction of the drug at this concentration to be very small. We can therefore assume that the density of the liquid solution is the same as that of pure benzene, which is 0.885 g cm^{-3}. As a basis for the calculation of the mole fraction, assume that we have 1 cm^3 of the liquid mixture. Then, for the drug mole fraction in benzene, we can write that

$$x_2 = \frac{\dfrac{0.04 \text{ g}}{cm^3} \times 1 \text{ cm}^3 \text{ solution} \times \dfrac{1 \text{ mole of drug}}{230 \text{ g}}}{\left(\dfrac{0.04 \text{ g}}{cm^3} \times \dfrac{1 \text{ mole of drug}}{230 \text{ g}} + \dfrac{0.885 - 0.04 \text{ g}}{cm^3} \times \dfrac{1 \text{ mole benzene}}{78 \text{ g}} \right) \times 1 \text{ cm}^3 \text{ solution}}$$

and we get that $x_2 = 0.0158$. Now using Equation 2.129:

$$\Delta T = T_m - T_f = \frac{1.987 \text{ cal mol}^{-1} K^{-1} \, 278.7^2 \text{ K}^2}{2350 \text{ cal mol}^{-1}} \times 0.0158 = 1.038 \text{ K}.$$

2.6.3.3 Equilibrium between a Solid and a Gas Phase

Now we consider a pure solid phase that is in equilibrium with a surrounding gas phase. Here, the question that can be addressed is, for a given T and P, what is the equilibrium composition of the gas phase? We let component 1 represent the gas and component 2 is the material that makes up the pure solid phase. We also assume that the gas has negligible solubility in the solid phase, so the solid phase is pure. Equilibrium requires that the solid phase fugacity of component 2 equals the fugacity of component 2 in the gas phase. We use Equation 2.71 to describe the fugacity of component 2 in the gas phase and obtain the following statement for equilibrium between the pure solid phase and the gas:

$$f_2^S = \hat{\phi}_2 \, y_2 \, P. \tag{2.130}$$

The mole fraction of component 2 in the gas phase is then given by

$$y_2 = \frac{f_2^S}{\hat{\phi}_2 \, P}. \tag{2.131}$$

If the pressure is significantly above atmospheric, then a mixture equation of state is needed to calculate the fugacity coefficient of component 2 in the gas phase, as described by Equation 2.74 (Poling et al. 2001). However, for most applications in biomedical engineering, the pressure will be near atmospheric and we can treat the gas phase as an ideal gas for which $\hat{\phi}_2 = 1$.

The fugacity of the pure solid is given by Equation 2.95, recognizing that we are now considering a subcooled or compressed solid rather than that of a liquid for which we derived Equation 2.95:

$$f_2^S = f_2^{Sat} \exp\left(\frac{V_2^{S,Sat}(P - P_2^{Sat})}{RT} \right) = \phi_2^{Sat} P_2^{Sat} \exp\left(\frac{V_2^{S,Sat}(P - P_2^{Sat})}{RT} \right), \tag{2.132}$$

where P_2^{Sat} is the sublimation pressure (or vapor pressure) wherein the solid will vaporize directly to a vapor without first forming a liquid. Since the sublimation pressure is generally very small at ambient temperatures, the saturated fugacity coefficient (ϕ_2^{Sat}) is equal to one. If the pressure is low and on the order of P_2^{Sat}, then the exponential term is also equal to unity. Therefore, for this case, we get that the pure component solid phase fugacity of component 2 is simply equal to its sublimation or vapor pressure. Hence, we can write that

$$y_2 = \frac{P_2^{Sat}}{P}. \tag{2.133}$$

Example 2.12

Estimate the gas phase equilibrium mole fraction of the drug considered in Example 2.10 at a temperature of 35°C. The pressure of the gas is 1 atm.

Solution

Since the pressure is 1 atm, we assume that the gas phase is ideal and set $\phi_2^{Sat} = 1$. We then calculate the fugacity of the drug as a solid using Equation 2.132:

$$f_2^S = P_2^{Sat} \exp\left(\frac{V^{S,Sat}(P - P_2^{Sat})}{RT} \right).$$

We also assume that the molar volume of the solid at saturation T and P is the same as the molar volume of the solid at ambient conditions. For P_2^{Sat}, we were given an expression in Example 2.10.

For the given T, we calculate that $P_2^{Sat} = 0.189$ mmHg. Using the previous equation, we then calculate the fugacity of the pure solid as

$$f_2^S = 0.189 \text{ mmHg}$$

$$\times \exp\left(\cfrac{\cfrac{1 \text{ cm}^3}{1.04 \text{ g}} \times \cfrac{1 \text{ L}}{1000 \text{ cm}^3} \times 230 \cfrac{\text{g}}{\text{mol}} \times (760 - 0.189) \text{ mmHg} \times \cfrac{1 \text{ atm}}{760 \text{ mmHg}}}{0.082 \cfrac{\text{atmL}}{\text{Kmol}} \times 308.15 \text{ K}}\right),$$

$$f_2^S = 0.189 \text{ mmHg} \times 1.0088 = 0.191 \text{ mmHg}.$$

As expected, because of the low pressure, we see in the previous calculation that the exponential correction to the solid fugacity is negligible and that the fugacity of the solid is nearly the same as its vapor pressure. Using Equation 2.133, we then find that the gas phase mole fraction of the drug is $y_2 = (0.191 \text{ mmHg}/760 \text{ mmHg}) = 0.00025$.

2.6.3.4 Solubility of a Gas in a Liquid

We now consider the calculation of the equilibrium solubility of a sparingly soluble gas (taken as component 2) in a liquid, i.e., $x_2 \to 0$. Our phase equilibrium relationship, i.e., Equation 2.84, requires for the solute gas that

$$\hat{f}_2^L = \hat{f}_2^G. \tag{2.134}$$

The definition of the activity coefficient in Equation 2.99, where $\gamma_2 = \hat{f}_2^L/x_2\,f_2^L$, provides an approach for calculating the fugacity of component 2 within the liquid phase:

$$\hat{f}_2^L = \gamma_2\,x_2\,f_2^L, \tag{2.135}$$

where f_2^L is the pure component liquid fugacity of mixture T and P and is given by Equation 2.95. Recall from Equation 2.71 that we also defined the component mixture fugacity coefficient as $\hat{\phi}_2 = \hat{f}_2^G/y_2 P$, which also says that the fugacity of component 2 within a vapor or gas mixture is given by

$$\hat{f}_2^G = y_2\,\hat{\phi}_2\,P \tag{2.136}$$

By Equation 2.84, Equations 2.135 and 2.136 are equal, so

$$y_2\,\hat{\phi}_2\,P = \gamma_2\,x_2\,f_2^L. \tag{2.137}$$

We then assume that the solvent has a negligible vapor pressure, hence $y_2 = 1$ and $\hat{\phi}_2 = \phi_2$. The gas solubility in the solvent expressed as a mole fraction is then given by

$$x_2 = \frac{\phi_2\,P}{\gamma_2\,f_2^L} = \frac{f_2^G}{\gamma_2\,f_2^L}. \tag{2.138}$$

TABLE 2.1
Physical Properties of Some Common Gases

Gas	Critical Properties T_C (K) and P_C (MPa)	Liquid Volume V^L (cm³ mol⁻¹)	Solubility Parameter δ (cal cm⁻³)¹ᐟ²
Oxygen	154.6, 5.046	33.0	4.0
Nitrogen	126.2, 3.394	32.4	2.58
Carbon dioxide	304.2, 7.376	55	6.0

Usually the temperature of interest will be much higher than the critical temperature of the soluble gas being considered, so we have that $T > T_{C2}$. The critical temperature and other properties for several common gases are summarized in Table 2.1.

Although Equation 2.138 provides a convenient means of calculating the gas solubility, it is encumbered by the fact that the soluble gas does not really exist as a liquid since $T > T_{C2}$. If by chance the soluble gas is below its critical temperature, then this problem of the gas existing as a hypothetical liquid is not an issue.

One approach to solving this problem involves using Equation 2.138 with the Scatchard–Hildebrand equation for the activity coefficient of the solute. When Equation 2.118 is substituted into Equation 2.138, one obtains

$$x_2 = \frac{f_2^G}{f_2^L} \exp\left[-\frac{V_2^L (\delta_1 - \delta_2)^2 \Phi_1^2}{RT} \right]. \tag{2.139}$$

Prausnitz and Shair (1961) used Equation 2.139 to correlate known solubility data for a number of gases in a variety of solvents and obtained the three parameters in Equation 2.139, which describe the soluble gas as a hypothetical liquid, i.e., the pure liquid fugacity (f_2^L), the liquid volume (V_2^L), and the gas solubility parameter (δ_2). Table 2.1 summarizes for several gases the values of (V_2^L) and (δ_2) that were found and Equation 2.140 provides the fugacity of the soluble gas as a hypothetical pure liquid at a pressure of 1.013 bar:

$$\ln\left(\frac{f_2^L}{P_{C2}}\right) = 7.81 - \frac{8.06}{\dfrac{T}{T_{C2}}} - 2.94 \ln\left(\frac{T}{T_{C2}}\right). \tag{2.140}$$

If the pressure is greater than 1.013 bar, then the liquid phase fugacity (f_2^L) calculated from Equation 2.140 needs to be corrected for the effect of the pressure, as shown in Equation 2.95, after changing the reference pressure from P^{Sat} to atmospheric pressure of 1.013 bar:

$$f_2^L(P) = f_2^L(1.013 \text{ bar}) \exp\left[\frac{V_2^L(P - 1.013 \text{ bar})}{RT} \right]. \tag{2.141}$$

Example 2.13

Estimate the solubility of oxygen and carbon dioxide in toluene at a gas partial pressure of 1 atm and 25°C. The solubility parameter for toluene is 8.91 (cal cm⁻³)¹ᐟ².

Solution

Since the pressure is atmospheric, we can neglect the pressure correction to the liquid fugacity. From Equation 2.140, we then calculate the fugacity of oxygen and carbon dioxide in the liquid state at 25°C. For carbon dioxide, we then have that

$$\ln\left(\frac{f_2^l}{P_{C2}}\right) = 7.81 - \frac{8.06}{\dfrac{298.15}{304.2}} - 2.94\ln\left(\frac{298.15}{304.2}\right) = -0.3545,$$

$$f_2^l = 7.376 \text{ MPa} \times \exp(-0.3545) \times \frac{10^6 \text{ Pa}}{1 \text{ MPa}} \times \frac{1 \text{ atm}}{101,325 \text{ Pa}} = 51.07 \text{ atm},$$

and for oxygen, we have that

$$\ln\left(\frac{f_2^l}{P_{C2}}\right) = 7.81 - \frac{8.06}{\dfrac{298.15}{154.6}} - 2.94\ln\left(\frac{298.15}{154.6}\right) = 1.70,$$

$$f_2^l = 5.046 \text{ MPa} \times \exp(1.70) \times \frac{10^6 \text{ Pa}}{1 \text{ MPa}} \times \frac{1 \text{ atm}}{101,325 \text{ Pa}} = 272.6 \text{ atm}.$$

Since the gas partial pressure is only 1 atm, we can treat the gas phase as an ideal gas and set the gas phase fugacity equal to the partial pressure. From Equation 2.139, we can then calculate the mole fraction of carbon dioxide and oxygen in toluene. Note that we have set $\Phi_1 = 1$ since we expect the solubility of these gases in toluene to be very small. For carbon dioxide, we then get that

$$x_2 = \frac{1 \text{ atm}}{51.07 \text{ atm}} \exp\left[-\frac{55 \text{ cm}^3 \text{ gmol}^{-1}(6-8.91)^2 \text{ calcm}^{-3} \times 1}{1.987 \text{ calmol}^{-1}\text{K}^{-1} \times 298.15 \text{ K}}\right] = 0.0089,$$

and for oxygen we obtain

$$x_2 = \frac{1 \text{ atm}}{272.6 \text{ atm}} \exp\left[-\frac{33 \text{ cm}^3 \text{ gmol}^{-1}(4-8.91)^2 \text{ calcm}^{-3} \times 1}{1.987 \text{ calmol}^{-1}\text{K}^{-1} \times 298.15 \text{ K}}\right] = 0.0010.$$

The reported solubility of carbon dioxide in toluene at 25°C and at a partial pressure of 1 atm is 0.010 and that for oxygen in toluene is 0.0009. Calculating the percent error, we see that the error for prediction of the carbon dioxide and oxygen solubility using the Scatchard–Hildebrand model is 11% for both gases.

Using Equation 2.139 for predicting the solubility of a gas in a liquid is limited by the assumptions made in the development of the Scatchard–Hildebrand model for estimating activity coefficients. Generally, the prediction will be better for nonpolar gases and liquids. Also, the accuracy will rapidly diminish as the absolute magnitude of the difference between the solubility parameter of the gas and the solvent increases. For example, consider the solubility of oxygen in ethanol ($\delta_{\text{ethanol}} = 12.8$ (cal cm^{-3})$^{1/2}$). The difference in the solubility parameters of oxygen and ethanol is quite large and we would therefore expect greater error in the prediction of the solubility. We can use the results from Example 2.13 and calculate the predicted oxygen solubility in ethanol as 4.91×10^{-5}, whereas the reported value is about 6×10^{-4}. So, in this case, we have underestimated the solubility of oxygen by over a factor of 10.

Because of these limitations on predicting gas solubility, a more empirical approach is often used. At a given temperature and pressure, it has been found that, for the most part, the gas phase fugacity is directly proportional to the gas solubility expressed as a mole fraction (x_2), i.e., $\hat{f}_2^G = y_2 \hat{\phi}_2 P = H x_2$. The proportionality constant H is called *Henry's constant*. If the gas phase is ideal, then $\hat{\phi}_2 = 1$, and $y_2 P$ is just the partial pressure of the solute gas, i.e., P_2 or pX, where X is the name of the solute gas. For example, the partial pressure of oxygen can be written as either P_{oxygen} or pO_2. So we can write that $P_2 = pX = H x_2$. It is important to remember though that the Henry constant for a specific gas will depend on the solvent, the temperature, and to some extent the pressure.

In some cases, Henry's constant is based on the solute concentration of the solute in the solvent. This is common, for example, in describing the solubility of oxygen in blood where we write that $pO_2 = HC_{oxygen}$. The partial pressure of oxygen in the gas phase is represented by pO_2 and C_{oxygen} is the molar concentration of dissolved oxygen in the blood. The value of Henry's constant (H) for describing the solubility of oxygen in blood is 0.74 mmHg µm^{-1}.

The thermodynamic basis for Henry's constant can be obtained from Equation 2.135, where we have the following expression for the liquid phase fugacity, i.e., $\hat{f}_2^L = \gamma_2 x_2 f_2^L$. The product of γ_2 and f_2^L as $x_2 \to 0$ is the Henry constant. Substituting H for γ_2 and f_2^L results in Henry's law for describing gas solubility, that is

$$\hat{f}_2^L = \gamma_2 x_2 f_2^L = H x_2. \tag{2.142}$$

For sparingly soluble gases, where $x_2 \to 0$, the activity coefficient approaches, in the limit of infinite dilution, a value that becomes independent of the solute mole fraction, i.e., $\gamma_2 \to \gamma_2^\infty$. Hence, from Equation 2.142, we would expect Henry's constant to be nearly independent of the solute mole fraction. Since H is also related to the pure component liquid fugacity, we would also expect H to be somewhat dependent on the temperature and only weakly dependent on the pressure.

Example 2.14

Calculate the concentration of oxygen dissolved in blood for a gas phase partial pressure of oxygen equal to 95 mmHg.

Solution

Using Henry's law and the value of H given above for oxygen and blood, we can write that

$$C_{oxygen} = \frac{pO_2}{H} = \frac{95 \text{ mmHg}}{0.74 \text{ mmHg µm}^{-1}} = 128.4 \text{ µm}.$$

2.6.3.5 Osmotic Pressure

Consider the situation illustrated in Figure 2.5 where a rigid membrane separates region A from region B. Region B contains only pure solvent, whereas region A contains the solvent and a solute. The membrane that separates region A from region B is only permeable to the solvent, hence the solute is completely retained in region A. Solvent will diffuse from region B into region A in an attempt to satisfy the requirement that at equilibrium (see Equations 2.82 and 2.84) the chemical

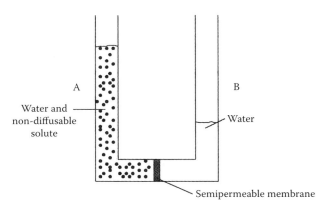

FIGURE 2.5 Concept of osmotic pressure.

potential or fugacity of the solvent on each side of the membrane must be equal. This movement of solvent from region B to region A is called *osmosis* and will tend to increase the level of region A until its increase in hydrostatic pressure stops the flow of solvent from region B and equilibrium is attained. The pressure difference between regions A and B at this point of equilibrium is the osmotic pressure of region A.

A quantitative understanding of osmosis is needed for a variety of calculations and concepts involving solutions and mass transport across semipermeable membranes. We can easily derive an expression for the osmotic pressure of a solution by starting with the fundamental requirement that at equilibrium the temperature and fugacity of the solvent in regions A and B must be equal:

$$\hat{f}^A_{\text{solvent}}(T, P^A) = f^B_{\text{solvent}}(T, P^B). \tag{2.143}$$

Note that at equilibrium the temperatures of regions A and B will be the same; however, the pressures will not be the same and the difference $(P^A - P^B)$ will be the osmotic pressure of region A. Equation 2.143 can be rewritten as follows using an activity coefficient model (see Equation 2.99) to describe the fugacity of the solvent in region A:

$$\gamma^A_{\text{solvent}} x^A_{\text{solvent}} f^A_{\text{solvent}}(T, P^A) = f^B_{\text{solvent}}(T, P^B). \tag{2.144}$$

The pure component solvent fugacity in region A at T and P^A can be related to that in region B at T and P^B through the use of Equation 2.95 after changing the reference pressure from P^{Sat} to P^B. Hence, we can write that

$$f^A_{\text{solvent}}(T, P^A) = f^A_{\text{solvent}}(T, P^B) \exp\left(\frac{V^L_{\text{solvent}}(P^A - P^B)}{RT}\right). \tag{2.145}$$

Substituting Equation 2.145 into Equation 2.144 results in Equation 2.146 for the osmotic pressure $(P^A - P^B)$. This osmotic pressure difference is given the symbol π and represents the osmotic pressure of the solution in region A relative to that of the pure solvent in region B:

$$\pi = (P^A - P^B) = \frac{-RT}{V^L_{\text{solvent}}} \ln\left(\gamma^A_{\text{solvent}} x^A_{\text{solvent}}\right) \tag{2.146}$$

Note that measurement of the osmotic pressure of a solution provides a convenient method through the use of Equation 2.146 to determine the activity coefficient of the solvent. For nonideal solutions, a variety of theoretical and empirical models exist for estimating the activity coefficient in Equation 2.146 (Poling et al. 2001; also see Problems 28 and 29 at the end of this chapter).

Equation 2.146 may be simplified for the special case where the solvent and solute form an ideal solution. In this case, $\gamma^A_{\text{solvent}} = 1$ and since x^A_{solvent} is close to unity, we can rewrite Equation 2.146 as

$$\pi = \frac{RT}{V^L_{\text{solvent}}} x^A_{\text{solute}} = RTC_{\text{solute}}, \tag{2.147}$$

where x^A_{solute} represents the mole fraction of the solute in region A. Since the solute mole fraction is generally quite small, we may approximate this as $x^A_{\text{solute}} = V^L_{\text{solvent}} C_{\text{solute}}$, where C_{solute} is the concentration of the solute in gram moles per liter of solution and V^L_{solvent} is the molar volume of the solvent. This ideal dilute solution osmotic pressure, described by Equation 2.147, is known as *van't Hoff's law*. If the solution contains N ideal solutes, then the total osmotic pressure of the solution is the summation of the osmotic pressure generated by each solute according to Equation 2.147:

$$\pi = RT \sum_{i=1}^{N} C_{\text{solute}_i}.$$ (2.148)

Osmotic pressure is not determined on the basis of the mass of the solute in the solution, but rather on the number of particles that are formed by a given solute. Each nondiffusing or retained particle in the solution contributes the same amount to the osmotic pressure regardless of the size of the particle. Thus, if we take 1/1000th of an Avogadro's number of glucose molecules or 1/1000th of an Avogadro's number of albumin molecules and form a 1 L solution of each with water, we have a 1 mM (millimole per liter) solution of each solute. According to Equation 2.147, the osmotic pressure of these two solutions is the same. However, the mass of glucose added to the solution is 180 mg, whereas the mass of albumin added to the solution is nearly 70 g, demonstrating that it is not the mass of solute in the solution that is important in determining the osmotic pressure.

Example 2.15

Consider the situation shown in Figure 2.5 where on side A of the membrane we have a solution with a water mole fraction of 0.99. On side B of the membrane we have pure water. If the temperature is 25°C and side B is at 1 atm, estimate the pressure needed on side A to stop the osmosis of water from region B. What will happen if the pressure on side A was increased to a value above this pressure?

Solution

Assuming that the solution in region A is an ideal solution, we can use Equation 2.147 to calculate the osmotic pressure of region A, that is

$$\pi = (P_A - P_B) = \frac{8.314 \dfrac{\text{Pa m}^3}{\text{mol K}} \times \dfrac{1\ \text{atm}}{101{,}325\ \text{Pa}} \times 298\ \text{K}}{18 \times 10^{-6} \dfrac{\text{m}^3}{\text{mol}}} \times 0.01 = 13.6\ \text{atm}.$$

Therefore, the pressure in region A will have to be 14.6 atm to prevent the osmosis of water from region B into region A. If the pressure in region A is greater than 14.6 atm, then water will move from region A into region B and this process is called *reverse osmosis*.

Since small pressures are easy to measure, the osmotic pressure is also useful for determining the molecular weight of macromolecules such as proteins. For example, letting m_{solute} represent the mass concentration (gram per liter) of the solute in the solution, then from Equation 2.147 we can write the molecular weight of the solute (MW_{solute}) in terms of the osmotic pressure of the solution as

$$MW_{\text{solute}} = \frac{RT\,m_{\text{solute}}}{\pi}.$$ (2.149)

Example 2.16

The osmotic pressure of a solution containing a macromolecule is equivalent to the pressure exerted by 8 cm of water. The mass concentration of protein in the solution is 15 g L^{-1}. Estimate the molecular weight of this macromolecule.

Solution

We can use Equation 2.149 to estimate the molecular weight as follows:

$$MW_{solute} = \frac{8.314 \dfrac{m^3\,Pa}{molK} \times 298\ K \times \dfrac{15\ g}{L} \times \dfrac{1\ L}{1000\ cm^3} \times \dfrac{(100\ cm)^3}{m^3} \times \dfrac{1\ atm}{101{,}325\ Pa}}{8\ cmH_2O \times \dfrac{1\ in}{2.54\ cm} \times \dfrac{1\ ft}{12\ in} \times \dfrac{1\ atm}{33.91\ ftH_2O}},$$

$$MW_{solute} = 47{,}400\ gmol^{-1}.$$

2.6.3.6 Distribution of a Solute between Two Liquid Phases

Many purification processes are based on the unequal distribution of a solute between two partially miscible liquid phases. For example, through the process of liquid extraction, a drug produced by fermentation can be extracted into a suitable solvent and purified from the aqueous fermentation broth. Factors to consider in the selection of the solvent include its toxicity, cost, degree of miscibility with the fermentation broth, and selectivity for the solute.

Consider the resulting equilibrium when N_1 moles of solute 1 are mixed with N_2 moles of solvent 2, and N_3 moles of solvent 3. The two solvents are partially miscible and form two liquid phases. The equilibrium distribution of these three components between the two resulting liquid phases (I and II) may be written for component i as follows using Equations 2.84 and 2.99:

$$\gamma_i^I\, x_i^I\, f_i^I(T,P) = \gamma_i^{II}\, x_i^{II}\, f_i^{II}(T,P). \tag{2.150}$$

Now the pure component fugacities of component i are the same in phase I and II since component i exists in the same state at the same temperature and pressure in each of the two phases. Therefore, Equation 2.150 becomes

$$\gamma_i^I\, x_i^I = \gamma_i^{II}\, x_i^{II}. \tag{2.151}$$

The ratio of the mole fractions of component i in the two phases is called the *distribution coefficient* (K_i) and is defined as follows after rearranging Equation 2.151:

$$K_i = \frac{x_i^I}{x_i^{II}} = \frac{\gamma_i^{II}}{\gamma_i^I}. \tag{2.152}$$

It is important to remember that in liquid–liquid equilibrium problems, the solvents that form the two partially miscible liquid phases form highly nonideal solutions. Therefore, the activity coefficients of each component in each phase tend to be strong functions of the composition and need to be determined by multicomponent activity coefficient models that have the capability to describe liquid–liquid equilibrium problems as discussed in Poling et al. (2001).

In this type of problem, it is desired to determine the values of the mole fractions of the three components in the two liquid phases and the amounts of each phase, that is N^I and N^{II}. For the three components and two phases, we therefore have eight unknowns, i.e., the six mole fractions and N^I and N^{II}. So, we need a total of eight relationships between these variables in order to obtain a solution. The equilibrium relationships (Equation 2.152) provide us with three equations and we can also write a mole balance for each component as

$$N_i = x_i^I\, N^I + x_i^{II}\, N^{II}, \tag{2.153}$$

providing us with an additional three equations. In addition, we can use the fact that the mole fractions in each phase must sum to unity, that is $\sum_{i=1}^{3} x_i^I = 1$ and $\sum_{i=1}^{3} x_i^{II} = 1$. Solution of these eight equations provides the desired solution. However, it should be pointed out that this is a somewhat challenging problem to solve because of the strong nonlinear dependence of the activity coefficients on the composition of each phase.

The previous problem can be simplified by assuming that the two solvents are immiscible. Usually, solvents are selected to miminize their mutual solubility, so in a practical sense this is a pretty good assumption. As given by Equation 2.152, the distribution coefficient for the solute taken as component 1 will depend on the composition of each of the two phases. In many biological applications the solute concentration is so low that the activity coefficients approach their infinite dilution values and the distribution coefficient is a constant. Also, since the solute is present in such a small amount, the value of N_2 and N_3 for solvents 2 and 3 is assumed to be constant and equal to N^I and N^{II}, respectively. With these assumptions, a mole balance on the solute can then be written as

$$N_1 = x_1^I N_2 + x_1^{II} N_3, \tag{2.154}$$

and with $x_1^I = K x_1^{II}$ from Equation 2.152, we can solve for the mole fraction of the solute in phase II as

$$x_1^{II} = \frac{N_1}{K N_2 + N_3}. \tag{2.155}$$

Example 2.17

The octanol–water partition coefficient ($K_{o/w}$) is frequently used to describe lipophilicity and to estimate the distribution of a drug between the aqueous and lipid regions of the body. $K_{o/w}$ is the same as the distribution coefficient described earlier with component 2 and phase I being the octanol phase and component 3 and phase II being the aqueous phase. Suppose we have 0.01 mole (N_1) of drug dissolved in 100 moles of water (N_3). The mole fraction of the drug in the aqueous phase is therefore equal to $0.01 \times 100^{-1} = 10^{-4}$. We then add 100 moles (N_2) of octanol to this phase. Estimate the mole fractions of the drug in the two phases once equilibrium has been attained. Also, determine the percent extraction of the drug from the aqueous phase. The octanol–water partition coefficient for the drug in this example is 89.

Solution

Using Equation 2.155, we can calculate the mole fraction of the drug in the aqueous phase once equilibrium has been reached

$$x_1^{II} = \frac{0.01}{89 \times 100 + 100} = 1.11 \times 10^{-6}.$$

The corresponding mole fraction of the drug in the octanol phase is given by Equation 2.152

$$x_1^I = 89 \times 1.11 \times 10^{-6} = 9.89 \times 10^{-5}.$$

The percent extraction of the drug from the aqueous phase is calculated as

$$\% \text{extraction} = \frac{N_1 - x_1^{II} N_3}{N_1} = \frac{0.01 - 1.11 \times 10^{-6} \times 100}{0.01} \times 100 = 98.89\%.$$

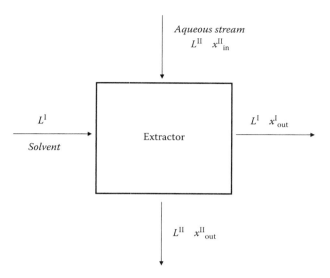

FIGURE 2.6 A single stage liquid–liquid extractor.

Now consider the situation shown in Figure 2.6 where, for example, a pure flowing solvent stream at molar flow rate L^I is contacted in an extractor with an aqueous stream flowing at L^{II} with a solute of mole fraction x^{II}_{in}. Also, the solvent and water are immiscible and we assume that the amount of solute is small so there is no change in either L^I or L^{II}. Our mole balance on the solute can then be written as

$$L^{II} x^{II}_{in} = L^I x^I_{out} + L^{II} x^{II}_{out}. \qquad (2.156)$$

The streams exiting the extractor are at equilibrium, so from Equation 2.152 we can write that $x^I_{out} = K x^{II}_{out}$. Using this relationship in Equation 2.156, we can solve for x^{II}_{out} as

$$\frac{x^{II}_{out}}{x^{II}_{in}} = \frac{1}{1+E}, \qquad (2.157)$$

where E is defined as the extraction factor, i.e., $E = L^I K / L^{II}$. The percent extraction of the solute from phase II is given by

$$\% \text{extraction} = \left(1 - \frac{x^{II}_{out}}{x^{II}_{in}}\right) \times 100. \qquad (2.158)$$

Example 2.18

Suppose a drug is in an aqueous stream flowing at 100 moles min^{-1} at a drug mole fraction of 0.01. The aqueous stream is then contacted within an extractor with a solvent that is flowing at 200 moles min^{-1}. The solvent has a distribution coefficient of 6 for this particular drug. What is the equilibrium mole fraction of the drug in the streams exiting the extractor? What is the percent extraction of the drug from the aqueous stream?

Solution

From the earlier data, we find that the extraction factor is 12. Then, from Equation 2.157 the mole fraction of the drug in the aqueous stream that leaves the extractor is

$$x^{II}_{out} = 0.01 \times \frac{1}{1+12} = 7.69 \times 10^{-4}.$$

The mole fraction of the drug in the exiting solvent stream is

$$x^I_{out} = 6 \times 7.69 \times 10^{-4} = 4.62 \times 10^{-3},$$

and the percent extraction of the solute works out to be 92.3%.

A percent extraction of 92.3% in Example 2.18 results in the potential for loss of valuable product. Therefore, extraction is frequently carried out in a countercurrent multistage extractor. Figure 2.7 illustrates this for N equilibrium stages of extraction. Since the phases are immiscible and the solute concentration is low, we will assume that L^I and L^{II} are constant within the extractor. In addition, the distribution coefficient K is also assumed to be constant. A solute mole balance around the dashed box enclosing stages 1 through n of the extractor provides Equation 2.159 for the mole fraction of the solute entering stage n in phase II:

$$x^{II}_{n+1} = \frac{L^I}{L^{II}} x^I_n + \frac{L^{II} x^{II}_1 - L^I x^I_0}{L^{II}}. \tag{2.159}$$

The streams exiting stage n will be in equilibrium, so from Equation 2.152 we can write that $x^I_n = K x^{II}_n$, where K is the distribution coefficient. Substituting this result into Equation 2.159 results in

$$x^{II}_{n+1} = E x^{II}_n + x^{II}_1 - \left(\frac{E}{K}\right) x^I_0. \tag{2.160}$$

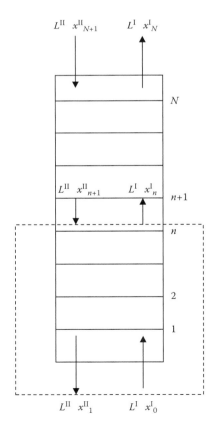

FIGURE 2.7 N stage liquid–liquid extractor.

We can now use Equation 2.160 in a stepwise manner to calculate successive values of x_{n+1}^{II} starting with stage 1. Note that in the following, we have also made use of the fact that we can replace x_0^I in Equation 2.160 with Kx_e^{II}, where x_e^{II} is the solute concentration in phase II that would be in equilibrium with the solute concentration that enters in phase I. So, for stage 1, we have that

$$n = 1: \quad x_2^{II} = E x_1^{II} + x_1^{II} - E x_e^{II} = (E+1)x_1^{II} - E x_e^{II}. \tag{2.161}$$

For stage 2, we can eliminate x_2^{II} by using Equation 2.161, which then gives

$$n = 2: \quad x_3^{II} = E x_2^{II} + x_1^{II} - E x_e^{II} = x_1^{II}(1 + E + E^2) - x_e^{II}(E + E^2),$$

and for stage 3, we obtain the following result after using the previous equation to eliminate x_3^{II}:

$$n = 3: \quad x_4^{II} = x_1^{II}(1 + E + E^2 + E^3) - x_e^{II}(E + E^2 + E^3).$$

This equation can then be generalized for the nth stage with the following result:

$$x_{n+1}^{II} = x_1^{II}(1 + E + E^2 + \cdots + E^n) - x_e^{II}(E + E^2 + \cdots + E^n). \tag{2.162}$$

Using Equation 2.162, we can then write for stage N that

$$x_{N+1}^{II} = x_1^{II}(1 + E + E^2 + \cdots + E^N) - x_e^{II}(E + E^2 + \cdots + E^N). \tag{2.163}$$

The summations in Equation 2.163 are geometric progressions wherein the ratio of each successive term to the previous term is a constant, in this case E. The first series is then given by $(1-E^{N+1})/(1-E)$ and the second series is given by $E(1-E^N)/(1-E)$. Using these relationships for the series, Equation 2.163 can then be written as

$$x_{N+1}^{II} = x_1^{II}\left(\frac{1 - E^{N+1}}{1 - E}\right) - x_e^{II} E\left(\frac{1 - E^N}{1 - E}\right). \tag{2.164}$$

In Equation 2.164, for example, phase II represents the aqueous phase and we would know the incoming mole fraction of the solute, i.e., x_{N+1}^{II}. Recall that $x_e^{II} = x_0^I/K$ and we would also know the amount of solute in the solvent feed stream, i.e., x_0^I. Therefore, for a given number of equilibrium stages (N), Equation 2.164 can be solved to provide the amount of solute present in phase II as it exits the extractor, i.e., x_1^{II}. If there is no solute in the entering solvent phase, then $x_0^I = 0$ and we have that

$$\frac{x_1^{II}}{x_{N+1}^{II}} = \frac{E - 1}{E^{N+1} - 1}. \tag{2.165}$$

The percent extraction of the solute from phase I is then given by

$$\%\,\text{extraction} = \left(1 - \frac{x_1^{II}}{x_{N+1}^{II}}\right) \times 100. \tag{2.166}$$

Equation 2.164 can also be rearranged and solved to give the number of equilibrium stages (N) for specified solute mole fractions.

$$N = \frac{\ln\left(\dfrac{x_{N+1}^{II} - x_N^{II}}{x_1^{II} - x_e^{II}}\right)}{\ln E}, \tag{2.167}$$

with $x_N^{II} = x_N^I/K = x_N^I$ and $x_e^{II} = x_0^I/K$. The number of actual stages will be greater than that calculated from Equation 2.167 since equilibrium is seldom achieved in each stage of the extractor. To find the actual number of stages, we can use the overall efficiency (η) of an equilibrium stage defined as

$$\eta = \frac{N}{N_{actual}}. \tag{2.168}$$

Equilibrium stage efficiencies may be obtained from manufacturer's information on their extractor or from experimental data.

Example 2.19

Rework the previous example and determine the number of equilibrium and actual stages needed to achieve a 99.99% extraction of the drug from the aqueous phase. Assume that the efficiency of an equilibrium stage is 60%. Let phase II represent the aqueous phase and phase I the solvent. No solute is present in the solvent stream that enters the extractor.

Solution

From Equation 2.166, we can calculate the mole fraction of the solute in the aqueous phase that exits the extractor

$$x_1^{II} = \left(1 - \%\,\text{extraction}/100\right)x_{N+1}^{II} = \left(1 - 0.9999\right)0.01 = 1 \times 10^{-6}.$$

Next, we can write an overall solute mole balance around the extractor that can be used to solve for the mole fraction of the solute in the exiting solvent stream:

$$x_N^I = x_0^I + \left(\frac{L^{II}}{L^I}\right)\left(x_{N+1}^{II} - x_1^{II}\right) = 0 + \left(\frac{100}{200}\right)\left(0.01 - 1\times10^{-6}\right) = 4.9995 \times 10^{-3}.$$

Next, we use Equation 2.167 to find the number of equilibrium stages:

$$N = \frac{\ln\left(\dfrac{x_{N+1}^{II} - \dfrac{x_N^I}{K}}{x_1^{II} - \dfrac{x_0^I}{K}}\right)}{\ln E} = \frac{\ln\left(\dfrac{0.01 - \dfrac{4.9995\times10^{-3}}{6}}{10^{-6}}\right)}{\ln 12} = 3.67 \approx 4.$$

So we need four equilibrium stages or $4/0.6 \approx 7$ actual stages.

2.6.3.7 Vapor–Liquid Equilibrium

Vapor–liquid equilibrium plays an important role in understanding and designing separation processes that involve distillation, gas absorption, and the removal of solutes from liquids by contact with another gas stream, also known as stripping. These separation processes are widely used in the chemical process industry, and the design of these processes falls within the realm of chemical engineering. Here, we will illustrate vapor–liquid equilibrium calculations by determining the bubble point (boiling point) and dew point (condensation) temperatures of mixtures. We will also use these concepts to develop an understanding of the flammability or explosive limits of combustible materials.

Consider now the equilibrium distribution of N components at a given T and P between a vapor and a liquid. Our phase equilibrium relationship, i.e., Equation 2.84, requires for component i that

$$\hat{f}_i^L = \hat{f}_i^V. \tag{2.169}$$

For nonideal liquid solutions, an activity coefficient model is once again required to describe the behavior of the components of the liquid phase. For a nonideal liquid solution, we can then use the activity coefficient defined by Equation 2.99, where $\gamma_i = \hat{f}_i^L / x_i f_i^L$, to calculate the fugacity of component i within the liquid phase:

$$\hat{f}_i^L = \gamma_i \, x_i \, f_i^L, \tag{2.170}$$

where f_i^L is the pure component liquid fugacity of mixture T and P and is given by Equation 2.95. In addition, at pressures considerably greater than atmospheric pressure, the vapor phase will deviate from that of an ideal gas and a mixture equation of state is also needed to calculate the mixture fugacity coefficient of each species in the vapor phase, as given by Equation 2.74 (Poling et al. 2001). Recall from Equation 2.71 that the component mixture fugacity coefficient is defined as $\hat{\phi}_i = \hat{f}_i^V / y_i P$, which also says that the fugacity of component i within a nonideal vapor mixture is given by

$$\hat{f}_i^V = y_i \, \hat{\phi}_i \, P. \tag{2.171}$$

By Equation 2.84, Equations 2.170 and 2.171 are equal, so we can write Equation 2.172, which expresses the equilibrium condition for each component i in the vapor–liquid mixture:

$$y_i \, \hat{\phi}_i \, P = \gamma_i \, x_i \, f_i^L. \tag{2.172}$$

If the total pressure (P) and the component vapor pressures are on the order of atmospheric pressure, then we can usually treat the vapor phase as an ideal gas and set $\hat{\phi}_i = 1$ and $\phi_i^{Sat} = 1$. Then, from Equation 2.95, we also have for these conditions that $f_i^L \approx P_i^{Sat}$. Therefore, we can write Equation 2.172 as

$$y_i \, P = \gamma_i \, x_i \, P_i^{Sat} = P_i, \tag{2.173}$$

with P_i equal to the partial pressure of component i. Once again, we also have the constraint on the mole fractions that $\sum_i x_i = \sum_i y_i = 1$.

For an ideal liquid solution, we also have $\gamma_i = 1$, and then we have that

$$y_i \, P = x_i \, P_i^{Sat} = P_i. \tag{2.174}$$

This equation is also known as *Raoult's law*, which shows that the partial pressure of a component ($P_i = y_i P$) in an ideal liquid solution is equal to the product of its mole fraction and its vapor pressure.

Now consider a liquid solution consisting of N components and mole fraction x_i for each component i. For a given pressure P, we can ask at what temperature would this liquid solution just start to boil, that is it first forms a vapor bubble? This temperature is known as the *bubble point temperature* and can be calculated as follows. When the liquid mixture is at its bubble point temperature, it is also referred to as a saturated liquid solution. Below the bubble point temperature, the liquid is a subcooled liquid.

First, we solve Equation 2.172 for the equilibrium mole fraction of component i in the vapor bubble that is formed when the solution reaches the bubble point temperature:

$$y_i = \frac{\gamma_i \, x_i \, f_i^L}{\hat{\phi}_i \, P}. \tag{2.175}$$

Since $\sum_i y_i = 1$, we then have Equation 2.176, which can be solved to find the bubble point temperature:

$$\sum_i \frac{\gamma_i x_i f_i^L}{\hat{\phi}_i} = P. \tag{2.176}$$

It is important to recognize that Equation 2.176 depends on temperature through the terms that include the component liquid phase fugacities and the vapor phase fugacity coefficient. The activity coefficient may also have a temperature dependence. One, therefore, solves the nonlinear algebraic equation given earlier for the temperature that makes the left-hand side of Equation 2.176 equal to the pressure. This temperature is then known as the bubble point temperature or the boiling temperature of the liquid mixture. The composition of the first vapor bubble that is formed is then given by Equation 2.175. If the liquid continues to boil, then the composition of the liquid and vapor phases will change as the more volatile components escape from the liquid to the vapor phase. This is the essence of distillation. The design of distillation processes falls within the study of staged processes and will not be considered here.

If the pressure is low, we can once again replace the liquid phase fugacity of each component with its vapor pressure and set the component mixture fugacity coefficient to one. If the liquid also forms an ideal solution, then the activity coefficients can also be set to unity. In this latter case, which is Raoult's law, we have for the bubble point temperature the requirement that

$$\sum_i x_i P_i^{Sat}(T) = P. \tag{2.177}$$

The only temperature dependence remaining in Equation 2.177 is in the component vapor pressures. This equation can then be solved for the bubble point temperature.

A similar analysis leads to the calculation of the dew point or condensation temperature of a vapor mixture of composition y_i for component i. When the vapor mixture is at its *dew point temperature*, it is also referred to as a saturated vapor. Above the dew point temperature, the vapor is referred to as superheated.

First, we solve Equation 2.172 for the equilibrium mole fraction of component i in the first drop of liquid that is formed when the vapor is cooled to its dew point temperature:

$$x_i = \frac{\hat{\phi}_i y_i P}{\gamma_i f_i^L}. \tag{2.178}$$

Since $\sum_i x_i = 1$, we then have Equation 2.179, which can be solved to find the dew point temperature:

$$\sum_i \frac{\hat{\phi}_i y_i}{\gamma_i f_i^L} = \frac{1}{P}, \tag{2.179}$$

where once again $\hat{\phi}_i$, γ_i, and f_i^L will depend on the temperature. The temperature that satisfies Equation 2.179 is then the dew point temperature of the vapor mixture. The composition of the liquid drop that is just formed is given by Equation 2.178.

If the pressure is low, we can once again replace the liquid phase fugacity of each component with its vapor pressure and set the component mixture fugacity coefficient to one. If the liquid also forms an ideal solution, then the activity coefficients can also be set to unity. In this latter case, which is Raoult's law, we have that

$$\sum_i \frac{y_i}{P_i^{\text{Sat}}(T)} = \frac{1}{P}. \tag{2.180}$$

The temperature dependence is only in the component vapor pressures and Equation 2.180 can then be solved for the dew point temperature.

Example 2.20

Estimate the normal (i.e., $P = 1$ atm) bubble point or boiling temperature of a liquid solvent solution consisting of 25 mol% benzene (C_6H_6, MW = 78.11, $T_B = 353.3$ K), 45 mol% toluene (C_7H_8, MW = 92.14, $T_B = 383.8$ K), and 30 mol% ethylbenzene (C_8H_{10}, MW = 106.17, $T_B = 409.3$ K), where the chemical formula, molecular weight, and normal boiling point are given in parenthesis after each of the components. What is the composition of the vapor bubble that is formed at the bubble point temperature? The vapor pressures of these components are given by the Antoine vapor pressure equation (Reid et al. 1977) with the vapor pressure in mmHg and the temperature in Kelvin:

$$\ln P_{\text{benzene}}^{\text{Sat}} = 15.9008 - \frac{2788.51}{T - 52.36},$$

$$\ln P_{\text{toluene}}^{\text{Sat}} = 16.0137 - \frac{3096.52}{T - 53.67},$$

$$\ln P_{\text{ethylbenzene}}^{\text{Sat}} = 16.0195 - \frac{3279.47}{T - 59.95}.$$

Solution

Since these molecules are all chemically similar, we will assume that they form an ideal solution. Equation 2.177 may therefore be used to calculate the bubble point temperature. The approach for solving this problem is as follows:

1. Assume a value of the bubble point temperature, T.
2. Calculate the vapor pressures for each component at T.
3. Calculate $\sum_i x_i P_i^{\text{Sat}} = \hat{P}$.
4. If $\hat{P} = P$, which in this problem is 1 atm, then T is the bubble point temperature.
5. If $\hat{P} > P$, then the assumed T is too high; decrease T and go back to step 2.
6. If $\hat{P} < P$, then the assumed T is too low; increase T and go back to step 2.

After following the above-mentioned procedure, we obtain the following results:

Bubble point temperature	= 377.81 K
y_{benzene}	= 0.503
y_{toluene}	= 0.379
$y_{\text{ethylbenzene}}$	= 0.118

Note the change in the distribution of the components between the liquid and the vapor bubble that is formed. Benzene, being the most volatile component, has a near twofold increase in its composition, whereas ethylbenzene, being considerably less volatile than benzene, has its mole fraction in the vapor decreased by nearly a factor of three as compared to its composition in the liquid phase.

Example 2.21

Repeat the previous calculation, but this time find the dew point temperature and the composition of the first drop of liquid that is formed.

Solution

Equation 2.180 may be used to calculate the dew point temperature. The approach for solving this problem is as follows:

1. Assume a value of the dew point temperature, T.
2. Calculate the vapor pressures for each component at T.
3. Calculate $\sum_i (y_i / P_i^{Sat}) = 1/\hat{P}$.
4. If $\hat{P} = P$, which in this problem is 1 atm, then T is the dew point temperature.
5. If $\hat{P} > P$, then the assumed T is too high; decrease T and go back to step 2.
6. If $\hat{P} < P$, then the assumed T is too low; increase T and go back to step 2.

After following the above-mentioned procedure, we obtain the following results:

Dew point temperature	$= 389.6$ K
$y_{benzene}$	$= 0.092$
$y_{toluene}$	$= 0.382$
$y_{ethylbenzene}$	$= 0.526$

Note that the dew point temperature is greater than the bubble point temperature. This makes sense since if we first consider that we have a subcooled liquid of the given composition, then as we heat this liquid mixture, we will reach the bubble point temperature that we found in Example 2.20. Then, as we continue to add heat at constant pressure, we will vaporize more and more of the components in the liquid phase until, when we reach the dew point temperature, we have vaporized all of the components and we have a saturated vapor of the same composition as the original liquid mixture. As we continue heating, the temperature rises above the dew point temperature and we have a superheated vapor.

2.6.3.8 Flammability Limits

Most manufacturing processes employ a variety of flammable substances. It is important that these materials be handled and processed in a safe manner in order to avoid fires and explosions, which can have devastating consequences. The ignition, combustion, or explosion of flammable substances in air only occurs within a narrow range of composition called the *flammability limits*. Below a certain concentration of the flammable substance in air, the mixture is too "lean" to burn, while above a certain concentration the mixture is too "rich" to burn. The flammable or explosive range lies within these limits. This is important in the design and operation of combustion engines, furnaces, and incinerators. It is also important from a safety viewpoint to prevent fires and explosions in manufacturing and other processes that involve the use of flammable compounds.

A fire or explosion requires three elements: the fuel, a source of oxygen, and an ignition source. This is referred to as the *fire triangle*. The fuel can be the flammable substances in your manufacturing process. Oxygen is in the air and air certainly exists outside your manufacturing equipment and may also be within your manufacturing process. Also, you can count on Mother Nature to provide

TABLE 2.2
Flammability Limits of Some Common Substances in Air at Atmospheric Pressure

Chemical	Lower Flammability Limit (vol% or mol%)	Upper Flammability Limit (vol% or mol%)
Acetylene	2.5	100.0
Acetic acid	5.4	16.0
Acetone	2.6	13.0
Ammonia	15.0	28.0
Benzene	1.4	8.0
Ethanol	3.3	19.0
n-Butane	1.8	8.4
Methanol	6.7	36.0
Diethyl ether	1.9	36.0
Propane	2.1	9.5
Octane	1.0	6.5
Gasoline	1.3	7.6
Toluene	1.3	7.0
Hydrogen	4.0	75.0

the ignition source, although other likely ignition sources in a manufacturing facility can include hot surfaces and flames, sparks, unstable chemicals, and static electricity. Flammability limits for a large number of organic chemicals are available in the literature.

These limits are provided as the lower and upper flammability (or explosion) limits. The upper (UFL) and lower flammability limit (LFL) is stated as the volume percent (which is the same as the mole percent) of the flammable chemical in air. If the flammable compound's volume percent or mole percent in air at a given T and P lies within the LFL to UFL range, then that mixture is ignitable or explosive. If the volume percent or mole percent of the flammable chemical in air is less than the LFL, or greater than the UFL, then that mixture is not ignitable or explosive.

Table 2.2 summarizes from a variety of sources the LFL and UFL for some common chemicals in air. Note that different sources may give slightly different numbers than those reported in Table 2.2.

For a flammable substance, it is important to remember that it is not the liquid that actually burns, but the vapors that are produced from that liquid. In many cases involving the processing or storage of flammable chemicals, we can assume that the vapor and liquid phases are in equilibrium. We can then use our vapor liquid equilibrium calculations, as discussed earlier, to determine the equilibrium composition of the vapor phase. This vapor phase composition can then be compared to the flammability limits to determine whether or not there may be an issue regarding a fire or explosion. This type of calculation is illustrated in the following example.

Example 2.22

Consider the storage at ambient conditions (25°C and 1 atm) of a solvent such as benzene in a large storage tank in a manufacturing facility. Determine whether or not the vapor phase composition in this tank lies within the flammability limits for benzene.

Solution

In a closed and partially full storage tank, the vapor phase may be assumed to be in equilibrium with the liquid benzene provided there have been no recent transfers of benzene into or out of the storage tank. At 25°C, the vapor pressure of benzene is 94.5 mmHg, using the vapor pressure equation previously used in Example 2.20. From Raoult's law:

$$P\, y_{benzene} = P_{benzene} = P_{benzene}^{Vap}\, x_{benzene} = P_{benzene}^{Vap},$$

and the partial pressure of benzene in the vapor phase is 94.5 mmHg. Since the total pressure in the vapor space must be atmospheric, this means that the partial pressure of air in the vapor space of the tank is 760 – 94.5 = 655.5 mmHg. From Raoult's law, we can then calculate the mole fraction of benzene in the vapor space as

$$y_{benzene} = \frac{P_{benzene}^{Vap}}{P} = \frac{94.5 \text{ mmHg}}{760 \text{ mmHg}} = 0.1243.$$

Therefore, the mole percent or volume percent of benzene in the vapor space is 12.4%. Comparing this value to the LFL and UFL values for benzene in Table 2.2 indicates that the vapor space in the benzene storage tank at ambient conditions is not flammable or explosive, assuming that the liquid and vapor phases are in equilibrium. However, nonequilibrium events, such as the removal of benzene from the tank, will result in the addition of fresh air that will temporarily decrease the concentration of benzene in the vapor space from the above value until equilibrium is reestablished. It is possible during these transfer periods for the benzene concentration to fall within the flammability limits. In addition, for outside storage tanks, seasonal variations of temperature should also be considered. For example, at 40°F or 4.4°C, the vapor pressure of benzene decreases to 33.5 mmHg, giving at equilibrium a vapor phase benzene concentration of 4.4 mol% or 4.4 vol%. Now, even at equilibrium the vapor phase lies well within the flammability range. If an ignition source is present, then there is the possibility that this vapor mixture will ignite and cause a fire or explosion.

Example 2.22 illustrates that the storage of flammable substances can result in vapor mixtures that lie within the flammability range. In large storage tanks under these conditions nitrogen gas is frequently supplied as a replacement rather than simply using ambient air. The use of nitrogen gas therefore eliminates the formation of flammable mixtures within a storage tank. For smaller tanks and portable containers for which this is not possible, it is important to make sure that these tanks and containers are properly grounded to eliminate static discharges that can serve as an ignition source. In addition, all vent openings into the tank or container can be fitted with flame arrestors to prevent the propagation of a fire from outside the tank to inside the tank. Commercial flame arrestors are relatively simple and inexpensive devices made of screens or mesh-like materials with small openings on the order of a millimeter or so. The design and construction of flame arrestors is based on the fact that flame propagation can be suppressed and eliminated if the flame has to pass through a large number of narrow openings. There exists a critical opening size below which the flame is quenched and this is called the *quenching distance*. Quenching distances for a variety of flammable substances may be found in the combustion literature.

2.6.3.9 Thermodynamics of Surfaces

So far, we have discussed the thermodynamic properties of solutions containing solids, liquids, and gases. However, when two phases are in contact, a surface is formed at the interface. The surface that is formed may have properties that are quite different from the bulk phases. Understanding the thermodynamics of surfaces is important in many biomedical applications that involve the wetting of surfaces, as well as for understanding such processes as capillary action.

Consider the differential increase in surface area (dA_S) of a fluid of arbitrary shape. The Gibbs free energy change for this differential increase in the surface area may be written as follows:

$$dG = -S\, dT + V\, dP + \gamma\, dA_S, \tag{2.181}$$

where the additional term $\gamma \, dA_S$ represents the increase in the Gibbs free energy due to the change in the surface area, dA_S. The *surface energy* is denoted by γ and is defined by

$$\gamma = \left(\frac{\partial G}{\partial A_S} \right)_{T,P}. \qquad (2.182)$$

For a pure component, γ is the same as the *surface tension*.

The surface tension acts as a restoring force that resists the addition of the new surface area. At constant T and P, the work (dW) required to change the surface area an amount dA_S is given by $dW = \gamma \, dA_S$. Typical units for the surface tension are dynes per centimeter and millinewton per meter. The surface tension of pure water in air at 25°C is about 72 mN m^{-1}.

In order to develop some useful relationships, let us first consider a soap bubble. A soap bubble consists of a very thin spherical film of liquid. The film is relatively thick on a molecular level, so the innermost portion of the film acts like the bulk liquid. This film has two surfaces that are exposed to the outside atmosphere and the air trapped within the soap bubble. According to Equation 2.181, the soap bubble can decrease its free energy by decreasing its surface area. Hence, the bubble will decrease in size, placing more of the liquid film molecules in the innermost portion of the film. However, as the bubble shrinks in size, the internal pressure (P^B) will increase until a point is reached where the soap bubble can shrink no further. An equilibrium state then occurs where the free energy decrease of the liquid film is equal to the work done compressing the air within the soap bubble. At equilibrium, the tension in the surface of the bubble is balanced by the difference in pressure between the inside of the bubble (P^B) and the outside of the bubble (P^A). Hence, we can write that

$$\gamma dA_S = (P^B - P^A) \, A_S dr. \qquad (2.183)$$

The left-hand side of Equation 2.183 represents the free energy gained by shrinking the soap bubble an amount dA_S. The right-hand side represents the net force acting on the surface A_S, and this force multiplied by the displacement of the interface, dr, is the work done against the pressure difference as a result of the change in area, dA_S.

Recall that for a sphere of radius r the surface area is $4\pi r^2$. For a soap bubble, there are two interfaces, the inside and outside surfaces of the liquid film, so dA_S in this case is equal to $16\pi r dr$. For a droplet or a gas bubble, where we have either a drop of liquid suspended in a gas or a gas bubble suspended in a liquid, there is only one interface and $dA_S = 8\pi r dr$. Substituting in for dA_S and A_S in Equation 2.183, we get Equations 2.184 and 2.185 for the pressure difference $P^B - P^A$:

$$\text{Two interfaces} \quad P^B - P^A = \frac{4\gamma}{r}, \qquad (2.184)$$

$$\text{One interface} \quad P^B - P^A = \frac{2\gamma}{r}. \qquad (2.185)$$

Equations 2.184 and 2.185 are known as the *Laplace–Young* equation and show that the excess pressure ($P^B - P^A$) is inversely proportional to the radius of the bubble or the droplet. The radius r in these equations is always considered to have its center of curvature in the phase in which P^B is measured. Hence, for a gas bubble in a liquid, P_B is the pressure within the bubble, which is then greater than the pressure (P_A) in the liquid phase.

Example 2.23

Consider a droplet of water, 1 μm in diameter, that is suspended in air at 1 atm and 25°C. What is the pressure of the water inside the droplet?

Solution

The center of curvature for the droplet of water lies within the droplet. Therefore, P^B represents the pressure within the droplet. Using Equation 2.185 for the droplet, we have

$$P^B - 1 \text{ atm} = \frac{2 \times 72 \times 10^{-3} \text{ Nm}^{-1}}{0.5 \times 10^{-6} \text{ m}} = 288,000 \text{ Pa} = 2.84 \text{ atm},$$

$$P^B = 3.84 \text{ atm}.$$

Now consider a liquid droplet suspended in its own vapor. The liquid and vapor phases are therefore in equilibrium. The Gibbs free energy for the droplet phase may be written as follows, assuming we have only a single component i:

$$dG = -S\,dT + V\,dP + \gamma\,dA_S + \mu_i^{(P)}\,dn_i. \tag{2.186}$$

The chemical potential in this case is defined as $\mu_i^{(P)} = (\partial G/\partial n_i)|_{T,P,A_S,n_j}$. Superscript P on the chemical potential denotes the fact that the transfer of dn_i moles does not change the surface area of the droplet, since in the above definition of the chemical potential, A_S is constant. The only way that this can occur is if $\mu_i^{(P)}$ refers to the chemical potential of component i as a planar surface.

For a spherical droplet, the surface area of the droplet will depend on how much material is in the droplet. With v_i the molar volume of liquid i, we can then write that

$$dV = v_i\,dn_i = d(\tfrac{4}{3}\pi r^3) = 4\pi r^2 dr. \tag{2.187}$$

Similarly, we can write for the surface area of the droplet that

$$dA_S = d(4\pi r^2) = 8\pi r\,dr = \frac{2\,dV}{r} = \frac{2v_i}{r}\,dn_i. \tag{2.188}$$

Using Equation 2.188, we can then rewrite Equation 2.186 as

$$dG = -S\,dT + V\,dP + \left(\mu_i^{(P)} + \frac{2v_i\gamma}{r}\right)dn_i. \tag{2.189}$$

The parenthetical term represents the chemical potential of component i as a droplet and takes into account the free energy change associated with changes in the droplet surface area.

Since the droplet is at equilibrium with its vapor at constant T and P, then we must have that

$$\mu_i^V = \mu_i^L = \mu_i^{(P)} + \frac{2v_i\gamma}{r}. \tag{2.190}$$

From Equation 2.56, we can write that

$$\mu_i^{L\,\text{ideal gas}} = C_i(T) + RT\ln(P_i^{VAP}) \text{ and } \mu_i^{(P)\text{ideal gas}} = C_i(T) + RT\ln(P_i^{(P)VAP}), \tag{2.191}$$

where $y_i P$ or the partial pressure for a pure component would be the same as the vapor pressure by Raoult's law. Substituting the relationships in Equation 2.191 into Equation 2.190 provides the following result:

$$\ln\left(\frac{P_i^{VAP}}{P_i^{(P)VAP}}\right) = \frac{2 v_i \gamma}{r RT}.$$ (2.192)

This equation, also known as the *Kelvin equation*, shows that the vapor pressure of a small droplet of liquid of radius r is higher than the corresponding value as a planar layer of liquid. If the surface is concave toward the vapor side, e.g., a gas or vapor bubble in a liquid phase, then use a minus sign on the right-hand side of Equation 2.192.

Equation 2.192 shows that a swarm of liquid droplets of uniform radius at equilibrium with its vapor is unstable in terms of its radius. For example, if one droplet has a slight increase in its radius, then by Equation 2.192 the vapor pressure of the droplet will decrease and there will be a corresponding transfer of mass from the vapor state to the liquid state, thereby increasing the size of the droplet and so on. On the other hand, if a droplet has a slight decrease in radius, then its vapor pressure will increase and there will be transfer of mass from the liquid state to the vapor state, thereby decreasing the size of the droplet and so on. This means that in a mixture of droplets of different sizes, the larger droplets will grow at the expense of the smaller droplets and this is the phenomenon that leads to the formation of clouds and rain.

Example 2.24

Consider a droplet of water, 1 μm in diameter, that is suspended in air at 1 atm and 25°C. What is the vapor pressure of the water droplet? At 25°C, the vapor pressure of planar water is 3.166 kPa and the molar volume of liquid water is 18.05 cm³ mol⁻¹.

Solution

We can use Equation 2.192 to calculate the vapor pressure of the droplet of water as

$$\ln\left(\frac{P_w^{VAP}}{P_w^{(P)VAP}}\right) = \frac{2 \times 18.05 \text{ cm}^3\text{mol}^{-1} \times 1 \text{ m}^3 \left(100 \text{ cm}\right)^{-3} \times 72 \times 10^{-3} \text{ Nm}^{-1}}{0.5 \times 10^{-6} \text{ m} \times 8.314 \text{ m}^3 \text{ Nm}^{-2}\text{mol}^{-1}\text{K}^{-1} \times 298 \text{ K}} = 0.0021,$$

$$P_w^{VAP} = e^{0.0021} \times 3.166 \text{ kPa} = 3.173 \text{ kPa}.$$

2.6.3.10 Equilibrium Dialysis

In the laboratory, equilibrium dialysis is a useful method for studying the binding of a small molecule (called the *ligand*) to a much larger macromolecule such as a protein. An important example is determining by this method what fraction of a drug is bound to the proteins found in blood. Another important example of this, which we will study later in Chapter 6, is the binding of oxygen with the protein hemoglobin that is found in the red blood cell. The equilibrium binding of oxygen with hemoglobin will depend on the partial pressure of oxygen ($P_{oxygen} = y_{oxygen} P$) in the gas phase. In order to understand the oxygenation of tissue and to design blood oxygenators used in heart surgery, we need to know for a given partial pressure of oxygen, how much oxygen is actually bound to the hemoglobin.

The simplest approach to performing equilibrium dialysis experiments is to form a membrane bag. The membrane bag can be formed by taking a membrane tube and tying off both ends. Usually, an aqueous solution containing the macromolecule of interest is placed within the membrane bag and then the membrane bag is immersed into a well-stirred solution containing the ligand. Stirring the solution facilitates the transport of the ligand and shortens the time to reach equilibrium. These experiments can also be performed within specially designed dialysis cells that consist of two well-stirred chambers separated by the dialysis membrane. Initially, one of these chambers holds a solution containing only the ligand and the other chamber contains a solution of the macromolecule.

Equilibrium dialysis experiments employ semipermeable membranes. The membranes have small pores that are large enough to allow the free passage of the ligand and the solvent while retaining the macromolecules. The ligand then diffuses across the dialysis membrane into the solution containing the macromolecule. The ligand will then bind with the macromolecule and this process continues until the system reaches a state of equilibrium. At equilibrium, we can use Equation 2.84 to express the equilibrium of the ligand molecules (L) inside and outside the membrane bag:

$$\hat{f}_L^{\text{inside}} = \hat{f}_L^{\text{outside}}. \tag{2.193}$$

Using Equation 2.99, we can express the ligand fugacity in the solution inside and outside the membrane as

$$\gamma_L^{\text{inside}} \, x_L^{\text{inside}} \, f_L = \gamma_L^{\text{outside}} \, x_L^{\text{outside}} \, f_L. \tag{2.194}$$

In this case, unlike for osmosis discussed earlier, the temperature and pressure inside and outside the membrane bag are the same, so the pure component ligand fugacity (f_L) is the same in both regions. Also, the usual assumption is that the ligand forms an ideal solution, therefore the ligand activity coefficients are equal to unity. Furthermore, assuming the ligand concentration is small, the ligand mole fractions may be approximated as $x_L^{\text{inside}} = V_{\text{solvent}}^L \, C_L^{\text{inside}}$ and $x_L^{\text{outside}} = V_{\text{solvent}}^L \, C_L^{\text{outside}}$. With these substitutions into Equation 2.194, the net result is that at equilibrium, $C_L^{\text{inside}} = C_L^{\text{outside}}$. The binding of the ligand to the macromolecule acts to concentrate the ligand inside the bag, even though at equilibrium the concentrations of the ligand inside and outside the membrane bag are the same.

The goal of these equilibrium dialysis experiments is to develop an expression that describes, for a given concentration of ligand, the amount of ligand that is bound to the macromolecule. This requires in general that we know the equilibrium constant that describes the binding of the ligand to the macromolecule, as well as the number of ligand binding sites per macromolecule. The simplest case is when the macromolecule has only a single site that can bind to the ligand. In this case, the binding of the ligand molecule (L) with the macromolecule (M) can be described by the following chemical equation:

$$L + M \overset{K}{\leftrightarrow} L \bullet M, \tag{2.195}$$

where L $\bullet$ M is the ligand–macromolecule complex.

Recall that the chemical equilibrium constant is defined in terms of the concentration of the ligand (C_L), the free or unbound macromolecule concentration (C_M), and the concentration of the ligand–macromolecule complex ($C_{L \bullet M}$) and is given by

$$K = \frac{C_{L \bullet M}}{C_L \, C_M}. \tag{2.196}$$

In order to calculate the value of K and the fraction of the macromolecules that are bound to the ligand, the above-mentioned concentrations need to be related to the measured ligand, macromolecule, and ligand–macromolecule complex concentrations. The total concentration of the macromolecule, i.e., C_M^{total}, is known at the start of the experiment and this value is conserved over time within the membrane bag. At any time during the experiment, C_M^{total} is given by the sum of the free macromolecule concentration and that bound to the ligand by

$$C_M^{\text{total}} = C_M + C_{L \bullet M}. \tag{2.197}$$

In a similar manner, the total amount of ligand inside the bag is the sum of the free ligand and that bound with the macromolecule as

$$C_L^{total} = C_L^{inside} + C_{L \cdot M}, \tag{2.198}$$

where C_L^{inside} is the concentration of unbound ligand inside the membrane bag, which at equilibrium is the same as that outside the membrane bag, i.e., $C_L = C_L^{inside} = C_L^{outside}$. In terms of the previous definitions, it can be shown that the equilibrium constant (K) is then given by

$$K = \frac{C_{L \cdot M}}{C_L \left(C_M^{total} - C_{L \cdot M} \right)}. \tag{2.199}$$

To facilitate the use of Equation 2.199, it is convenient to define the fraction of macromolecules (f) that are bound to the ligand. This is given by

$$f \equiv \frac{C_{L \cdot M}}{C_{L \cdot M} + C_M}. \tag{2.200}$$

Combining Equations 2.199 and 2.200, the following equations are obtained for the relationship between f and K:

$$K = \frac{f}{(1-f)C_L} \quad \text{and} \quad f = \frac{K C_L}{1 + K C_L}. \tag{2.201}$$

These equations then provide the relationship between the ligand concentration, the equilibrium constant, and the fraction of macromolecules that are bound to the ligand at equilibrium. The previous expressions can also be rearranged into a convenient form for data analysis as shown by the following equation:

$$\frac{f}{C_L} = K(1-f) = K - Kf. \tag{2.202}$$

This equation is known as the *Scatchard equation* and shows that for a given ligand concentration (C_L), if one determines the fraction of macromolecules that are bound to the ligand (f) then a plot of f/C_L vs. f should be linear with a slope equal to $-K$ and an intercept equal to K.

The use of Equation 2.202 is illustrated in Example 2.25 for the binding of oxygen to myoglobin. Myoglobin is an oxygen-binding protein found in muscles that store oxygen molecules and release the oxygen when the oxygen partial pressure becomes low, e.g., during strenuous exercise. Myoglobin has only one site available for the binding of oxygen. The absorption spectra of myoglobin bound to oxygen is different from that for free myoglobin and this can be used to determine at a given partial pressure of oxygen, the fraction of myoglobin that is bound to oxygen.

Example 2.25

The following data express the fractional binding of oxygen to myoglobin as a function of the partial pressure of oxygen. From these data, determine the value of the equilibrium constant for the binding of oxygen to myoglobin. Also, determine P_{50}, which is the partial pressure of oxygen where the fractional binding of the myoglobin molecules is 50%.

Oxygen Partial Pressure (pO$_2$, mmHg)	Fraction of Myoglobin Bound to Oxygen (f)
0.5	0.25
1.0	0.40
1.5	0.50
2.0	0.61
2.5	0.65
4.5	0.76
6.0	0.8

Source: Tinoco et al. (2001)

Solution

Since the pO$_2$ of oxygen is related to the dissolved oxygen concentration (C_{oxygen}) by Henry's law (see Example 2.14, i.e., pO$_2 = HC_{oxygen}$), we can first rewrite Equation 2.202 as follows, where H is the Henry's constant:

$$\frac{f}{pO_2} = \frac{K}{H} - \frac{K}{H}f.$$

P_{50} is that value of the partial pressure of oxygen where one-half of the myoglobin molecules are bound to oxygen (i.e., $f = 0.5$). In terms of P_{50}, it is easy to show using the previous equation that $(K/H) = 1/P_{50}$. Our regression equation in terms of P_{50} then becomes

$$\frac{f}{pO_2} = \frac{1}{P_{50}} - \frac{f}{P_{50}} \quad \text{and} \quad f = \frac{pO_2}{P_{50} + pO_2}.$$

Performing a linear regression on the data given in the table, according to the previous equation on the left, we find that P_{50} based on the average of the values obtained from the slope and the intercept is equal to 1.53 mmHg. We see in Figure 2.8 that the previous equation on the right for f with the value of P_{50} found from a regression of the data provides an excellent fit to these data on oxygen binding to myoglobin.

In many cases, a macromolecule may have several binding sites for the ligand. In the general case, there are N sites available and it is also assumed that these sites are independent and have the same affinity for the ligand. This means that the binding of a ligand at a particular site has no effect on the binding of a ligand at the other sites on the macromolecule. Each of these sites are, therefore, independent and they will have the same ligand equilibrium constant K. The binding of a ligand to the ith site on the macromolecule is given by the following equilibrium equation:

$$L + M_i \overset{K}{\leftrightarrow} L \bullet M_i. \tag{2.203}$$

We can then let ϕ_i be defined as the fraction of site i on the macromolecules that are bound to the ligand. This is given by

$$\phi_i = \frac{C_{L \bullet M_i}}{C_{M_i} + C_{L \bullet M_i}}. \tag{2.204}$$

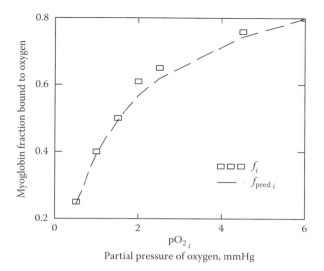

FIGURE 2.8 Binding of oxygen to myoglobin.

The equilibrium constant for the binding of the ligand to site i is given by

$$K = \frac{C_{\text{L} \cdot \text{M}_i}}{C_{\text{L}} C_{\text{M}_i}}. \tag{2.205}$$

Equations 2.204 and 2.205 can then be combined to give Equation 2.206 for the fraction of site i that is bound to the ligand:

$$\phi_i = \frac{K C_{\text{L}}}{1 + K C_{\text{L}}}. \tag{2.206}$$

This equation can then be summed over all N sites on the macromolecule, recognizing that $f = \sum_{i=1}^{N} \phi_i$, where f is given by Equation 2.200, and defines the moles of ligand bound per mole of macromolecule, which is also the same as the average number of ligand molecules that are bound to a macromolecule. The result is that f is given by Equation 2.207 for the case of a macromolecule containing N identical and independent ligand binding sites:

$$f = \frac{N K C_{\text{L}}}{1 + K C_{\text{L}}}. \tag{2.207}$$

For ease of data analysis, this equation can be rearranged as

$$\eta = \frac{f}{C_{\text{L}}} = K(N - f) = K N - K f. \tag{2.208}$$

This is the *Scatchard equation* for N binding sites on a macromolecule, which are all identical and independent. Equation 2.208 shows that for a given ligand concentration (C_{L}), if one determines the moles of ligand that are bound to a mole of macromolecule (f), then a plot of $\eta = f/C_{\text{L}}$ vs. f should be linear with a slope equal to $-K$ and a y-intercept equal to $N K$ or an x-intercept of N.

   If the plot of the data based on Equation 2.208 is not linear, then this is an indication that the binding sites for the ligand are not identical or not independent. For example, an important case where the Scatchard equation does not apply is the binding of oxygen to hemoglobin. Although hemoglobin has four oxygen-binding sites, these sites are not independent and exhibit what is known as *cooperative binding.* As we will discuss in Chapter 6, the binding of oxygen to hemoglobin enhances the binding of additional oxygen molecules. In Chapter 6, we will develop an equation that can be used to describe cooperative binding.

The use of Equation 2.208 is illustrated in the following example for the binding of Mg–ATP (a complex of Mg and ATP) to the enzyme tetrahydrofolate synthetase.

Example 2.26

The data shown in the following table express the binding of Mg–ATP complex to the enzyme tetrahydrofolate synthetase in terms of f and f/C_{Mg-ATP}. The enzyme consists of four identical subunits. From these data, determine the value of the equilibrium constant (K) for the binding of Mg–ATP complex to this enzyme and show that the number of binding sites (N) on the enzyme is consistent with there being one Mg–ATP binding site on each subunit of the enzyme.

f (Fraction of Mg–ATP Bound to Enzyme)	$\dfrac{f}{C_{Mg-ATP}}$
0.8	45,500
1.0	49,000
1.1	40,000
1.7	32,000
1.75	37,000
2.0	24,000
2.05	29,000
2.8	19,000
3.1	15,500
3.2	12,000
3.45	10,000
3.55	12,500
3.7	9500
3.9	8000

Source: Tinoco et al. (2001)

Solution

Performing a linear regression on the data given in the table according to Equation 2.208, we find that the value of K based on the slope is equal to 1.29×10^4 M^{-1}. The number of binding sites N on the enzyme for the Mg–ATP complex is 4.33, which is consistent with the enzyme consisting of four identical subunits. We also see in Figure 2.9 that regression of the data according to Equation 2.208 provides an excellent fit to these data on the binding of Mg–ATP to the enzyme.

2.6.3.11 Gibbs–Donnan Effect

Now we can consider the situation of equilibrium dialysis where the macromolecule has a concentration of C_M and a net charge of Z_M. To simplify the following discussion, assume that the macromolecule is contained inside a semipermeable membrane bag in an aqueous solution containing a strong electrolyte,† such as NaCl. The strong electrolyte and its ions do not bind to the macromolecule. Only the ligand and the ions are free to cross the semipermeable dialysis membrane. However,

---

* See Section 6.2.
† A strong electrolyte is a substance that completely dissociates into charged ions when placed in water.

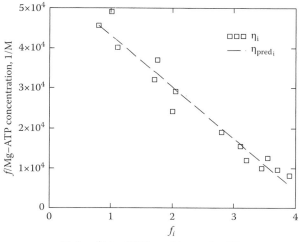

FIGURE 2.9   Binding of Mg–ATP to tetrahydrofolate synthetase.

because the solutions on each side of the semipermeable dialysis membrane must be electrically neutral, it is observed that if the ligand has a charge opposite to that of the macromolecule, then there will be an increase in the binding of the ligand to the macromolecule, whereas if the ligand has the same charge, then there will be a decrease in the binding of the ligand. By using a high concentration of a strong electrolyte like NaCl, this effect can be minimized, since the charge of the macromolecule becomes insignificant in comparison to the charge carried by the ions formed by the dissociation of the strong electrolyte.

To understand this phenomena, consider the simplest case, which is the dialysis of a charged macromolecule in an aqueous solution containing NaCl. The solution that is opposite or outside the solution containing the macromolecule (i.e., the solution inside the membrane bag) must be electrically neutral. If $C_{NaCl}^{outside}$ is the total concentration of NaCl in the outside solution, then electrical neutrality of this solution requires

$$C_{NaCl}^{outside} = C_{Na^+}^{outside} = C_{Cl^-}^{outside}. \tag{2.209}$$

The dialysis membrane is permeable to NaCl, so when equilibrium occurs between the inside and outside solutions, it is convenient to use Equation 2.82 and write this equilibrium for NaCl as

$$\mu_{NaCl}^{inside} = \mu_{NaCl}^{outside}. \tag{2.210}$$

Since NaCl is a strong electrolyte, Equation 2.210 can also be rewritten using Equation 2.70 in terms of Na^+ and Cl^-, where use is made of the fact that $\mu_{NaCl} = \mu_{Na^+} + \mu_{Cl^-}$, as

$$\left(G_{Na^+}^{inside} + G_{Cl^-}^{inside}\right) + RT \ln\left(\frac{\hat{f}_{Na^+}^{inside}}{f_{Na^+}}\right) + RT \ln\left(\frac{\hat{f}_{Cl^-}^{inside}}{f_{Cl^-}}\right)$$

$$= \left(G_{Na^+}^{outside} + G_{Cl^-}^{outside}\right) + RT \ln\left(\frac{\hat{f}_{Na^+}^{outside}}{f_{Na^+}}\right) + RT \ln\left(\frac{\hat{f}_{Cl^-}^{outside}}{f_{Cl^-}}\right). \tag{2.211}$$

In this equation, the pure component value of the Gibbs free energy and the fugacity for each species are the same on the inside and the outside solutions, since both are at the same temperature and pressure. We can also make use of the fact from Equation 2.99 that $\hat{f}_i = \gamma_i x_i f_i$. Hence, Equation 2.211 becomes

$$\ln\left(x_{Na^+}^{inside}\, \gamma_{Na^+}^{inside}\right) + \ln\left(x_{Cl^-}^{inside}\, \gamma_{Cl^-}^{inside}\right) = \ln\left(x_{Na^+}^{outside}\, \gamma_{Na^+}^{outside}\right) + \ln\left(x_{Cl^-}^{outside}\, \gamma_{Cl^-}^{outside}\right). \tag{2.212}$$

If it is also assumed that $\gamma_i^{inside} \approx \gamma_i^{outside}$ and $x_i^{inside} = V_{solvent}^L\, C_i^{inside}$ and $x_i^{outside} = V_{solvent}^L\, C_i^{outside}$, then Equation 2.212 can be shown to simplify to the following result:

$$C_{Na^+}^{inside}\, C_{Cl^-}^{inside} = C_{Na^+}^{outside}\, C_{Cl^-}^{outside} = \left(C_{NaCl}^{outside}\right)^2. \tag{2.213}$$

This equation can then be solved for the Cl$^-$ concentration in the solution containing the macromolecule as

$$C_{Cl^-}^{inside} = \frac{\left(C_{NaCl}^{outside}\right)^2}{C_{Na^+}^{inside}}. \tag{2.214}$$

If the macromolecule is positively charged, then electroneutrality of the inside solution will require that the following equation be satisfied:

$$C_{Cl^-}^{inside} = Z_M C_M + C_{Na^+}^{inside}, \tag{2.215}$$

whereas if the macromolecule is negatively charged, then

$$C_{Cl^-}^{inside} = -Z_M C_M + C_{Na^+}^{inside}. \tag{2.216}$$

Implicit in these equations is that $Z_M > 0$. Equations 2.214 and 2.215 can then be combined to obtain the following expression for the equilibrium concentration of Na$^+$ in the solution containing the macromolecule:

$$C_{Na^+}^{inside} = \frac{-Z_M C_M + \sqrt{\left(Z_M C_M\right)^2 + 4\left(C_{NaCl}^{outside}\right)^2}}{2}. \tag{2.217}$$

Of particular interest is the ratio of ion concentrations for the inside and outside solutions. By letting $r = \left(C_{Na^+}^{inside}\big/C_{Na^+}^{outside}\right) = \left(C_{Na^+}^{inside}\big/C_{NaCl}^{outside}\right)$, Equation 2.217 may be rewritten in terms of r:

$$r = \frac{-Z_M C_M}{2C_{NaCl}^{outside}} + \sqrt{\left(\frac{Z_M C_M}{2C_{NaCl}^{outside}}\right)^2 + 1}. \tag{2.218}$$

The ratio of Cl$^-$ concentrations inside and outside can also be shown to be related to r where $\left(C_{Cl^-}^{inside}\big/C_{Cl^-}^{outside}\right) = 1/r$.

Equation 2.218 shows that for a macromolecule with no charge, the value of $r = 1$ and the Na$^+$ and Cl$^-$ ion concentrations inside and outside at equilibrium are the same as expected. However, for a macromolecule with a very large positive charge, Equation 2.218 shows that for this case, $r \to 0$, which means that the Na$^+$ ion concentrations are excluded from the inside solution due to the excessive positive charge carried by the macromolecule. This also means that the Cl$^-$ concentration in the inside solution is much larger than that on the outside, and this occurs in order to offset the positive charge of the macromolecule and make the inside solution electrically neutral. In general, for a positively charged macromolecule within the inside solution, $r < 1$ and $C_{Na^+}^{inside} < C_{Na^+}^{outside}$ and $C_{Cl^-}^{inside} > C_{Cl^-}^{outside}$. This unequal distribution of Na$^+$ and Cl$^-$ at equilibrium is due to the presence of the charged macromolecule and this is called the *Gibbs–Donnan effect*.

A similar analysis for a negatively charged macromolecule contained inside the membrane bag leads to the following expression for the value of r:

$$r = \frac{Z_M C_M}{2 C_{NaCl}^{outside}} + \sqrt{\left(\frac{Z_M C_M}{2 C_{NaCl}^{outside}}\right)^2 + 1},\tag{2.219}$$

where $Z_M > 0$. Also, the ratio of Cl^- concentrations inside and outside the membrane bag is given by $\left(C_{Cl^-}^{inside}/C_{Cl^-}^{outside}\right) = 1/r$. Once again, if the macromolecule contains no charge, then from Equation 2.219, $r = 1$ and the Na^+ and Cl^- ion concentrations inside and outside at equilibrium are the same as expected. However, for a macromolecule with a very large negative charge, Equation 2.219 shows for this case that $r \gg 1$, which means that the Na^+ ions are concentrated in the inside solution due to the excessive negative charge carried by the macromolecule, and this occurs in order to offset the negative charge of the macromolecule and make the inside solution electrically neutral. This also means that the Cl^- concentration in the inside solution is much smaller than that on the outside. In general, for a negatively charged macromolecule within the inside solution, $r > 1$ and $C_{Na^+}^{inside} > C_{Na^+}^{outside}$ and $C_{Cl^-}^{inside} < C_{Cl^-}^{outside}$.

Example 2.27

The concentration of a macromolecule in an aqueous solution contained inside a membrane bag is 0.005 M. The macromolecule carries a net positive charge of + 15. If the NaCl concentration in the solution outside the bag at equilibrium is 0.15 M, calculate the ratio of the concentration of Na^+ inside the bag to that outside the bag. Also, find the ratio of the concentration of Cl^- inside the bag to that outside the bag.

Solution

With $Z_M = 15$, $C_M = 0.005$ M, and $C_{NaCl}^{outside} = 0.15$ M, then from Equation 2.218 it is found that

$$r = \frac{-15 \times 0.005 \text{ M}}{2 \times 0.15 \text{ M}} + \sqrt{\left(\frac{15 \times 0.005 \text{ M}}{2 \times 0.15 \text{ M}}\right)^2 + 1} = 0.781.$$

Therefore, the ratio of the Na^+ concentration inside the membrane bag to that outside the bag is 0.781 and the ratio of Cl^- inside the bag to that outside the bag is $1/r = 1.281$.

Although the solutions on the inside and the outside of the membrane bag are electrically neutral, there is, as shown in Example 2.27, an unequal distribution of the ions, which is caused by the charged macromolecules inside the membrane bag. For example, in Example 2.27, the Cl^- ions are concentrated inside the bag to counter the positive charges on the macromolecules. The Na^+ ions are repulsed by the positive charges on the macromolecules and are concentrated in the solution that is outside the bag. This means that there is a concentration gradient of Cl^- that would tend to cause these ions to diffuse from inside the bag to the outside. Similarly, a concentration gradient of Na^+ is also formed, which would tend to cause these ions to diffuse from outside the bag to the inside. To counter these concentration gradients, at equilibrium there exists an attractive electrical force caused by the positive charges of the macromolecules, which counters the loss of these Cl^- ions due to the presence of their concentration gradient. These same positive charges on the macromolecule oppose the diffusion of Na^+ into the bag. This transmembrane electrical potential is known as the *Donnan potential* and is described in the next section.

2.6.3.12 Donnan Potential

We can obtain an equation that relates the equilibrium membrane potential (V), i.e., the difference in electrical potential between the inside (V_{inside}) and the outside ($V_{outside}$) of the membrane bag, to the equilibrium concentrations of the ions on either side of the membrane. The presence of an electrical field across the membrane will produce a force on the ions as a result of their charge. Hence, the ions will move and their motion causes a current to flow. Recall that voltage (V) times current (I) is equal to the power and power is the rate of doing work. Electrical work is therefore done on the ion and is given by

$$W_{electrical} = -VIt. \tag{2.220}$$

If we are only infinitesimally removed from equilibrium, then $W_{electrical}$ becomes $W_{reversible\ electrical}$ and from our earlier discussion in Section 2.3.3.1, the reversible electrical work (maximum work at constant T and P) is equal to the change in the Gibbs free energy. Also, the product of current and time in Equation 2.220 is equal to the total charge that is transferred. Faraday's constant (F) is defined as the charge carried by a gram mole of ions of unit positive valency and is equal to 9.649×10^4 coulombs mol^{-1} or 2.306×10^4 cal V^{-1} mol^{-1}. We can now rewrite Equation 2.220 for a gram mole of ions as

$$W_{reversible\ electrical} = \Delta G_{electrical} = -zFV, \tag{2.221}$$

where z is the charge on the ion being considered. For example, for a sodium ion (Na^+) $z = +1$, for a calcium ion (Ca^{2+}) $z = 2$, and for a chloride ion (Cl^-) $z = -1$.

In addition to the movement of these ions by the electrical field, there is also movement of the ions as a result of the concentration gradient. The change in the Gibbs free energy resulting from the transport of a solute from the inside of the cell to the outside of the cell:

$$-\Delta G_{transport} = \bar{G}_{inside} - \bar{G}_{outside} = \mu_{inside} - \mu_{outside}. \tag{2.222}$$

Recall from Equation 2.75 that for an ideal solution, we can write that

$$\mu_{inside}^{ideal\ solution} = G'_{inside} + RT \ln C_{inside} \quad and \quad \mu_{outside}^{ideal\ solution} = G'_{outside} + RT \ln C_{outside}, \tag{2.223}$$

where we have replaced the mole fraction with the concentration and the G''s are pure component constants. Also, since the state of the ion is the same in the two regions, then $G'_{inside} = G'_{outside}$, and we can write that

$$\Delta G_{transport} = RT \ln \frac{C_{outside}}{C_{inside}}. \tag{2.224}$$

The total change in the Gibbs free energy for movement of the ion by transport and the electrical field is given by the sum of Equations 2.221 and 2.224. At equilibrium, the total change in the Gibbs free energy is zero:

$$\Delta G_{total} = RT \ln \frac{C_{outside}}{C_{inside}} - zFV = 0. \tag{2.225}$$

Solving for V, we obtain Equation 2.226, which is known as the *Nernst equation*. This equation can be used to calculate the equilibrium membrane potential for a particular ion in equilibrium dialysis involving a charged macromolecule:

$$V = \frac{RT}{zF} \ln \frac{C_{outside}}{C_{inside}} = -\frac{RT}{zF} \ln \frac{C_{inside}}{C_{outside}}. \tag{2.226}$$

In this equation, R represents the gas constant (1.987 cal mol^{-1} K^{-1}), T is the temperature in Kelvin, z is the charge on the ion, and F is Faraday's constant (2.3×10^4 cal V^{-1} mol^{-1}). At 25°C for a univalent ion, $|RT/zF|$ is equal to 25.68 mV, whereas at 37°C the value is 26.71 mV.

Example 2.28

With the results from Example 2.27, calculate the transmembrane electrical potential based on the Na$^+$ concentration and the Cl$^-$ concentration inside and outside the membrane bag.

Solution

In Equation 2.226, we can replace $C_{inside}/C_{outside}$ with r, where we have shown earlier that $r = \left(C_{Na^+}^{inside} / C_{Na^+}^{outside} \right) = \left(C_{Na^+}^{inside} / C_{NaCl}^{outside} \right) = 0.781$ for Na$^+$ and $\left(C_{Cl^-}^{inside} / C_{Cl^-}^{outside} \right) = 1/r = 1/0.781$ for Cl$^-$. Hence, for Na$^+$ with $z = +1$, we have that

$$V = -\frac{RT}{zF}\ln\frac{C_{inside}}{C_{outside}} = -25.68 \text{ mV}\ln 0.781 = 6.35 \text{ mV},$$

and for Cl$^-$ with $z = -1$, we obtain the same answer as given above, that is

$$V = -\frac{RT}{zF}\ln\frac{C_{inside}}{C_{outside}} = 25.68 \text{ mV}\ln\frac{1}{0.781} = 6.35 \text{ mV}.$$

The Donnan potential or the transmembrane voltage is positive in this case because the positively charged macromolecule is contained on the inside of the membrane bag.

2.6.3.13 Chemical Equilibrium in Ideal Aqueous Solutions

In the previous discussion on the Gibbs–Donnan effect, the aqueous solution contained a strong electrolyte such as NaCl. Recall that a strong electrolyte in water completely dissociates into its constituent ions. Oftentimes, in biological systems our aqueous solutions will also contain weak electrolytes, such as weak acids or weak bases. Weak acids and bases only partially dissociate. An important example of a weak acid is acetic acid (CH$_3$COOH or HOAc) and an example of a weak base is aqueous ammonia or ammonium hydroxide (NH$_4$OH). Weak acids and weak bases, because of their partial dissociation, will affect the concentration of H$^+$ and as a result the pH of the solution.

In addition, water can also dissociate into H$^+$ and OH$^-$. At 25°C, the dissociation equilibrium constant for water (K_W) is given by

$$K_W = C_{H^+} C_{OH^-} = 10^{-14} \text{ M}^2. \tag{2.227}$$

Since water must remain electrically neutral, this means that $C_{H^+} = C_{OH^-}$ and using Equation 2.227, this means that $(C_{H^+})^2 = 10^{-14}$ M^2, which says that for pure water at 25°C, $C_{H^+} = C_{OH^-} = 10^{-7}$ M.

The pH of an aqueous solution is defined in terms of the H$^+$ concentration (M) and is given by

$$pH \equiv -\log_{10} C_{H^+}. \tag{2.228}$$

For pure water at 25°C, we showed that $C_{H^+} = 10^{-7}$ M; hence from Equation 2.228 for pure water, pH = 7, which is referred to as neutral since this pH value just expresses the normal concentration of H$^+$ due to the equilibrium dissociation of water. As the pH decreases below 7, e.g., by adding an acid to water, the concentration of H^+ increases from the neutral value where pH = 7 ($C_{H^+} = 10^{-7}$ M), and if the pH increases above 7, e.g., by adding a base to water, the concentration of H^+ decreases.

If a weak electrolyte such as acetic acid is added to pure water, then additional hydrogen ions will be added to the solution due to the partial dissociation of the weak acid, i.e., HOAc $\leftrightarrow$ H^+ + OAc$^-$, where OAc$^-$ denotes the acetate anion. The release of H^+ from the acetic acid in solution will lower the pH of the solution at equilibrium. In solving a chemical equilibrium problem such as this, involving the addition of acetic acid to water, the goal is usually to determine the concentrations of all the species that are present in the solution at equilibrium. For example, for the case of adding acetic acid (HOAc) to water, this means finding the concentrations at equilibrium of HOAc, OAc$^-$, H^+, OH$^-$, and HOH. This requires that we have the same number of independent equations as there are unknowns.

The equations used to find these unknowns are of three general types and include those of chemical equilibrium for each dissociation reaction, conservation of mass on the N species involved in the M reactions, and conservation of electrical charge. The equilibrium dissociation constants (K_j) are specific for each reaction (j) and depend on the temperature and can be affected by the presence of other species. For most reactions of interest, these values are tabulated in chemistry handbooks or may be found by Internet searches. Since the equilibrium dissociation constant for weak electrolytes can span many orders of magnitude, oftentimes the equilibrium constant for a particular reaction j is given as the base 10 logarithm of the equilibrium constant. Hence, the dissociation of a weak electrolyte by reaction j is given as pK_j, where pK_j is defined in terms of K_j by Equation 2.229. Note that pK_j is *not* the same as the dissociation equilibrium constant:

$$pK_j \equiv -\log_{10} K_j. \tag{2.229}$$

Note that pK_W for water is $-\log_{10}(10^{-14})$ and is equal to 14, whereas for a weak acid like acetic acid, the pK_{HOAc} is equal to 4.745.

It is also convenient in these calculations to express the differential change in the moles of each species i, i.e., dn_i, in terms of the differential *extent of reaction j*, i.e., $d\varepsilon_j$. The extent of reaction j ($d\varepsilon_j$) will have units of moles. We also let v_{ij} represent the stoichiometric coefficient of species i in reaction j. With these definitions, Equation 2.230 relates the differential change in the moles of species i as a result of its participation in each of the M reactions being considered. If species i does not participate in reaction j, then for that reaction $v_{ij} = 0$:

$$dn_i = \sum_{j=1}^{M} v_{ij}\, d\varepsilon_j. \tag{2.230}$$

This equation can then be integrated from the initial state where $\varepsilon_j = 0$ and $n_i = n_i^0$ to the equilibrium state of n_i and ε_j:

$$n_i = n_i^0 + \sum_{j=1}^{M} v_{ij}\, \varepsilon_j. \tag{2.231}$$

This equation then provides the moles of each species at equilibrium (n_i) in terms of the extent of each reaction (ε_j). The total moles of all species at equilibrium can be found by summing Equation 2.231 over i as

$$n = \sum_{i=1}^{N} n_i^0 + \sum_{i=1}^{N}\sum_{j=1}^{M} v_{ij}\, \varepsilon_j. \tag{2.232}$$

The mole fraction of species i in the solution is then given by $x_i = n_i/n$. Note that the concentrations of the respective species and the total concentration can be obtained by dividing Equations 2.231 and 2.232 by the solution volume.

The solution of these equations for the concentrations of the species will involve a lot of algebra as well as clever approximations that simplify the process of finding a solution. Example 2.29 illustrates the solution of a chemical equilibrium problem involving the addition of acetic acid to water.

Example 2.29

Determine the equilibrium concentrations of all species in a 0.05 M solution of acetic acid. Also find the pH of the solution.

Solution

There are two dissociation reactions that must be considered, one is for acetic acid and the other is for water. These are shown below along with their corresponding equations for chemical equilibrium. The equilibrium constants are from Tinoco et al. (2001):

$$(1)\ HOAc \leftrightarrow H^+ + OAc^- \quad \text{and} \quad K_{HOAc} = \frac{C_{H^+} C_{OAc^-}}{C_{HOAc}} = 1.8 \times 10^{-5}\,M$$

$$(2)\ HOH \leftrightarrow H^+ + OH^- \quad \text{and} \quad K_W = C_{H^+} C_{OH^-} = 10^{-14}\,M^2.$$

It is convenient to first take as a basis 1 L of solution and let n^0_{HOAc} represent the initial moles of HOAc added to make the 1 L solution. Hence, in this example to make a 1 L solution, $n^0_{HOAc} = 0.05$ mol. In terms of the extent of each of the previous reactions, we can make use of Equations 2.231 and 2.232 and write the following equations that express conservation of mass for each of the species that are present in the solution:

$$(3)\ n_{HOAc} = n^0_{HOAc} - \varepsilon_1$$

$$(4)\ n_{OAc^-} = \varepsilon_1$$

$$(5)\ n_{H^+} = \varepsilon_1 + \varepsilon_2$$

$$(6)\ n_{OH^-} = \varepsilon_2$$

$$(7)\ n_{HOH} = n^0_{HOH} - \varepsilon_2.$$

Note that in writing these equations, there is no OAc^-, OH^-, and H^+ present initially in the solution, since we start with undissociated water and acetic acid. If we now divide these equations by the 1 L volume of solution taken as our basis, we can then express n_i in terms of concentration C_i where the extents now have units of moles per liter or molar:

$$(8)\ C_{HOAc} = C^0_{HOAc} - \varepsilon_1$$

$$(9)\ C_{OAc^-} = \varepsilon_1$$

$$(10)\ C_{H^+} = \varepsilon_1 + \varepsilon_2$$

$$(11)\ C_{OH^-} = \varepsilon_2$$

$$(12)\ C_{HOH} = C^0_{HOH} - \varepsilon_2.$$

Addition of Equations 8 and 9 expresses the fact that at equilibrium $C_{HOAc}^0 = C_{HOAc} + C_{OAc^-}$. We can also insert the previous equations into the equilibrium relationships given by Equations 1 and 2 to give two equations in terms of the two unknown extents, ε_1 and ε_2:

$$(13) \; \frac{(\varepsilon_1 + \varepsilon_2)\varepsilon_1}{0.05 - \varepsilon_1} = 1.8 \times 10^{-5}$$

$$(14) \; (\varepsilon_1 + \varepsilon_2)(\varepsilon_2) = 10^{-14}.$$

Since we are dealing with the dissociation of an acid in water, we expect that at equilibrium $C_{H^+} \gg C_{OH^-}$. This means from Equation 11 that $C_{H^+} \gg \varepsilon_2$ and from Equation 10 this is the same as saying that $\varepsilon_1 \gg \varepsilon_2$. As a result of these assumptions, Equation 13 can be simplified to give

$$(15) \; \frac{\varepsilon_1^2}{0.05 - \varepsilon_1} = 1.8 \times 10^{-5} \quad \text{or} \quad \varepsilon_1^2 + 1.8 \times 10^{-5}\varepsilon_1 - 9.0 \times 10^{-7} = 0.$$

The quadratic form on the right-hand side of Equation 15 can then be solved for $\varepsilon_1 = 9.4 \times 10^{-4}$ M. Note that the other root is invalid since it is negative. Equation 14 can then be solved for $\varepsilon_2 = 1.064 \times 10^{-11}$ M. Note that $\varepsilon_1 \gg \varepsilon_2$ as we assumed earlier. With the extents of the reactions known, we can now use Equations 8 through 12 to calculate the concentrations of the species at equilibrium, as shown below. Note that the concentration of pure water, i.e., C_{HOH}^0, is equal to 1000 g $L^{-1} \times$ 1 mol 18^{-1} g^{-1} = 55.56 M and for practical purposes is constant:

$$(16) \; C_{HOAc} = C_{HOAc}^0 - \varepsilon_1 = 0.05 - 9.4 \times 10^{-4} = 0.0491 \text{ M}$$

$$(17) \; C_{OAc^-} = \varepsilon_1 = 9.4 \times 10^{-4} \text{ M}$$

$$(18) \; C_{H^+} = \varepsilon_1 + \varepsilon_2 = 9.4 \times 10^{-4} \text{ M}$$

$$(19) \; C_{OH^-} = \varepsilon_2 = 1.064 \times 10^{-11} \text{ M}$$

$$(20) \; C_{HOH} = C_{HOH}^0 - \varepsilon_2 = 55.6 \text{ M}.$$

The pH of the solution at equilibrium is $- \log_{10} (9.4 \times 10^{-4}) = 3.03$. Also, electroneutrality requires that $C_{H^+} = C_{OAc^-} + C_{OH^-}$ and we see from the previous solution that this requirement is also satisfied.

In biological systems it is also very important to control the pH at some particular value, and buffers are substances that are employed to make the solution pH less sensitive to changes in the concentration of H^+. These buffers usually consist of a weak acid or a weak base and their corresponding salt. For example, in a solution made from a particular acid, the addition of the salt of this acid will buffer the solution by consuming excess H^+, thereby keeping the pH of the solution relatively constant. This is illustrated in the following example.

Example 2.30

Determine the amount of sodium acetate (NaOAc) needed to form a buffer solution of pH = 4.5 in a 1 L solution containing 0.05 M acetic acid.

Solution

In Example 2.29, it was shown that the pH of a 0.05 M solution of acetic acid is 3.03. NaOAc acts as a strong electrolyte dissociating into Na$^+$ and OAc$^-$. The acetate anions will tend to consume the H^+ and the problem is to find how much NaOAc is needed to raise the pH to 4.5. Now we have three dissociation reactions:

(1) $NaOAc \rightarrow Na^+ + OAc^-$ completely dissociates, $K_{NaOAc} = \infty$

(2) $HOAc \leftrightarrow H^+ + OAc^-$ and $K_{HOAc} = \dfrac{C_{H^+} C_{OAc^-}}{C_{HOAc}} = 1.8 \times 10^{-5}$ M

(3) $HOH \leftrightarrow H^+ + OH^-$ and $K_W = C_{H^+} C_{OH^-} = 10^{-14}$ M^2.

As in the previous example, we can express the concentrations of each species in terms of the extent of each of the previous reactions as shown by the following equations:

(4) $C_{NaOAc} = C_{NaOAc}^0 - \varepsilon_1$

(5) $C_{Na^+} = \varepsilon_1$

(6) $C_{HOAc} = C_{HOAc}^0 - \varepsilon_2$

(7) $C_{OAc^-} = \varepsilon_1 + \varepsilon_2$

(8) $C_{H^+} = \varepsilon_2 + \varepsilon_3$

(9) $C_{OH^-} = \varepsilon_3$

(10) $C_{HOH} = C_{HOH}^0 - \varepsilon_3$.

Since NaOAc is a strong electrolyte and completely dissociates, then from Equation 4 $C_{NaOAc} = 0$ and $\varepsilon_1 = C_{NaOAc}^0$. Also by Equations 5 and 7, we have that $C_{Na^+} = \varepsilon_1 = C_{NaOAc}^0$ and $C_{OAc^-} = \varepsilon_1 + \varepsilon_2 = C_{NaOAc}^0 + \varepsilon_2$. Adding Equations 6 and 7, we also get that $C_{HOAc} + C_{OAc^-} = C_{HOAc}^0 + \varepsilon_1 = C_{HOAc}^0 + C_{NaOAc}^0 = C_A^0 + C_S^0$. Here we generalize the result by denoting C_A^0 and C_S^0 as the initial concentrations of the acid and the salt in the solution that is formed.

A charge balance can also be written by summing Equations 5 and 8 for the cations and Equations 7 and 9 for the anions, and we observe that these are equal, giving the following result:

(11) $C_{Na^+} + C_{H^+} = C_{OAc^-} + C_{OH^-}$.

To form a buffer solution, the amount of NaOAc or salt that must be added is usually significant. Therefore, we expect that $C_{Na^+} \gg C_{H^+}$ and $C_{OAc^-} \gg C_{OH^-}$. This means that $\varepsilon_1 \gg \varepsilon_2 + \varepsilon_3$ as well as $\varepsilon_1 \gg \varepsilon_3$. From the charge balance given by Equation 11:

(12) $C_{Na^+} = C_{OAc^-} = C_{NaOAc}^0 = \varepsilon_1$.

In addition, as shown in Example 2.29, the dissociation of the weak acid, in this case acetic acid by Equation 6, is very small, hence $C_{HOAc} \approx C_{HOAc}^0$. Therefore, the equilibrium constant for HOAc can be written as follows:

(13) $K_{HOAc} = \dfrac{C_{H^+} C_{OAc^-}}{C_{HOAc}} \approx \dfrac{C_{H^+} C_{NaOAc}^0}{C_{HOAc}^0} = \dfrac{C_{H^+} C_S^0}{C_A^0}$.

This equation can then be solved for C_{H^+} as follows:

(14) $C_{H^+} = K_{HOAc} \dfrac{C_A^0}{C_S^0}$

(15) $\log_{10} C_{H^+} = \log_{10} K_{HOAc} + \log \dfrac{C_A^0}{C_S^0}$.

The previous equations can then be rewritten in terms of the pH and the pK of the acid (A) as shown below. This equation is known as the *Henderson–Hasselbach equation*:

$$(16) \quad pH = pK_A + \log_{10} \frac{C_S^0}{C_A^0}$$

This equation is a general result that can be used to prepare buffer solutions. Since $C_S^0 \approx C_A^0$, one usually selects a weak acid with a pK_A close to the desired pH. Recall that subscript S refers to the salt of the weak acid A and pK_A is for the weak acid. For this example, we have that $pK_A = 4.745$ and $C_A^0 = 0.05$ M. Also, the solution pH is to be 4.5. The Henderson–Hasselbach equation can then be solved for the concentration of salt, i.e., NaOAc in this example, as shown below:

$$4.5 = 4.745 + \log_{10} \frac{C_S^0}{0.05 \text{ M}}, \quad \text{which gives } C_S^0 = 0.0284 \text{ M}.$$

Since the solution formed is 1 L, this means that 0.0284 mol of NaOAc is needed. The molecular weight of NaOAc is 82 gmol^{-1}, so the amount of NaOAc needed is 0.0284 mol × 82 gmol^{-1} = 2.33 g. In obtaining this solution, we have assumed that the solutions are ideal. In actual practice the pH of the solution may need some minor tweaking to account for nonideal solution behavior.

PROBLEMS

1. Assuming that air is an ideal gas, calculate the change in entropy and enthalpy when 100 mol of air at 25°C and 1 atm is heated and compressed to 300°C and 10 atm. Use the following equation for C_P:

$$\frac{C_P}{R} = A + B \cdot T + C \cdot T^{-2},$$

   where $A = 3.355$, $B = 0.575 \times 10^{-3}$, and $C = -0.016 \times 10^5$, with T in K.

2. Calculate the work required to compress adiabatically (i.e., $Q = 0$) 60 mol of carbon dioxide initially at 25°C and 1 atm to 350°C and 10 atm. Assume that $C_P = 37.1$ J K^{-1} mol^{-1}. If the process is carried out reversibly, how much work is required? Assume that carbon dioxide is an ideal gas.

3. A toy rocket consists of a Styrofoam rocket snugly fitted into a long tube of volume equal to 0.5 L. The pressure in the tube is increased to a final pressure of 3 atm and a temperature of 30°C at which time the rocket takes off. If the rocket weighs 50 g, estimate the maximum initial velocity of the rocket and the maximum expected height of the rocket. Assume air is an ideal gas and has a $C_V = (5/2)R$. Explain why the rocket would not be expected to achieve this elevation.

4. Air at 400 kPa and 400 K passes through a turbine. The turbine is well insulated. The air leaves the turbine at 125 kPa. Find the maximum amount of work that can be obtained from the turbine and the exit temperature of the air leaving the turbine. Assume air is an ideal gas with a $C_P = (7/2)R$.

5. Consider the slow adiabatic expansion of a closed volume of gas for which $C_P = (7/2)R$. If the initial gas temperature is 825 K, and the ratio of the final pressure to the initial pressure is 1/3, what is the change in enthalpy of the gas, the change in internal energy, the heat transferred Q, and the work W? Assume a basis of 1 mole of gas.

6. Air at 125 kPa and 350 K passes through a compressor. The compressor is well insulated. The air leaves the compressor at 600 kPa and 650 K. Find the work required to do this compression. Is this change of state of the air possible? Assume air is an ideal gas with a $C_P = (7/2)R$.

7. Consider the slow adiabatic compression of a closed volume of gas for which $C_P = 3.5\ R$. If the initial gas temperature is 320 K and the ratio of the final pressure to the initial pressure is 4, what is the change in enthalpy of the gas, the change in internal energy, the heat transferred Q, and the work W? Assume a basis of 1 mole of gas.

8. One mole of an ideal gas, initially at 30°C and 1 atm, undergoes the following reversible changes. It is first compressed isothermally to a point such that when it is heated at constant volume to 150°C, its final pressure is 12 atm. Calculate Q, W, ΔE, and ΔH for the process. Take $C_p = 3.5\ R$ and $C_V = 2.5\ R$.

9. A rigid vessel of 75 L volume contains an ideal gas at 400 K and 1 atm. If heat in the amount of 15,000 J is transferred to the gas, determine the entropy change of the gas. Assume $C_V = 2.5\ R$ and $C_p = 3.5\ R$.

10. An osmotic pump (see Figure 7.7) is a device used to deliver a drug at a constant rate within the body. An excess of NaCl is used to form a continuously saturated solution that is contained within a compartment, called the *osmotic engine*, which is located between a semipermeable membrane that is exposed to the interstitial fluid and a piston that acts on a reservoir containing the drug. In operation, water from the surrounding tissue is brought into the osmotic engine by osmosis. The resulting expansion of the osmotic engine drives the piston that forces the drug from the drug reservoir out of the osmotic pump at a constant rate. The drug delivery rate (R) can be shown to be proportional to the osmotic pressure of the fluid within the osmotic engine and the concentration (C) of the drug in the reservoir. The delivery rate of the drug is therefore given by

$$R = k\pi C.$$

At 37°C, the saturation concentration of NaCl in water is equal to 5.4 M. For a particular osmotic pump, it was found that the drug release rate (R) was 125 µg day^{-1}. The drug concentration (C) was 370 mg mL^{-1}. What is the value of the proportionality constant (Kelvin, in milliliters per day per atmosphere) for this osmotic pump?

11. The Navy is considering an osmotic device in their submarines in order to desalinate water when the submarine is submerged. When the submarine is at the appropriate depth, the osmotic device converts seawater into pure water that would be available for use on the submarine. The density of seawater is 1024 kg m^{-3} and the composition of the seawater is equivalent to a 0.5 M NaCl solution. At what cruising depth would this proposed desalination process work?

12. One mole of an ideal gas initially at 25°C and 1 atm is heated and allowed to expand against an external pressure of 1 atm until the final temperature is 400°C. For this gas, $C_V = 20.8$ J mol^{-1} K^{-1}.
   a. Calculate the work done by the gas during the expansion.
   b. Calculate the change in internal energy and the change in enthalpy of the gas during the expansion.
   c. How much heat is absorbed by the gas during the expansion?
   d. What is the change in the entropy of the gas?

13. At 50°C, the vapor pressure of A and B as pure liquids is 268.0 and 236.2 mmHg, respectively. At this temperature, calculate the total pressure and the composition of the vapor, which is in equilibrium with the liquid containing a mole fraction of A of 0.25.

14. When 0.013 mole of urea is dissolved in 1000 g of water, the freezing point of water decreases by 0.024 K. How does this measured freezing point depression compare to that based on the ideal solution theory?

15. Gas in a cylinder expands slowly by pushing on a frictionless piston. The following data show what happens to the gas in the cylinder. How much work (W) is done by the gas? Assuming $C_V = 2.5$ R, what is the change in the internal energy (ΔE) of the gas? How much heat was transferred during the process (Q)?

Gas Pressure (bar)	Volume of Gas (L)	Temperature of Gas (°C)
10	15	70
7	17	23
5	20	10
3	24	50
6	28	150
8	35	450

16. Cubic equations of state are commonly used for *PVT* calculations. Most of the modern cubic equations of state derive from that first developed by van der Waals in 1873. The van der Waals equation of state is given by

$$\left(P + \frac{a}{V^2}\right)(V - b) = RT,$$

where *a* is called the *attraction parameter* and *b* is referred to as the *repulsion parameter* or the *effective molecular volume*. Note that with *a* and *b* equal to zero, this equation simplifies to the ideal gas law. These constants are found by either fitting the earlier equation to actual *PVT* data, or as we shall see next, we can use the critical point to define them in terms of the component critical properties.

   a. Show that the previous equation can also be written in the following cubic forms in terms of volume or the compressibility factor:

$$V^3 - \left(b + \frac{RT}{P}\right)V^2 + \frac{a}{P}V - \frac{ab}{P} = 0,$$

$$z^3 - \left(\frac{bP}{RT} + 1\right)z^2 + \frac{aP}{(RT)^2}z - \frac{abP^2}{(RT)^3} = 0.$$

   For isotherms (*T*) below the critical isotherm and $P = P^{Sat}(T)$ (also refer to Figure 2.3), the solution to either of the previous equations will produce three roots for either *V* or *z*. The smallest of these roots will correspond to the saturated liquid value, i.e., point B in Figure 2.3, and the largest of these roots will correspond to the saturated vapor value, i.e., point D in Figure 2.3. The middle root has no physical meaning. At the critical point, point C in Figure 2.3, these three roots become equal, i.e., $V_L^{Sat} = V_V^{Sat} = V_C = RT_C/P_C$, where V_C is the critical volume. Hence, at the critical point we have that $(V - V_C)^3 = 0$ or when this is expanded we have that $V^3 - 3V_C V^2 + 3V_C^2 V - V_C^3 = 0$.

   b. Comparing the like powers of the expanded volume equation at the critical point to those in the volume explicit form of the van der Waals equation, show that the van der Waals parameters are given by the following equations in terms of the critical properties:

$$a = \frac{27 R^2 T_C^2}{64 P_C}, \quad b = \frac{RT_C}{8 P_C}, \quad R = \frac{8 P_C V_C}{3 T_C}, \quad z_C = 0.375.$$

   Although the value for the gas constant *R* is found to depend on the component critical properties, the usual practice is to use the universal value of the gas constant (shown in Table 1.5) in *PVT* calculations. In this way, the van der Waals equation of state becomes the ideal gas law at low pressures.

   c. For water at 250°C the vapor pressure is 3977.6 kPa and the specific volumes of the saturated vapor and liquid are 50.04 and 1.251 cm³ g⁻¹. Compare these specific volumes to those predicted by the van der Waals equation of state using the following critical properties for water:

$$T_C = 647.1 \text{ K}, \quad P_C = 220.55 \text{ bar} \quad (1 \text{ bar} = 100 \text{ kPa}).$$

17. A new drug has a molecular weight of 625 and a melting temperature of 310°C. Estimate the solubility of this drug in benzene at 25°C. Use the Scatchard–Hildebrand equation to estimate the activity coefficient of the drug. Additional information needed to solve this problem is shown subsequently and in Example 2.9:

Heat of fusion of the drug	5850 cal mol^{-1}
Density of the drug	1.028 g cm^{-3} at 25°C
Vapor pressure of the solid drug	$\ln P^{\text{Sat}}\,(\text{mmHg}) = 26.3 - \dfrac{8780}{T\,(\text{K})}$

18. Estimate the gas phase equilibrium mole fraction of the drug considered in Problem 17 at a temperature of 45°C. The pressure of the gas is 1 atm.

19. A protein solution containing 0.75 g of protein per 100 mL of solution has an osmotic pressure of 22 mmH$_2$O at 25°C. What is the molecular weight of the protein? (1 mmHg = 13.65 mmH$_2$O and 1 atm = 760 mmHg, also $R = 0.082$ L atm K^{-1} mol^{-1})

20. Consider the situation where a semipermeable membrane separates a bulk fluid (region A) with an osmotic pressure of 1750 mmHg from another *enclosed* fluid (region B) with an osmotic pressure of 10,000 mmHg. The hydrodynamic pressure of region A is 760 mmHg. The solvent in both regions A and B is water. Assuming the container and membrane enclosing region B is rigid, what is the equilibrium hydrodynamic pressure in region B, that is the situation wherein there is no net flow of water across the membrane?

21. When 0.5 mole of sucrose is dissolved in 1000 g of water, the osmotic pressure at 20°C is found to be 12.75 atm. Calculate the activity coefficient of water.

22. It is desired to extract from water a drug with a mole fraction of 0.02. A single equilibrium stage extractor is to be used. The flow rate of the aqueous stream to the extractor is 50 mol min^{-1} and the flow rate of the solvent to the extractor is 150 mol min^{-1}. If 90% of the drug is to be removed from the aqueous stream, what should the distribution coefficient of the solvent for this drug be?

23. Using the results from Problem 22, what would the percent extraction of the drug be if four equilibrium extraction stages were used?

24. The activity coefficients of ethanol (1) and water (2) can be described by the van Laar equation, where

$$\ln\gamma_1 = A_{12}\left(\frac{A_{21}\,x_2}{A_{12}\,x_1 + A_{21}\,x_2}\right)^2,$$

$$\ln\gamma_2 = A_{21}\left(\frac{A_{12}\,x_1}{A_{12}\,x_1 + A_{21}\,x_2}\right)^2.$$

The infinite dilution activity coefficient (γ_1^∞) of ethanol in water is 4.66 and that for water in ethanol (γ_2^∞) is 2.64. At 25°C the vapor pressure of water is 3.166 kPa and the vapor pressure of ethanol is 7.82 kPa. Assuming a liquid phase contains 50 wt% ethanol, estimate the composition of the vapor phase that is in equilibrium with the liquid phase. Would this vapor phase composition be explosive? The density of pure water at 25°C is 0.9971 g cm^{-3} and the density of pure ethanol is 0.7851 g cm^{-3}. The density of a 50 wt% solution of ethanol in water is 0.9099 g cm^{-3}.

25. Explain why a water bug can walk on water.

26. A droplet of water at 25°C has a diameter of 0.1 μm. Calculate the pressure within the droplet of water.

27. What is the vapor pressure of the water within a droplet of water at 25°C that has a diameter of 10 nm?

28. Consider the situation of a liquid mixture where the solute molecules are much larger than those of the solvent. An example of such a mixture would be a polymer in a solvent or a protein solution. In the solution model developed by Flory and Huggins (Sandler 1989), the entropy change as a result of mixing a solvent (1) with a much larger solute (2) is given by the following expression:

$$\Delta S^{\text{mix}} = -R\left(x_1 \ln\phi_1 + x_2 \ln\phi_2\right),$$

where x_1 and x_2 are the mole fractions and ϕ_1 and ϕ_2 are the volume fractions of the solvent and the solute, respectively. The volume fractions are given by

$$\phi_1 = \frac{x_1}{x_1 + r\,x_2} \quad \text{and} \quad \phi_2 = \frac{r\,x_2}{x_1 + r\,x_2},$$

with $r = V_2/V_1$, where V_1 and V_2 are the molar volumes of each species. Using Equations 2.44 and 2.96 that define the property change of mixing and the excess property, respectively, along with the result from Example 2.6 that $\overline{S}_i^{\text{ideal solution}} = S_i - R \ln x_i$, show that the excess entropy of the solution is given by

$$S^{\text{E}} = -R\left(x_1 \ln\frac{\phi_1}{x_1} + x_2 \ln\frac{\phi_2}{x_2} \right).$$

Note that if the volumes of the solvent and solute are comparable in size, then from the above expression, $S^{\text{E}} = 0$, and the solution is then an ideal solution. So, we see that it is the difference in the size of the solvent and the solute that leads to the nonideal solution behavior in the model proposed by Flory and Huggins. The excess enthalpy (H^{E}) is then expressed by

$$H^{\text{E}} = \chi\,RT\left(x_1 + r\,x_2 \right)\phi_1\,\phi_2,$$

where χ is an adjustable parameter, known as the Flory–Huggins interaction parameter. The Flory–Huggins interaction parameter for nonpolar systems can be shown to be related to the solubility parameters of the solvent and the solute by the following equation (Prausnitz et al. 1986):

$$\chi = \frac{V_1}{RT}\left(\delta_1 - \delta_2 \right)^2.$$

A good solvent for the polymer or macromolecule is one for which χ is very small or $\delta_1 \sim \delta_2$. In addition, to ensure that the solvent and polymer are completely miscible, the largest value of χ is equal to 0.5. If $\chi > 0.5$, the solvent and solute are only partially miscible.

Since $G^{\text{E}} = H^{\text{E}} - TS^{\text{E}}$, show that the previous two equations for S^{E} and H^{E} can be combined to give the Flory–Huggins model for the excess Gibbs free energy of the solution:

$$\frac{G^{\text{E}}}{RT} = \left(x_1 \ln\frac{\phi_1}{x_1} + x_2 \ln\frac{\phi_2}{x_2} \right) + \chi\left(x_1 + r\,x_2 \right)\phi_1\,\phi_2.$$

Using Equation 2.100, show (if you can) that the Flory–Huggins activity coefficients for solvent (1) and solute (2) are given by the following expressions:

$$\ln\gamma_1 = \ln\frac{\phi_1}{x_1} + \left(1 - \frac{1}{r} \right)\phi_2 + \chi\,\phi_2^2,$$

$$\ln\gamma_2 = \ln\frac{\phi_2}{x_2} + \left(r - 1 \right)\phi_1 + \chi\,\phi_1^2.$$

Recall from Equation 2.146 that the osmotic pressure of the solution containing our solvent (1) and solute (2) is given by

$$\pi = -\frac{RT}{V_1}\ln\left(\gamma_1\,x_1 \right).$$

Show that if the activity coefficient of solvent (1) is described by the Flory–Huggins activity coefficient model, then the osmotic pressure is given by the following expression. Note that for a dilute solute solution, we can expand the solvent and solute volume fraction in an infinite series where $\ln \phi_1 = \ln(1 - \phi_2) \approx -\phi_2 - (\phi_2^2 / 2) - (\phi_2^3 / 3) - \cdots$

$$\pi = \frac{RT}{V_1} \left[\frac{\phi_2}{r} + \left(\frac{1}{2} - \chi \right) \phi_2^2 + \frac{1}{3} \phi_2^3 + \cdots \right]$$

From the definition of the solute volume fraction given above, we can then write for a dilute solution that $\phi_2 = (x_2 V_2 / V_1) = V_2 C_2$, where C_2 is the molar concentration of the solute. Substituting this result for ϕ_2 into the previous equation for π, show that in a general sense the osmotic pressure for a nonideal solution can be described by a power series in the molar concentration of the solute (or virial series) as

$$\pi = RT C_2 \left(1 + \bar{B} C_2 + \bar{C} C_2^2 + \cdots \right),$$

where $\bar{B}$ and $\bar{C}$ are known as the second and third virial coefficients, respectively. Note that for an ideal solution, $\bar{B}$ and $\bar{C}$ are equal to zero and $\pi = RT C_2$, which is also the result shown in Equation 2.147. Usually with polymers and macromolecules, the solute concentration is expressed in weight concentration, e.g., grams per liter. Letting c_2 represent the weight concentration of the solute, then $C_2 = c_2 / MW_2$ and $\phi_2 = V_2 (c_2 / MW_2)$. Substituting this result for ϕ_2 into the previous equation for π and neglecting terms higher than the second order in ϕ_2, show that the following result is obtained for the osmotic pressure:

$$\pi = \frac{RT c_2}{MW_2} \left[1 + \frac{MW_2}{V_1 \rho_2^2} \left(\frac{1}{2} - \chi \right) c_2 \right].$$

Also show that the second virial coefficient is equal to $(MW_2 / V_1 \rho_2^2)((1/2) - \chi)$.

29. The following data for the osmotic pressure of hemoglobin (MW = 68,000 Da) in an aqueous solution at 0°C are presented by Freeman (1995).

Hemoglobin Concentration (g 100^{-1} cm^{-3} solution)	Osmotic Pressure (mmHg)
2.5	8
5	15
8	25
10	37
12	42
15	65
19.5	100
23.4	150
24.5	167
27.5	229
28.6	254

Make a graph of the above-mentioned data and compare the data to regression fits of the following virial expression derived in Problem 28 for the osmotic pressure, i.e., $\pi = (RT c_2 / MW_2) [1 + \bar{B} c_2 + \bar{C} c_2^2]$. Do three regressions where the first regression assumes an ideal solution (i.e., $\bar{B}$ and $\bar{C}$ equal zero), the second regression ($\bar{C} = 0$) truncates the osmotic pressure equation after the second virial coefficient, and the third regression includes $\bar{B}$ and $\bar{C}$.

30. What is the predicted temperature of the seawater surrounding the polar ice caps, assuming that the seawater and the ice are in equilibrium? Assume the freezing point of pure water is 0°C and that the heat of fusion of pure water is 6012 J mol^{-1}.

31. A mountain climber has reached an elevation where the observed boiling point of water is 97°C. Estimate how high above sea level the climber is. Carefully state your assumptions and any references that you have consulted.

32. The concentration of a macromolecule in an aqueous solution contained inside a semi-permeable membrane bag is 0.002 M. The macromolecule carries a net negative charge of −8. If the NaCl concentration in the solution outside the bag at equilibrium is 0.05 M, calculate the ratio of the concentration of Na$^+$ inside the bag to that outside the bag. Also find the ratio of the concentration of Cl$^-$ inside the bag to that outside the bag. What is the transmembrane voltage at equilibrium based on the Na$^+$ concentration and the Cl$^-$ concentration inside and outside the membrane bag?

33. For Example 2.27, calculate the osmotic pressure of the solution inside the membrane bag relative to that outside the bag. How does the Donnan effect affect the osmotic pressure?

34. Find the concentrations of the species at equilibrium for the case when a 0.10 M solution of sodium acetate (NaOAc) is made. NaOAc is the salt of acetic acid (HOAc). Also determine the pH of the resulting solution. Assume that sodium acetate as a salt is a strong electrolyte and that it completely dissociates in water to form Na$^+$ and OAc$^-$. In this case, the high concentration of OAc$^-$ will tend to consume H$^+$ with the result that HOAc will be formed and we expect the resulting solution to be basic. The water and HOAc dissociation equilibrium constants may be found in Example 2.29.

35. *Tris*, i.e., *tris*-(hydroxymethyl) aminomethane (MW = 121.14 gmol^{-1}) is a weak base (i.e., removes H$^+$) and a very important buffer used to make biological solutions. Although it is a weak base, it is still convenient in solving buffer problems to work instead with its acid dissociation reaction, which can be written as follows: $Tris - H^+ \leftrightarrow Tris + H^+$. $Tris$-H$^+$ has a pK$_A$ = 8.3 at 20°C. If a 1 L solution contains 0.05 M of *Tris* and 0.03 M of hydrochloric acid (HCl), what is the pH of this solution? Hydrochloric acid is a strong acid and will completely dissociate in an aqueous solution. Suppose after this solution is formed that an additional millimole of HCl is added. What is the pH now? Note that the addition of 1 mM of HCl to 1 L of pure water gives a solution pH of 3.0.

36. A protein with a molecular weight of 67396 gmol^{-1} is added to water such that the concentration of the protein is 4 g 100^{-1} cm^{-3} of solution. Estimate the osmotic pressure (mmHg) of the resulting solution at a temperature of 0°C.

37. The osmotic pressure (mmHg) of a protein solution at a temperature of 30°C was found to be 10 mmHg. If the concentration of the protein is 4 g 100^{-1} cm^{-3} of solution, estimate the molecular weight of the protein molecule.

38. A peptide solution containing 0.35 g of protein per 100 mL of solution has an osmotic pressure of 36 mmH$_2$O at 25°C. What is the molecular weight of the protein? (1 mmHg = 13.65 mmH$_2$O)

39. What is the osmotic pressure of a solution containing a solute with a molecular weight of 75,000 at a concentration of 30 g L^{-1} and a solute with a molecular weight of 20,000 at a concentration of 10 g L^{-1}? The solute with a molecular weight of 75,000 is unstable and completely dissociates into four solutes of equal size.

40. A protein solution with an osmotic pressure of 45 mm of water needs to be prepared. If the molecular weight of the protein is 50,000 gmol^{-1}, how many grams of the protein need to be in 100 mL of the solution at 25°C?

3 Physical Properties of the Body Fluids and the Cell Membrane

3.1 BODY FLUIDS

To begin our study of transport phenomena in biomedical engineering, we must first examine the physical properties of the fluids within the human body. Many of our engineering calculations and the development and design of new procedures, devices, and treatments will either involve or affect the fluids within the human body. Therefore, we will focus our initial attention on the types and characteristics of the fluids that reside within the body.

Body fluids can be classified into three types: *extracellular*, *intracellular*, and *transcellular* fluids. As shown in Table 3.1, nearly 60% of the body weight for an average 70 kg male is composed of these body fluids, resulting in a total fluid volume of about 40 L.

The largest fraction of the fluid volume, about 36% of the body weight, consists of *intracellular fluid*, which is the fluid contained within the body's cells, e.g., the fluid found within red blood cells, muscle cells, and liver cells. *Extracellular fluid* consists of the *interstitial fluid* that comprises about 17% of the body weight and the *blood plasma* that comprises around 4% of the body weight. Interstitial fluid circulates within the spaces (*interstitium*) between cells. The interstitial fluid space represents about one-sixth of the body volume. The interstitial fluid is formed as a filtrate from the plasma within the *capillaries*. The capillaries are the smallest element of the cardiovascular system and represent the site where the exchange of vital substances occurs between the blood and the tissue surrounding the capillary. We shall see that the interstitial fluid composition is very similar to that of plasma.

The blood volume of a 70 kg male is 5 L, with 3 L consisting of plasma and the remaining 2 L representing the volume of cells in the blood, primarily the red blood cells. The 2 L of cells found in the blood are filled with intracellular fluid. The fraction of the blood volume due to the red blood cells is called the *hematocrit*. The red blood cell volume fraction is found by centrifuging a given volume of blood in order to find the "packed" red cell volume. Since a small amount of plasma is trapped between the packed red blood cells, the true hematocrit (H) is about 96% of this measured hematocrit (Hct). The true hematocrit is about 40% for a male and about 36% for a female. The *transcellular fluids* are those fluids found only within specialized compartments and include the cerebrospinal, intraocular, pleural, pericardial, synovial, sweat, and digestive fluids. Some of these compartments have membrane surfaces that are in proximity to one another and have a thin layer of lubricating fluid between them.

Measurement of these fluid volumes can be achieved by using "tracer materials," which have the unique property of remaining in specific fluid compartments. The fluid volume of a specific compartment can then be found by adding a known mass of a tracer to a specific compartment and, after an appropriate period of time for dispersal, measuring the concentration (mass/volume) of the tracer in the fluid compartment. The compartment volume is then given by the ratio of the tracer mass that was added and the measured tracer concentration, i.e., mass/mass/volume = volume.

Examples of tracers used to measure fluid volumes include radioactive water for measuring *total body water* and radioactive sodium, radioactive chloride, and inulin for measuring *extracellular fluid volume*. Tracers that bind strongly with plasma proteins may be used to measure the *plasma*

TABLE 3.1
Body Fluids

Fluid	Fluid Volume (L)	Wt %
Intracellular	25	36
Extracellular	15	21
Interstitial	12	17
Plasma	3	4
Transcellular	–	–
Total	40	57

Source: Data from Guyton, A.C., *Textbook of Medical Physiology*, W.B. Saunders Co., Philadelphia, PA, 1991.

volume. Interstitial fluid volume may then be found by subtracting the plasma volume from the extracellular fluid volume. Subtraction of the extracellular fluid volume from the total body water provides the *intracellular fluid volume*.

3.2   FLUID COMPOSITIONS

The compositions of the body fluids are presented in Table 3.2 in terms of concentration in units of *milliosmolar* (mOsmole/liter of solution or mOsM). We provide a definition of milliosmolar in Section 3.4.1. Since the solutes listed in Table 3.2 do not dissociate, these concentrations are the same as millimolar (mmol L^{-1}). Of particular interest is the fact that nearly 80% of the total osmolarity of the interstitial fluid and plasma is produced by sodium and chloride ions. As we discussed earlier, the interstitial fluid arises from filtration of plasma through the capillaries. Therefore, we would expect the composition of these two fluids to be very similar. This is shown in Table 3.2. We find this to be true with the exception that the protein concentration in the interstitial fluid is significantly lower in comparison to its value in the plasma.

3.3   CAPILLARY PLASMA PROTEIN RETENTION

The retention of proteins by the walls of the capillary during filtration of the plasma is readily explained by comparing the molecular sizes of typical plasma protein molecules to the sizes of the pores within the capillary wall. Figure 3.1 illustrates the relative sizes of various solutes as a function of their molecular weight. The wall of a capillary, illustrated in Figure 3.2, consists of a single layer of *endothelial cells* surrounded by a *basement membrane*. The basement membrane is a mat-like cellular support structure, or an extracellular matrix, that consists primarily of a protein called *type IV collagen* and is joined to the cells by the glycoprotein called *laminin*. The basement membrane is about 50–100 nm thick. The total thickness of the capillary wall is about 0.5 μm.

As shown in Figure 3.2, there are several mechanisms that allow for the transport of solutes across the capillary wall. These include the *intercellular cleft* and the *pinocytotic vesicles* and *channels*. The intercellular cleft is a thin slit or slit pore that is formed at the interface between adjacent endothelial cells. The size of the openings or pores in the slit is about 6–7 nm, just sufficient to retain plasma proteins such as albumin and other larger proteins. The collective surface area of these openings represents less than 1/1000th of the total capillary surface area.

Plasma proteins are generally larger than the capillary slit pore. Although ellipsoidally shaped proteins, e.g., the clotting protein fibrinogen, may have a minor axis that is smaller than that of the capillary slit pore, the streaming effect caused by the fluid motion within the capillary orients the

TABLE 3.2
Osmolar Solutes Found in the Extracellular and Intracellular Fluids

Solute	Plasma (mOsm L^{-1})	Interstitial (mOsm L^{-1})	Intracellular (mOsm L^{-1})
Na$^+$	143	140	14
K$^+$	4.2	4.0	140
Ca^{2+}	1.3	1.2	0
Mg^{2+}	0.8	0.7	20
Cl$^-$	108	108	10
HCO$_3^-$	24	28.3	10
HPO$_4^-$, H$_2$PO$_4^-$	2	2	11
SO$_4^-$	0.5	0.5	1
Phosphocreatine			45
Carnosine			14
Amino acids	2	2	8
Creatine	0.2	0.2	9
Lactate	1.2	1.2	1.5
Adenosine triphosphate			5
Hexose monophosphate			3.7
Glucose	5.6	5.6	
Protein	1.2	0.2	4
Urea	4	4	4
Others	4.8	3.9	11
Total (mOsm L^{-1})	302.8	301.8	302.2
Corrected osmolar activity (mOsm L^{-1})	282.5	281.3	281.3
Total osmotic pressure at 37°C (mmHg)	5450	5430	5430

Source: Data from Guyton, A.C., *Textbook of Medical Physiology*, W.B. Saunders Co., Philadelphia, PA, 1991.

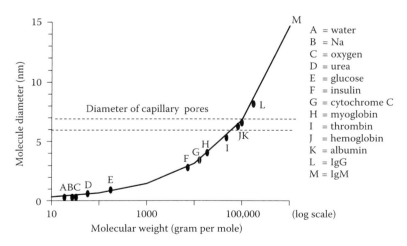

FIGURE 3.1 Approximate diameter of molecules as a function of their molecular weight.

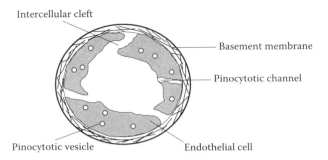

FIGURE 3.2 Cross section of a capillary.

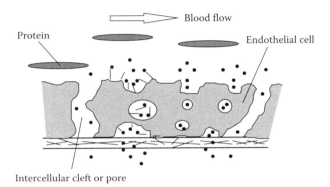

FIGURE 3.3 Orientation of ellipsoidal-shaped proteins by streaming prevents their entry into the capillary pores.

major axis of the proteins parallel to the flow axis and prevents their entry into the slit pore. This streaming effect is shown in Figure 3.3. Therefore, the smaller substances found in the plasma, such as ions, glucose, and metabolic waste products, will readily pass through the slit pores of the capillary wall, whereas the plasma proteins will be retained in the lumen of the capillary. The composition of the plasma and interstitial fluid should therefore only differ in their protein content. This is shown in Table 3.2.

3.4 OSMOTIC PRESSURE

Retention of proteins in the plasma in comparison to the interstitial fluid creates an *osmotic pressure* between the plasma and the interstitial fluid. We discussed the thermodynamics of osmosis in Section 2.6.3.5 in Chapter 2. Since the plasma proteins are the only constituents of the plasma that do not readily pass through the capillary wall, it is the plasma proteins that are responsible for the formation of the osmotic pressure between the interstitial and plasma fluids. The osmotic pressure created by these proteins is given the special name of *colloid osmotic pressure* or *oncotic pressure*. For human plasma, the colloid osmotic pressure is about 28 mmHg, with 19 mmHg caused by the plasma proteins and 9 mmHg caused by the cations within the plasma that are also retained through electrostatic interaction with the negative surface charges of these proteins (see the Gibbs–Donnan Effect in Section 2.6.3.11 in Chapter 2).

The colloid osmotic pressure is actually quite small in comparison to the osmotic pressure that develops when a cell is placed in pure water. In this case, it is assumed that all of the species present within the intracellular fluid are retained by the cell membrane. As shown in Table 3.2, the total osmotic pressure of the intracellular fluid in this case would be 5430 mmHg at 37°C.

3.4.1 OSMOLARITY

Recognizing that it is the number of nondiffusing solute molecules that contributes to the osmotic pressure, we now must make the distinction between a nondiffusing substance that dissociates and one that does not dissociate. The term *nondiffusing* is used to describe solutes that are retained by the membrane that separates the solutions of interest. For example, compounds such as NaCl are strong electrolytes. In water, they completely dissociate to form two ions (i.e., Na^+ and Cl^-). Each ion or particle formed will exert its own osmotic pressure. It is important to note that the charge of the ion has no effect (assuming an ideal solution) on the osmotic pressure, so a sodium ion with a charge of +1 is equivalent to a calcium ion with a charge of +2. Substances such as glucose do not dissociate and their osmotic pressure is based on their nondissociated concentration only.

The term *osmole* has been introduced to account for the effect of a dissociating solute. One osmole is therefore defined as one mole of a nondiffusing and nondissociating substance. Therefore, one mole of a dissociating substance such as NaCl is equivalent to two osmoles, or a 1 M solution of NaCl is equivalent to a two osmolar (OsM) solution. However, one mole of glucose is the same as one osmole since glucose does not dissociate in solution. *Osmolarity* simply defines the number of osmoles per liter of solution.

If a cell is placed within a solution that has a lower concentration of solutes or osmolarity, then the cell is in a *hypotonic solution*, and the establishment of osmotic equilibrium requires the osmosis of water into the cell. This influx of water into the cell results in swelling of the cell and a subsequent decrease in its osmolarity. On the other hand, if the cell is placed in a solution with a higher concentration of solutes or osmolarity, i.e., *hypertonic*, then osmotic equilibrium requires the osmosis of water out of the cell, concentrating the intracellular solution and resulting in shrinkage of the cell. An *isotonic* solution is a fluid that has the same osmolarity as the cell. When cells are placed in an isotonic solution, there is neither swelling nor shrinkage of the cell. A 0.9 wt% solution of sodium chloride or a 5 wt% solution of glucose is just about isotonic with respect to a cell.

3.4.2 CALCULATING THE OSMOTIC PRESSURE

Recall from our discussion in Chapter 2 on osmosis that for the special case where the solvent and solute form an ideal solution, the osmotic pressure given by Equation 2.147 may be written as

$$\pi = \frac{RT}{V^L_{solvent}} x^A_{solute} = RTC_{solute}. \tag{3.1}$$

Here, x^A_{solute} represents the mole fraction of the solute in the solution, which is region A in Figure 2.5. Recall that since the solute mole fraction is generally quite small, we may approximate this as $x^A_{solute} = V^L_{solvent} C_{solute}$, where C_{solute} is the concentration of the solute in moles per liter of solution and $V^L_{solvent}$ is the molar volume of the solvent. This ideal dilute solution osmotic pressure, described by Equation 3.1, is also known as *van't Hoff's law*.

If the solution contains *N* ideal nondissociating solutes, then the total osmotic pressure of the solution will be the summation of the osmotic pressure generated by each solute according to Equation 3.1:

$$\pi = RT \sum_{i=1}^{N} C_{solute_i}. \tag{3.2}$$

As we discussed in Chapter 2, remember that osmotic pressure is not determined on the basis of the mass of the solute in the solution, but rather on the number of particles that are formed by a given solute. Each nondiffusing particle in the solution contributes the same amount to the osmotic pressure regardless of the size of the particle.

For physiological solutions, it is convenient to work in terms of milliosmoles (mOsm) or milliosmolar (mOsM). At a physiological temperature of 37°C, Equation 3.2 may be written as follows to give the osmotic pressure in mmHg when the solute concentration of each nondiffusing species is expressed in milliosmolar:

$$\pi = 19.33 \sum_{i=1}^{N} C_{\text{solute}_i}. \tag{3.3}$$

3.4.3 OTHER FACTORS THAT MAY AFFECT THE OSMOTIC PRESSURE

The previous discussion assumed that the mixture of solutes formed an ideal solution. However, the osmotic pressure should take into account the various secondary solute interactions that occur within the solution due to their charge, size, shape, and other effects. These effects will either increase or decrease the osmotic activity of a particular solute and result in the corrected osmolar activity shown in Table 3.2. We see in Table 3.2 that the ratio of the corrected osmolar concentration to the total osmolar concentration is about 0.93 for each of the body fluids. This value of 0.93 represents the overall average activity coefficient. In most cases, it is conventional practice to disregard the calculation of solute activity coefficients and calculate the osmotic pressure on the basis of solute concentration only, as just discussed. Since we will primarily be concerned with differences in the osmotic pressure and the generation of fluid flow across a membrane due to a difference in osmotic pressure, this constant of 0.93 will become absorbed in other constants that generally need to be determined by experiment. Thus, for our purposes, little error is introduced by simply calculating, using Equation 3.2, the osmotic pressure on the basis of solute concentration only.

3.5 FORMATION OF THE INTERSTITIAL FLUID

Recall that, in general, a *flow* of something is proportional to a *driving force* and inversely proportional to the flow *resistance*. The flow of fluid across the capillary wall, or, for that matter, any porous semipermeable membrane, is driven by a difference in pressure across the capillary wall or membrane, which is the driving force. This pressure difference arises not only from hydrodynamic effects but also from the difference in osmotic pressure between the fluids separated by the membrane. The properties of the fluid as well as the nature of the capillary wall or membrane produce a resistance to this flow.

Figure 3.4 shows the hydrodynamic and osmotic pressures in the capillary and the surrounding interstitial fluid. The arrows indicate for each pressure the corresponding direction of fluid flow induced by that pressure. We may write that the volumetric fluid transfer rate (Q) across the capillary membrane is directly proportional to the *effective pressure drop* ($\overline{\Delta P}$) across the capillary membrane, as given by

$$\frac{J}{L_\text{p}S} = [(P_\text{C} - P_\text{IF}) - (\pi_\text{C} - \pi_\text{IF})] = \overline{\Delta P}. \tag{3.4}$$

In this equation, L_p represents a proportionality constant that is inversely related to the flow resistance of the capillary membrane and is called the *hydraulic conductance*. The hydraulic conductance is usually best determined by experiment, although a model for its prediction is shown below. For example, Renkin (1977) reports values for L_p as low as 3×10^{-14} m² sec kg⁻¹ for the tight junctions between the endothelial cells found in the capillaries of the rabbit brain, 5×10^{-12} m² sec kg⁻¹ for nonfenestrated or continuous capillaries, and as high as 1.5×10^{-9} m² sec kg⁻¹ for the capillaries in the glomeruli of the kidney. In Equation 3.4, S is the total circumferential surface area of the capillary membrane or the total membrane surface area.

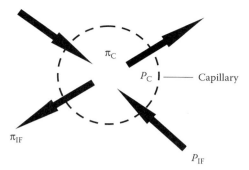

FIGURE 3.4   Forces acting to cause a flow of fluid across the capillary wall. Arrows indicate the direction of flow induced by each pressure.

The first difference term in the brackets of Equation 3.4 represents the difference between the hydrodynamic pressure of the capillary (P_C) and the interstitial fluid (P_{IF}). When this term is positive, fluid will leave the capillary. When this term is negative, fluid will flow from the interstitial fluid into the capillary. The second difference term within the brackets represents the difference in the osmotic pressure between the capillary (π_C) and the interstitial fluid (π_{IF}). When this parenthetical term is positive, there will be an osmotic flow of water from the interstitial fluid into the capillary, whereas if this difference is less than zero, there will be an osmotic flow of water out of the capillary. If the entire term in brackets in Equation 3.4 is positive, then there is a net flow of fluid from the capillary into the interstitium. If the bracketed term is negative, then there will be a net flow of fluid from the interstitium into the capillary.

Equation 3.4 represents the fluid flow across the capillary membrane. It can also be used to describe the flow across any semipermeable membrane regardless of whether or not it is of biological or synthetic origin. We will find, for example, that this equation applies to the dialysis membranes used in an artificial kidney.

We can also develop a model for describing how the hydraulic conductance (L_p), or the resistance to fluid flow across the membrane, depends on the membrane pore geometry and the physical properties of the filtrate fluid. First, we model the porous structure of the capillary wall as a series of parallel cylindrical pores. We will show in Chapter 4 that Poiseuille's equation (Equation 4.11) provides a relationship between flow (Q) and the pressure drop (P_0-P_L) in a cylindrical tube or, in this case, a cylindrical pore:

$$Q = \frac{\pi R^4 (P_0 - P_L)}{8\mu L}, \tag{3.5}$$

where μ is the viscosity (a measure of the flow resistance) of the fluid, R is the radius of the cylindrical tube or pore, and L is the length of the tube or pore. Equation 3.5 provides the flow rate across a single pore in the capillary wall or in a semipermeable membrane. If we have a total of N pores in a membrane of total surface area S, then we can combine Equation 3.5 for N pores with Equation 3.4 to obtain:

$$Q = L_p S \overline{\Delta P} = N \left(\frac{\pi R^4}{8\mu \tau t_m} \right) \overline{\Delta P}, \tag{3.6}$$

where we have replaced (P_0-P_L) in Poiseuille's equation with the effective pressure drop $\overline{\Delta P}$. The actual pore length L is usually longer than the thickness of the capillary wall or membrane (t_m) since the pores are not straight but tortuous. Hence, we replaced L with the product of the tortuosity (τ)

and the capillary wall or membrane thickness, t_m. The tortuosity is a correction factor and for the capillary wall it is about two.

The total cross-sectional area of the pores, i.e., A_P, is equal to $N\pi R^2$. Substituting this into Equation 3.6 allows us to solve for the hydraulic conductance as

$$L_P = \left(\frac{A_P}{S}\right)\frac{R^2}{8\mu\tau t_m}. \tag{3.7}$$

Equation 3.7 provides a model that allows one to understand the factors that affect the hydraulic conductance. We see that the hydraulic conductance is directly proportional to the *porosity* of the capillary wall, i.e., A_P/S, and inversely proportional to the flow resistance of the fluid, i.e., its viscosity (μ), and the thickness of the capillary wall or membrane. In addition, the hydraulic conductance is directly proportional to the square of the pore radius.

Example 3.1

Calculate the filtration flow rate (cm³ sec⁻¹) of a pure fluid across a 100 cm² membrane. Assume the viscosity (μ) of the fluid is 1.8 cP. The porosity of the membrane is 40% and the thickness of the membrane is 500 μm. The pores run straight through the membrane, and these pores have a radius of 0.225 μm. The pressure drop applied across the membrane is 75 psi. (Recall that 14.7 psi = 1 atm = 101325 Pa (or N m⁻²), that 1 N = 1 kg m sec⁻², and from Chapter 4 we have for the viscosity that 1 cP = 0.001 N sec m⁻² = 0.001 Pa sec.)

Solution

To find the filtration flow rate, we will use Equations 3.4 and 3.7. Since there are no retained solutes, there are no osmotic effects and the effective pressure drop across the membrane is equal to the applied pressure drop of 75 psi. From the information about the membrane, we first calculate the value of the hydraulic conductance as

$$\frac{A_P}{S} = 0.4,\ R = 0.225\ \mu m = 2.25\times10^{-5}\ cm,\ t_m = 500\ \mu m = 0.05\ cm,$$

$$\mu = 1.8\ cP = 0.0018\ Pa\,sec,\ \text{and}\ \overline{\Delta P} = 75\ psi\times\frac{1\ atm}{14.7\ psi}\times\frac{101325\ Pa}{1\ atm} = 5.17\times10^5\ Pa.$$

Now, substituting these values into Equation 3.7:

$$L_P = \frac{0.40\times(2.25\times10^{-5}\ cm)^2}{8\times0.0018\ Pa\,sec\times0.05\ cm} = 2.813\times10^{-7}\ \frac{cm}{Pa\,sec},$$

and from Equation 3.4:

$$Q = 2.813\times10^{-7}\ \frac{cm}{Pa\,sec}\times100\ cm^2\times5.17\times10^5\ Pa$$

$$Q = 14.54\ \frac{cm^3}{sec}.$$

3.6 NET CAPILLARY FILTRATION RATE

We can use Equation 3.4 to estimate the *net capillary filtration rate* for the human body. To perform this calculation, we will first need to define nominal values for the pressures in Equation 3.4

(note: these are gauge pressures or relative to an atmospheric pressure of 760 mmHg). At the arterial end of the capillary, the capillary pressure is about 30 mmHg, and at the venous end of the capillary, the pressure is about 10 mmHg. The mean capillary pressure is considered to be about 17.3 mmHg and its bias to the lower end is based on the larger volume of the venous side of the capillaries in comparison to the arterial side of the capillaries. Surprisingly, the interstitial fluid pressure has been found to be subatmospheric and the accepted value is −3 mmHg. As mentioned before, the colloid osmotic pressure for human plasma is 28 mmHg and the value for the interstitial fluid is about 8 mmHg.

At the arterial end of the capillary, we can calculate the flow rate of fluid across the capillary wall by using Equation 3.4. Therefore, $Q/L_pS = [(30 - - 3)–(28–8)] = 13$ mmHg, which is positive, indicating that there is a net flow of fluid from the capillary into the interstitium. At the venous end of the capillary, $Q/L_pS = [(10 - - 3)–(28–8)] = –7$ mmHg, which is less than zero, indicating a net reabsorption of fluid from the interstitium back into the capillary.

This flow of fluid out of the capillary at the arterial end, and its reabsorption at the venous end, is called the *Starling flow*, after E.H. Starling, who first described it more than 100 years ago. Thus, Equation 3.4 is also known as the *Starling equation*. If we base the filtration rate on the mean capillary pressure, then $Q/L_pS = [(17.3 - -3)–(28–8)] = 0.3$ mmHg. Hence, on average, there is a slight imbalance in pressure that results in more filtration of fluid out of the capillary than is reabsorbed.

Approximately 90% of the fluid that leaves at the arterial end of the capillary is reabsorbed at the venous end. However, the 10% that is not reabsorbed by the capillary collects within the interstitium and enters the lymphatic system. The amount of this net filtration of fluid from the circulation for the human body can be estimated as illustrated in the following example.

Example 3.2

Calculate the normal rate of net filtration for the human body. Assume that the capillaries have a total surface area of 500 m² and that the slit pore surface area is 1/1000th of the total capillary surface area.

Solution

We model the porous structure of the capillary wall as a series of parallel cylindrical pores with a diameter of 7 nm. Plasma filtrate may be considered a Newtonian fluid with a viscosity of 1.2 cP. The mean net filtration pressure or the effective pressure drop for the capillary was just calculated previously to be 0.3 mmHg. Using Equation 3.7, we can calculate the hydraulic conductance and the net filtration rate:

$$L_p = \left(\frac{0.5 \text{ m}^2}{500 \text{ m}^2} \right) \frac{(0.5 \times 7 \times 10^{-9} \text{ m})^2}{8 \times 1.2 \text{ cP} \times \dfrac{0.01 \text{ P}}{\text{cP}} \times \dfrac{1 \text{ g}}{\text{cm sec P}} \times \dfrac{100 \text{ cm}}{\text{m}} \times \dfrac{1 \text{ kg}}{1000 \text{ g}} \times 2 \times 0.5 \times 10^{-6} \text{ m}}$$

$$= 1.28 \times 10^{-12} \frac{\text{m}^2 \text{ sec}}{\text{kg}}.$$

Note that this prediction of L_p is consistent with the values reported by Renkin (1977) for the hydraulic conductance of continuous capillaries:

$$Q = 1.28 \times 10^{-12} \frac{\text{m}^2 \text{ sec}}{\text{kg}} \times 500 \text{ m}^2 \times 0.3 \text{ mmHg} \times \frac{1 \text{ atm}}{760 \text{ mmHg}} \times 101.33 \frac{\text{kPa}}{\text{atm}} \times \frac{1 \text{ Nm}^{-2}}{\text{Pa}}$$

$$\times \frac{1000 \text{ Pa}}{\text{kPa}} \times \frac{\text{kg m sec}^{-2}}{N} \times \frac{60 \text{ sec}}{\text{min}} \times \frac{(100 \text{ cm})^3}{\text{m}^3}$$

$$Q = 1.54 \frac{\text{cm}^3}{\text{min}}.$$

We find that the *total net filtration rate* due to the pressure imbalance at the capillaries for the human body is on the order of 2 mL min^{-1}.

We can define the *filtration coefficient* as the ratio of Q to $\overline{\Delta P}$, which is equal to $L_p S$. For the capillary wall, using the results from the previous example, we obtain

$$\frac{Q}{\overline{\Delta P}} = L_p S = \frac{5.1 \text{ mL}}{\min \text{mmHg}}. \tag{3.8}$$

For the capillary wall or a particular membrane, the filtration coefficient is a constant and represents the properties of the filtrate fluid, the membrane porosity, and the dimensions of the pores. The net filtration rate for any average pressure imbalance at the capillaries is given by simply multiplying the filtration coefficient by the pressure imbalance. For example, if the pressure imbalance across a capillary increased from 0.3 to 1.5 mmHg, then the net filtration rate would increase by a factor of 5 to about 10 mL min^{-1}.

3.7   LYMPHATIC SYSTEM

We must now address where the net filtration of fluid from the capillaries, about 120 mL hr^{-1} for a human, ultimately goes. It clearly cannot continue to collect within the interstitium since this would lead to *edema* or excess fluid (swelling) within the tissues of the body.

Two types of edema can occur. The first type involves excess extracellular fluid and is caused either by too much filtration of fluid from the capillaries as just discussed, a failure to drain this excess fluid from the interstitium, or retention of salt and water as a result of impaired kidney function. Edema can also be caused by intracellular accumulation of fluid as a result of cellular metabolic problems or inflammation. In these cases, either the sodium ion pumps are impaired (see Section 3.10) or the cell membrane permeability to sodium is increased. In either case, the excess sodium in the cell causes the osmosis of water into the cell.

The *lymphatic system* is an accessory flow or circulatory system in the body that drains excess fluid from the interstitial spaces and returns it to the blood. The lymphatics consist of a system of lymphatic capillaries and ducts that empty into the venous system at the junctures of the left and right internal jugular and subclavian veins. This system is also responsible for the removal of large proteins and other matter that cannot be reabsorbed into the capillary from the interstitial space. Like the capillaries of the vascular system, the lymphatic capillaries are also formed by endothelial cells. However, the lymphatic endothelial cells have much larger intercellular junctions and the cells overlap in such a manner to form a valve-like structure. Interstitial fluid can force the valve open to flow into the lymphatic capillary; however, backflow out of the lymphatic fluid from the lymphatic capillary is prevented.

3.8   SOLUTE TRANSPORT ACROSS THE CAPILLARY ENDOTHELIUM

Lipid-soluble substances, such as oxygen and carbon dioxide, can diffuse directly through the endothelial cells that line the capillary wall without the use of the slit pores. Accordingly, their rate of transfer across the capillary wall is significantly higher than water-soluble, but lipid-insoluble substances, such as sodium ions, chloride ions, and glucose, for which the cell membrane of the endothelial cell is essentially impermeable. The transport of these latter substances across the capillary wall is through the use of the capillary slit pores.

In addition to the slit pores, there are two other pathways that can provide an additional route for the transport of large lipid-insoluble solutes, such as proteins, across the endothelium of the capillary wall. These pathways are called *pinocytosis* and *receptor-mediated transcytosis* (Lauffenberger and Linderman 1994).

Pinocytosis is not solute specific and, as shown in Figure 3.5, involves the ingestion by the cell of the surrounding extracellular fluid and its associated solutes. A small portion of the cell's plasma

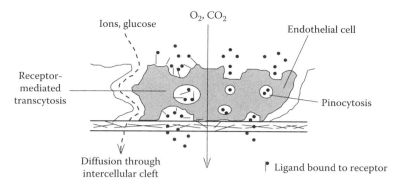

FIGURE 3.5 Mechanisms for solute transport across the capillary endothelium.

membrane forms a pocket containing the extracellular fluid. This pocket grows in size and finally pinches off to form the intracellular pinocytotic vesicles. This ingestion of extracellular material by the cell is also known as *endocytosis*. The pinocytotic vesicles are either processed and their contents used internally, or they migrate through the cell and reattach to the cell membrane on the opposite side, where they release their contents to the surrounding milieu. This release of material from the pinocytotic vesicle is known as *exocytosis*. Because of its nonspecific nature, pinocytosis is usually not a significant solute transport mechanism.

However, receptor-mediated transcytosis can provide for significant transport of specific solutes across the capillary endothelium. This process is also shown in Figure 3.5. The solute, referred to as a *ligand*, first binds with complementary receptors that are located on the surface of the cell membrane. Unlike the nonspecific process of pinocytosis, receptor-mediated transcytosis can concentrate a particular ligand by many orders of magnitude. This solute-concentrating mechanism by the cell-surface receptors is responsible for the significant transport rates that can be achieved by this process. The ligand–receptor complexes are then endocytosed, forming transcytotic vesicles. These vesicles are either processed internally or they can move through the cell and reattach themselves to the opposite side of the cell. The transcytotic vesicle then releases its contents by exocytosis.

The importance of receptor-mediated transcytosis of large lipid-insoluble solutes is provided by the following example. In vitro studies of insulin transport across vascular endothelial cells have shown that 80% of the insulin was transported by receptor-mediated processes. The remaining 20% was transported either through the slit pores between the endothelial cells or by nonspecific pinocytosis (Hachiya et al. 1988).

3.9 CELL MEMBRANE

The cell membrane, illustrated in Figure 3.6, is composed mainly of a *lipid bilayer*. Lipid molecules are insoluble in water, but they readily dissolve in organic solvents such as benzene. Three classes of lipids are found in cellular membranes: *phospholipids*, *cholesterol*, and *glycolipids*. The lipid bilayer results because the lipid molecule has a head and tail configuration, as shown in Figure 3.6. The head of the lipid molecule is polar and thus hydrophilic, whereas the tail of the lipid molecule is nonpolar and hydrophobic (typically derived from a fatty acid). Such molecules are also called *amphipathic* because the molecule has both hydrophilic and hydrophobic properties. The term *lipid bilayer* indicates their tendency to form bimolecular sheets when surrounded on all sides by an aqueous environment, as is the case for a cell membrane. Thus, the hydrophilic heads of the lipid molecules face into the aqueous environment, and the hydrophobic tails are sandwiched between the heads of the lipid molecules.

The lipid bilayer forms the basic structure of the cell membrane. However, other molecules, such as proteins, are scattered throughout the lipid bilayer of the cell membrane and serve many

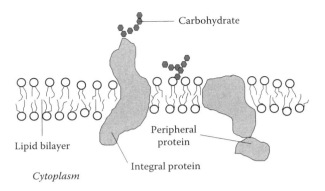

FIGURE 3.6 Cell membrane structure.

important functions. For example, special proteins allow for the transport of specific molecules across the cell membrane. Other proteins have catalytic activity and mediate chemical reactions that occur within the cell membrane. These proteins are known as *enzymes*. Still other proteins provide structural support to the cell or provide connections to surrounding cells or other extracellular materials. Some proteins found in the cell membrane act as receptors to extracellular substances or chemical signals, and through the transduction of these signals, they control intracellular events. Other proteins present foreign materials to the immune system or identify the cell as self.

Protein molecules associated with the cell membrane can be classified into two broad categories. The *transmembrane proteins* are also amphipathic and extend through the lipid bilayer. They typically have hydrophobic regions that may travel across the membrane several times and hydrophilic ends that are exposed to water on either side of the membrane. Integral membrane proteins are transmembrane proteins that are held tightly within the cell membrane through chemical linkages with other components of the cell membrane. Integral proteins have major functions related to the transport of water-soluble but lipid-insoluble substances across the cell membrane. The *peripheral membrane proteins* are not located within the plasma membrane but associate on either side of the membrane with transmembrane or integral proteins. Peripheral proteins mostly function as enzymes.

For the most part, the cell membrane is impermeable to polar or other water-soluble molecules. Hence, large neutral polar molecules, like glucose, have very low cell membrane permeabilities. Charged molecules and ions, such as H^+, Na^+, K^+, and Cl^-, also have very low permeabilities. Hydrophobic molecules, such as oxygen and nitrogen, readily dissolve in the lipid bilayer and show very high permeabilities. Smaller neutral polar molecules, such as CO_2, urea, and water, are able to permeate the lipid bilayer because of their much smaller size and neutral charge.

The transport of essential water-soluble molecules across the cell membrane is achieved through the use of special transmembrane proteins that have a high specificity for a certain type or class of molecules. These membrane transport proteins come in two basic types: *carrier proteins* and *channel proteins*. The carrier proteins bind to the solute and then undergo a change in shape or conformation (*ding* to *dong*; see Figure 3.7), which allows the solute to traverse the cell membrane. The carrier protein, therefore, changes between two shapes, alternately presenting the solute-binding site to either side of the membrane. Channel proteins actually form water-filled pores that penetrate across the cell membrane. Solutes that cross the cell membrane by either carrier or channel proteins are said to be passively transported.

Figure 3.7 illustrates the passive transport of solutes through the cell membrane by either carrier proteins or channel proteins. Carrier proteins can transport only a single solute across the membrane, a process called *uniport*, or there can be transport of two different solutes, called *coupled transport*. The passive transport of glucose into a cell by glucose transporters is an example of a

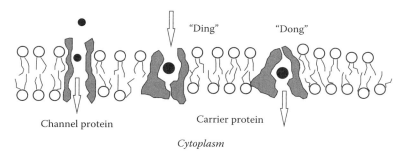

FIGURE 3.7 Membrane transport proteins. Carrier protein exists in two conformational states, "Ding" and "Dong."

uniport. In a coupled transport, the transfer of one solute occurs in combination with the transport of another solute. This coupled transport of the two solutes can occur with both solutes transported in the same direction, *symport*, or in opposite directions, *antiport*. An example of a coupled transport is the sodium ion gradient-driven symport of glucose and sodium ions.

In general, the driving force for the passive transport of these solutes is due to the combined effect of their concentration gradient and the electrical potential difference that exists across the membrane. Neutral molecules diffuse from regions of high concentration to regions of low concentration. However, if the molecule carries an electrical charge, then both the concentration gradient and the electrical potential difference, or voltage gradient, across the cell membrane will affect the transport of the molecule. The *electrochemical gradient* is the term used to describe the combined effect of charge and the solute concentration on the transport of a molecule. The voltage gradient for a cell membrane is such that the inside of the cell membrane is negative in comparison to the outside. This membrane potential (V_M) for cells at rest is about –90 mV.

The flow of charged molecules through channels in the cell membrane is responsible for the creation of the membrane potential. For example, the higher concentration of potassium ions within the cell relative to the surroundings will tend to cause a leakage of these ions out of the cell through the potassium ion leak channels. The loss of these positive ions will make the interior of the cell negative in charge. This creates an electric field that is called the membrane potential. The growth of this membrane potential with continued loss of potassium ions will reach a point where the negative charge created within the cell begins to retard the loss of the positively charged potassium ions due to the difference in potassium concentration. When these two forces balance each other, there is no net flow of ions. This balance or equilibrium of the concentration and voltage gradients for an ion is known as the *equilibrium membrane potential* for the cell. Recall from Chapter 2 that the following equation (Equation 2.226), known as the *Nernst equation*, can be used to calculate the equilibrium membrane potential for a particular ion:

$$V = \frac{RT}{zF} \ln \frac{C_{\text{outside}}}{C_{\text{inside}}} = -\frac{RT}{zF} \ln \frac{C_{\text{inside}}}{C_{\text{outside}}}. \tag{3.9}$$

In this equation, R represents the gas constant (1.987 cal gmol^{-1} K^{-1}), T is the temperature in kelvin, z is the charge on the ion, and F is Faraday's constant (2.3×10^4 cal V^{-1} gmol^{-1}). At 25°C for a univalent ion, $|RT/zF|$ is equal to 25.68 mV, whereas at 37°C the value is 26.71 mV. Considering potassium ions, the intracellular concentration from Table 3.2 is 140 mOsM and the interstitial concentration is 4 mOsM. Therefore, the equilibrium membrane potential calculated from Equation 3.9 for this ion is equal to about –95 mV. Since this value is less than zero, this indicates that there are more negative charges within the cell than outside the cell.

The actual resting membrane potential is less than this value because of two opposing effects. First, there is a very slight leakage of sodium ions into the cell, which will add a positive charge of about +9 mV. Second, the sodium–potassium pump (Section 3.10) has the effect of removing a positive charge equivalent to about −4 mV. These two effects increase the net resting membrane potential by about +5 to −90 mV.

We shall soon see that the cell must expend cellular energy in order to maintain these potassium and sodium ion gradients, thereby maintaining its resting membrane potential. Although the chloride ion concentration is very high outside the cell, these ions are repelled by the negative charges within the cell. This is shown by the fact that the equilibrium membrane potential for chloride ions is the same as the resting membrane potential, and hence there is no net driving force for the transport of chloride ions into the cell.

In nerve and muscle cells, the resting membrane potential can change very rapidly. This rapid change in the membrane potential is called an *action potential* and provides for the conduction of a nerve signal from neuron to neuron or the contraction of a muscle fiber. Any stimulus to the cell that raises the membrane potential above a threshold value will lead to the generation of a self-propagating action potential. It is also important to note that the action potential is an all-or-nothing response. The development of an action potential is dependent on the presence of voltage-gated sodium and potassium channels. The voltage-gated sodium channels only open or become active when the membrane potential is less negative than during the resting state. They typically begin to open when the membrane potential is about −65 mV. These sodium gates remain open for only a few tenths of a millisecond, after which time they close or become inactive. The voltage-gated sodium channels remain in this inactive or closed state until the membrane potential has returned to near its resting value. The voltage-gated potassium channels also open when the membrane potential becomes less negative than during the resting state; however, unlike the sodium channels, they open more slowly and become fully opened only after the sodium channels have closed. The potassium channels then remain open until the membrane potential has returned to near its resting potential. Figure 3.8 illustrates the events during the generation of an action potential.

The first stage of an action potential is a rapid *depolarization* of the cell membrane. The cell membrane becomes very permeable to sodium ions because of the opening of the voltage-gated sodium channels. This rapid influx of sodium ions, a positive charge, increases the membrane potential in the positive direction, and in some cases, can result in a positive membrane potential (overshoot) for a brief period of time. This depolarization phase may last only a few tenths of a millisecond. Following the depolarization of the membrane, the sodium channels close and the potassium channels, which are now fully opened, allow for the rapid loss of positively charged potassium ions from

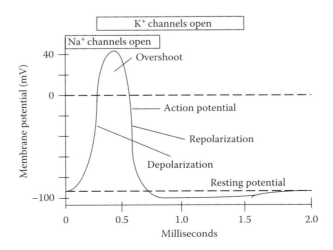

FIGURE 3.8   Action potential.

TABLE 3.3

Physical Constants Used to Describe the Electrical Properties of the Cell Membrane

Basic Electrical Properties	Units
Charge	Coulomb (C), charge carried by 6.2×10^{18} univalent ions
Electric potential	Volt (V), potential caused by separation of charges
Current	Ampere (A), flow of charge, C sec^{-1}
Capacitance	Farad (F), amount of charge needed on either side of a membrane to produce a given potential, C V^{-1}
Conductance	Siemens (S), ability of a membrane to conduct a flow of charge, A V^{-1}

Electrical Properties of Cells	Typical Values
Membrane potential	-20 to -200 mV
Membrane capacitance	~ 0.01 pF μm^{-2} of cell membrane surface area
Conductance of a single ion channel	1–150 pS
Number of specific ion channels	~ 75 μm^{-2} of cell membrane surface area

Other Relationships	
Milli (m) = 10^{-3}, micro (μ) = 10^{-6}	Nano (n) = 10^{-9}, pico (p) = 10^{-12}

Source: Data from Alberts, B., Bray, D., Lewis, J., Raff, M., Roberts, K., and Watson, J. D., *Molecular Biology of the Cell*, Garland Publishing, New York, 1989.

the cell, thus re-establishing within milliseconds the normal negative resting potential of the cell membrane (repolarization). However, for a brief period of time following an action potential, the sodium channels remain inactive and they cannot open again, regardless of external stimulation, for several milliseconds. This is known as the *refractory period*.

The net driving force for the transport of an ion due to the combined effect of the concentration gradient and the membrane potential, i.e., the electrochemical gradient, is proportional to the difference between the actual membrane potential, V_M, and the ion's equilibrium potential, V. If the quantity ($V_M - V$) is greater than zero, then the ion will be transported out of the cell through either the channel protein pores or by membrane carrier proteins. However, if this quantity is less than zero, then the ion will be transported into the cell. The proportionality constant for the transport of an ion due to its electrochemical gradient is referred to as the *membrane conductance* (g), which is the inverse of membrane resistance.

The transport of ions generates a current (i) that is given by the product, g ($V_M - V$). This current, or flow of charge can be related to the flow of the ions themselves, using the definitions summarized in Table 3.3. The following example illustrates the calculation of the flow of ions through a membrane channel.

Example 3.3

Calculate the flow of sodium ions through the voltage-gated sodium channels in a cell membrane during depolarization. Assuming the cell membrane has a surface area of 1 micron^2, how long would it take to change the membrane potential by 100 mV? Assume the equilibrium membrane potential for sodium ions is 62 mV (see Table 3.2 and Equation 3.9), that the threshold membrane potential for the sodium channels is -65 mV, and that at the peak of the action potential the membrane potential is 35 mV. In addition, the membrane conductance for the Na ions is 4×10^{-12} S channel^{-1}.

Solution

Since the membrane potential rapidly changes during depolarization from the threshold value of -65 mV to the peak value of 35 mV, assume for the calculation of the sodium ion flow that the "average" membrane potential during this phase is −15 mV. The flow or current of sodium ions may then be calculated from

$$i_{Na} = g_{Na}(V_M - V_{Na}),$$

$$i_{Na} = \frac{4 \times 10^{-12}\,S}{channel} \times \frac{1\,A\,V^{-1}}{S} \times \frac{1\,C\,sec^{-1}\,V^{-1}}{A} \times \frac{6.2 \times 10^{18}\,\oplus}{C} \times \frac{1\,V}{1000\,mV} \times \frac{1\,Na}{\oplus} \times \frac{75\,channels}{\mu m^2}$$

$$\times (-15 - 62)mV = -1.43 \times 10^8 \; Na\,sec^{-1}\,\mu m^{-2}.$$

This flow of sodium ions into the cell is also equivalent to a current of −23 pA μm^{-2}. The flow of sodium ions transfers the charge across the membrane, thereby changing the membrane potential. We can relate this change in charge and membrane potential to the membrane capacitance by the following relationship, $C_{membrane} = i_{Na}t/\Delta V_M$. Recall from Table 3.3 that the membrane capacitance is about 0.01 pF μm^{-2}. For the calculated sodium ion current and the 100 mV change in the membrane potential, we can then solve for the time required to achieve this change in membrane potential:

$$t = \frac{0.01\,\dfrac{pF}{\mu m^2} \times \dfrac{1\,C}{VF} \times \dfrac{1\,V}{1000\,mV} \times 100\,mV}{23\,\dfrac{pA}{\mu m^2} \times \dfrac{1\,C}{sec\,A} \times \dfrac{1\,sec}{1000\,msec}} = 0.043\;msec.$$

This example shows that the membrane potential is rapidly depolarized during the initial phase of the action potential.

3.10 ION PUMPS

Cells also have the ability to "pump" certain solutes against their electrochemical gradient. This process is known as *active transport* and involves the use of special carrier proteins. Since this is an "uphill" process, active transport requires the expenditure of cellular energy.

The large differences in the concentrations of sodium and potassium ions across the cell membrane will result in some leakage of potassium ions out of the cell and leakage of sodium ions into the cell. However, as we have discussed, proper functioning of the cell requires that these concentration differences be maintained in order to preserve the cell's resting membrane potential. Substances cannot diffuse against their own electrochemical gradient without the expenditure of energy. Therefore, to compensate for the loss of potassium ions by leakage through the cell membrane, the cell must have a "pump" mechanism to shuttle potassium ions from the external environment into the cell. On the other hand, sodium ions leak into the cell and, similarly, the cell needs a "pump" to remove these ions. This process of shuttling substances across the cell membrane against their electrochemical gradient and at the expense of cellular energy is called active transport.

The energy for active transport is provided by the cellular energy storage molecule called *adenosine triphosphate* (ATP). ATP is a nucleotide consisting of three components: a base called adenine, a ribose sugar, and a triphosphate group. Through the action of the enzyme *ATPase*, a molecule of ATP can be converted into a molecule of adenosine diphosphate (ADP) and a free high-energy phosphate bond that can cause conformational changes in special cell membrane carrier proteins.

The best example of active transport is the *sodium–potassium pump* present in all cells. The Na–K pump transports sodium ions out of the cell and, at the same time, transfers potassium ions into the cell. Figure 3.9 illustrates the essential features of the Na–K pump. The carrier protein protrudes through both sides of the cell membrane. Within the cell, the carrier protein has three receptor sites for binding sodium ions and also has ATPase activity. On the outside of the cell membrane, the carrier protein has two receptor sites for binding potassium ions. When these ions are

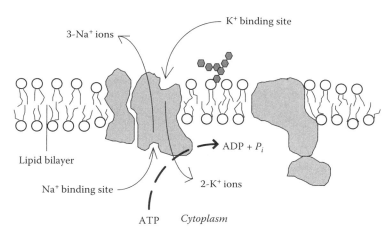

FIGURE 3.9 The sodium–potassium pump.

bound to their respective receptor sites, the ATPase becomes activated, liberating the high-energy phosphate bond (P_i) from ATP. The energy in the phosphate bond causes a shape or conformational change in the carrier protein that allows for the passage of the sodium and potassium ions. The Na–K ATPase pump is electrogenic, that is, it removes a net positive charge from the cell equivalent to about –4 mV. Similar active transport carrier proteins are also available to transfer other ions, such as Cl^-, Ca^{+2}, Mg^{+2}, and HCO_3^-. The only difference in the carrier proteins would be their preference for a specific ion.

Active transport of a solute can also be driven by ion gradients, a process referred to as *secondary active transport*. For example, the higher concentration of sodium ions outside the cell can lead to a conformational change in a carrier protein that favors the symport of another solute. The other solute is therefore pumped into the cell against its own electrochemical gradient. The sodium ion gradient that drives this pump is maintained by the Na–K ATPase pump discussed earlier. Sodium ion antiports are also used to control intracellular pH. In this case, the removal of excess hydrogen ions generated by acid-forming reactions within the cell is coupled with the influx of sodium ions.

PROBLEMS

1. Derive Equation 3.3.
2. A membrane has pure water on one side and a protein solution on the other side. The membrane has pores equivalent in size to a spherical molecule with a molecular weight of 100,000. The protein solution consists of albumin (40 g L^{-1} ; MW = 69,000), globulins (70 g L^{-1} ; MW = 150,000), and fibrinogen (60 g L^{-1} ; MW = 340,000). Assuming that the proteins form an ideal solution, what is the osmotic pressure of the protein solution? What would be the osmotic pressure of the protein solution if the membrane was completely impermeable to the proteins?
3. Explain why osmotic pressure is based on the molar concentration of the solute and not the solute mass concentration?
4. Explain the difference between a mole and an osmole. Cite some examples.
5. You are concentrating a solution containing a polypeptide of modest molecular weight by pressure filtration through a membrane. The solute concentration on the feed side of the membrane is 0.10 M and the temperature is 25°C. The applied pressure on the feed side of the membrane is 6 atm and the pressure on the opposite side of the membrane is atmospheric. If the polypeptide solute is completely rejected by the membrane, what is the "effective" pressure drop across the membrane? If the hydraulic conductance of the membrane is 3 mL hr^{-1} m^{-2} $mmHg^{-1}$, what is the filtration rate of the solvent? Assume that the membrane has a total surface area of 1 m^2. $R = 0.082$ L atm mol^{-1} K^{-1}.

6. The following clean water flow rates were reported for a particular series of ultrafiltration membranes.

Nominal Molecular Weight Cutoff (NMWCO)	Clean Water Flow Flux (mL min⁻¹ cm⁻² at 50 psi)
10,000	0.90
30,000	3.00
100,000	8.00

From these data, calculate the hydraulic conductance for each case (L_p). Assuming the membranes have similar porosity and thickness, is there a relationship between the value of L_p and the NMWCO? Hint: Base your analysis on Equation 3.7.

7. Perform a literature search and write a short paper on the biomedical applications of liposomes.

8. Look up in a biochemistry text (e.g., Stryer 1988) the chemical structures of collagen, laminin, cholesterol, and the various lipids found in the cell membrane.

9. Search the literature and write a short paper on the biomedical applications of osmotic pumps.

10. How many potassium ions must a cell lose in order to produce a membrane potential of −95 mV for potassium? How does this compare to the number of potassium ions within the cell?

11. Consider a membrane that is permeable to Ca^{2+} ions. On one side of the membrane the Ca^{2+} concentration is 100 mM and on the other side of the membrane the Ca^{2+} concentration is 1 mM. The electrical potential of the high concentration side of the membrane relative to the low concentration side of the membrane is +10 mV. How much reversible work is required to move each mole of Ca^{2+} from the low concentration side of the membrane to the high concentration side of the membrane? What is the equilibrium membrane potential?

12. The cell membrane is permeable to many different ions. Therefore, the equilibrium membrane potential for the case of multiple ions will depend not only on the concentrations of the ions within and outside the cell, but also on the permeability (P) of the cell membrane to each ion. The Goldman equation may be used to calculate the equilibrium membrane potential for the case of multiple ions. For example, considering Na^+ and K^+ ions, the Goldman equation may be written as

$$V = 26.71 \times \ln\left(\frac{P_{Na}C_{Na,0} + P_K C_{K,0}}{P_{Na}C_{Na,i} + P_K C_{K,i}}\right).$$

For a cell at rest, P_{Na} is much smaller than P_K; typically, P_{Na} is about 0.01 P_K. Calculate the equilibrium membrane potential for a cell under these conditions. During the depolarization phase of an action potential, the sodium ion permeability increases dramatically due to the opening of the voltage-gated sodium ion channels. Under these conditions, P_K is about 0.5 P_{Na}. Under these conditions, recalculate the value of the equilibrium membrane potential.

13. Composite membranes are often used in membrane filtration and in hemodialysis (Clark and Gao 2002). One of these membranes is used to provide structural support and can be tens or even hundreds of microns thick. This structural membrane is usually microporous to minimize its flow resistance and will therefore have pore diameters on the order of a micron. Attached to this structural membrane is a much thinner permselective membrane skin on the order of a micron in thickness that will have pores whose diameters are defined in terms of their nominal molecular weight cutoff (NMWCO; see Equation 5.5, which can be used to relate the molecular radius of a solute to its molecular weight). Hence, solutes whose molecular weights are less than the NMWCO of the selective membrane will travel across the membrane through the pores, whereas those solutes whose molecular weights

are larger than the NMWCO will not cross the permselective membrane. Since the filtration flow rate (see Equation 3.4, where L_p and $\overline{\Delta P}$ are, respectively, the hydraulic conductance and the effective pressure drop across the ith membrane) through each membrane layer has to be the same, show that the overall hydraulic conductance for three membranes that are stacked together is given by the equation shown below. Note that in this equation, L_{P_i} is the hydraulic conductance for the ith membrane as calculated by Equation 3.7 or as found by independent filtration flow measurements for that particular membrane:

$$L_{P_{composite}} = \frac{1}{\displaystyle\sum_{i=1}^{3} \frac{1}{L_{P_i}}}.$$

14. Consider a composite membrane that is being used to filter plasma from blood. Plasma has a viscosity of 1.2 cP. The composite membrane consists of a microporous spongy-like material that provides structural support. This membrane is 25 μm thick and has pores that are 2 μm in diameter. These pores are also tortuous and have a tortuosity (i.e., τ) of 1.67. The porosity of this membrane (i.e., A_p/S) is also equal to 0.60. Attached to this microporous membrane is a thin permselective skin, 3.23 μm thick, that has an NMWCO of 1000. The pores in this membrane skin are therefore about 0.0015 μm in diameter. The tortuosity of the pores in the membrane skin is also equal to 1.67, and the porosity of the membrane skin is equal to 0.60. Use the formula found in Problem 13 to predict the overall hydraulic conductance for this composite membrane. Find the total filtration flow rate (milliliters per hour) across this composite membrane assuming that the total surface area of the membrane is 1 m² and that the overall effective pressure drop across the composite membrane is 160 mmHg.

15. Consider a solution of glucose on one side of a membrane that is impermeable to the transport of glucose. The temperature of the glucose solution is 20°C. What pressure drop must be applied across the membrane to stop the flow of water into the glucose solution? Assume that the glucose concentration is 1 mg mL^{-1} and that the molecular weight of glucose is 180 gmol^{-1}.

16. A hollow fiber membrane cartridge is being evaluated for use in an aquapheresis system. In one experiment using blood, a cartridge with a surface area of 1.5 m² had a filtration flow across the hollow fiber membranes of 1000 mL h^{-1}. The average pressure of the blood flowing inside the tubes of the hollow fiber membranes was 120 mmHg and the suction pressure on the filtrate side of the hollow fibers averaged −150 mmHg. Assuming that the plasma proteins were totally retained on the blood side of the hollow fiber membrane, estimate the hydraulic conductance of these membranes in milliliters per hour per square meter per mmHg.

4 The Physical and Flow Properties of Blood and Other Fluids

4.1 PHYSICAL PROPERTIES OF BLOOD

Blood is a viscous fluid mixture consisting of plasma and cells. Table 4.1 summarizes the most important physical properties of blood. Recall that the chemical composition of plasma was previously shown in Table 3.2. Proteins represent about 7–8 wt% of plasma. The major proteins found in plasma are albumin (MW = 69,000; 4.5 g 100 mL^{-1}), globulins (MW = 35,000–1,000,000; 2.5 g 100 mL^{-1}), and fibrinogen (MW = 400,000; 0.3 g 100 mL^{-1}). *Albumin* has a major role in regulating the pH and the colloid osmotic pressure. The so-called *alpha* and *beta globulins* are involved in solute transport, whereas the *gamma globulins* are the antibodies that fight infection and form the basis of the humoral component of the immune system. *Fibrinogen*, through its conversion to long strands of *fibrin*, has a major role in the process of blood clotting. *Serum* is simply the fluid remaining after blood is allowed to clot. For the most part, the composition of serum is the same as that of plasma, with the exception that the clotting proteins, primarily fibrinogen, and the cells have been removed.

4.2 CELLULAR COMPONENTS

The cellular component of blood consists of three main cell types. The most abundant cells are the red blood cells (RBCs) or *erythrocytes* comprising about 95% of the cellular component of blood. Their major role is the transport of oxygen by the hemoglobin contained within the RBC. Note from Table 4.1 that the density of an RBC is higher than that of plasma. Therefore, in a quiescent fluid, the RBCs will tend to settle. The RBC volume fraction is called the *hematocrit* and typically varies between 40% and 50%. The true hematocrit (H) is about 96% of the measured hematocrit (Hct).

The RBC has a unique shape, described as a *biconcave discoid*. Figure 4.1 illustrates the size of the RBC and Table 4.2 summarizes its typical dimensions. RBCs can form stacked coin-like structures called *rouleaux*. Rouleaux can also clump together to form larger RBC structures called *aggregates*. Both rouleaux and aggregates break apart under conditions of increased blood flow or higher shear rates.

Platelets are the next most abundant cell type, comprising about 4.9% of the cell volume. Platelets are major players in blood coagulation and *hemostasis*, which is the prevention of blood loss. The remaining 0.1% of the cellular component of blood consists of the white blood cells (WBCs) or *leukocytes*, which form the basis of the cellular component of the immune system. Since the WBCs and platelets only comprise about 5% of the cellular component of blood, their effect on the macroscopic flow characteristics of blood is negligible.

4.3 RHEOLOGY

The field of rheology concerns the deformation and flow behavior of fluids. The prefix *rheo-* is from Greek and refers to something that flows. Due to the particulate nature of blood, we expect the rheological behavior of blood to be somewhat more complex than a simple fluid such as water.

TABLE 4.1
Physical Properties of Adult Human Blood

Property	Value
Whole Blood	
pH	7.35–7.40
Viscosity (37°C)	3.0 cP (at high shear rates)
Specific gravity (25/4°C)	1.056
Venous hematocrit—male	0.47
Venous hematocrit—female	0.42
Whole blood volume	~78 mL kg^{-1} body weight
Plasma or Serum	
Colloid osmotic pressure	~330 mm H$_2$O
pH	7.3–7.5
Viscosity (37°C)	1.2 cP
Specific gravity (25/4°C)	1.0239
Formed Elements	
Erythrocytes (RBCs)	
Specific gravity (25/4°C)	1.098
Count—male	5.4×10^9 mL^{-1} whole blood
Count—female	4.8×10^9 mL^{-1} whole blood
Average life span	120 days
Production rate	4.5×10^7 mL^{-1} whole blood/day
Hemoglobin concentration	0.335 g mL^{-1} of erythrocyte
Leukocytes	
Count	~7.4×10^6 mL^{-1} whole blood
Diameter	7–20 μm
Platelets	
Count	~2.8×10^8 mL^{-1} whole blood
Diameter	~2–5 μm

Source: Data from Cooney, D.O., *Biomedical Engineering Principles*, Marcel Dekker, New York, 1976.

To understand the flow behavior of blood, we must first define the relationship between *shear stress* (τ) and *shear rate* ($\dot{\gamma}$). To develop this relationship, consider the situation shown in Figure 4.2. A fluid is contained between two large parallel plates, both of area A. The plates are separated by a small distance equal to h. Initially, the system is at rest. At time $t = 0$, the lower plate is set in motion in the x direction at a constant velocity V. As time proceeds, momentum is transferred in the y direction to successive layers of fluid from the lower plate that is in motion in the x direction. Momentum, therefore, "flows or diffuses" from a region of high velocity to a region of low velocity. After a sufficient length of time, a steady state velocity profile is obtained that is a linear function of y, i.e., $v_x(y) = (V/h)(h-y)$. Recall that velocity is a vector and v_x represents the component of the fluid velocity in the x direction. For the situation shown in Figure 4.2, the other components of the velocity vector, i.e., v_y and v_z, are equal to zero; v_y is zero because at steady state there is no flow in the y direction and v_z is zero because the plates are very large in the z direction in comparison to the distance h that separates them.

At steady state, a constant force (F) must be applied to overcome the resistance of the fluid and maintain the motion of the lower plate. For the situation shown in Figure 4.2, the shear stress on

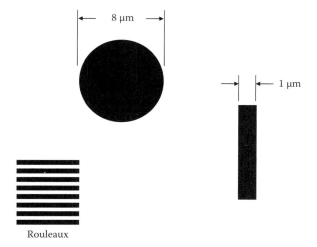

Rouleaux

FIGURE 4.1 The dimensions of the red blood cell.

TABLE 4.2
Dimensions of the Normal Red Blood Cell

Property	Value
Diameter	8.1 ± 0.43 μm
Greatest thickness	2.7 ± 0.15 μm
Least thickness	1.0 ± 0.3 μm
Surface area	138 ± 17 μm²
Volume	95 ± 17 μm³

Source: Data from Burton, A.C., *Physiology and Biophysics of the Circulation*, Year Book Medical Publishers, Chicago, 1972.

the lower plate is defined as *F/A* and is given the symbol τ_{yx}, where *yx* denotes the viscous flux* of *x* momentum in the *y* direction (Bird et al. 2002). Note that the momentum flux (momentum/area/time) of the fluid adjacent to the plate that is in motion is given by $\rho\,V\,V$, where ρ is the fluid density and the area in this case is perpendicular to the *y* direction. The shear rate at any position *y* in the fluid is defined as $-(dv_x(y)/dy) = V/h = \dot{\gamma}$. The shear rate is given the symbol $\dot{\gamma}$. Notice that the shear rate has units of reciprocal time.

Newton's law of viscosity states that for *laminar flow* (nonturbulent) the shear stress is proportional to the shear rate. The proportionality constant is called the *viscosity*, μ, which is a property of the fluid and is a measure of the flow resistance of the fluid. Viscosity is usually expressed in the following units where 1 P = 100 cP = 1 g cm⁻¹ sec⁻¹ = 1 dyn sec cm⁻² = 0.1 newton sec m⁻² = 0.1 Pa sec. Also 1 cP = 0.001 Pa sec. For the situation shown in Figure 4.2, this may be stated as

$$\frac{F}{A} = \tau_{yx} = \mu\frac{V}{h} = \mu\dot{\gamma}. \tag{4.1}$$

---

* A flux is a flow that has been normalized with respect to the area that is perpendicular to the direction of the flow.

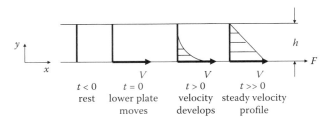

FIGURE 4.2 Velocity profile development for flow of a fluid between two parallel plates.

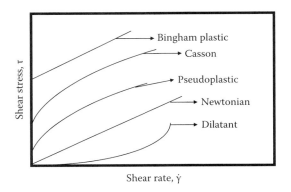

FIGURE 4.3 Types of shear stress–shear rate relationships.

Most simple homogeneous liquids and gases obey this law and are called *Newtonian fluids*.

For more complicated geometries, the steady state velocity profile is not linear. However, Newton's law of viscosity may be stated at any point in the laminar flow field as

$$\tau_{yx} = -\mu \frac{dv_x}{dy} = \mu \dot{\gamma}, \tag{4.2}$$

where v_x is the velocity in the x direction at position y. This equation tells us that the momentum flows in the direction of decreasing velocity. The velocity gradient is therefore the driving force for momentum transport, much like the temperature gradient is the driving force for heat transfer, and the concentration gradient is the driving force for mass transport.

A fluid whose *shear stress–shear rate relationship* does not follow Equation 4.2 is known as a *non-Newtonian fluid*. Figure 4.3 illustrates the types of shear stress–shear rate relationships that are typically observed for non-Newtonian fluids. The Newtonian fluid is shown for comparison. Note from Equation 4.2 that the shear stress–shear rate relationship for a Newtonian fluid is linear with the slope equal to the viscosity. For non-Newtonian fluids, the slope of the line at a given value of the shear rate is the *apparent viscosity*. Hence, from Equation 4.2 for given values of τ_{yx} and $\dot{\gamma}$ the apparent viscosity ($\mu_{apparent}$) is given by $\tau_{yx}/\dot{\gamma}$. A *dilatant* fluid thickens or has an increase in apparent viscosity as the shear rate increases. A *pseudoplastic* fluid, on the other hand, tends to thin out, or its apparent viscosity decreases, with an increase in shear rate. An example of a dilatant fluid is a solution of cornstarch and water, which allows one to "walk on water." As a person quickly walks or runs across this fluid, each step induces a high shear rate in the fluid surrounding the foot, increasing the fluid viscosity, and the resulting shear stress that is generated supports the person and he/she does not sink into the fluid. Paint is an example of a pseudoplastic fluid since it thins out as it is brushed or quickly sheared as it is applied to a surface.

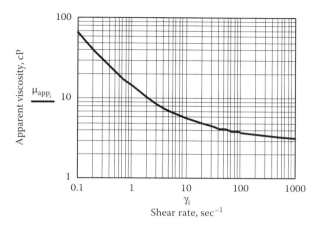

FIGURE 4.4 Apparent viscosity of blood at 37°C.

Heterogeneous fluids that contain a particulate phase that forms aggregates at low rates of shear exhibit a yield stress, τ_y. Two types of fluids that exhibit this behavior are the *Bingham plastic* and the *Casson fluid*. The yield stress must be exceeded in order to get the material to flow. The Casson fluid has been used to describe fluids that contain particulates that also aggregate, forming large complex structures. An example of this type of fluid is the ink used in ballpoint pens. When the pen is not being used, the ink thickens and cannot flow out of the pen. As one writes, the rotating ball at the point of the pen shears the fluid, thinning it out so that it can be applied to the paper. In the case of the Bingham plastic, once the yield stress is exceeded, the fluid behaves as if it were Newtonian. For the Casson fluid, we see that as the shear rate increases, the apparent viscosity decreases, indicating that the particulate aggregates are getting smaller and smaller and, at some point, the fluid behaves as a Newtonian fluid. Blood is a heterogeneous fluid, with the particulates consisting primarily of RBCs. As mentioned earlier and shown in Figure 4.1, the RBCs form rouleaux and aggregates at low shear rates. We will see in the ensuing discussion that blood follows the curve shown for the Casson fluid.

The apparent viscosity of blood as a function of shear rate is shown in Figure 4.4 at a temperature of 37°C. At low shear rates, the apparent viscosity of blood is quite high due to the presence of rouleaux and aggregates. However, at shear rates above about 100 sec^{-1}, only individual cells exist, and blood behaves as if it were a Newtonian fluid. We then approach the asymptotic high shear rate limit for the apparent viscosity of blood, which is about 3 cP.

4.4 RELATIONSHIP BETWEEN SHEAR STRESS AND SHEAR RATE

The simplest approach to examining the shear stress–shear rate behavior of blood or other fluids is through the use of the capillary viscometer shown in Figure 4.5. The diameter of the capillary tube is typically on the order of 500 μm. To eliminate entrance effects, the ratio of the capillary length to its radius should be greater than 100 (Rosen 1993). For a given flow rate of blood, Q, the pressure drop across the viscometer, ΔP, is measured. From this information, it is possible to deduce an analytical expression for the shear stress–shear rate relationship, i.e., $\dot{\gamma} = \dot{\gamma}(\tau_{rz})$.

Analysis of the capillary flow illustrated in Figure 4.5 requires the use of cylindrical coordinates (r, θ, and z) and the following simplifying assumptions: the length of the tube (L) is much greater than the tube radius (R) (i.e., $L/R \gg 100$) to eliminate entrance effects; steady incompressible (i.e., constant density) and isothermal flow (i.e., constant temperature); no external forces acting on the fluid; no holes in the tube so that there is no radial velocity component, v_r; axisymmetric flow or no

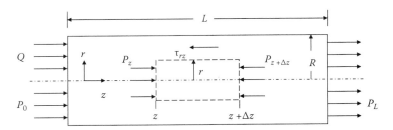

FIGURE 4.5 Forces acting on a cylindrical fluid element within a capillary viscometer.

swirls so that the tangential velocity, v_θ, is also zero; and the no-slip condition at the wall $(r = R)$ requiring $v_z = 0$. Continuity or conservation of mass for an incompressible fluid with these assumptions therefore requires that only an axial velocity component exists and it will be a function of the tube radius only, therefore $v_z = v_z(r)$.

Now consider in Figure 4.5 a cylindrical volume of fluid of radius r and length Δz. For steady flow, the viscous force acting to retard the fluid motion, i.e., $\tau_{rz} 2\pi r \Delta z$, must be balanced by the force developed by the pressure drop acting on the volume of fluid of length Δz, i.e., $-\pi r^2(P|_{z+\Delta z} - P|_z)$. Equating these forces, dividing by Δz, and taking the limit as $\Delta z \to 0$, the following equation is obtained for the shear stress distribution for the fluid flowing within the capillary tube:

$$\tau_{rz}(r) = -\frac{r}{2}\frac{dP}{dz} = \frac{(P_0 - P_L)r}{2L}. \tag{4.3}$$

We note that the shear stress vanishes at the centerline of the capillary and achieves its maximum value, τ_w, at the wall where $r = R$. The *wall shear stress* is easily calculated by the following equation in terms of the measured pressure drop $(P_0 - P_L)$ and the tube dimensions $(R$ and $L)$:

$$\tau_w = \tau_{rz}(R) = -\frac{R}{2}\frac{dP}{dz} = \frac{(P_0 - P_L)R}{2L}. \tag{4.4}$$

It should also be pointed out that Equations 4.3 and 4.4 hold whether the fluid is Newtonian or non-Newtonian. Nothing has been said so far about the relationship between a particular fluid's shear stress and shear rate. Using Equation 4.4, one may rewrite Equation 4.3 to give the shear stress in terms of the wall shear stress and the fractional distance from the centerline of the capillary tube, i.e., r/R:

$$\tau_{rz}(r) = \tau_w \frac{r}{R}. \tag{4.5}$$

What we want to find through the use of data obtained from the capillary viscometer is a general relationship between the shear rate and some function of the shear stress in terms of the measurable quantities Q, $P_0 - P_L$, L, and R:

$$\dot{\gamma} = -\frac{dv_z}{dr} = \dot{\gamma}(\tau_{rz}). \tag{4.6}$$

For the special case of a Newtonian fluid, we have

$$\dot{\gamma}(\tau_{rz}) = \frac{\tau_{rz}}{\mu}. \tag{4.7}$$

A general relationship between $\dot{\gamma}$ and τ_{rz}, and the measurable quantities of the capillary viscometer, can be obtained by the following development. First, we write the total flow rate, Q, in terms of the axial velocity profile as

$$Q = 2\pi \int_0^R v_z(r) r\, dr. \tag{4.8}$$

Next, we integrate Equation 4.8 by parts to obtain the following equation

$$Q = -\pi \int_0^R r^2 \left(\frac{dv_z(r)}{dr} \right) dr. \tag{4.9}$$

Using Equations 4.5 and 4.6, one can show that the following equation is obtained. This is called the *Rabinowitsch equation*:

$$Q = \frac{\pi R^3}{\tau_w^3} \int_0^{\tau_w} \dot{\gamma}(\tau_{rz}) \tau_{rz}^2\, d\tau_{rz}. \tag{4.10}$$

For data obtained from a given capillary viscometer, the experiments will provide Q as a function of the wall shear stress, τ_w, which is related, by Equation 4.4, to the observed pressure drop, i.e., $P_0 - P_L$, and the capillary dimensions R and L. The appropriate shear rate–shear stress relationship, i.e., Equation 4.6, is the one that best fits the data according to Equation 4.10. This will be illustrated for blood in the following discussion.

4.5 HAGEN–POISEUILLE EQUATION

Let us first consider the case of a Newtonian fluid. Equation 4.7 provides the relationship for $\dot{\gamma}(\tau_{rz})$. Substituting this equation into Equation 4.10, one readily obtains the following result for the volumetric flow rate:

$$Q = \frac{\pi R^4 (P_0 - P_L)}{8\mu L}. \tag{4.11}$$

This is the *Hagen–Poiseuille law* for laminar flow of a Newtonian fluid in a cylindrical tube. It provides a simple relationship between the volumetric flow rate in the tube given by Q, and the pressure drop between the entrance and the exit of the tube, i.e., $P_0 - P_L$, for the given dimensions of the tube R and L, and the fluid viscosity μ. Using the analogy that flow is proportional to a driving force divided by a resistance, we see that the driving force is the pressure drop, $P_0 - P_L$, and the resistance is given by $8\mu L / \pi R^4$. This resistance term is very important since it shows that for laminar flow within blood vessels, small changes in the radius of the vessel can have a significant effect on the blood flow rate for a given pressure drop. To achieve this control of the blood flow rate, the arterioles, the smallest elements of the arterial system with diameters less than 100 μm, consist of an endothelial cell lining that is surrounded by a layer of vascular smooth muscle cells. Contraction of the smooth muscle layer provides a reactive method for controlling arteriole diameter and hence the blood flow rate within organs and tissues.

Using Equations 4.3, 4.6, and 4.7, we can also write for a Newtonian fluid in cylindrical tube flow:

$$\tau_{rz} = -\mu \frac{dv_z}{dr} = \frac{(P_0 - P_L) r}{2L}. \tag{4.12}$$

This equation can be easily integrated with the boundary conditions that at $r = R$, $v_z = 0$, and that at $r = 0$, $dv_z/dr = 0$. The result of this integration provides an equation that describes the velocity profile for laminar tube flow of a Newtonian fluid. The following equation predicts that the velocity profile will have a parabolic shape:

$$v_z(r) = \frac{(P_0 - P_L)R^2}{4\mu L}\left[1 - \left(\frac{r}{R}\right)^2\right].$$ (4.13)

Note that the maximum velocity ($V_{maximum}$) occurs at the centerline of the cylindrical tube, where $r = 0$; hence the maximum velocity is given by

$$V_{maximum} = \frac{R^2(P_0 - P_L)}{4\mu L}.$$ (4.14)

The average velocity ($V_{average}$) of the fluid flowing within a cylindrical tube is defined as the ratio of the volumetric flow rate (Q) to the cross-sectional area (i.e., $A = \pi R^2$) of the tube normal to the flow direction, i.e., Q/A. Using Equation 4.11 for Q, the average velocity is then given by

$$V_{average} = \frac{R^2(P_0 - P_L)}{8\mu L} = \frac{1}{2}V_{maximum}.$$ (4.15)

4.6   OTHER USEFUL FLOW RELATIONSHIPS

Some other useful relationships for Newtonian flow may be obtained by combining Equations 4.4 and 4.11. First, we obtain the following expression that relates the wall shear stress to the volumetric flow rate:

$$\tau_w = \left(\frac{4\mu}{\pi}\right)\frac{Q}{R^3}.$$ (4.16)

Also, by writing Equation 4.7 at the tube wall and using the above result, it is found that the shear rate at the wall is given by

$$\dot{\gamma}_w = \frac{4Q}{\pi R^3} = \frac{4V_{average}}{R}.$$ (4.17)

The reduced average velocity $\bar{U}$ is related to the wall shear rate and is defined as the ratio of the average velocity and the tube diameter:

$$\bar{U} = \frac{V_{average}}{D} = \frac{4Q}{\pi D^3}.$$ (4.18)

Example 4.1

A large artery can have a diameter of about 0.5 cm, whereas a large vein may have a diameter of about 0.8 cm. The average blood velocity in these arteries and veins is, respectively, on the order of 40 and 20 cm sec^{-1}. Calculate for these blood flows the wall shear rate, $\dot{\gamma}_w$. What would the shear rate be at the centerline of these blood vessels?

Solution

We may use Equation 4.17 to calculate the wall shear rates.

For the large artery:

$$\dot{\gamma}_w = \frac{4 \times 40 \frac{cm}{sec}}{0.25 \text{ cm}} = 640 \frac{1}{sec},$$

and for the large vein:

$$\dot{\gamma}_w = \frac{4 \times 20 \frac{cm}{sec}}{0.40 \text{ cm}} = 200 \frac{1}{sec}.$$

From Equation 4.12, the shear rate in both cases is zero at the centerline, $r = 0$.

Example 4.2

Nutrient media is flowing at a rate of 1.5 L min^{-1} in a tube that is 5 mm in diameter. The walls of the tube are covered with antibody-producing cells, and these cells are anchored to the tube wall by their interaction with a special coating material that was applied to the surface of the tube. If one of these cells has a surface area of 500 μm^2, what is the amount of force that each cell must resist as a result of the flow of this fluid in the tube? Assume the viscosity of the nutrient media is 1.2 cP = 0.0012 Pa sec.

Solution

We can use Equation 4.16 to find the shear stress acting on the cells that cover the surface of the tube:

$$\tau_w = \left(\frac{4 \times 0.0012 \text{ Pa sec}}{\pi} \right) \frac{1500 \text{ cm}^3}{\text{min}(0.25 \text{ cm})^3} \times \frac{1 \text{ min}}{60 \text{ sec}} = 2.44 \text{ Pa} = 2.44 \text{ Nm}^{-2}.$$

Multiplying τ_w by the surface area of a cell gives the force acting on that cell as a result of the fluid flow. Therefore

$$F_{cell} = 500 \text{ μm}^2 \times \frac{(10^{-6} \text{ m})^2}{\text{μm}^2} \times 2.44 \frac{N}{m^2} \times \frac{10^{12} \text{ pN}}{N} = 1220 \text{ pN}.$$

4.7 RHEOLOGY OF BLOOD

As shown in Figure 4.4, blood is a non-Newtonian fluid except at shear rates above about 100 sec^{-1}, where the viscosity is seen to approach its asymptotic limit of about 3 cP. Therefore, as shown in Example 4.1, blood flow in arteries and veins as well as smaller vessels is Newtonian near the vessel wall, since the wall shear rate is significantly higher than 100 sec^{-1}. However, as one gets closer to the centerline of the vessel, the shear rate approaches zero, and blood exhibits non-Newtonian behavior. Therefore, in general, blood cannot be expected to follow the relationships summarized in Table 4.3 for Newtonian fluids.

Figure 4.6 provides the shear stress–shear rate relationship for blood under a variety of conditions, as obtained by Replogle et al. (1967) using a capillary viscometer at ambient temperatures. This log–log plot clearly shows that below a shear stress of about 100 sec^{-1}, the flow of blood (open circle) becomes non-Newtonian. Note in this figure the rapid rise in the viscosity of blood as the

TABLE 4.3
Summary of Key Flow Equations

Relationship	Non-Newtonian	Newtonian
Shear stress $\tau_{rz} =$	$\dfrac{(P_0 - P_L)r}{2L} = \tau_w \dfrac{r}{R}$	$\dfrac{(P_0 - P_L)r}{2L} = \tau_w \dfrac{r}{R}$
Wall shear stress $\tau_w =$	$\dfrac{(P_0 - P_L)R}{2L}$	$\dfrac{(P_0 - P_L)R}{2L}$
Shear rate $\dot{\gamma} =$	$-\dfrac{dv_z}{dr} = \dot{\gamma}(\tau_{rz})$	$-\dfrac{dv_z}{dr} = \dfrac{\tau_{rz}}{\mu}$
Volumetric flow rate $Q =$	$\dfrac{\pi R^3}{\tau_w^3} \displaystyle\int_0^{\tau_w} \dot{\gamma}(\tau_{rz})\tau_{rz}^2 \, d\tau_{rz}$	$\dfrac{\pi R^4 \Delta P}{8\mu L}$
Wall shear stress $\tau_w =$	$-$	$\dfrac{4\mu Q}{\pi R^3}$
Wall shear rate $\lvert \dot{\gamma}_{\cdot w} \rvert =$	$-$	$\dfrac{4Q}{\pi R^3}$
Reduced average velocity $\bar{U} =$	$\dfrac{4Q}{\pi D^3}$	$\dfrac{4Q}{\pi D^3}$

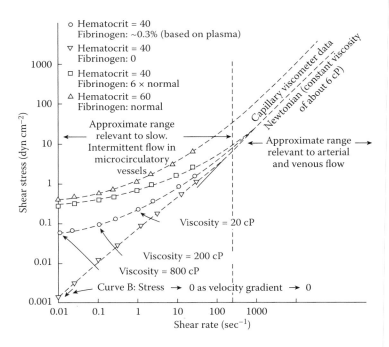

FIGURE 4.6 Shear stress versus shear rate for blood. (From Replogle, R.L., Meiselman, H.J., and Merrill, E.W., *Circulation*, 36, 148–60, 1967. With permission.)

shear rate decreases. Above a shear rate of 100 sec^{-1}, the fluid is Newtonian with a constant viscosity of 6 cP for these ambient conditions; however, at 0.01 sec^{-1} the viscosity is 800 cP. If the clotting protein fibrinogen (typically about 0.3 wt% in plasma) is removed, keeping the hematocrit constant, the resulting RBC suspension (upside down triangle) behaves nearly like a Newtonian fluid for shear rates as low as 0.01 sec^{-1}. Fibrinogen and its effect of making the RBCs stick together is therefore primarily responsible for the non-Newtonian flow behavior of blood at low shear rates. Thus, the other plasma proteins, such as albumin and the globulins, do not contribute to the non-Newtonian behavior of blood, although their concentration will affect the viscosity of the plasma. It should also be pointed out that serum and plasma alone are Newtonian fluids.

Increasing the hematocrit has the effect of increasing the flow resistance of blood (compare open circle $H = 40\%$ with open triangle $H = 60\%$). The same effect is also seen when the fibrinogen concentration is increased (compare open circle with open square). Therefore, the rheology of blood can be significantly affected by the hematocrit and the plasma protein concentrations.

4.8 CASSON EQUATION

The shear stress–shear rate relationship for blood may be described by the following empirical equation, known as the *Casson equation*:

$$\tau^{1/2} = \tau_y^{1/2} + s\dot{\gamma}^{1/2}. \tag{4.19}$$

In this equation, τ_y is the yield stress and s is a constant, both of which must be determined from viscometer data. The yield stress represents the fact that a minimum force must be applied to stagnant blood before it will flow. This was illustrated earlier in Figure 4.3. The yield stress for normal blood at 37°C is about 0.04 dyn cm^{-2}. It is important to point out, however, that the effect of the yield stress on the flow of blood is small, as the following example will show.

Example 4.3

Estimate the pressure drop in a small blood vessel that is needed to just overcome the yield stress.

Solution

One may use Equation 4.4 to solve for the pressure drop needed to overcome the yield stress ($\tau_y = \tau_w$):

$$(P_0 - P_L)_{min} = \frac{2L\tau_y}{R}.$$

Using the yield stress of 0.04 dyn cm^{-2} and an L/R of 200 for a blood vessel, the pressure drop required to just initiate flow is

$$(P_0 - P_L)_{min} = 2 \times 200 \times 0.04 \frac{dyn}{cm^2} \times \frac{1\ bar}{10^6\ \frac{dyn}{cm^2}} \times \frac{750.061\ mmHg}{1\ bar} = 0.012\ mmHg.$$

This result is considerably less than the mean blood pressure, which is on the order of 100 mmHg.

At large values of the shear rate, the Casson equation should approach the asymptotic viscosity of the Newtonian fluid, as shown in Figure 4.4. Equation 4.19 at large shear rates thus provides the interpretation that the parameter s is the square root of this asymptotic Newtonian viscosity. Table 4.1 provides the asymptotic viscosity of blood to be 3 cP. Therefore, the parameter s is $\sqrt{3\,cP} = 1.732$ $(cP)^{1/2}$ or 0.173 (dyn sec cm^{-2})$^{1/2}$. It should be stressed that the values of the yield stress and the parameter s will also depend on the plasma protein concentrations, hematocrit, and temperature. Therefore, one must be careful to properly calibrate the Casson equation to the actual blood and the conditions used in the viscometer or the situation under study.

4.9 USING THE CASSON EQUATION

We are now ready to use the Casson equation to represent the shear stress–shear rate relationship for blood. Consider the data shown in Figure 4.7 as obtained by Merrill et al. (1965) for the flow of blood in tubes of various diameters. The figure provides data for the pressure drop as a function of the flow rate for five tube diameters ranging from 288 to 850 μm. The hematocrit is 39.3 and the temperature is about 20°C.

If the flow were Newtonian, we could use the Hagen–Pouiselle law, i.e., Equation 4.11, to express the relationship between the pressure drop and the volumetric flow rate. In a log–log system of coordinates, the Hagen–Pouiselle equation becomes

$$\log \Delta P = \log Q + \log\left(\frac{8\mu L}{\pi R^4}\right). \tag{4.20}$$

The data shown in the log–log plot of ΔP vs. Q in Figure 4.7 should therefore have a slope of unity. However, at the lower flow rates, we see significant deviation from this requirement as we would expect since for a given tube radius, Equation 4.17 predicts a decrease in the wall shear rate with decreasing volumetric flow. Thus, there is a transition from Newtonian to non-Newtonian flow at the lower flow rates and the slope is no longer unity.

To facilitate analysis of the data shown in Figure 4.7, it is convenient to express the pressure drop in terms of the wall shear stress according to Equation 4.4 and to express the volumetric flow rate in terms of the reduced average velocity using Equation 4.18.

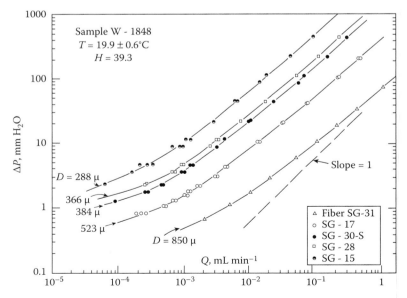

FIGURE 4.7 Pressure-flow curves for blood. (From Merrill, E.W., Benis, A.M., Gilliland, E.R., Sherwood, T.K., and Salzman, E.W., *J. Appl. Physiol.*, 20, 954–67, 1965. With permission.)

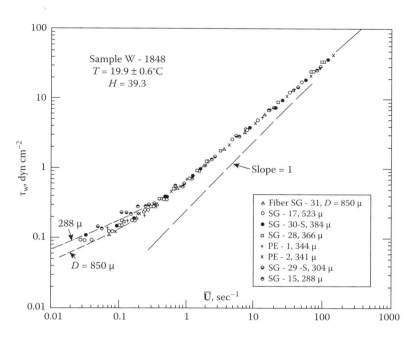

FIGURE 4.8 Wall shear stress versus the reduced average velocity. (From Merrill, E.W., Benis, A.M., Gilliland, E.R., Sherwood, T.K., and Salzman, E.W., *J. Appl. Physiol.*, 20, 954–67, 1965. With permission.)

Figure 4.8 provides a plot of the data shown in Figure 4.7 as a plot of τ_w vs. $\bar{U}$. We observe that all the data shown in Figure 4.7 now fall nearly along a single line. This result is expected when the Rabinowitsch equation (4.10) is recast as $\bar{U}$ vs. τ_w:

$$\bar{U} = \frac{4Q}{\pi D^3} = \frac{1}{2\tau_w^3} \int_0^{\tau_w} \dot{\gamma}(\tau_{rz})\tau_{rz}^2 \, d\tau_{rz}. \tag{4.21}$$

This equation predicts that the reduced average velocity should only be a function of the wall shear stress and this result is generally supported by the data presented in Figure 4.8. This result also applies whether or not the fluid is Newtonian or non-Newtonian.

The shear rate or $\dot{\gamma}(\tau_{rz})$ may be written as follows for the Casson equation (4.19):

$$\dot{\gamma}(\tau_{rz}) = \frac{\left[\tau_{rz}^{1/2} - \tau_y^{1/2}\right]^2}{s^2}. \tag{4.22}$$

This equation may be substituted into Equation 4.21 and the resulting equation integrated to obtain the following fundamental equation that describes the tube flow of the Casson fluid. This equation depends on only two parameters, τ_y and s, which may be determined from experimental data:

$$\bar{U} = \frac{1}{2s^2}\left[\frac{\tau_w}{4} - \frac{4}{7}\sqrt{\tau_y}\sqrt{\tau_w} - \frac{1}{84}\frac{\tau_y^4}{\tau_w^3} + \frac{\tau_y}{3}\right]. \tag{4.23}$$

If the above equation is written in the form of Equation 4.11, this equation for the Casson fluid becomes

$$Q = \frac{\pi R^4 (P_0 - P_L)}{8\mu_\infty L}\left[1 - \frac{16}{7}\sqrt{\frac{\tau_y}{\tau_w}} + \frac{4}{3}\frac{\tau_y}{\tau_w} - \frac{1}{21}\left(\frac{\tau_y}{\tau_w}\right)^4\right], \tag{4.24}$$

where μ_∞ is equal to s^2 and is the viscosity of the fluid at high shear rates; τ_w is given by Equation 4.4. Equation 4.24 is the Casson fluid equivalent to the Hagen–Poiseuille equation.

Merrill et al. (1965) provide the following values for these parameters at about 20°C: $\tau_y = 0.0289$ dyn cm^{-2} and $s = 0.229$ (dyn sec cm^{-2})$^{1/2}$. It is left as an exercise (see Problem 4 at the end of this chapter) to show that the Casson equation with these values for the parameters provides an excellent fit to the data shown in Figure 4.8.

4.10 VELOCITY PROFILE FOR TUBE FLOW OF A CASSON FLUID

We can also use the Casson equation to obtain an expression for the axial velocity profile for the flow of blood in a cylindrical tube or vessel. The maximum shear stress, τ_w, is at the wall. If the yield stress, τ_y, is greater than τ_w, then there will be no flow of the fluid. On the other hand, if τ_w is greater than τ_y, there will be flow; however, there will be a critical radius ($r_{critical}$) at which the local shear stress will equal τ_y. From the tube centerline to this critical radius, this core fluid will have a flat velocity profile, i.e., $v_z(r) = v_{core}$, and will move as if it were a solid body or in what is known as "plug flow." For the region from the critical radius to the tube wall ($r_{critical} \leq r \leq R$), Equation 4.3 describes once again the shear stress distribution as a function of radial position. We can set this equation equal to the shear stress relation provided by the Casson equation (4.19) and rearrange to obtain the following equation for the shear rate.

$$\dot{\gamma} = -\frac{dv_z(r)}{dr} = \left\{ \frac{P_0 - P_L}{2L} r - 2\tau_y^{1/2} \left[\frac{(P_0 - P_L)}{2L} r \right]^{1/2} + \tau_y \right\} \frac{1}{s^2}. \tag{4.25}$$

This equation may be integrated to find $v_z(r)$ using the boundary condition that at $r = R$, $v_z(R) = 0$, thereby obtaining the following result:

$$v_z(r) = \frac{R\tau_w}{2s^2} \left\{ \left[1 - \left(\frac{r}{R}\right)^2 \right] - \frac{8}{3}\sqrt{\frac{\tau_y}{\tau_w}} \left[1 - \left(\frac{r}{R}\right)^{3/2} \right] + 2\left(\frac{\tau_y}{\tau_w}\right)\left(1 - \frac{r}{R}\right) \right\}. \tag{4.26}$$

Equation 4.26 applies for the region defined by $r_{critical} \leq r \leq R$. At this point, however, we need to determine the value of $r_{critical}$. Since the shear stress at $r_{critical}$ must equal the yield stress, i.e., $\tau_{rz} = \tau_y$, it is easy to then show from Equation 4.5 that $r_{critical} = R(\tau_y/\tau_w)$. For locations from the centerline to the critical radius, the velocity of the core would be given by Equation 4.26 after setting $r/R = r_{critical}/R = \tau_y/\tau_w$.

4.11 TUBE FLOW OF BLOOD AT LOW SHEAR RATES

Note in Figure 4.8 the scatter in the data at lower shear rates, i.e., $\bar{U}$ less than about 1 sec^{-1}. The data start to show significant deviations from Equation 4.21, which predicts that τ_w vs. $\bar{U}$ should be independent of the tube diameter. Clearly, at these lower shear rates, RBC aggregates are starting to form, and their characteristic size is becoming comparable to that of the tube diameter. The assumption of blood being a homogeneous fluid is no longer correct. However, for practical purposes, blood flow in the body and through medical devices will be at significantly higher shear rates. Therefore, we really need not concern ourselves with the limiting case of blood flow at these very low shear rates.

4.12 EFFECT OF DIAMETER AT HIGH SHEAR RATES

Tube flow of blood at high shear rates (i.e., >100 sec^{-1}) shows two anomalous effects that involve the tube diameter. These are the *Fahraeus effect* and the *Fahraeus–Lindqvist effect* (Barbee and Cokelet 1971a, 1971b; Gaehtgens 1980). To understand the effect of tube diameter on blood

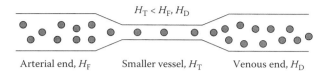

FIGURE 4.9   The Fahraeus effect.

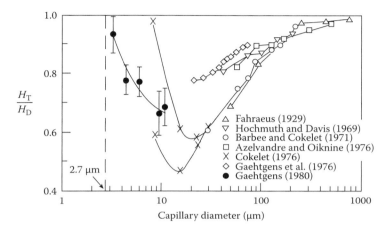

FIGURE 4.10   The Fahraeus effect. (From Gaehtgens, P., *Biorheology,* 17, 183–89, 1980. Data sources may be found in this reference. With permission.)

flow, consider the conceptual model illustrated in Figure 4.9. In this figure, blood flows from a larger vessel, such as an artery, into progressively smaller vessels, ultimately reaching the size of a capillary, after which the vessels progress in size to a vein. We make the distinction between the hematocrit found in the feed to a smaller vessel represented by H_F, the hematocrit in the vessel of interest H_T, and the discharge hematocrit, H_D. All these hematocrits are averaged over the entire cross-sectional area of the tube. In tubes with diameters from about 15 to 500 μm, it is found that the tube hematocrit (H_T) is actually less than that of the discharge hematocrit (at steady state, $H_D = H_F$, to satisfy the mass balance on the RBCs). This is called the Fahraeus effect and is shown in Figure 4.10, where we see H_T/H_D plotted as a function of tube diameter (D) in microns. We observe a pronounced decrease in the tube hematocrit in comparison to the discharge hematocrit (or feed hematocrit). For tubes less than about 15 μm in diameter, the ratio of H_T to H_D starts to increase, as shown in Figure 4.10. This is because H_D is now less than H_F. One reason for this is plasma skimming. When the tube size is smaller than about 15 μm, it becomes physically very difficult for the RBCs to even enter the smaller vessel. Hence, the orientation of the branch as well as that of the RBCs will affect the tube hematocrit as well as the resulting discharge hematocrit.

To explain the Fahraeus effect, it has been shown both by in vivo and in vitro experiments that the cells do not distribute themselves evenly across the tube cross section. Instead, the RBCs tend to accumulate along the tube axis forming, in a statistical sense, a thin cell-free layer along the tube wall. This is illustrated in Figure 4.11. Recall from Equation 4.13 that the fluid velocity is maximal along the tube axis. The axial accumulation of RBCs, in combination with the higher fluid velocity near the tube axis, maintains the RBC balance ($H_F = H_D$) even though the hematocrit in the tube is reduced. The thin cell-free layer along the tube wall is called the *plasma layer*. The thickness of the plasma layer depends on the tube diameter and the hematocrit and is typically on the order of several microns.

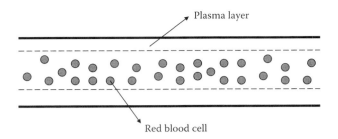

FIGURE 4.11 Axial accumulation of red blood cells.

Because of the Fahraeus effect, it is found that as the tube diameter decreases below about 500 μm, the *apparent viscosity* of blood also decreases. The decreased viscosity is a direct result of the decrease in the tube hematocrit and the presence of the cell-free plasma layer near the wall of the tube. The reduction in the viscosity of blood is known as the Fahraeus–Lindqvist effect.

Since the rheology of blood can also depend on the tube diameter, the simple relationship (Equation 4.11) developed to describe tube flow of blood at high shear rates no longer applies. However, since for tube flow of blood, we can still measure both Q and $\Delta P/L$, we can use these measurements to calculate from Equation 4.11 for a given tube diameter, the apparent viscosity of blood. This apparent viscosity can be written in terms of the observed values of Q and $\Delta P/L$ by rearranging Equation 4.11 (assuming the flow of blood is laminar):

$$\mu_{\text{apparent}} = \frac{\pi R^4 \Delta P}{8LQ}. \tag{4.27}$$

Figure 4.12 illustrates the Fahraeus–Lindqvist effect, where we see the ratio of the apparent viscosity to its large tube value (μ_F) plotted as a function of the tube diameter. We observe a significant decrease in the apparent viscosity of blood between tube diameters from about 500 to 4–6 μm. The decrease in the apparent viscosity in the smaller tubes is explained by the fact that the viscosity in the plasma layer near the tube wall is less than that of the central core that contains the cells.

As the tube diameter becomes less than about 4–6 μm, the apparent viscosity increases dramatically as shown in Figure 4.12 (Fung 1993). At this point, it becomes very difficult for an RBC to enter the tube. The RBC is about 8 μm in diameter and is deformable to some extent and can enter tubes somewhat smaller than this characteristic dimension. A tube diameter of about 2.7 μm is about the smallest size that an RBC can enter. However, by Equation 4.11, in order to maintain the flow rate under these conditions, the pressure drop must increase to overcome the resistance presented by the RBC; hence, an increase in the apparent viscosity is observed under these conditions.

4.13 MARGINAL ZONE THEORY

The marginal zone theory proposed by Haynes (1960) may be used to characterize the Fahraeus–Lindqvist effect in the range from 4–6 to 500 μm. Using this theory, it is possible to obtain an expression for the apparent viscosity in terms of the plasma layer thickness, tube diameter, and the hematocrit. Development of the marginal zone theory makes use of the relationships for a Newtonian fluid that we developed earlier. These are summarized in Table 4.3.

As shown in Figure 4.11, the blood flow within a tube or vessel is divided into two regions; a central core that contains the cells with a viscosity μ_c, and the cell-free marginal or plasma layer that consists only of plasma with a thickness of δ, and a viscosity equal to that of the plasma given by the symbol μ_p. In each of these regions, the flow is considered to be Newtonian and Equation 4.12 applies to each. For the core region, we may then write

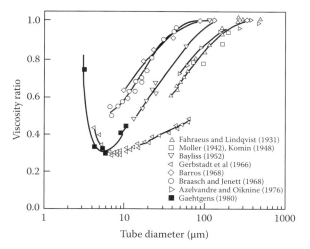

FIGURE 4.12 The Fahraeus–Lindqvist effect. (From Gaehtgens, P., *Biorheology,* 17, 183–89, 1980. Data sources may be found in this reference. With permission.)

$$\tau_{rz} = \frac{(P_0 - P_L)r}{2L} = -\mu_c \frac{dv_z^c}{dr},$$

BC1: $r = 0$, $\dfrac{dv_z^c}{dr} = 0$,

(4.28)

BC2: $r = R - \delta$, $\tau_{rz}|_c = \tau_{rz}|_p$.

The first boundary condition (BC1) expresses the fact that the axial velocity profile is symmetric about the axis of the tube and the velocity is also maximum at the centerline of the tube. The second boundary condition (BC2) derives from the fact that the transport of momentum must be continuous across the interface between the core and the plasma layer.

For the plasma layer, the following equations apply:

$$\tau_{rz} = \frac{(P_0 - P_L)r}{2L} = -\mu_p \frac{dv_z^p}{dr},$$

BC3: $r = R - \delta$, $v_z^c = v_z^p$,

(4.29)

BC4: $r = R$, $v_z^p = 0$.

The third boundary condition (BC3) states the requirement that the velocity in each region must be the same at their interface. The last boundary condition (BC4) simply requires that the axial velocity is zero at the tube wall.

Equations 4.28 and 4.29 may be readily integrated to give the following expressions for the axial velocity profiles in the core and plasma regions:

$$v_z^p(r) = \frac{(P_0 - P_L)R^2}{4\mu_p L}\left[1 - \left(\frac{r}{R}\right)^2\right] \quad \text{for } R - \delta \le r \le R.$$

(4.30)

$$v_z^c(r) = \frac{(P_0 - P_L)R^2}{4\mu_p L}\left\{1 - \left(\frac{R-\delta}{R}\right)^2 - \frac{\mu_p}{\mu_c}\left(\frac{r}{R}\right)^2 + \frac{\mu_p}{\mu_c}\left(\frac{R-\delta}{R}\right)^2\right\} \quad \text{for } 0 \le r \le R - \delta.$$

(4.31)

The plasma and core volumetric flow rates are given by

$$Q_p = 2\pi \int_{R-\delta}^{R} v_z^p(r) r\, dr,$$

$$Q_c = 2\pi \int_{0}^{R-\delta} v_z^c(r) r\, dr. \tag{4.32}$$

Integration of Equation 4.32 with the values of v_z^p and v_z^c from Equations 4.30 and 4.31 provides the following result for the volumetric flow rates of the plasma layer and the core:

$$Q_p = \frac{\pi(P_0 - P_L)}{8\mu_p L} \left[R^2 - (R - \delta)^2 \right]^2. \tag{4.33}$$

$$Q_c = \frac{\pi(P_0 - P_L)R^2}{4\mu_p L} \left[(R-\delta)^2 - \left(1 - \frac{\mu_p}{\mu_c}\right)\frac{(R-\delta)^4}{R^2} - \frac{\mu_p}{\mu_c}\frac{(R-\delta)^4}{2R^2} \right]. \tag{4.34}$$

The total flow rate of blood within the tube would equal the sum of the flow rates in the core and plasma regions. After adding the Equations 4.33 and 4.34, we obtain the following expression for the total flow rate:

$$Q = \frac{\pi R^4 (P_0 - P_L)}{8\mu_p L} \left[1 - \left(1 - \frac{\delta}{R}\right)^4 \left(1 - \frac{\mu_p}{\mu_c}\right) \right]. \tag{4.35}$$

Comparing Equation 4.35 with Equation 4.27 allows one to develop a relationship for the apparent viscosity based on the marginal zone theory. We then arrive at the following expression for the apparent viscosity in terms of δ, R, μ_c, and μ_p:

$$\mu_{apparent} = \frac{\mu_p}{1 - \left(1 - \dfrac{\delta}{R}\right)^4 \left(1 - \dfrac{\mu_p}{\mu_c}\right)}. \tag{4.36}$$

As $\delta/R \to 0$, then $\mu_{apparent} \to \mu_c \to \mu$, which is the bulk viscosity of blood in a large tube at high shear rates, i.e., a Newtonian fluid, as one would expect. We can use Equations 4.35 and 4.36 for blood flow calculations in tubes less than 500 μm in diameter provided that we know the thickness of the plasma layer, δ, and the viscosity of the core, μ_c. The next section shows how to determine the thickness of the plasma layer, δ, and the viscosity of the core, μ_c, using apparent viscosity data like that shown in Figure 4.12.

4.14   USING THE MARGINAL ZONE THEORY

We can now use Equation 4.35 based on the marginal zone theory to fit the apparent viscosity data shown in Figure 4.12. This will allow us to obtain values of the plasma layer thickness and the core hematocrit as a function of the tube diameter. In order to do this, we must first develop the relationship for how the core hematocrit (H_C) depends on the feed hematocrit (H_F) and the thickness of the plasma layer.

We will also need an equation to describe the dependence of the blood viscosity on the hematocrit since the value of H_C will be larger than H_F because of the axial accumulation of the RBCs.

This relative increase in the core hematocrit will make the blood in the core have a higher viscosity than the blood in the feed.

The following equation may be used to express the dependence of the viscosity of blood at high shear rates on the hematocrit and temperature (Charm and Kurland 1974):

$$\mu = \mu_p \frac{1}{1 - \alpha H}, \quad \text{for } H \le 0.6,$$

$$\text{where } \alpha = 0.070 \exp\left[2.49H + \frac{1107}{T}\exp(-1.69H)\right].$$

(4.37)

In the above equations, temperature (T) is in Kelvin. These equations may be used to a hematocrit of 0.60 and the stated accuracy is within 10%. If a subscript in the following discussion occurs on the viscosity, then the corresponding values of H and α in Equation 4.37 will carry the same subscript. For example, if we are talking about the core region, then $\mu = \mu_c$, $H = H_C$, and $\alpha = \alpha_c$.

Returning to Equation 4.35, we can rewrite the total volumetric flow rate Q using Equation 4.37 to express the viscosity ratio μ_p/μ_c as $(1 - \alpha_c H_c)$; we then obtain

$$Q = \frac{\pi R^4 (P_0 - P_L)}{8L\mu_p}\left[1 - \left(1 - \frac{\delta}{R}\right)^4 \alpha_c H_c\right].$$

(4.38)

Continuity or conservation of the cells allows us to write the following expression that relates the feed hematocrit to the core hematocrit: $Q_F H_F = Q_c H_C$. Hence, on rearranging this equation, we have that $H_C/H_F = Q_F/Q_c$. We also have that the volumetric flow rate of the feed must equal that of the core and the plasma layer, or $Q_F = Q_c + Q_p$. This can then be used to give the following relationship for the ratio of the core and feed hematocrit, $H_C/H_F = Q_F/Q_c = 1 + Q_p/Q_c$. We can now use Equations 4.33 and 4.34 for Q_p and Q_c along with $\sigma \equiv 1 - \delta/R$, to obtain an expression for the hematocrit of the core in terms of the plasma layer thickness, δ, and the core viscosity, μ_c:

$$\frac{H_C}{H_F} = 1 + \frac{(1 - \sigma^2)^2}{\sigma^2\left[2(1 - \sigma^2) + \sigma^2 \dfrac{\mu_p}{\mu_c}\right]}.$$

(4.39)

Equation 4.39 is nonlinear in H_C (since $\mu_c = \mu_c(H_C)$ by Equation 4.37) and also depends for a given tube diameter, D, on the thickness of the plasma layer since $\delta = (1 - \sigma)R$, which at this point is not known. Therefore, we have two unknowns, H_C and δ, which must be found, so we need another equation in addition to Equation 4.39 to obtain our solution.

The viscosity ratio for a given tube diameter provides the other relationship that may be used, along with Equation 4.39, to determine H_C and δ. The viscosity ratio, which is also the ordinate of Figure 4.12, may be determined by dividing both sides of Equation 4.36 by the feed or large tube viscosity, μ_F, and using Equation 4.37 to eliminate μ_p/μ_F in favor of the term $(1 - \alpha_F H_F)$ and μ_p/μ_c in terms of $(1 - \alpha_c H_C)$:

$$\frac{\mu_{apparent}}{\mu_F} = \frac{1 - \alpha_F H_F}{[1 - \sigma^4 \alpha_C H_C]}.$$

(4.40)

For a given tube diameter and a corresponding value of $\mu_{apparent}/\mu_F$, Equations 4.36 and 4.37 can be solved to find δ and H_C. This is illustrated in the following example for some of the data shown in Figure 4.12. The tube hematocrit (i.e., averaged over the entire tube cross section) is related to the core hematocrit by the expression $H_T = (1 - (\delta/R))^2 H_C = \sigma^2 H_C$.

Example 4.4

Find values of δ and H_C as a function of tube diameter for the data of Bayliss (upside-down open triangle) and Gaehtgens (filled square) shown in Figure 4.12.

Solution

To obtain the solution, we need to solve Equations 4.39 and 4.40 simultaneously. These two algebraic equations are nonlinear, meaning that they cannot be explicitly solved for δ (or $\sigma = 1-(\delta/R)$) and H_C. Numerical techniques based on the Newton–Raphson method can be used to solve these equations in an iterative manner. For example, to solve these equations by the Newton–Raphson method, one must first rearrange Equations 4.38 and 4.40 as

$$f(\sigma, H_C) = \frac{H_C}{H_F} - 1 - \frac{(1-\sigma^2)^2}{\sigma^2 \left[2(1-\sigma^2) + \sigma^2 \dfrac{\mu_p}{\mu_c} \right]},$$

$$g(\sigma, H_C) = \frac{\mu_{apparent}}{\mu_F} - \frac{1-\alpha_F H_F}{\left[1-\sigma^4 \alpha_C H_C\right]}.$$

When the problem is recast in terms of $f(\sigma, H_C)$ and $g(\sigma, H_C)$, the goal is to find the values of σ and H_C that make these functions equal to zero. To find the solution using the Newton–Raphson method, a Taylor series truncated after the second term is written around the ith iteration values of σ_i and H_{Ci}, which allows us to predict the next best values $(i+1)$ of σ and H_C based on the current values of σ_i and H_{Ci}, which are

$$f\left(\sigma_{i+1}, H_{Ci+1}\right) \approx f\left(\sigma_i, H_{Ci}\right) + \left.\frac{\partial f}{\partial \sigma}\right|_{\sigma_i, H_{Ci}} \left(\sigma_{i+1} - \sigma_i\right) + \left.\frac{\partial f}{\partial H_C}\right|_{\sigma_i, H_{Ci}} \left(H_{Ci+1} - H_{Ci}\right),$$

$$g\left(\sigma_{i+1}, H_{Ci+1}\right) \approx g\left(\sigma_i, H_{Ci}\right) + \left.\frac{\partial g}{\partial \sigma}\right|_{\sigma_i, H_{Ci}} \left(\sigma_{i+1} - \sigma_i\right) + \left.\frac{\partial g}{\partial H_C}\right|_{\sigma_i, H_{Ci}} \left(H_{Ci+1} - H_{Ci}\right).$$

We want to choose σ_{i+1} and H_{Ci+1} such that both $f(\sigma_{i+1}, H_{Ci+1})$ and $g(\sigma_{i+1}, H_{Ci+1})$ equal zero. Letting the left-hand side of the above equations equal zero then allows for the solution of $\varepsilon_i = \sigma_{i+1} - \sigma_i$ and $\tau_i = H_{Ci+1} - H_{Ci}$, as

$$\begin{bmatrix} \varepsilon_i \\ \tau_i \end{bmatrix} = - \begin{bmatrix} f_i \\ g_i \end{bmatrix} \begin{bmatrix} \left.\dfrac{\partial f}{\partial \sigma}\right|_i & \left.\dfrac{\partial f}{\partial H_C}\right|_i \\ \left.\dfrac{\partial g}{\partial \sigma}\right|_i & \left.\dfrac{\partial g}{\partial H_C}\right|_i \end{bmatrix}^{-1}.$$

Solution of this matrix equation then provides the values of ε_i and δ_i from which $\sigma_{i+1} = \sigma_i + \varepsilon_i$ and $H_{Ci+1} = H_{Ci} + \tau_i$. If σ_{i+1} and H_{Ci+1} differ significantly from the previous values, i.e., σ_i and H_{Ci}, then convergence has not been obtained and one repeats the calculation by letting $\sigma_i = \sigma_{i+1}$ and $H_{Ci} = H_{Ci+1}$. Fortunately, many scientific and engineering software packages have built-in routines for simultaneously solving sets of nonlinear algebraic equations. Table 4.4 summarizes the results of these calculations for tube diameters from 130 to 25 μm using the Haynes marginal zone layer theory. Tubes smaller than 25 μm in diameter give a core hematocrit greater than 0.60, which is beyond the range allowed for by Equation 4.37. Knowing how δ and H_C vary with tube diameter for a particular data set allows one to then calculate the apparent viscosity from Equation 4.40 and the flow rate, for a given pressure drop, from Equation 4.27.

TABLE 4.4
Calculation of H_C and δ

d (μm)	$\dfrac{\mu_{apparent}}{\mu_f}$	δ (μm)	H_C	H_T	$\dfrac{\delta}{d}$
130	1.0	0	0.40	0.40	0
100	0.93	1.15	0.40	0.38	0.012
80	0.90	1.4	0.40	0.38	0.018
60	0.82	2.27	0.41	0.35	0.038
40	0.77	2.22	0.43	0.34	0.055
30	0.69	2.92	0.48	0.31	0.093
25	0.63	3.87	0.59	0.28	0.155

Source: Data taken from Figure 4.12 for results obtained by
Bayliss (1952) and Gaehtgens (1980).

4.15 BOUNDARY LAYER THEORY

Describing the flow of a fluid near a surface is extremely important in a wide variety of engineering problems. Generally, the effect of the fluid viscosity is such that the fluid velocity changes from zero at the surface to the free stream value over a narrow region near the surface, which is referred to as the *boundary layer*. It is the presence of this boundary layer that affects the rates of mass transfer and heat transfer between the surface and the fluid.

Analysis of these types of problems using boundary layer theory for relatively simple cases can provide a great deal of insight into how the flow of the fluid affects the transport of mass and energy leading to the rational development of correlations to describe mass and energy transfer in more complex geometries and flow systems. We will use these correlations in later chapters in problems that involve the transport of mass across a bounding surface into a flowing fluid.

4.15.1 FLOW NEAR A WALL THAT IS SET IN MOTION

Consider the situation shown in Figure 4.13. A semi-infinite quantity of a viscous fluid is contacted from below by a flat, horizontal plate. For $t < 0$, the plate and the fluid are not moving. At $t = 0$, the plate is set in motion with a constant velocity (V) to the right, as shown in Figure 4.13. There are no pressure gradients or gravitational forces acting on the fluid, so the motion of the fluid is solely due to the momentum transferred from the plate to the fluid. The flow is also laminar, meaning there is no mixing of fluid elements in the y direction. Since we have no holes in the plate, v_y is zero, and the flow is laminar. Since the fluid and the plate are assumed to be infinite in the z direction, v_z is zero. The only component of the fluid velocity is therefore $v_x(y)$.

As time progresses, momentum is transferred further into the fluid. This creates a velocity profile in the y direction. We can arbitrarily define the boundary layer thickness (δ) for this problem as the distance perpendicular to the plate surface where the fluid has just been set in motion, and this distance is defined as that point where the local velocity is equal to 1% of V. Our goal here is to determine the velocity profile $v_x(y,t)$ and the boundary layer thickness, $\delta(t)$.

The concept of a shell balance can be used to analyze this problem. The shell balance is an important technique for developing mathematical models to describe the transport of such quantities as momentum, mass, and energy. The shell balance approach is conceptually easy to use and is based on the application of the generalized balance equation (Equation 1.8) to a given finite volume of interest.

Consider a small volume element of the fluid $\Delta x\, \Delta y\, W$, where W is the width of the plate in the z direction (normal to the page) and is assumed to be very large. Recall that momentum is

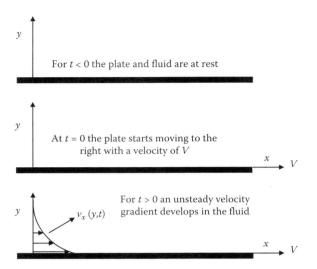

FIGURE 4.13   Flow of a fluid near a flat plate that is set in motion.

(mass) × (velocity) and we can write the momentum per unit volume of the fluid as ρv_x. The rate of accumulation of momentum within this volume element of the fluid is equal to $\rho(\partial v_x/\partial_t)\Delta_x\Delta_yW$. This term has units of force and is also equal to the sum of the forces acting on the volume element of the fluid. The only forces acting on this volume element of fluid are the shear forces acting on the surfaces at y and $y + \Delta y$, i.e., $\tau_{yx}|_y$ and $\tau_{yx}|_{y+\Delta y}$. These terms also, respectively, represent the flux of x momentum in the y direction at y and $y + \Delta y$, respectively. Hence, our momentum shell balance on the volume element can be written as

$$\rho\frac{\partial v_x}{\partial t}\Delta x\Delta yW = \left(\tau_{yx}|_y - \tau_{yx}|_{y+\Delta y}\right)\Delta x\,W. \tag{4.41}$$

Eliminating Δx and W, and then dividing by Δy and taking the limit as $\Delta y \to 0$, we obtain the following partial differential equation:

$$\rho\frac{\partial v_x}{\partial t} = -\frac{\partial \tau_{yx}}{\partial y}. \tag{4.42}$$

The above equation is valid for any fluid, Newtonian or non-Newtonian. For the special case of the Newtonian fluid, we can use Equation 4.2 for τ_{yx} and obtain

$$\frac{\partial v_x}{\partial t} = \nu\frac{\partial^2 v_x}{\partial y^2}, \tag{4.43}$$

where ν is the *kinematic viscosity* and is defined as μ/ρ. The initial condition (IC) and boundary conditions (BC) are

$$\text{IC: } t = 0, \ \ v_x = 0 \text{ for all values of } y, \tag{4.44}$$

$$\text{BC1: } y = 0, \ \ v_x = V \text{ for all values of } t > 0,$$

$$\text{BC2: } y = \infty, \ \ v_x = 0 \text{ for all values of } t > 0.$$

Equation 4.44 can easily be solved using the Laplace transform technique. Recall that the Laplace transform of a function $f(t)$ is defined by

$$L[f(t)] = \overline{f}(s) = \int_0^\infty e^{-st} f(t) dt. \tag{4.45}$$

Tables of Laplace transforms can be found in a variety of calculus textbooks and mathematical handbooks. Table 4.5 summarizes some of the more commonly used Laplace transforms.

Taking the Laplace transform of Equation 4.43 results in the following equation:

$$s\overline{v}_x - v_x(t = 0) = \nu \frac{d^2 \overline{v}_x}{dy^2}, \tag{4.46}$$

where $\overline{v}_x$ denotes the Laplace transform of the velocity. From the initial condition (Equation 4.44), we have that $v_x(t = 0) = 0$, and the other boundary conditions, when transformed, become

$$\text{BC1:} \quad y = 0, \ \overline{v}_x = \frac{V}{s},$$

$$\text{BC2:} \quad y = \infty, \ \overline{v}_x = 0. \tag{4.47}$$

Equation 4.46 can then be written as

$$\frac{d^2 \overline{v}_x}{dy^2} - a^2 \overline{v}_x = 0, \tag{4.48}$$

with $a^2 = s/\nu$. The above equation is a homogeneous second order differential equation having the general solution:

$$\overline{v}_x = C_1 e^{ay} + C_2 e^{-ay}. \tag{4.49}$$

The constants C_1 and C_2 can then be found from the transformed boundary conditions given by Equation 4.47. Using these boundary conditions, we find that $C_1 = 0$ and $C_2 = V/s$. Hence, our solution in the Laplace transform space is

$$\frac{\overline{v}_x}{V} = \frac{1}{s} e^{-ay} = \frac{1}{s} e^{-\frac{y}{\sqrt{\nu}}\sqrt{s}}. \tag{4.50}$$

Now inverting Equation 4.50, we find the function $v_x(t)$ whose Laplace transform is the above equation. From Table 4.5, we use transform pair 25 and then find that the inverse of Equation 4.50 provides the following solution for $v_x(t)$:

$$\frac{v_x(y,t)}{V} = \text{erfc}\left(\frac{y}{\sqrt{4\nu t}}\right) = 1 - \text{erf}\left(\frac{y}{\sqrt{4\nu t}}\right). \tag{4.51}$$

The above solution is in terms of a new function called the *error function* and the *complementary error function*, which are abbreviated as "erf" and "erfc," respectively. The error function is defined by the following equation:

$$\text{erf}(x) = \frac{2}{\sqrt{\pi}} \int_0^x e^{-x^2} dx \ \text{ and } \ \text{erfc}(x) = 1 - \text{erf}(x). \tag{4.52}$$

TABLE 4.5
Some Commonly Used Laplace Transforms

ID	Function	Transform
1	$C_1 f(t) + C_2 g(t)$	$C_1 \bar{f}(s) + C_2 \bar{g}(s)$
2	$\dfrac{df(t)}{dt}$	$s\bar{f}(s) - f(0+)$
3	$\dfrac{\partial^n f(x_i, t)}{\partial x_i^n}$	$\dfrac{\partial^n \bar{f}(x_i, s)}{\partial x_i^n}$
4	$\displaystyle\int_0^t f(\tau)\, d\tau$	$\dfrac{1}{s}\bar{f}(s)$
5	$f(\alpha t)$	$\dfrac{1}{\alpha}\bar{f}\left(\dfrac{s}{\alpha}\right)$
6	$e^{-\beta t} f(t)$	$\bar{f}(s+\beta)$
7	$\displaystyle\int_0^t f(t-\tau)g(\tau)\, d\tau$	$\bar{f}(s)\bar{g}(s)$
8	1	$\dfrac{1}{s}$
9	t	$\dfrac{1}{s^2}$
10	$\dfrac{1}{(\pi t)^{1/2}}$	$\dfrac{1}{s^{1/2}}$
11	$-\dfrac{1}{2\pi^{1/2} t^{3/2}}$	$s^{1/2}$
12	$e^{-\alpha t}$	$\dfrac{1}{s+\alpha}$
13	$\dfrac{e^{-\beta t} - e^{-\alpha t}}{\alpha - \beta}$	$\dfrac{1}{(s+\alpha)(s+\beta)}$
14	$te^{-\alpha t}$	$\dfrac{1}{(s+\alpha)^2}$
15	$\dfrac{(\gamma-\beta)e^{-\alpha t} + (\alpha-\gamma)e^{-\beta t} + (\beta-\alpha)e^{-\gamma t}}{(\alpha-\beta)(\beta-\gamma)(\gamma-\alpha)}$	$\dfrac{1}{(s+\alpha)(s+\beta)(s+\gamma)}$
16	$\dfrac{1}{2}t^2 e^{-\alpha t}$	$\dfrac{1}{(s+\alpha)^3}$
17	$\dfrac{\alpha e^{-\alpha t} - \beta e^{-\beta t}}{\alpha - \beta}$	$\dfrac{s}{(s+\alpha)(s+\beta)}$
18	$\dfrac{(\beta-\gamma)\alpha e^{-\alpha t} + (\gamma-\alpha)\beta e^{-\beta t} + (\alpha-\beta)\gamma e^{-\gamma t}}{(\alpha-\beta)(\beta-\gamma)(\gamma-\alpha)}$	$\dfrac{s}{(s+\alpha)(s+\beta)(s+\gamma)}$

TABLE 4.5 (Continued)
Some Commonly Used Laplace Transforms

ID	Function	Transform
19	$\sin \alpha t$	$\dfrac{\alpha}{s^2 + \alpha^2}$
20	$\cos \alpha t$	$\dfrac{s}{s^2 + \alpha^2}$
21	$\sinh \alpha t$	$\dfrac{\alpha}{s^2 - \alpha^2}$
22	$\cosh \alpha t$	$\dfrac{s}{s^2 - \alpha^2}$
23	$\dfrac{x}{2(\pi a t^3)^{1/2}} e^{-x^2/4at}$	$e^{-qx},\ q = (s/a)^{1/2}$
24	$\left(\dfrac{a}{\pi t}\right)^{1/2} e^{-x^2/4at}$	$\dfrac{e^{-qx}}{q},\ q = (s/a)^{1/2}$
25	$\mathrm{erfc}\left(\dfrac{x}{2(at)^{1/2}}\right)$	$\dfrac{e^{-qx}}{s},\ q = (s/a)^{1/2}$
26	$2\left(\dfrac{at}{\pi}\right)^{1/2} e^{-x^2/4at} - x\,\mathrm{erfc}\left(\dfrac{x}{2(at)^{1/2}}\right)$	$\dfrac{e^{-qx}}{qs},\ q = (s/a)^{1/2}$
27	$\left(t + \dfrac{x^2}{2a}\right)\mathrm{erfc}\left(\dfrac{x}{2(at)^{1/2}}\right) - x\left(\dfrac{t}{\pi a}\right)^{1/2} e^{-x^2/4at}$	$\dfrac{e^{-qx}}{s^2}$

Source: From Arpaci, V.S., *Conduction Heat Transfer*, Addison-Wesley, Reading, MA, 1966.

Note that the integral of e^{-x^2} cannot be integrated analytically. Since this integral is quite common in the solution of many engineering problems, this function has been tabulated in mathematical handbooks and mathematical software and can be treated as a known function, much like logarithmic and trigonometric functions.

Recall that we defined the boundary layer thickness as that distance y from the surface of the plate where the velocity has decreased to a value of 1% of V. The complementary error function of $y/\sqrt{4vt} = 1.821$ provides a value of v_x/V, which is equal to 0.01. Hence, we can define the boundary layer thickness, $\delta(y,t)$, as

$$\delta(y,t) = 3.642\sqrt{vt} \approx 4\sqrt{vt}. \qquad (4.53)$$

The value of δ can also be interpreted as the distance to which momentum from the moving plate has penetrated into the fluid at time t.

Example 4.5

Calculate the boundary layer thickness 1 sec after the plate has started to move. Assume the fluid is water for which $v = 1 \times 10^{-6}$ m² sec⁻¹.

Solution

Using Equation 4.53, we can calculate the thickness of the boundary layer as

$$\delta = 4\sqrt{10^{-6} \text{ m}^2 \text{ sec}^{-1} \times 1 \text{ sec}} = 0.004 \text{ m} = 4 \text{ mm}.$$

4.15.2 Laminar Flow of a Fluid along a Flat Plate

Now consider the situation shown in Figure 4.14 of the steady laminar flow of a fluid along a flat plate. The plate is assumed to be semi-infinite and the fluid approaches the plate at a uniform velocity, V. Since the fluid velocity at the surface of the plate is zero, a boundary layer is formed near the surface of the plate, and within this boundary layer, v_x increases from zero to its free stream value of V.

In the following discussion, we will obtain an approximate solution for the boundary layer flow over a flat plate. The approximate solution is very close to the exact solution. Details on exact solutions to the boundary layer equations can be found in Schlichting (1979).

Consider the shell volume shown in Figure 4.14 and located from x to $x + \Delta x$ and from $y = 0$ to $y = \delta(x)$. We first perform a steady state mass balance on this shell volume, which is given by

$$W \int_0^\delta \rho v_x \, dy \bigg|_x - W \int_0^\delta \rho v_x \, dy \bigg|_{x+\Delta x} - \rho v_y \big|_\delta W \Delta x = 0. \tag{4.54}$$

The first two terms in Equation 4.54 provide the net rate (i.e., In–Out) at which mass is being added to the shell volume. The third term accounts for the loss of mass from the shell volume at the top of the boundary layer due to flow in the y direction. Eliminating W, dividing by Δx, then taking the limit as $\Delta x \to 0$, provides the following equation for $v_y \big|_\delta$:

$$v_y \big|_\delta = -\frac{d}{dx} \int_0^\delta v_x \, dy. \tag{4.55}$$

In a similar manner, we can write an x-momentum balance on the shell volume as

$$W \int_0^{\delta(x)} \rho v_x v_x \, dy \bigg|_x - W \int_0^{\delta(x)} \rho v_x v_x \, dy \bigg|_{x+\Delta x} - \rho v_y \big|_\delta VW \Delta x + W \Delta x \, \tau_{yx} \big|_{y=0} = 0. \tag{4.56}$$

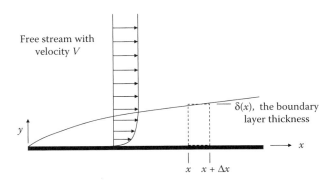

FIGURE 4.14 Laminar boundary layer flow of a fluid over a flat plate.

The first two terms in the above equation represent the net rate at which x-momentum is being added to the shell volume. The third term represents the rate at which x-momentum at the top of the boundary layer is being lost due to the flow of fluid out of the boundary layer in the y direction. The last term in Equation 4.56 represents the loss of momentum as a result of the shear stress generated by the fluid at the surface of the plate. After eliminating W and dividing by Δx and taking the limit as $\Delta x \rightarrow 0$, we can write Equation 4.56 as

$$-\tau_{yx}\big|_{y=0} = -\frac{d}{dx}\left(\int_0^\delta \rho v_x v_x \, dy\right) - \rho v_y\big|_{y=\delta} V. \tag{4.57}$$

Now we can use Equation 4.55 to eliminate v_y in the previous equation and obtain the result as

$$-\tau_{yx}\big|_{y=0} = \frac{d}{dx}\left(\int_0^\delta \rho(V - v_x)v_x \, dy\right). \tag{4.58}$$

For a Newtonian fluid, we can use Equation 4.2 once again and obtain

$$\mu \frac{\partial v_x}{\partial y}\bigg|_{y=0} = \frac{d}{dx}\left(\int_0^\delta \rho(V - v_x)v_x \, dy\right). \tag{4.59}$$

Equation 4.59 is also known as the *von Karman* momentum balance equation and forms the basis for obtaining an approximate solution to the boundary layer flow over a flat plate. To obtain an approximate solution, we first need to approximate the shape of the velocity profile within the boundary layer, i.e., $v_x(x,y)$. The simplest function that reasonably approximates the shape of the velocity profile is a simple cubic equation:

$$v_x(y) = a + by + cy^2 + dy^3. \tag{4.60}$$

The velocity profile assumed above also has to satisfy the following boundary conditions:

$$\text{BC1: } y = 0, \ v_x = 0, \tag{4.61}$$

$$\text{BC2: } y = \delta(x), \ v_x = V,$$

$$\text{BC3: } y = \delta(x), \ \frac{\partial v_x}{\partial y} = 0,$$

$$\text{BC4: } y = 0, \ \frac{\partial^2 v_x}{\partial y^2} = 0.$$

The first boundary condition is referred to as the "no slip" boundary condition, which requires that the velocity of the fluid at the surface of the plate be the same as the velocity of the plate that, in this case, is zero. The last boundary condition expresses the fact that the stress at the surface of the plate only depends on x and not on y. Boundary conditions two and three state that beyond the boundary layer, the velocity is constant and equal to the free stream value, V.

When the above boundary conditions are imposed on Equation 4.60, the following expression is obtained for the velocity profile within the boundary layer:

$$\frac{v_x(x,y)}{V} = \frac{3}{2}\left(\frac{y}{\delta(x)}\right) - \frac{1}{2}\left(\frac{y}{\delta(x)}\right)^3. \tag{4.62}$$

Equation 4.62 indicates that v_x depends on $\delta(x)$, which is still not known. However, we can insert Equation 4.62 into the von Karman momentum balance equation (Equation 4.59) and after some simplification we obtain

$$\delta\frac{d\delta}{dx} = \frac{140}{13}\left(\frac{\mu}{\rho V}\right), \tag{4.63}$$

with the condition that at $x = 0$, $\delta = 0$. Integration of Equation 4.63 results in the following expression for the boundary layer thickness, $\delta(x)$:

$$\delta(x) = 4.64\sqrt{\frac{\nu x}{V}}. \tag{4.64}$$

This equation shows that the boundary layer thickness grows in proportion to the square root of the distance from the upstream leading edge of the plate. Also, Equation 4.64 indicates that the boundary layer thickness is inversely proportional to the square root of the free stream velocity. We can also rearrange Equation 4.64 into the following form:

$$\frac{\delta(x)}{x} = \frac{4.64}{\sqrt{\frac{\rho V x}{\mu}}} = 4.64 Re_x^{-1/2}. \tag{4.65}$$

In the above equation, $Re_x = \rho V x / \mu$ is defined as the local value (i.e., at location x) of the *Reynolds number*. The Reynolds number is a very important dimensionless number in the field of fluid mechanics. Physically, the Reynolds number represents the ratio of the inertial forces ($\rho V \times VL^2 = \rho V^2 L^2$) acting on the fluid to the viscous forces ($\mu(V/L) \times L^2 = \mu VL$) acting on the fluid, where L is a characteristic dimension. Hence, in the previous equation, the characteristic dimension is the local position along the flat plate in the direction of the flow, which is x. A high Reynolds number indicates that the inertial forces dominate the viscous forces. High Reynolds number flow tends to be turbulent. On the other hand, a low Reynolds number means viscous forces are comparable to the inertial forces and the flow tends to be laminar. The Reynolds number, therefore, provides insight into when the fluid transitions from uniform laminar flow to turbulent flow. This critical Reynolds number for transition from laminar to turbulent flow depends on the geometry of the flow being considered. For example, for boundary layer flow over the flat plate, experiments show that the flow is laminar provided $Re_x < 300,000$, whereas flow in a cylindrical tube the flow is laminar if $Re_D < \sim 2100$ and can transition to turbulent flow at $Re_D > \sim 4000$. In this case, the characteristic dimension is the diameter of the cylindrical tube (D).

The approximate velocity profile within the boundary layer for laminar flow on a flat plate is then given by the combination of Equations 4.62 and 4.64 as

$$\frac{v_x(x,y)}{V} = 0.3233\left(y\sqrt{\frac{V}{\nu x}}\right) - 0.005\left(y\sqrt{\frac{V}{\nu x}}\right)^3. \tag{4.66}$$

With the velocity profile given by Equation 4.66, we can also calculate the drag force exerted by the fluid on the plate. For a plate of length L and width W, the force that acts on the surface of both sides of the plate in the positive x direction is given by

$$F_x = 2 \int_0^W \int_0^L \left(\mu \frac{\partial v_x}{\partial y} \bigg|_{y=0} \right) dx\,dz = 1.293\sqrt{\rho\mu LW^2 V^3}. \tag{4.67}$$

The exact solution as well as the experimental data for the drag force on the flat plate is about 3% greater than that predicted by the above approximate solution to the flat plate boundary layer problem. The constant in Equation 4.67 being 1.328 for the exact solution. Hence, we see that this approximate solution to the flat plate boundary layer problem is quite good.

We can also calculate the power needed to overcome the drag force. Recall that power is defined as (force × velocity), so after multiplying Equation 4.67 by V, the power is given by

$$P = VF_x = 1.293\sqrt{\rho\mu LW^2 V^5}. \tag{4.68}$$

The *friction factor* (f) is defined as the ratio of the shear stress at the wall and the kinetic energy per volume of the fluid based on the free stream velocity. The local value of the friction factor is then given by

$$f_x = \frac{-\tau_{yx}|_{y=0}}{\frac{1}{2}\rho V^2} = \frac{\mu \frac{\partial v_x}{\partial y}\big|_{y=0}}{\frac{1}{2}\rho V^2} = \frac{3\mu}{\rho V \delta(x)} = \frac{0.646}{\sqrt{Re_x}}. \tag{4.69}$$

For a plate of length L and width W, the average value of the friction factor (f) is given by integrating the above equation as

$$f = \frac{\int_0^W \int_0^L \frac{0.646}{\sqrt{Re_x}}\,dx\,dz}{WL} = 1.293\frac{1}{\sqrt{Re_L}}, \tag{4.70}$$

where $Re_L = \rho V L / \mu$. This definition of the friction factor means that the force acting on both sides of the plate (i.e., F_x) is given by the product of the kinetic energy per volume of the fluid $((1/2)\rho V^2)$, the area of the flat plate ($2\,LW$), and the friction factor (f), and is given by

$$F_x = \frac{1}{2}\rho V^2 \times (\text{area}) \times f = \frac{1}{2}\rho V^2 \times (2LW) \times 1.293\frac{1}{\sqrt{Re_L}} = 1.293\sqrt{\rho\mu LW^2 V^3}. \tag{4.71}$$

This result for F_x is, as expected, the same as that given by Equation 4.67. As shown by Equations 4.70 and 4.71, the friction factor is a convenient method for finding the force acting on a surface as a result of fluid motion. When the flow is across or over an object, the friction factor is also known as the *drag coefficient* ($f = C_D$) since it allows for the calculation of the drag force exerted on the object by the flowing fluid.

Example 4.6

Water ($\mu = 0.001$ Pa sec, $\rho = 1000$ kg m^{-3}) is flowing over a flat plate at a speed of 3 m sec^{-1}. Over what length of the plate (centimeters) is the boundary layer flow laminar? What is the boundary layer thickness (millimeters) at the position where the flow becomes turbulent? If the plate has a width of 4 cm, what is the drag force acting on the flat plate over this length?

Solution

The flow will transition to turbulent flow once the Reynolds number reaches 300,000. Letting this distance be where $x = L$, the Reynolds number can be solved for the value of L as

$$L = \frac{Re_L \mu}{\rho V} = \frac{300{,}000 \times 0.001 \dfrac{\text{Nsec}}{\text{m}^2} \times \dfrac{\text{kgm}}{\text{sec}^2 \text{N}}}{1000 \dfrac{\text{kg}}{\text{m}^3} \times 3 \dfrac{\text{m}}{\text{sec}}} = 0.10 \text{ m} = 10 \text{ cm.}$$

From Equation 4.64, the thickness of the boundary layer at this location can be found:

$$\delta(L) = 4.64\sqrt{\frac{\mu L}{\rho V}} = 4.64\sqrt{\frac{0.001 \dfrac{\text{Nsec}}{\text{m}^2} \times 0.1 \text{ m} \times \dfrac{\text{kgm}}{\text{sec}^2 \text{N}}}{1000 \dfrac{\text{kg}}{\text{m}^3} \times 3 \dfrac{\text{m}}{\text{sec}}}} = 0.00085 \text{ m} = 0.85 \text{ mm.}$$

The drag force acting on the plate can be found from Equation 4.67:

$$F_{\text{drag}} = 1.293\sqrt{1000 \frac{\text{kg}}{\text{m}^3} \times 0.001 \frac{\text{Nsec}}{\text{m}^2} \times 0.1 \text{ m} \times 0.04^2 \text{ m}^2 \times 3^3 \frac{\text{m}^3}{\text{sec}^3} \times \frac{1 \text{ Nsec}^2}{\text{kgm}}} = 0.085 \text{ N.}$$

Example 4.7

Consider a manta ray swimming along in the ocean at 0.75 m sec^{-1}. Assuming that the ray approximates a rectangular shape with a length of 0.30 m and a width of 0.75 m, how much power is the ray expending to move through the water as a result of the drag force of the water on its surface? Assume the density of the water is 1000 kg m^{-3} and the viscosity of the water is 1 cP = 0.001 Pa sec = 0.001 kg m^{-1} sec^{-1}.

Solution

Assume the geometry of the manta ray approximates that of a flat plate and that we can assume that we have laminar flow of the seawater over a flat plate. Using Equation 4.68, we can calculate the power requirement as

$$P = 1.293\sqrt{1000 \text{ kg m}^{-3} \times 0.001 \text{ kg m}^{-1} \text{ sec}^{-1} \times 0.3 \text{ m} \times 0.75^2 \text{ m}^2 \times 0.75^5 \text{ m}^5 \text{ sec}^{-5}}$$

$$= 0.26 \text{ kg m sec}^{-2} \text{ m sec}^{-1} = 0.26 \text{ J sec}^{-1} = 0.26 \text{ W.}$$

As shown below, the Reynolds number calculation shows that the flow is laminar over the surface of the manta ray since it is less than 300,000 at $x = L = 0.30$ m:

$$Re_L = \frac{1000 \dfrac{\text{kg}}{\text{m}^3} \times 0.75 \dfrac{\text{m}}{\text{sec}} \times 0.30 \text{ m}}{0.001 \dfrac{\text{kg}}{\text{msec}}} = 225{,}000.$$

4.16 GENERALIZED MECHANICAL ENERGY BALANCE EQUATION

Equation 4.11 provides a useful relationship for describing the laminar flow of a fluid in a cylindrical tube of constant cross section. Oftentimes, we will need a more generalized relationship that can handle turbulent flow and account for not only the effect of the pressure drop on fluid flow, but also the effect of changes in elevation, tube cross section, changes in fluid velocity, sudden contractions or expansions, pumps, as well as the effect of fittings such as valves. This general relationship for the flow of a fluid is called the *Bernoulli equation* and is valid for laminar or turbulent flow.

The Bernoulli equation accounts for the effect of changes in fluid pressure, potential energy, and kinetic energy on the flow of the fluid. It also accounts for the energy added to the fluid by pumps and accounts for energy losses due to a variety of frictional effects. For steady state flow of an incompressible fluid (density ρ = constant), such as blood from inlet station "1" to exit station "2," the Bernoulli equation may be written as follows (McCabe et al. 1985; Bird et al. 2002):

$$\frac{P_1}{\rho} + gZ_1 + \frac{\alpha_1 V_1^2}{2} + W_{\text{device}} = \frac{P_2}{\rho} + gZ_2 + \frac{\alpha_2 V_2^2}{2} + h_{\text{friction}}. \tag{4.72}$$

In this equation, P represents the pressure, Z the elevation relative to a reference plane, and V the average fluid velocity. Gravitational acceleration is represented by g and in SI units is equal to 9.8 m sec^{-2}. The αs are kinetic energy correction terms that account for the shape of the velocity profile. For laminar flow in a cylindrical tube, that is for $Re = (\rho D_{\text{tube}} V/\mu) < 2100$, $\alpha = 2.0$, and for turbulent flow, $\alpha = 1.0$. The work effect per unit mass of fluid is W_{device}. If $W_{\text{device}} > 0$, then work is done on the fluid, e.g., by a pump. The actual work required to achieve these changes in the fluid properties will be greater than W_{device} due to frictional losses and mechanical inefficiencies within the pump. If $W_{\text{device}} < 0$, then the fluid produces work, e.g., by a turbine. The actual amount of work that is generated will be less than this value because of frictional losses and mechanical inefficiencies within the turbine. Other frictional effects between positions 1 and 2 due to the tube itself, contractions, expansions, and fittings are accounted for by the term h_{friction}.

The units of each term in Equation 4.72 must be consistent with one another and, in general, are expressed in terms of energy per unit mass of fluid. In SI units, each term in Equation 4.72 will have units of square meters per square second. If these units are multiplied by kilograms per kilogram, then in the following equation we see that these units are the same as a Joule per kilogram:

$$\frac{m^2}{\sec^2} = \frac{m^2}{\sec^2} \times \frac{kg}{kg} = \frac{kg\,m}{\sec^2} \times \frac{m}{kg} = \frac{N\,m}{kg} = \frac{J}{kg}. \tag{4.73}$$

In addition to the Bernoulli equation, we also need to write a mass balance on the fluid. At steady state, this simply says that the rate of mass leaving the system, or a region of interest at position 2, must equal the rate at which mass enters the system at position 1. For steady state flow, this can be expressed by the mass conservation or continuity equation

$$\dot{m} = (\rho VS)_1 = (\rho VS)_2 = \text{constant}. \tag{4.74}$$

In this equation, $\dot{m}$ is the mass flow rate and S represents the cross-sectional area of the tube.

The frictional effects are described by

$$h_{\text{friction}} = \left(4 \sum_i f_i \frac{L_i V_i^2}{2D_i} + \sum_j \frac{V_j^2}{2} K_{\text{fitting}_j} \right). \tag{4.75}$$

In this equation, f_i is the friction factor and accounts for the loss in fluid energy per unit mass of fluid in tube segment i of length L_i and diameter D_i; V_i represents the average velocity (V_i = volumetric flow rate/$(\pi D_i^2/4)$) within tube section i. The friction factor f, also known as the *Fanning friction factor*, is defined as the ratio of the wall shear stress (τ_w) and the average kinetic energy per unit mass of fluid, also called the *velocity head*, that is $\rho V^2/2$. For laminar flow, it is easy to show from Equation 4.11 that $f = 16/Re$. For turbulent flow ($Re > {\sim}2100$) in smooth tubes, the friction factor may be evaluated from either of the following equations (McCabe et al. 1985; Bird et al. 2002):

$$\frac{1}{\sqrt{f}} = 4.07\log\left(Re\sqrt{f}\right) - 0.60,$$

$$f = \frac{0.0791}{Re^{1/4}}.$$

(4.76)

The second equation is valid up to $Re = 100,000$ and is more convenient to use since f is explicit in Re. If the flow is turbulent and the walls are not smooth, then the friction factor charts in McCabe et al. or Bird et al. should be consulted (i.e., graphs of f vs. Re with surface roughness as a parameter).

The term in the second summation of Equation 4.75, represented by K_{fitting}, accounts for energy loss in the fluid due to tube contractions, expansions, or fittings such as valves. The average velocity in this summation is usually for the fluid just downstream of the contraction, expansion, or fitting. A K_{fitting} for such items as valves is highly dependent on the valve's degree of openness and on the type of valve used. It is generally best to consult the manufacturer's literature for the particular valve being considered. As examples, a gate valve that is wide open has a K_{fitting} of about 0.2 and a value of about 6 when half-open. A globe valve that is wide open may have a value as high as 10. Simple fittings, such as a tee, has a value of 2.0, and a 90° elbow has a value of about 1.0. For sudden contractions and expansions of the fluid, the following equations may be used to estimate a K_{fitting}. Once again, S refers to the cross-sectional area of the tube. If for an expansion ($S_{\text{downstream}}/S_{\text{upstream}})^{-1} = 0$, then $K_{\text{fitting}} = 1$ and the velocity upstream of the expansion is used in Equation 4.75:

$$K_{\text{fitting}} = 0.45\left(1 - \frac{S_{\text{downstream}}}{S_{\text{upstream}}}\right), \text{ sudden contraction, turbulent flow,}$$

$$K_{\text{fitting}} = \left(\frac{S_{\text{downstream}}}{S_{\text{upstream}}} - 1\right)^2, \text{ sudden expansion, turbulent flow.}$$

(4.77)

For the most part, these fitting losses tend to be negligible for laminar flow of a fluid.

If the tube through which a fluid flows is not circular, then the *hydraulic diameter* can be used in the above calculations. The hydraulic diameter for flow in tubes of noncircular cross section is defined as four times the *hydraulic radius* (i.e., $4 \times r_H$). The hydraulic radius is defined as the ratio of the cross-sectional area of the flow channel to the wetted perimeter of the tube. The factor of four is needed so that $D_H = D$ for a circular tube:

$$D_H = 4 \times r_H = \frac{4 \times (\text{cross-sectional area})}{(\text{wetted perimeter})}.$$

(4.78)

The hydraulic diameter is then used in the calculation of the Reynolds number, i.e., $Re = \rho V D_H/\mu$, and in Equation 4.75. However, the average velocity (V) is still defined as the volumetric flow rate (Q) divided by the cross-sectional area normal to the flow. It is also important to note that the hydraulic diameter should only be used for turbulent flow.

For flow in an annulus with an inner tube of outer diameter D_i and an outer tube with inner diameter D_o, the hydraulic diameter is found from Equation 4.78 to be ($D_o - D_i$). For a rectangular tube completely filled with fluid of dimensions L and W, the hydraulic diameter is equal to

$2LW/(L + W)$. Note that for a very wide rectangular duct where $L \gg W$, the hydraulic diameter for this slit flow is $2W$.

The following examples illustrate the use of the Bernoulli equation and the above relationships.

Example 4.8

Consider the flow of a fluid through an orifice located at the bottom of a tank. The cross-sectional area of the tank is A_1 and the area of the orifice located at the bottom surface of the tank is A_2. The fluid leaving through the orifice will tend to contract (called the *vena contracta*) such that at a short distance downstream of the orifice its cross-sectional area will be about 0.64 times that of the area of the orifice or $A_{\text{vena contracta}} = 0.64\, A_2$. If the pressure in the tank is maintained at P_1, and if the fluid leaving through the orifice is at atmospheric pressure, develop an expression for the average fluid velocity at the vena contracta, assuming the height of the fluid in the tank is given by h.

Solution

First we write the Bernoulli equation (4.72) for this situation where the index 1 denotes the surface of the fluid within the tank. Note that there are no work devices and we ignore any frictional effects and assume the flow leaving the tank through the orifice is turbulent. We also set $g(Z_1 - Z_2) = gh$. Hence, for this situation we can write the Bernoulli equation as

$$\frac{P_1}{\rho} + gh + \frac{V_1^2}{2} = \frac{P_{\text{vena contracta}}}{\rho} + \frac{V_{\text{vena contracta}}^2}{2}.$$

Since $P_{\text{vena contracta}}$ is the same as the atmospheric pressure (P_{atm}), the previous equation can be rewritten as

$$V_{\text{vena contracta}}^2 = V_1^2 + 2\frac{\left(P_1 - P_{\text{vena contracta}}\right)}{\rho} + 2gh.$$

Mass conservation given by Equation 4.74 allows for the calculation of the velocity of the fluid surface within the tank in terms of the velocity at the vena contracta, that is

$$V_1 = V_{\text{vena contracta}}\frac{A_{\text{vena contracta}}}{A_1}.$$

Using the above equation to eliminate V_1 in the previous equation for $V_{\text{vena contracta}}$ allows for the calculation of the average velocity of the fluid in the vena contracta as

$$V_{\text{vena contracta}} = \sqrt{\frac{2\left(\dfrac{P_1 - P_{\text{atm}}}{\rho}\right) + 2gh}{\left[1 - \left(\dfrac{A_{\text{vena contracta}}}{A_1}\right)^2\right]}}.$$

For flow of a fluid from a tank through an orifice, we usually have that $A_1 \gg A_{\text{vena contracta}}$. Hence, in this case:

$$V_{\text{vena contracta}} = \sqrt{2\left(\frac{P_1 - P_{\text{atm}}}{\rho}\right) + 2gh}.$$

And if the tank is open to the atmosphere, then $P_1 = P_{atm}$ and we obtain what is known as the *Torricelli equation*:

$$V_{vena\,contracta} = \sqrt{2gh}.$$

Based on this result, the volumetric flow rate of the fluid, that is $Q = A_{vena\,contracta} \times V_{vena\,contracta}$, is given by the next result:

$$Q = A_{vena\,contracta}\sqrt{2gh} \approx 0.64\,A_{orifice}\sqrt{2gh}.$$

If the flow through the orifice is driven by the pressure difference and not by the potential energy of the fluid, that is if $((P_1 - P_{atm})/\rho) \gg 2gh$, then we can show that

$$V_{vena\,contracta} = \sqrt{\frac{2(P_1 - P_{atm})}{\rho}}.$$

And in this case, Q is given by

$$Q = A_{vena\,contracta}\sqrt{\frac{2(P_1 - P_{atm})}{\rho}} \approx 0.64\,A_{orifice}\sqrt{\frac{2(P_1 - P_{atm})}{\rho}}.$$

Example 4.9

Consider a patient being evaluated for stenosis (narrowing) of his/her aortic valve. Catheterization of the heart gave a cardiac output of 5000 mL min^{-1}, a mean systolic pressure drop across the aortic valve of 50 mmHg, a heart beat of 80 beats per minute, and a systolic ejection period of 0.33 sec. From these data, estimate the cross-sectional area of the aortic valve.

Solution

We can use the last equation developed in the previous example to determine the cross-sectional area of the aortic valve ($A_{orifice} = A_{aortic\,valve}$) from these data. From Table 4.1, the density of blood is 1.056 g cm^{-3} = 1056 kg m^{-3}. From the cardiac output and the number of heart beats per minute, we find that each beat of the heart moves a volume of blood, given by (5000 cm^3/min) × (1 min/80 beats) = 62.5 (cm^3/beat). If the aortic valve remains open for 0.33 sec for each beat, then each beat of the heart gives a flow rate of blood through the valve of (62.5 cm^3/beat) × (1 beat/0.33 sec) = 189.4 (cm^3/sec). Solving now for the area of the aortic valve:

$$A_{aortic\,valve} = \frac{Q}{0.64\sqrt{\dfrac{2\Delta P_{systolic}}{\rho}}} = \frac{189.4\,\dfrac{cm^3}{sec}}{0.64\sqrt{\dfrac{2 \times 50\,mmHg}{1056\,\dfrac{kg}{m^3}} \times \dfrac{101{,}325\,Pa}{760\,mmHg} \times \dfrac{N}{m^2\,Pa} \times \dfrac{kg\,m}{sec^2\,N} \times \dfrac{10^4\,cm^2}{m^2}}}$$

$$A_{aortic\,valve} = 0.83\,cm^2.$$

We can also calculate the Reynolds number at the vena contracta of the blood flowing out of the aortic valve. At the vena contracta, the cross-sectional area is 0.64 × 0.83 cm^2 = 0.53 cm^2, which gives a diameter for the vena contracta of 0.82 cm. The average velocity of the blood at the vena contracta is given by

$$V = \frac{189.4 \text{ cm}^3}{\text{sec}} \times \frac{1}{0.64 \times 0.83 \text{ cm}^2} = 356.6 \frac{\text{cm}}{\text{sec}}.$$

From Table 4.1, we have that the viscosity of blood is 3 cP. Therefore, we can calculate the Reynolds number as

$$Re = \frac{\rho V D_{\text{vena contracta}}}{\mu_{\text{blood}}} = \frac{1.056 \frac{\text{g}}{\text{cm}^3} \times 356.6 \frac{\text{cm}}{\text{sec}} \times 0.82 \text{ cm}}{3 \text{ cP} \times 0.01 \frac{\text{g}}{\text{cP cm sec}}} = 10,293.$$

Hence, we conclude that the flow of the blood through the aortic valve under these conditions is turbulent and that the assumptions used to calculate the aortic valve cross-sectional area that were based on the results of the previous example are valid.

Example 4.10

A *Pitot tube* (see Figure 4.15) is a simple device for measuring the velocity of a fluid that is passing nearby. This device is used to measure the airspeed on airplanes and can be inserted into pipes to measure the velocity of the flowing fluid. As shown in Figure 4.15, the Pitot tube is aligned with the axis of the flow. At the tip of this tube (1) there is a small opening into an inner tube that leads to a pressure measuring system. In this case, the pressure measuring system is a simple manometer that contains a denser fluid such as mercury with density ρ_m. The fluid that impinges at the tip of the Pitot tube (1) stagnates or has a zero velocity (i.e., $V_1 = 0$) and its kinetic energy is converted according to the Bernoulli equation into what is called the *stagnation pressure*, P_1. Surrounding or coaxial to this tube is another tube and along the sides of this tube are small holes that are also connected to the pressure measuring system. These small holes along the side of this outer tube sense the static pressure (P_2) of the surrounding fluid that passes by at a velocity of V_2. From this information, develop an expression for the velocity of the fluid flowing near a Pitot tube in terms of the pressure difference ($P_1 - P_2$). Also, write this expression for the velocity in terms of the manometer reading (h) for a manometer fluid of density ρ_m.

Solution

For the fluid flowing past the Pitot tube there is no change in elevation. Also, since in this case we are measuring the local velocity of the fluid and not the average velocity, there is no kinetic energy correction term, hence α is equal to one. If we also neglect any work and frictional effects,

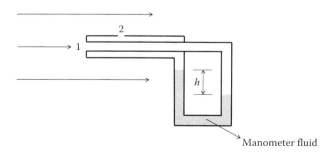

FIGURE 4.15   The pitot tube.

the Bernoulli equation (4.72) can be written between the stagnation point (1) (note $V_1 = 0$) and the static holes (2) as

$$\frac{P_1}{\rho} = \frac{P_2}{\rho} + \frac{V_2^2}{2}.$$

Note that this equation says that the stagnation pressure (P_1) is equal to the static pressure (P_2) plus the dynamic pressure, or the velocity head of the fluid, that is $1/2\rho V^2$, hence $P_1 = P_2 + 1/2\rho V^2$. This equation can then be rearranged and solved to give the velocity of the fluid passing near the static holes along the sides of the Pitot tube:

$$V_2 = \sqrt{\frac{2(P_1 - P_2)}{\rho}}.$$

If a manometer is used as shown in Figure 4.15 to measure the pressure difference ($P_1 - P_2$), then the Bernoulli equation can be used to express this pressure difference in terms of the difference in height (h) between the legs of the manometer fluid. At the surface of the upper leg of the manometer fluid the pressure is P_2 and the velocity of this surface is equal to zero. Similarly, at the surface of the lower leg of the manometer fluid the pressure is P_1 and the velocity of this surface is also zero. Neglecting work and frictional effects, the Bernoulli equation for this situation becomes

$$\frac{P_1}{\rho_m} + gZ_1 = \frac{P_2}{\rho_m} + gZ_2.$$

Letting $Z_2 - Z_1 = h$, the previous equation can be solved for ($P_1 - P_2$) in terms of h for a manometer fluid of density ρ_m. This result is also known as the *manometer equation*:

$$(P_1 - P_2) = \rho_m gh.$$

Using this result, we can express the velocity for the Pitot tube of Figure 4.15 as

$$V_2 = \sqrt{\frac{2\rho_m gh}{\rho}}.$$

Example 4.11

Farmer Jones needs to pump water from the Maumee River to irrigate his soybean fields. He figures that he will need 250 ft of 2 in inside diameter steel pipe and four 90° elbows to bring the water from the river to his water tower where the water is then stored. The total change in elevation from the river surface to where the water flows into the top of the water tower is 100 ft. Farmer Jones wants the water flow rate in this system to be 100 gal min⁻¹. Assuming the pump down by the river has an efficiency of 80%, estimate the required horsepower of the pump that is needed.

Solution

We assume that the storage tank is vented to the atmosphere and since the pipe enters at the top of the storage tank the water flows unimpeded into storage tank. With these assumptions the pressure at the surface of the river is the same as that inside the storage tank, hence we have $P_1 = P_2$. We also have that since the surface of the river is so large, $V_1 = 0$. With these assumptions, we can then write the Bernoulli equation (4.72) from the surface of the lake (1) to where the pipe enters the storage tank (2):

$$gZ_1 + W_{device} = gZ_2 + \frac{\alpha_2 V_2^2}{2} + h_f.$$

We now use Equation 4.75 for h_f to account for the resistance of the pipe of length L and diameter D and the fittings. The previous equation then becomes

$$W_{device} = g(Z_2 - Z_1) + \frac{\alpha_2 V_2^2}{2} + \frac{4fLV_2^2}{2D} + \sum_j \frac{V_j^2}{2} K_{fitting_j}.$$

The second last term in the previous equation accounts for the loss of energy per unit mass as the water flows through the 2 in pipe (D) that is 250 ft in length (L). The friction factor (f) in this term can be calculated after we find the Reynolds number and determine whether the flow in the pipe is laminar or turbulent. To find the Reynolds number, we first need to determine the velocity, V_2, in the pipe as

$$V_2 = \frac{4Q}{\pi D^2} = \frac{4 \times 100 \frac{\text{gal}}{\text{min}} \times \frac{1 \text{ min}}{60 \text{ sec}} \times \frac{0.13368 \text{ ft}^3}{\text{gal}} \times \frac{(12 \text{ in})^3}{\text{ft}^3} \times \frac{(2.54 \text{ cm})^3}{\text{in}^3}}{\pi \times (2 \text{ in})^2 \times \frac{(2.54 \text{ cm})^2}{\text{in}^2}} = 311.3 \frac{\text{cm}}{\text{sec}}.$$

Assuming that water has a density of 1 g cm^{-3} and a viscosity of 1 cP, we can now calculate the value of the Reynolds number:

$$Re = \frac{\rho DV}{\mu} = \frac{1 \frac{\text{g}}{\text{cm}^3} \times 2 \text{ in} \times \frac{2.54 \text{ cm}}{\text{in}} \times 311.3 \frac{\text{cm}}{\text{sec}}}{1 \text{ cP} \times \frac{1 \text{ g}}{\text{cm sec} 100 \text{ cP}}} = 158,140.$$

The flow in the pipe is therefore turbulent and $\alpha_2 = 1.0$. We can also use Equation 4.76 to find the friction factor, which is found to be equal to 0.0041.

Next, the effect of the four elbows and the contraction of the fluid as it enters the pipe and the expansion of the fluid into the storage tank is found as

$$\sum_j \frac{V_j^2}{2} K_{fitting_j} = \frac{1}{2} V_2^2 \times (4 \times K_{elbow}) + 1 \times \frac{1}{2} V_2^2 + 0.45 \left(1 - \frac{S_{pipe}}{S_{river}}\right) \times \frac{1}{2} V_2^2 = 2.725 V_2^2.$$

The above result is obtained when the value of K_{elbow} is set equal to 1. Also, we assume that the tank cross-sectional area is much larger than that of the pipe, hence $K_{expansion} = 1$ and we use the pipe velocity in the calculation of the expansion loss. We also assume for the contraction from the river into the pipe that $S_{river} \gg S_{pipe}$. The Bernoulli equation then becomes

$$W_{device} = g(Z_2 - Z_1) + \left(\frac{1}{2} + \frac{4fL}{2D} + 2.725\right) V_2^2.$$

Now inserting the numerical values where the elevation change was given as 100 ft (30.48 m), the pipe length is 250 ft (76.2 m), the diameter of the pipe is 2 in (0.051 m), along with f equal to 0.0041, and $V_2 = 311.3$ cm sec^{-1} = 3.11 m sec^{-1}, we then obtain

$$W_{device} = 9.8 \frac{\text{m}}{\text{sec}^2} \times 30.48 \text{ m} + \left(3.225 + \frac{4 \times 0.0041 \times 76.2 \text{ m}}{2 \times 0.051 \text{ m}}\right) \frac{3.11^2 \text{ m}^2}{\text{sec}^2} = 448.4 \frac{\text{m}^2}{\text{sec}^2}.$$

If this result is then multiplied by kilograms per kilogram (see Equation 4.73), then $W_{device} = 448.4$ J kg^{-1}. Next, we multiply this result by the mass flow rate of the water, which is calculated:

$$\dot{m} = \frac{\pi D^2}{4} \rho V_2 = \frac{\pi \times 0.051^2 \text{ m}^2}{4} \times 1000 \frac{\text{kg}}{\text{m}^3} \times 3.11 \frac{\text{m}}{\text{sec}} = 6.35 \frac{\text{kg}}{\text{sec}}.$$

So, the total amount of work required to pump the water is given by

$$W_{total} = \dot{m} W_{device} = 6.35 \frac{kg}{sec} \times 448.4 \frac{J}{sec} = 2847.3 \frac{J}{sec} \times \frac{1\,W}{J\,sec^{-1}} = 2847.3\,W.$$

Because of the inefficiencies of the pump, this value needs to be divided by the efficiency of the pump (here 0.80) to find the actual work required. This would be the rating of the pump. So, dividing the above result by the efficiency gives a value of $W_{actual} = 3.56$ kW, since 1 kW = 1000 W. In terms of horsepower (hp), which is a common unit still used in the United States, we multiply this result by the conversion factor of 1.341 hp kW^{-1} and obtain $W_{actual} = 4.77$ hp.

The next example illustrates a pseudo-steady state application* of the Bernoulli equation to estimate the draining time of an intravenous (IV) bag.

Example 4.12

The simplest patient infusion system is that of gravity flow from an IV bag. A 500 mL IV bag containing an aqueous solution is connected to a vein in the forearm of a patient. Venous pressure in the forearm is about 8 mmHg (gauge pressure†). The IV bag is placed on a stand such that the entrance to the tube leaving the IV bag is exactly 1 m above the vein into which the IV fluid enters. The length of the IV bag is 30 cm. The IV is fed through an 18 gauge tube (internal diameter = 0.953 mm) and the total length of the tube is 2 m. Calculate the flow rate of the IV fluid. Also, estimate the time needed to empty the bag.

Solution

We apply Bernoulli's equation from the surface of the fluid in the IV bag ("1") to the entrance to the vein ("2"). We expect the flow of the fluid through the bag and the tube to be laminar and therefore neglect the contraction at the entry to the feed tube and the expansion at the vein. The pressure at the surface of the fluid in the bag will be atmospheric ($P_1 = 760$ mmHg absolute) since the bag collapses as the fluid leaves the bag. The venous pressure, or P_2, is 8 mmHg gauge or 768 mmHg absolute. Because the fluid drains slowly from the bag, we neglect the velocity of the surface of the fluid in the bag in comparison to the fluid velocity at the exit of the tube. Therefore, we assume $V_1 = 0$ and $V_2 \gg V_1$. We also set the reference elevation as the entrance to the patient's arm, hence $Z_2 = 0$. Therefore, Z_1 is equal to the elevation of the bag relative to the position where the fluid enters the patient's arm. This would equal 1 m plus the 30 cm length of the bag. Since there are no work devices in the system, $W_{device} = 0$. We can now write the Bernoulli equation for this particular problem as

$$\frac{P_1}{\rho} + gZ_1 = \frac{P_2}{\rho} + V_2^2 + \left(4f \frac{L}{D}\right) \frac{V_2^2}{2}.$$

Recall that the friction factor for laminar flow in a cylindrical tube is equal to 16/Re. We can substitute this relationship into the previous equation to obtain the following quadratic equation that can be solved for the exiting velocity, V_2:

$$V_2^2 + \left(\frac{32\mu L}{\rho D^2}\right) V_2 - \left[gZ_1 + \frac{1}{\rho}(P_1 - P_2)\right] = 0.$$

---

* Problem 7 at the end of this chapter discusses an unsteady state solution.
† Recall that gauge pressure is that pressure relative to the local atmospheric pressure. Absolute pressure is gauge pressure plus local atmospheric pressure.

This equation may now be solved for the exit velocity, recognizing that this quantity must be positive:

$$V_2 = \frac{-\left(\frac{32\,\mu L}{\rho D^2}\right)+\left\{\left(\frac{32\,\mu L}{\rho D^2}\right)^2+4\left[gZ_1+\frac{1}{\rho}\left(P_1-P_2\right)\right]\right\}^{1/2}}{2}.$$

Assuming the IV fluid has the same properties as water, and substituting the appropriate values for the parameters in the previous equation, the exit velocity is calculated as shown in the following equation. Note that P_1-P_2 equals −8 mmHg, which is equal to −0.0105 atm or −1066.58 kg m^{-1} sec^{-2}:

$$V_2 = \frac{-\left(\dfrac{32\times0.001\frac{kg\,m\,sec}{sec^2\,m^2}\times2.0\,m}{1000\frac{kg}{m^3}\times0.000953^2\,m^2}\right)+\left[\left(\dfrac{32\times0.001\frac{kg\,m\,sec}{sec^2\,m^2}\times2.0\,m}{1000\frac{kg}{m^3}\times0.000953^2\,m^2}\right)^2+4\left(9.8\frac{m}{sec^2}1.3\,m-\dfrac{1066.58\frac{kg}{m\,sec^2}}{1000\frac{kg}{m^3}}\right)\right]^{1/2}}{2},$$

$$V_2 = 0.1652\frac{m}{sec}\times\frac{100\,cm}{m}\times\frac{60\,sec}{min}=991.49\frac{cm}{min},$$

$$Q = 991.49\frac{cm}{min}\times\frac{\pi}{4}(0.0953\,cm)^2\times\frac{mL}{cm^3}=7.072\frac{mL}{min}.$$

The time to empty the 500 mL bag based on this fluid flow rate of 7.072 mL min^{-1} is then given by

$$t_{empty} \approx \frac{V_{bag}}{Q}=\frac{500\,mL}{7.072\frac{mL}{min}}=70.7\ min.$$

With the exit velocity of the fluid now estimated, we need to check the Reynolds number to see if our assumption of laminar flow is valid:

$$Re = \frac{\rho D V_2}{\mu}=\frac{1000\frac{kg}{m^3}\times0.000953\,m\times0.1652\frac{m}{sec}}{0.001\frac{kg}{m\,sec}}=157.$$

Since $Re < 2100$, our assumption of laminar flow is correct.

4.17 CAPILLARY RISE AND CAPILLARY ACTION

Numerous processes depend on capillary action, that is the ability of liquids to penetrate freely into small pores, cracks, and openings. Capillary action is responsible for transporting water to the uppermost parts of tall trees and has a variety of applications in the fields of printing, textiles, agriculture, cleaning and sanitation products, and medical devices.

4.17.1 EQUILIBRIUM CAPILLARY RISE

Consider the situation shown in Figure 4.16. A small capillary tube is placed within a liquid. The liquid is drawn into the capillary as a result of the surface forces acting on the liquid wetting the inside surfaces of the capillary tube. These surface forces cause a curvature, called a *meniscus*, in the liquid surface as shown at position 3 in Figure 4.16, which, according to the *Laplace–Young equation* (i.e., Equation 2.185), lowers the pressure there relative to that outside the capillary tube.

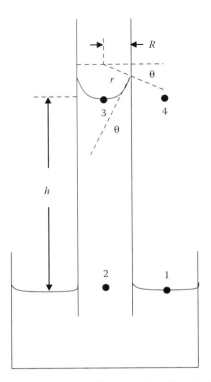

FIGURE 4.16   Capillary rise of a liquid in a small diameter tube of radius R.

This creates a suction that draws the fluid into the capillary tube. The angle (θ) at the meniscus between the liquid surface and the wall of the capillary is called the *contact angle*. The radius of curvature of the meniscus (r) is related to the radius of the capillary tube (R) and the contact angle by the following equation:

$$r = \frac{R}{\cos\theta}.$$
(4.79)

The liquid continues to increase in the capillary until the forces tending to draw up the liquid are balanced by the downward force of gravity acting on the fluid. The equilibrium capillary rise (h) can be found from an analysis of the pressures at points 1, 2, 3, and 4. From the Laplace–Young equation that we developed in Chapter 2 (i.e., Equation 2.185), we can write at the meniscus interface that

$$P_4 - P_3 = \frac{2\gamma}{r}.$$
(4.80)

The pressure at point 2 is greater than the pressure at point 3 by an amount equal to $\rho_L g h$ and the pressure at point 1 is greater than the pressure at point 4 by $\rho_V g h$, hence we can write that $P_1 = P_4 + \rho_V g h$, $P_2 = P_3 + \rho_L g h$. We also have the requirement that at equilibrium, $P_1 = P_2$. Therefore, $P_4 - P_3 = (\rho_L - \rho_V)g h$. Using this result, Equation 4.80 can be solved for the capillary rise as shown in the following equation, recognizing that $\rho_L \gg \rho_V$:

$$h = \frac{2\gamma\cos\theta}{\rho_L R g}.$$
(4.81)

Note that the previous equation also provides a simple means to determine the surface tension of a liquid by measuring its capillary rise.

Example 4.13

Calculate the capillary rise for water in a tube with a diameter of 1 mm. Assume that $\cos \theta \approx 1$ and that the surface tension of water is 72 mN m^{-1}.

Solution

Using Equation 4.81, we can calculate the capillary rise:

$$h = \frac{2 \times 72 \times 10^{-3} \text{ Nm}^{-1} \times 1}{0.0005 \text{ m} \times 1000 \text{ kgm}^{-3} \times 9.8 \text{ msec}^{-2}} = 0.0294 \text{ m} = 29.4 \text{ mm} = 1.16 \text{ in.}$$

4.17.2 DYNAMICS OF CAPILLARY ACTION

Suppose it is desired to estimate the rate at which the fluid enters the capillary, i.e., find $Q(t)$ and $h(t)$. The goal is then to determine an expression for $Q(t)$ and $h(t)$. For the situation shown in Figure 4.16, the liquid is drawn into the capillary by forces arising from the surface tension and this *capillary force* is retarded by the *inertial force* due to the mass of the rising fluid, the *viscous force*, and the *gravitational force* acting on the fluid. In a general sense, this is a very difficult problem; however an approximate solution can be obtained if we assume that the flow in the capillary tube is laminar ($Re < 2300$) and that the velocity profile maintains the same parabolic shape (i.e., Equation 4.13) as the liquid is drawn into the capillary tube.

The capillary force draws fluid into the tube as a result of the suction pressure developed between positions 4 and 3 of Figure 4.16. This is given by the following expression, where we have also made use of Equations 4.79 and 4.80:

$$F_{\text{capillary force}} = \pi R^2 (P_4 - P_3) = \pi R^2 \frac{2\gamma}{r} = 2\gamma R \cos\theta. \tag{4.82}$$

The gravitational force acts on the mass of fluid ($\rho_L \pi R^2 h(t)$) within the tube at time t. This is given by the equation below where ρ_L is the density of the fluid:

$$F_{\text{gravitational force}} = \rho_L g \pi R^2 h(t). \tag{4.83}$$

The viscous force arises as a result of the flow of the fluid as it is drawn into the capillary tube by the capillary force. It is assumed that the flow is laminar and that the Hagen–Poiseuille law (Equation 4.11) can be used to describe this unsteady flow. The driving force for this flow of fluid into the capillary tube is the pressure drop between points 2 and 3 as shown in Figure 4.16. Since the fluid is incompressible, the average velocity of the fluid (i.e., V) is the same as the observed meniscus velocity, which is $dh(t)/dt$. With these assumptions, the volumetric flow rate of the fluid at time t is given by

$$Q(t) = \pi R^2 V(t) = \pi R^2 \frac{dh(t)}{dt} = \frac{\pi R^4 (P_2 - P_3)}{8\mu h(t)}. \tag{4.84}$$

The wall shear stress (i.e., τ_w) is also related to (P_2–P_3) through Equation 4.4, so we can solve this equation for (P_2–P_3) in terms of the wall shear stress as

$$(P_2 - P_3) = \frac{2h(t)\tau_w}{R}. \tag{4.85}$$

Substituting Equation 4.85 into 4.84 and solving for $dh(t)/dt$ gives the next result:

$$\frac{dh(t)}{dt} = \frac{2\pi R h(t) \tau_w}{8\pi\mu h(t)}. \tag{4.86}$$

The viscous force is the wall shear stress times the circumferential area, which is the numerator of the right-hand side of Equation 4.86. Therefore, we obtain

$$F_{viscous\,force} = 8\pi\mu h(t)\frac{dh(t)}{dt}. \tag{4.87}$$

The inertial force is the mass of the fluid in the capillary tube multiplied by its acceleration. This is given by Newton's second law:

$$F_{inertial\,force} = \lim_{t \to \infty}\frac{mV|_{t+\Delta t} - mV|_t}{\Delta t} = \frac{d(mV)}{dt} = \frac{d}{dt}\left(m\frac{dh(t)}{dt}\right) = \rho_L \pi R^2 \frac{d}{dt}\left(h(t)\frac{dh(t)}{dt}\right). \tag{4.88}$$

The inertial force is then equal to the sum of all the forces acting on the fluid as it rises in the tube through capillary action. The capillary force draws the fluid into the capillary tube and the viscous and gravitational forces work in opposition. Using the expressions developed above, we can then write that

$$\rho_L \pi R^2 \frac{d}{dt}\left(h(t)\frac{dh(t)}{dt}\right) = 2\pi\gamma R\cos\theta - 8\pi\mu h(t)\frac{dh(t)}{dt} - \rho_L g\pi R^3 h(t). \tag{4.89}$$

When the Equation 4.89 is rearranged as shown in Equation 4.90, it is known as the *Bosanquet equation* (Zhmud et al. 2000; Kornev and Neimark 2001).

$$\frac{d}{dt}\left(h(t)\frac{dh(t)}{dt}\right) + \left(\frac{8\mu}{\rho R^2}\right)h(t)\frac{dh(t)}{dt} = \frac{2\gamma\cos\theta}{\rho_L R} - gh(t). \tag{4.90}$$

This equation can then be solved for the rise of the fluid in the capillary tube as a function of time, provided suitable initial conditions can be defined. The initial conditions can be found by considering a solution to Equation 4.90 that is valid for short contact times when penetration of the fluid just begins. In this case, only the inertial and capillary forces are dominant and Equation 4.90 becomes

$$\frac{d}{dt}\left(h(t)\frac{dh(t)}{dt}\right) = \frac{2\gamma\cos\theta}{\rho_L R}. \tag{4.91}$$

Integration of Equation 4.91 with the initial condition that $h(0) = 0$ gives

$$h(t)\frac{dh(t)}{dt} = \left(\frac{2\gamma\cos\theta}{\rho_L R}\right)t. \tag{4.92}$$

The left-hand side of Equation 4.92 is proportional to the fluid momentum and shows that as $t\rightarrow0$, the momentum of the fluid approaches zero. Since $h(0) = 0$, this implies for $t\rightarrow0$ that there is a finite velocity during the initial fluid entry phase, which is known as the *Bosanquet velocity* (U_B) that is defined by

$$\text{For } t \rightarrow 0, \frac{dh(t)}{dt} \approx U_B \quad \text{or} \quad h(t) \approx U_B t. \tag{4.93}$$

Equation 4.93 predicts that during the initial time of fluid penetration, the capillary rise increases linearly with time. On substitution of the above results into Equation 4.92, we can solve for the initial fluid velocity as it enters the capillary tube due to the capillary force:

$$U_B = \left(\frac{2\gamma\cos\theta}{\rho_L R}\right)^{1/2}. \tag{4.94}$$

Hence, from Equations 4.93 and 4.94 the initial conditions for Equation 4.90 are $h(0) = 0$ and $dh(0)/dt = U_B$.

At long times the fluid in the capillary reaches a stationary level, which represents a balance between the capillary forces and the gravitational forces. At this equilibrium, $dh(t)/dt = 0$ and Equation 4.90 simplifies to the following equation for the capillary rise, which is the same as Equation 4.81 that was found earlier:

$$h_{\text{equilibrium}} = \frac{2\gamma\cos\theta}{\rho_L R g}. \tag{4.95}$$

For intermediate times, the first term in Equation 4.90 can be neglected and we then obtain the *Lucas–Washburn equation*. This equation describes the rise of the fluid after the initial entry of the fluid into the tube since the fluid acceleration is decreasing and the inertial force is much smaller for these times than the viscous and gravitational forces:

$$\left(\frac{8\mu}{\rho_L R^2}\right)h(t)\frac{dh(t)}{dt} = \frac{2\gamma\cos\theta}{\rho_L R} - gh(t). \tag{4.96}$$

If the capillary rise is not large then the gravitational force can also be neglected and Equation 4.96 can be integrated to give the following result for the capillary rise as a function of time:

$$h(t) = \sqrt{\frac{R\gamma\cos\theta t}{2\mu}}. \tag{4.97}$$

This equation predicts that the capillary rise is directly proportional to $\sqrt{t}$. Also, since $V(t) = dh(t)/dt$ and $Q(t) = \pi R^2 V(t)$, we obtain the following equation after differentiating Equation 4.97 with respect to t:

$$V(t) = \sqrt{\frac{\gamma R\cos\theta}{8\mu t}} \quad \text{and} \quad Q(t) = \pi R^2\sqrt{\frac{\gamma R\cos\theta}{8\mu t}}. \tag{4.98}$$

Equation 4.98 predicts that for long times V and Q decrease in proportion to $1/\sqrt{t}$. Notice also that Equation 4.98 at $t = 0$ gives the result that V and Q are infinite. This is a result of the neglect of the inertial forces when the fluid is first being drawn into the capillary tube.

Example 4.14

Estimate the time for water at 25°C to reach a height of 15 mm in a capillary tube that has a diameter of 1 mm. Assume that $\cos\theta \approx 1$ and that the surface tension of water is 72 mN m^{-1}.

Solution

Using Equation 4.97, we can solve for the time for the fluid to reach a particular height.

$$t = \frac{2\mu h^2}{R\gamma \cos\theta} = \frac{2 \times 0.001 \text{ Nsecm}^{-2} \times 0.015^2 \text{ m}^2}{0.0005 \text{ m} \times 72 \times 10^{-3} \text{ Nm}^{-1} \times 1} = 0.0125 \text{ sec} = 12.5 \text{ msec}.$$

PROBLEMS

1. Derive Equations 4.10 and 4.23.
2. Derive Equation 4.26.
3. Derive Equations 4.30 and 4.31.
4. Using Equation 4.23 with a yield stress $\tau_y = 0.0289$ dyn cm^{-2} and $s = 0.229$ (dyn sec cm^{-2})$^{1/2}$, show that this equation provides an excellent fit to the data given in Figure 4.8.
5. Estimate the shear rate at the wall of a capillary.
6. Obtain an expression for the radial dependence of the shear rate for a Newtonian fluid. What is the ratio of the average shear rate to the wall shear rate?
7. In Example 4.12, we obtained a simple estimate of the time needed to drain the IV bag. However, the IV bag has a length of 30 cm, and as the fluid drains from the bag, the potential energy (Z_1) of the remaining fluid that drives the flow will change with time. Therefore, to obtain a better estimate of the time to drain the bag, we also need to include how Z_1 changes with time. This can be obtained by combining the expression for the exit velocity (V_2) derived in Example 4.12 with a mass balance on the bag itself. Letting $M(t)$ denote the mass of IV fluid remaining in the bag at any time t, we can write the IV bag mass balance as

$$\frac{dM(t)}{dt} = \rho_L S_{bag} \frac{d(Z_1(t) - H)}{dt} = -\rho\pi \frac{D_{tube}^2}{4} V_2.$$

In this equation, H represents the height of the tube leaving the bottom of the bag relative to where this tube then enters the patient's arm. For this problem, H is equal to 1 m and $Z_1 - H$ is then the depth of the fluid in the bag. S_{bag} is the cross-sectional area of the IV bag and D_{tube} is the diameter of the tube. Use this equation and the expression from Example 4.12 for V_2 to obtain the time to just drain the bag. How does this time compare to the pseudo-steady state estimate of drain time obtained in Example 4.12?
8. Derive Equation 4.11. Start with Equation 4.10 and the assumption of a Newtonian fluid. Also show that Q may be obtained by integrating $v_z(r)$ in Equation 4.13 using Equation 4.8.
9. The cardiac output in a human is normally about 6 L min^{-1}. Blood enters the right side of the heart at a pressure of about 0 mmHg gauge and flows via the pulmonary arteries to the lungs at a mean pressure of 11 mmHg gauge. Blood returns to the left side of the heart through the pulmonary veins at a mean pressure of 8 mmHg gauge. The blood is then ejected from the heart through the aorta at a mean pressure of 90 mmHg gauge. Use the Bernoulli equation to obtain an estimate of the total work performed by the heart. Carefully state any assumptions and express your answer in watts (recall that 1 W = 1 J sec^{-1}, where J = N m = kg m^2 sec^{-2}).
10. Use the Bernoulli equation to describe the expected velocity and pressure changes upstream, within, and downstream of an arterial stenosis. A stenosis is a partial blockage or narrowing of an artery by the formation of plaque (atherosclerosis).
11. Blood is flowing through a bundle of hollow fiber tubes that are each 50 μm in diameter. There are 10,000 tubes in the bundle. The hollow fiber tube length is 12 cm and the pressure drop across each tube is found to be 250 mmHg. The hematocrit of the blood is 0.40. The blood flow rate for these conditions is 80 μL h^{-1} in each tube. Estimate the marginal zone thickness, the hematocrit of the core, and the tube hematocrit. (Hint: you will need to solve Equations 4.39 and 4.40.)

12. Blood enters a hollow fiber unit (hemodialysis) that is used as an artificial kidney. The unit consists of 10,000 hollow fibers arranged in a shell and tube configuration. Blood flows from an artery in the patient's arm through a tube and is uniformly distributed to the fibers via an arterial head space region at the entrance of the unit. The blood then leaves each fiber through the venous head space region of the unit and is returned to a vein in the patient's arm. Each hollow fiber has an inside diameter of 220 μm and a length of 25 cm. The fiber void volume with the shell of the unit is 50%. Assuming the maximum available pressure drop across the hollow fiber unit is 90 mmHg, estimate the total flow rate of blood through the hollow fiber unit.

13. You are designing a hollow fiber unit. The flow rate of blood is assumed to be at a high enough shear rate that the blood behaves as a Newtonian fluid. The fiber diameter is 800 μm and its length is 30 cm. You want a flow rate of 8 mL min^{-1} for each fiber. What should be the pressure drop in mmHg (where 760 mmHg = 101,325 N m^{-2}) across each fiber length to achieve this flow rate?

14. Using the data shown in Figure 4.4, find the best values of s and τ_y in the Casson equation (see Equation 4.19) that fit these data. Recall that the shear stress can be found from the data shown in Figure 4.4 from the following relationship, $\tau = \mu_{apparent}\,\dot{\gamma}$. Also from the Casson equation, note that a plot of $\tau^{1/2}$ vs. $\dot{\gamma}^{1/2}$ should be linear with a slope equal to s and an intercept of $\tau_y^{1/2}$. Express the units of s as (dyn sec cm^{-2})$^{1/2}$ and τ_y in (dyn cm^{-2}).

15. The following values were obtained for the apparent viscosity of blood ($H = 40\%$) in tubes of various diameters. Estimate the thickness of the marginal or plasma layer (δ) in microns and the core viscosity in centipoise from these data. The viscosity of the plasma is 1.09 cP. Here, we can employ an approximate method based on Equation 4.36. From the binomial series, we know that:

$$(1 - \delta/R)^4 \approx 1 - 4(\delta/R) \text{ when } (\delta/R)^2 \ll 1.$$

Then show, using the above approximation, that Equation 4.36 simplifies to the following equation:

$$\mu_{apparent} = \frac{\mu_c}{1 + 4\left(\dfrac{\delta}{R}\right)\left(\dfrac{\mu_c}{\mu_p} - 1\right)}.$$

Make a plot of $1/\mu_{apparent}$ vs. $1/R$ and use the previous equation to find μ_c and δ.

Tube Radius (R) (μm)	Apparent Viscosity (cP)
20	1.68
40	2.25
60	2.49
100	2.88
300	3.00

16. A bioartificial liver has a plasma flow of 1000 mL min^{-1} through a hollow fiber unit that contains hepatocytes on the shell side. The hollow fiber unit contains 10,000 fibers. The fiber length is 75 cm and the inside diameter of the fibers is 300 μm. What is the pressure drop across each fiber in mmHg? (1 atm = 101,300 N m^{-2} = 760 mmHg)

17. A design for a novel aortic cannula consists of a smooth thin-walled polyethylene tube 7 mm in diameter and 40 cm in length. For a flow rate of blood of 5 L min^{-1} through the cannula, estimate the pressure drop (mmHg) over the length of the cannula. Carefully state and justify any assumptions that you make.

18. For a cell-free layer of 3 μm and a hematocrit of 40%, calculate the apparent viscosity for blood flowing in a 100 μm diameter tube. Assume the plasma viscosity is 1.093 cP and the core viscosity is about 3.7 cP.

19. You are designing a small implantable microfluidic pump for the continuous delivery of a drug. The pump is a two compartment cylindrical chamber; one compartment containing the drug dissolved within a solvent and the other compartment is the pump engine. These two compartments are separated by a movable piston that pushes on the drug compartment as the pump engine operates. At the other end of the drug compartment there is an exit tube through which the drug solution flows. The exit tube of the pump through which the drug leaves the pump has an internal diameter of 10 μm and a total length of 15 cm. What gauge pressure is needed (mmHg) within the drug compartment to maintain a flow rate of the drug solution of 350 micrograms day^{-1}? You may assume that the drug solution has a density of 1 g cm^{-3} and a viscosity of 3 cP.

20. You are part of a team developing an osmotic pump for the delivery of a drug. An osmotic pump has two compartments; one compartment, the osmotic engine, contains an osmotic agent that is retained by a membrane and imbibes water when placed within the body. This compartment also has a piston that expands against another compartment containing the drug solution as water is imbibed from the surroundings. The movement of the piston then forces the drug solution out into the body. A question has been raised as to what would be the maximum pressure within the device if, after implantation, the exit tube that delivers the drug becomes blocked? Assume the interstitial fluid pressure is −3 mmHg and its osmotic pressure is 8 mmHg. If the concentration of the osmotic agent is 0.05 Osm, what would be the maximum hydrodynamic pressure within the device assuming the delivery tube becomes blocked?

21. A viscometer has been used to measure the viscosity of a fluid at 20°C. The data of the shear stress vs. the shear rate when plotted on a log–log graph are linear. Is this fluid Newtonian? Explain your answer.

22. Blood flows through a bundle of hollow fibers at a total flow rate of 250 mL min^{-1}. There are a total of 7500 fibers, the diameter of a fiber is 75 μm and the length of a fiber is 15 cm. The viscosity ratio for blood under these conditions is estimated to be 0.70. What is the pressure drop across a fiber in mmHg?

23. Commercially available spermicidal or contraceptive gels have been developed for the purpose of preventing sperm transport and thus blocking fertilization (from Owen et al. 2000). Current interest has also led to the possible use of contraceptive gels as a means to reduce the spread of sexually transmitted diseases, such as AIDS, and has led to interest in developing formulations of these gels for both prophylaxis and contraception. The physical properties of these gels must be such that, when applied, they spread to coat the vaginal epithelia and then stay in place long enough to provide contraception as well as adequate protection from disease-causing agents such as bacteria and viruses. This is accomplished by gel formulations that deliver topically bioactive compounds, such as microbiocides, and also by the physical barrier to infection provided by the coating layer. The spreading and retention of intravaginal contraceptive formulations are fundamental to their efficacy and these performance characteristics are governed in part by their rheological properties.

In vivo, these contraceptive gels will experience a wide range of shear rates as a result of movements of the vaginal epithelial surfaces, gravity, capillary flow, and sex. It is estimated that these shear rates may range from as low as 0.1/sec to as high as 100/sec during sex. The polymeric nature of these gels suggests that they will exhibit non-Newtonian rheological behavior.

A popular model for describing the rheological behavior of non-Newtonian gels is that of the two-parameter power law model. This model is shown by the following equation for flow in a cylindrical tube:

$$\tau_{rz} = -m\dot{\gamma}^{n-1}\frac{dv_z}{dr} = -\mu_{apparent}\frac{dv_z}{dr} = \mu_{apparent}\dot{\gamma}.$$

In this equation, m and n are constants characterizing the fluid, and $\dot{\gamma}$ is the shear rate equal to $-dv_z/dr$. From the previous equation, what is the relationship for the apparent viscosity's ($\mu_{apparent}$) dependence on the shear rate of the fluid? Using the previous equation, derive an expression for the axial velocity profile, $v_z(r)$, and the mass flow rate ($\dot{m} = \rho_L Q$) of a fluid described by the power law model.

The table below shows data obtained from a rheometer for the commercially available gel, *Conceptrol*.

Apparent Viscosity of Conceptrol vs. Shear Rate

Shear Rate (sec⁻¹)	Viscosity (Pa sec)
0.01	6000
0.05	2000
0.1	800
0.5	400
1	100
5	80
10	40
50	15
100	5
500	0.80
1000	0.2

Perform a regression analysis of the data in the above table and find the power law parameters m and n. Compare your model predictions to the data shown in the above table. Carefully state the units of these parameters have.

24. Bush et al. (1997) studied the flow of urine in an anatomical model of the human female urethra as shown in Figure 4.17. They obtained the data shown in the table below from their experimental model.

Flow Rate (cm³ sec⁻¹)	Pressure Difference (ΔP, cm of H_2O)
5	2.0
8	4.5
10	8.0
13	14.0
15	21.0
17	28.0
20	38.0
23	52.0
25	60.0

The pressure difference represents the reservoir head (H) shown in Figure 4.17b; however, the total pressure head driving the urine flow also includes the additional 2 cm (see Figure 4.17a) from the bottom of the reservoir to the exit of the urethra. Use the Bernoulli equation (Equation 4.72) to predict the pressure difference for each of the flow rates given in the above table. Show your results as a graphical comparison between the data and the model. Assume that the diameter of the urethra is 3.25 mm and that its length is 4 cm. Determine whether the flow is laminar or turbulent. If the model does not fit the data, what

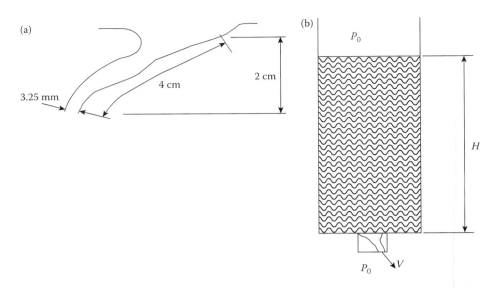

FIGURE 4.17 (a) A longitudinal profile of the model urethra adapted from a video cystogram. The lateral cross section is taken to be circular. (b) The experimental apparatus showing the reservoir head, H, and the distal velocity, V. The atmospheric pressure is P_0. The apparatus permitted a maximum head of 65 cm H_2O to be applied to the urethra. (Bush, et al., *J. Biomech.*, 30, 967–969, 1997. With permission.)

     parameters in the model can you change to improve the fit, e.g., what is the effect of the urethral diameter?

25. Consider the steady laminar flow of a Newtonian fluid in the thin channel formed between two large parallel and horizontal plates of length L and width W. The plates are separated by a distance of $2H$ and $H \ll L$ and W. Show that the velocity profile $v_z(y)$ and the volumetric flow rate are given by the following expressions:

$$v_z(y) = \frac{(P_0 - P_L)H^2}{2\mu L}\left[1 - \left(\frac{y}{H}\right)^2\right]$$

$$Q = \frac{2}{3}WH^3\frac{(P_0 - P_L)}{\mu L},$$

    with P_0 and P_L being the inlet and exit pressures.

26. Using the above result, show that the following expression approximates the penetration of liquid, $L(t)$, by capillary action into a slit channel used in a diagnostic device:

$$L(t) = 2\left[\frac{H\gamma\cos\theta}{3\mu}\right]^{1/2}t^{1/2}.$$

    A diagnostic device makes use of a thin rectangular channel to draw in a sample of blood. Assuming the blood sample has a viscosity of 3 cP and that the plates forming the channel are separated by a distance of 1 mm, estimate the time for the sample of blood to travel a distance of 15 mm in the channel. Assume the blood has a surface tension of 0.06 N m⁻¹ and that the contact angle is 70°.

27. The following table shows data for the measured velocity profile for the laminar boundary layer flow of a fluid across a flat plate (from Schlicting 1979). Make a plot of these data and compare it to the approximate velocity profile given by Equation 4.66.

$\dfrac{v_x(x,y)}{V}$	$y\sqrt{\dfrac{V}{nx}}$
0.25	0.1
0.50	0.175
1.0	0.34
1.5	0.495
2.0	0.63
2.5	0.745
3.0	0.85
3.5	0.92
4.0	0.955
5.0	0.98

28. Zhmud et al. (2000) obtained the following data for the capillary rise of dodecane in a 200 μm diameter capillary tube. Compare these results to those predicted by the Lucas–Washburn equation, i.e., Equation 4.97. The physical properties for dodecane are as follows: viscosity $= 1.7 \times 10^{-3}$ Pa sec, surface tension $= 2.5 \times 10^{-2}$ N m^{-1}, and density $= 750$ kg m^{-3}. What is the value of the contact angle that gives the best fit to these data using the Lucas–Washburn equation?

Time (sec)	$h(t)$ (mm)
0.03	2
0.06	4.3
0.1	6.1
0.13	7.9
0.16	9
0.20	10
0.23	11
0.26	12
0.30	12.8
0.33	13.5
0.36	14.2
0.40	14.8
0.42	15.5
0.46	16.1
0.50	16.3

29. In the paper by Zhmud et al. (2000), the capillary rise for diethyl ether in a 1 mm diameter capillary tube was found to be 8.6 mm at equilibrium. How does this value compare with the value predicted by Equation 4.81? The physical properties for diethyl ether are as follows: surface tension $= 1.67 \times 10^{-2}$ N m^{-1}, density $= 710$ kg m^{-3}, and contact angle $= 26°$.

30. Prove that for laminar flow in a cylindrical tube, the kinetic energy correction factor, α, in the Bernoulli equation is equal to 2.

31. Show, using Equation 4.11, that $f = 16/Re$ for laminar flow in a cylindrical tube.

32. A small airplane is flying at 3000 m above sea level. The density of air at this altitude is 0.83 kg m^{-3}. A Pitot tube gives a difference between the stagnation pressure and the static pressure of 50 mmHg. Based on this information, how fast is the airplane travelling in miles per hour?

33. The viscosity of a fluid (μ_{test}) may be found in terms of the viscosity of another reference fluid whose viscosity is known ($\mu_{reference}$) by measuring the time it takes for the fluid (t_{test}) to

drain by gravity a certain distance within a vertical tube of constant cross section and then comparing that time to the time ($t_{reference}$) for the reference fluid whose viscosity is known. Show that for either fluid the velocity of the fluid exiting the tube is related to the change in height (z) of the fluid by the following equation:

$$\frac{dz}{dt} = -V_2,$$

where V_2 is the velocity of the fluid exiting the bottom of the tube. Next, apply the Bernoulli equation to the fluid in the tube from the top of the fluid surface (1) to the bottom of the tube (2). Assuming laminar flow of the fluid with $f = 16/Re$ and using the previous equation for V_2, show that dz/dt is also given by the following equation:

$$\frac{dz}{dt} = -\frac{\rho D^2 g}{32\mu},$$

where D is the inside diameter of the tube and g is the acceleration of gravity. Now integrate the previous equation and show that the time for the fluid to drain a distance equal to Δz is given by

$$\Delta z = \frac{\rho D^2 g t}{32\mu}.$$

Now, if both fluids drain the same distance, i.e., $\Delta z_{test} = \Delta z_{reference}$, show that the following equation may be used to relate their viscosities in terms of their drain times and densities:

$$\frac{\mu_{test}}{\mu_{reference}} = \frac{\rho_{test} t_{test}}{\rho_{reference} t_{reference}}.$$

Below is shown some data for the drain time of different concentrations of chitosan in water, which was obtained by one of my former PhD students, Prasanjit Das. Pure water was used as the reference fluid with a viscosity of 0.001 Pa sec or 1 cP and a density of 997 kg m^{-3}. The time ($t_{reference}$) for the water to flow by gravity a defined distance ($\Delta z_{test} = \Delta z_{reference}$) in a capillary tube was found to be 0.120 sec. Calculate the viscosity (cP) of the chitosan solution for each of the concentration values given in the following table.

Chitosan Concentration (ppm)	Chitosan Solution Density (kg m^{-3})	Drain Time (t_{test}) (sec)
10	997	0.120
100	997	0.120
200	997	0.230
500	997	0.290
1000	998	0.480

34. The formation of a small droplet at the tip of a capillary tube can be used to determine the surface tension of a fluid. This is known as the hanging droplet method. The droplet can grow in size until the gravitational force exceeds the surface tension force that holds the droplet to the periphery of the tube, that is

$$\gamma \pi D_{tube} = \frac{4}{3} \pi R^3 \rho_L g,$$

where γ is the surface tension of the fluid and R is the radius of the droplet formed. Show that the surface tension is then given by the following equation:

$$\gamma = \frac{0.1667 D^3 \rho_L g}{D_{\text{tube}}},$$

where D is the diameter of the droplet that just releases itself from the tube. Below is shown data that were obtained by my PhD student Prasanjit Das on droplets formed at the tip of a 1 mm outside diameter capillary tube for different concentrations of chitosan in water. From these data, calculate the surface tension (millinewtons per meter) of the chitosan solution for each of the concentration values given in the following table.

Chitosan Concentration (ppm)	Chitosan Solution Density (kg m^{-3})	Droplet Diameter (mm)
10	997	3.48
100	997	3.46
200	997	3.43
500	997	3.40
1000	998	3.338

35. Bazilevsky et al. (2003) measured the entry flow rate of water into a 0.65 mm diameter capillary tube. The average entry flow rate of water was found to be 220 mm^3 sec^{-1}. What is the entrance velocity of the water and how does this compare to that predicted by the Bosanquet equation? Assume that the surface tension of water is 0.071 N m^{-1} and its viscosity is 0.001 Pa sec. Also $\theta = 0°$.

36. A concentrated solution of sugar dissolved in water is flowing through a capillary tube with an inside diameter of 2 mm. The capillary tube will be used to find the viscosity of this solution. The length of the capillary tube is 10 cm. The density of the solution is 1200 kg m^{-3}. For a flow rate of 60 cm^3 min^{-1} the pressure drop per length of tube was found to be 1.0 mmHg cm^{-1}. What is the viscosity (pascal-second) of the fluid?

37. A polymeric fluid having a viscosity of 0.40 Pa sec, a density of 800 kg m^{-3}, and a surface tension of 0.02 N m^{-1}, is drawn by capillary action into a glass tube of radius 0.025 cm. If the contact angle is such that cos $\theta \sim 1$, estimate the time required for the oil to reach a height equal to 90% of its equilibrium rise. Also, find the initial flow rate of this fluid into the tube in cubic centimeters per second.

38. The capillary rise, $h(t)$, as a function of time, t, for a biofluid was measured in a capillary tube of radius 0.3 mm. When the data were plotted as $h(t)$ vs. $t^{1/2}$, the data were found to be linear with a slope equal to 0.06 m sec$^{-1/2}$. If the fluid has a viscosity of 1.2×10^{-3} Pa sec, and the contact angle $\sim 0°$, estimate the surface tension of the fluid in newtons per meter.

39. A particular fluid has a shear stress of 0.005 N m^{-2} at a shear rate of 1 sec^{-1} and a shear stress of 2 N m^{-2} at a shear rate of 50 sec^{-1}. Present an argument as to whether this fluid is Newtonian or non-Newtonian.

40. Blood is flowing at a flow rate of 7 L min^{-1} through a tube that is 6 mm in diameter and 50 cm in length. Estimate the pressure drop of the blood over this length of tubing in mmHg.

41. The apparent viscosity of blood flowing in a 100 μm diameter tube was found to be 2.6 cP. Assuming that the core viscosity of blood is about 3.7 cP, estimate the thickness of the marginal zone layer. Assume the plasma viscosity is 1.1 cP.

42. Blood flows through a 100 μm diameter glass tube that is 0.1 cm in length. Assuming that the blood behaves like the data obtained by Fahraeus and Lindqvist (1931) as shown in Figure 4.12, estimate the volumetric flow rate of the blood in cubic centimeters per hour if the pressure difference across the capillary length is 6000 Pa. Assume blood at high shear rates has a viscosity of 3 cP.

43. You are designing a hollow fiber unit with 10,000 fibers. The flow rate of blood is assumed to be at a high enough shear rate that the blood behaves as a Newtonian fluid. The fiber diameter is 1000 μm and its length is 50 cm. You want a flow rate of 100 L min^{-1} for the entire unit. What should be the pressure drop (mmHg) across the unit to achieve this flow rate?

44. The fermentation broth from a continuous bioreactor has a density of 1.02 g cm^{-3} and a viscosity of 1.8 cP. The broth is being pumped into the bottom of a feed tank that leads into a filtration system. The level of the broth in the feed tank is 20 feet higher than the level of the broth in the fermentor. The filtration feed tank is continuously feeding a downstream filtration system at the same flow rate as the feed from the bioreactor. The pressure in the bioreactor is maintained at 1 atm and the pressure in the receiving tank is maintained at 6 atm in order to facilitate the downstream filtration process. The pipe connecting the bioreactor and the receiving tank is equivalent to 75 ft of pipe with an inside diameter of 3 in, hence this length includes the additional resistance of any valves and pipe fittings. The desired flow rate of the fermentation broth as it is pumped to the storage tank is 100 gal min^{-1}. What power (kilowatts) must be delivered by the pump to the fluid in order to effect this transfer of broth from the bioreactor to the storage tank?

45. Water at 20°C is pumped through 3000 cm of pipe with an internal diameter of 7.8 cm into an overhead storage tank that vents to the surroundings. The total elevation change is 1000 cm. The valves and other pipe fittings are equivalent to an additional pipe length equal to 15 pipe diameters. What outlet pressure in atmospheres of the pump is needed to move the water at a flow rate of 70 L min^{-1}? (At 20°C the viscosity of water is 1.002 cP and the density is 0.9982 g cm^{-3}.)

46. The pressure drop of a nutrient fluid flowing through the hollow fibers of a bioreactor cannot exceed 50 mmHg. There are 10,000 fibers in the bioreactor and the total flow rate of the nutrient media to the bioreactor is 50 L min^{-1}. Estimate the allowable length of these fibers assuming the following conditions:

    Fiber radius = 0.1 cm
    Fluid viscosity = 0.05 Pa sec

47. You are studying in the laboratory the laminar boundary layer flow of a fluid across a very thin flat plate. The flat plate is suspended vertically from a spring in the range where Hooke's law applies, i.e., $F = k \Delta\lambda$, where $\Delta\lambda$ is the amount of spring extension from the unloaded position of the spring and k is the spring constant. Under conditions of no air flow, the extension of the spring due to the weight of the flat plate is $\Delta\lambda_{plate} = 5$ cm. When there is an upward flow of air, the resulting drag opposes the weight of the flat plate. Find the air flow in meters per second for which the spring extension is zero, i.e., $\Delta\lambda = 0$. Use the following data to find your answer:

    Mass of the plate, $M = 0.001$ kg
    Plate length, $L = 0.2$ m
    Plate width, $W = 0.1$ m
    Air speed, $V =$ find this
    Air density, $\rho = 1.2$ kg m^{-3}
    Air kinematic viscosity, $\upsilon = 1.51 \times 10^{-5}$ m^2 sec^{-1}

48. Estimate the maximum power that is generated in megawatts from a wind turbine whose blades are 300 ft in diameter. Assume that the wind speed averages 20 miles per hour and the density of air is 1.2 kg m^{-3}. Carefully state your assumptions.

49. Estimate the pressure drop in mmHg of blood flowing in the annulus formed by two concentric cylindrical tubes. The inner tube has an outer diameter of 8 mm and the outer tube has an inner diameter of 15 mm. The blood flow rate is 15,000 mL min^{-1} and the overall length of the tube section being considered is 40 cm.

50. The following data were obtained for the flow of a fluid through a small tube that was 0.9 mm in diameter. The flow rate was 0.3 cm^3 sec^{-1} and the pressure drop over the length of the 50 cm fiber was found to be 50 kPa. Estimate the viscosity of this fluid in pascal-second.

51. Below is shown viscometry data obtained for a 5% aqueous solution of hydroxyethyl-cellulose (HEC). Choose the best answer that describes the rheological behavior of this solution:
    a. Newtonian
    b. Non-Newtonian
    c. Non-Newtonian power law solution (i.e., $\mu = m\dot{\gamma}^{n-1}$)
    d. None of these choices

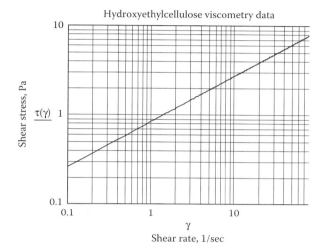

Hydroxyethylcellulose viscometry data

52. The equilibrium rise height of a rather viscous fluid in a small capillary tube of diameter equal to 0.1 mm was found to be 65 mm. The contact angle was also measured to be 75°. Estimate the surface tension of this fluid if its viscosity is $\mu = 0.75$ Pa-sec and $\rho_L = 800$ kg/m³.

53. Water at 70°F enters a pump through a 3 in schedule 40 pipe at atmospheric pressure and is being pumped at a rate of 100 gal min⁻¹ (1 gal = 3.7853 L) through a pipe system made up of 500 ft of 3 in schedule 40 steel pipe (internal diameter = 3.068 in). The pipe circuit includes four 90 degree elbows. The total change in elevation from the pump entrance to the discharge of the water into the atmosphere is equal to 400 ft. Find the horsepower (hp) required for the pump assuming that the pump efficiency is 70%. The viscosity of water under these conditions is 0.96 cP and the density of water is 1 g cm⁻³.

5 Solute Transport in Biological Systems

5.1 DESCRIPTION OF SOLUTE TRANSPORT IN BIOLOGICAL SYSTEMS

In this chapter, we will focus our discussion on the transport of solutes in biological systems. Solute transport occurs both through bulk fluid motion, also known as *convection*, and by solute diffusion due to the presence of solute concentration gradients. In biological as well as synthetic membrane systems, the diffusion of a solute will also be affected by the presence of a variety of heterogeneous structures. For example, solutes will need to diffuse through porous structures, such as the capillary wall or a polymeric membrane, around or through cells within the extravascular space, and through the interstitial fluid containing a variety of macromolecules.

5.2 CAPILLARY PROPERTIES

Our primary focus in this chapter concerns solute transport through the capillary wall as a representative porous semipermeable membrane. Therefore, the following development is also generally applicable to synthetic membranes that are used in a variety of biomedical device applications. To begin our discussion of capillary wall solute transport, we first need to define the physical properties of a typical capillary. These properties are summarized in Table 5.1. Note that capillaries are very small, having diameters of about 8–10 μm and lengths that are less than 1 mm. The residence time of blood in a capillary is also only on the order of 1–2 sec. Therefore, each capillary can only supply nutrients and remove waste products from a very small volume of tissue that surrounds each capillary.

There are three types of capillaries. They are referred to as continuous, fenestrated, and discontinuous. The *continuous capillaries* are found in the muscle, skin, lungs, fat, the nervous system, and in connective tissue. The capillary lumen lies within a circumferential ring of several endothelial cells, as shown in Figure 3.2. *Fenestrated capillaries* are much more permeable to water and small solutes in comparison to continuous capillaries. These capillaries are found in tissues that are involved in the exchange of fluid or solutes such as hormones. For example, within the kidney they are found in the glomerulus and allow for a high filtration rate of plasma. The endothelium of these capillaries is perforated by numerous small holes called *fenestrae*. The fenestrae are sometimes covered by a thin membrane that provides selectivity with regard to the size of solutes that are allowed to pass through. *Discontinuous capillaries* have large endothelial cell gaps that readily allow the passage of proteins and even red blood cells.

5.3 CAPILLARY FLOW RATES

Some of the blood plasma that enters the capillary will be carried or filtered across the capillary wall by the combined effect of the hydrodynamic and oncotic pressure differences that exist between the capillary and the surrounding interstitial fluid. This perfusion of plasma across the capillary wall is also known as *plasmapheresis*. We can determine the total flow rate of this fluid across the capillary

TABLE 5.1
Capillary Characteristics

Property	Value
Inside diameter (D_c)	10 μm
Length (L)	0.1 cm
Wall thickness (t_m)	0.5 μm
Average blood velocity (V)	0.05 cm sec^{-1}
Pore fraction	0.001
Wall pore diameter (d_p)	6–7 nm
Inlet pressure	30 mmHg
Outlet pressure	10 mmHg
Mean pressure (P_c)	17.3 mmHg
Colloid osmotic pressure (π_p)	28 mmHg
Interstitial fluid pressure (P_{if})	–3 mmHg
Interstitial fluid colloid osmotic pressure (π_{if})	8 mmHg

wall using the relationships developed in Chapter 3. Recall from Equation 3.7 that the value of the hydraulic conductance, L_P, is given by

$$L_P = \left(\frac{A_P}{S}\right)\frac{r^2}{8\mu t_m \tau},$$ (5.1)

where S represents the circumferential surface area of a given capillary and A_P is the total cross-sectional area of the pores of radius r in the capillary wall. The ratio, A_P/S, is also referred to as the porosity of the capillary wall (or membrane) and is often given the symbol ε.

Example 5.1

Calculate the convective flow rate of plasma across the capillary wall.

Solution

Using the capillary properties provided in Table 5.1, and a plasma viscosity of 1.2 cP, the value of the hydraulic conductance, L_P, can be shown as in Example 3.2 to be equal to 0.61 cm^3 h^{-1}m^{-2} mmHg^{-1} (or 1.28×10^{-12} m^2 sec kg^{-1}). We can then calculate the filtration rate as follows using Equation 3.4:

$$Q_{filtration} = \frac{0.61\ \text{cm}^3}{\text{hm}^2\,\text{mmHg}} \times \pi \times 10 \times 10^{-6}\ \text{m} \times 0.001\ \text{m}$$

$$\times \left[(17.3 - -3) - (28 - 8)\right]\text{mmHg} = 5.75 \times 10^{-9}\ \text{cm}^3\,\text{h}^{-1}$$

$$= 5.75 \times 10^{-6}\ \mu\text{L h}^{-1}.$$

We can compare this value of the plasma filtration flow rate across the capillary wall to the total flow rate of blood entering the capillary, i.e., $Q_{capillary} = (\pi/4)\,D_c^2\,V$:

$$Q_{capillary} = \frac{\pi}{4}(10 \times 10^{-6}\ \text{m})^2 \times \frac{0.05\ \text{cm}}{\text{sec}} \times \left(\frac{100\ \text{cm}}{\text{m}}\right)^2 \times \frac{3600\ \text{sec}}{\text{h}}$$

$$= 1.41 \times 10^{-4}\ \text{cm}^3\,\text{h}^{-1} = 0.14\ \text{mL h}^{-1}.$$

We find for the special case of a capillary in Example 5.1 that the volumetric flow rate of blood entering the capillary is significantly higher than the filtration flow of plasma across the capillary wall. We may therefore assume that the blood flow, $Q_{capillary}$, is constant along the length of the capillary.

5.4 SOLUTE DIFFUSION

In addition to the convective transport of solute, e.g., as carried along by the bulk motion of the fluid, solute can also diffuse down its own concentration gradient. From Equation 2.70, this is also equivalent to stating that a solute diffuses from a region of high chemical potential to a region of lower chemical potential.

5.4.1 FICK'S FIRST LAW AND DIFFUSIVITY

Consider the situation shown in Figure 5.1. The surface of a semi-infinite plate of length L contains a solute that maintains a constant concentration along the surface of the plate. At $t = 0$, this surface is contacted with a quiescent fluid or a solid material that initially does not contain solute. As time progresses, solute diffuses from the surface of the plate into the quiescent fluid or solid material. If the fluid has solute at a concentration of C_∞, then the following analysis still applies; however, the concentration would need to be defined as $\overline{C} = C - C_\infty$.

Figure 5.1 shows a thin shell of thickness Δy. The rate at which the solute enters and leaves this thin shell by diffusion at y and $y + \Delta y$ is proportional to the solute concentration gradient at these locations. The diffusion rate of this solute can therefore be described by *Fick's first law*:

$$J_S = -DS \frac{dC}{dy},$$ (5.2)

where C represents the concentration of the solute and typical units are moles per cubic centimeter or moles per liter. The solute diffusion rate is given by J_S and typically has units of moles per second. The surface area normal to the y-direction of solute diffusion is S, and D is the *solute diffusivity* or the *diffusion coefficient*. The solute diffusivity generally depends on the size of the solute and the physical properties of the fluid or material in which the solute is diffusing. Since the solute is assumed to be diffusing through a homogeneous medium, this solute diffusivity is sometimes referred to as the *bulk diffusivity*. The diffusivity, D, has units of square centimeters per second or square meters per second.

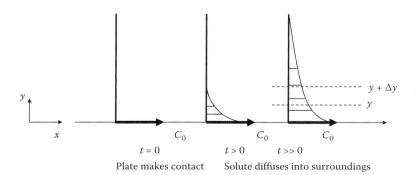

FIGURE 5.1   Solute concentration in the vicinity of a flat plate of constant surface concentration.

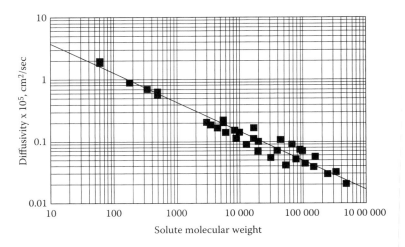

FIGURE 5.2 Solute diffusivity in water at 37°C. (Based on data from Renkin, E.M. and Curry, F.E., *Membrane Transport in Biology*, Springer-Verlag, New York, 1979.)

Figure 5.2 presents diffusivity data (Renkin and Curry 1979) for a variety of solutes in dilute aqueous solutions at 37°C as a function of solute molecular weight (MW). Note that the diffusivity for the data shown in Figure 5.2 has been multiplied by 10^5. The solid line through the data is the result of a linear least squares regression. The following empirical equation, based on the data in Figure 5.2, provides a useful relationship for estimating the diffusivity of a solute in water at 37°C, knowing only the MW of the solute:

$$D = 1.013 \times 10^{-4} \, (MW)^{-0.46} \, cm^2 \, sec^{-1}. \tag{5.3}$$

Cussler (1984) and Tyn and Gusek (1990) provide additional data on diffusivity for a variety of solutes.

The diffusivity of solutes in dilute solutions can also be estimated from the *Stokes–Einstein equation* (Bird et al. 2002; Cussler 1984):

$$D = \frac{RT}{6\pi\mu\, a N_A}. \tag{5.4}$$

This equation at best is only accurate to about 20% (Cussler 1984). In Equation 5.4, R is the ideal gas constant (8.314 J mol^{-1} K^{-1}), T is the temperature in K, a is the solute radius, N_A is Avogadro's number (6.023×10^{23} mol^{-1}), and μ is the solution viscosity. The viscosity of water at 37°C is 0.76 cP and that of plasma is 1.2 cP (recall that 1 cP = 0.01 g cm^{-1} sec^{-1}).

If the diffusivity of a solute is known, then Equation 5.4 may be used to obtain an estimate of the molecular radius (a) of the solute. If the diffusivity and size of the solute are not known, the solute size can first be estimated from Equation 5.5. This equation assumes that the solute of a given MW is a sphere with a density ($\rho \approx 1$ g cm^{-3}) equal to that of the solute in the solid phase:

$$a = \left(\frac{3 \, MW}{4\pi\rho N_A} \right)^{1/3}. \tag{5.5}$$

Example 5.2

Estimate the diffusivity of a spherically shaped protein with a MW of 36,000 in water at 37°C.

Solution

From Figure 5.2, we see that for a solute of this size, $D = 8 \times 10^{-7}$ cm² sec⁻¹. From Equation 5.3, we find that $D = 8.12 \times 10^{-7}$ cm² sec⁻¹. The Stokes–Einstein equation (Equation 5.4) can also be used, but first, using Equation 5.5, we need to estimate the radius of this molecule as follows:

$$a = \left(\frac{3 \times 36000 \,\frac{g}{mol}}{4\pi \times 1 \frac{g}{cm^3} \times 6.023 \times 10^{23} \frac{1}{mol}} \right)^{1/3} = 2.43 \times 10^{-7} \text{ cm} \times \frac{1 \text{ m}}{100 \text{ cm}} \times \frac{10^9 \text{ nm}}{m} = 2.43 \text{ nm}.$$

Next, using the Stokes–Einstein equation, we have

$$D = \frac{8.314 \frac{J}{molK} \times 310 \, K \times \frac{kg\,mm}{sec^2\,J} \times \frac{(100 \text{ cm})^2}{m^2}}{6\pi \times 0.76 \text{ cP} \times \frac{0.01 \frac{g}{cm\,sec}}{cP} \times 2.43 \times 10^{-7} \text{ cm} \times 6.023 \times 10^{23} \frac{1}{mol} \times \frac{1 \text{ kg}}{1000 \text{ g}}},$$

$$D = 1.23 \times 10^{-6} \text{ cm}^2 \text{ sec}^{-1}.$$

Note that the diffusivity obtained from the Stokes–Einstein equation is about 40% larger than that estimated from the actual data shown in Figure 5.2. Very large molecules like proteins can be solvated or hydrated, making the radius of the actual solute–solvent complex larger than the radius estimated from the solute MW alone using Equation 5.5.

5.4.2 Fick's Second Law

Now, if we perform an unsteady solute balance across the shell of thickness, Δy, as shown in Figure 5.1, we can then write, using Equation 1.8 as our guide, that

$$S\Delta y \frac{\partial C}{\partial t} = -DS \frac{\partial C}{\partial y}\bigg|_{y} - - DS \frac{\partial C}{\partial y}\bigg|_{y+\Delta y}. \tag{5.6}$$

The term on the left side of the equal sign in Equation 5.6 represents the accumulation of solute mass within the control volume $S \Delta y$. The term on the right side is the net rate at which solute enters the control volume by diffusion, according to Fick's first law. We can eliminate S in Equation 5.6 and after dividing by Δy and taking the limit as $\Delta y \to 0$, we obtain the following result, which is known as *Fick's second law*:

$$\frac{\partial C}{\partial t} = D \frac{\partial^2 C}{\partial y^2}. \tag{5.7}$$

Solution of Equation 5.7 for the situation shown in Figure 5.1 requires the following initial condition and boundary conditions for the solute within the fluid or solid material region:

$$\text{IC:} \quad t = 0, C = 0,$$

$$\text{BC1:} \, y = 0, C = C_0, \tag{5.8}$$

$$\text{BC2:} \, y = \infty, C = 0.$$

5.4.3 Solution for the Concentration Profile for Diffusion from a Flat Plate into a Quiescent Fluid

Equations 5.7 and 5.8 are analogous to the problem we examined in Chapter 4 for the flat plate that is set in motion within a semi-infinite fluid that is initially at rest, i.e., Equations 4.43 and 4.44. In that case, we used Laplace transforms to obtain a solution for the unsteady velocity profile within the fluid, i.e., $v_x(y,t)$. We can therefore use that result here by simply recognizing that we can replace v_x with C, and V with C_0, in Equation 4.51. Note also in Equation 4.51 that kinematic viscosity, v, is replaced by diffusivity, D. Hence, we obtain the following result for the concentration profile within the quiescent fluid or solid material at any location y and time t:

$$\frac{C(y,t)}{C_0} = \text{erfc}\left(\frac{y}{\sqrt{4Dt}}\right) = 1 - \text{erf}\left(\frac{y}{\sqrt{4Dt}}\right). \tag{5.9}$$

We can also define a concentration boundary layer thickness, δ_C, as that distance where the concentration has decreased to 1% of the value at the surface of the plate. Recall from Chapter 4 that the complementary error function of $y/\sqrt{4Dt} = 1.821$ provides a value of C/C_0 that is equal to 0.01. Hence, we can define the concentration boundary layer thickness, $\delta_C(y,t)$, as

$$\delta_C(y,t) = 3.642\sqrt{Dt} \approx 4\sqrt{Dt}. \tag{5.10}$$

The value of δ_C can also be interpreted as the distance to which the solute from the plate has penetrated into the fluid at time t.

Example 5.3

Calculate the concentration boundary layer thickness 1 sec after the plate has made contact with the fluid. Assume that the fluid is water and that the solute diffusivity is $D = 1 \times 10^{-5}$ cm^2 sec^{-1}.

Solution

Using Equation 5.10, we can calculate the thickness of the concentration boundary layer as

$$\delta_C = 4\sqrt{10^{-5} \text{ cm}^2 \text{sec}^{-1} \times 1 \text{ sec}} = 0.0126 \text{ cm} = 126 \text{ } \mu\text{m}.$$

Example 5.4

A polymeric material is being used as a barrier for a protective garment. The polymeric material has a thickness of 0.075 in. For a particular toxic agent, the diffusivity in this material was found to be equal to 1×10^{-6} cm^2 sec^{-1}. If the protective garment comes into contact with the toxic agent, estimate the time it will take for the toxic agent to just penetrate the garment.

Solution

We can use Equation 5.9 to estimate the breakthrough time, assuming that $C/C_0 = 0.001$ is a reasonable criterion for breakthrough of the toxic agent. Hence, we have that

$$\text{erf}\left(\frac{y}{\sqrt{4Dt}}\right) = 1 - 0.001 = 0.999.$$

From this equation, we have for this case that erf $(2.327) = 0.999$. Therefore, $y/\sqrt{4Dt} = 2.327 = (0.075 \text{ in} \times (2.54 \text{ cm/in}))/\sqrt{4 \times 1 \times 10^{-6} \text{ (cm}^2/\text{sec}) \times t}$. Solving for the time, we get that $t = 1675$ sec or 28 min.

5.4.4 Definition of the Solute Flux

The flux of solute diffusing at any location y is defined as the moles of solute per unit time per unit area normal to the direction of diffusion, i.e., $j_s = J_s/S$. The flux at the surface of the plate in Figure 5.1 is therefore given by

$$j_s = -D \left. \frac{\partial C}{\partial y} \right|_{y=0}. \tag{5.11}$$

For the situation shown in Figure 5.1, we can find $\partial C / \partial y|_{y=0}$ by differentiating Equation 5.9 with respect to y and evaluating this result at $y = 0$. We find for this case that

$$j_s = \frac{DC_0}{32\sqrt{\pi Dt}}. \tag{5.12}$$

Equation 5.12 shows that the solute flux at the surface of the plate is inversely proportional to the square root of the contact time of the plate with the fluid.

5.4.5 Definition of the Mass Transfer Coefficient

Solution of mass transfer problems is oftentimes facilitated by defining the *mass transfer coefficient*, k_m. The mass transfer coefficient can be thought of as the proportionality constant that relates the molar flux of the solute (j_s) to the overall concentration driving force, i.e., $j_s = k_m(C_{High} - C_{Low})$. The mass transfer coefficient is often given in terms of the dimensionless group, known as the *Sherwood number* (*Sh*). The Sherwood number is the ratio of the transport rate of the solute by convection to that by diffusion and is defined as $Sh = k_m L/D$, where L is a characteristic length. For example, for convective mass transfer within a cylindrical tube, the characteristic length is the tube diameter.

We can write, in general, Equation 5.13, which defines the mass transfer coefficient in terms of the flux of solute at the surface of an object, using Fick's first law:

$$j_s = -D \left. \frac{\partial C}{\partial y} \right|_{y=0} = k_m (C_{High} - C_{Low}). \tag{5.13}$$

For the problem just considered in Figure 5.1, $C_{High} = C_0$ and $C_{Low} = 0$. Comparing Equation 5.13 with Equation 5.10, we see that the mass transfer coefficient is then given by

$$k_m = \frac{D}{32\sqrt{\pi Dt}} = \frac{D}{8\sqrt{\pi}\,\delta_C(t)}. \tag{5.14}$$

From Equation 5.14, we also see that the mass transfer coefficient is directly proportional to the solute diffusivity and inversely proportional to the concentration boundary layer thickness, i.e., $k_m \infty\ D/\delta_C$. For the unsteady diffusion problem shown in Figure 5.1, the mass transfer coefficient, however, is not constant, but decreases as the concentration boundary layer thickness increases over time.

Example 5.5

Calculate the value of the mass transfer coefficient for the situation described in Example 5.3.

Solution

Using Equation 5.14 and a concentration boundary layer thickness of 0.0126 cm, after 1 sec of contact we can calculate the mass transfer coefficient as follows:

$$k_m = \frac{10^{-5} \text{ cm}^2\text{sec}^{-1}}{8\sqrt{\pi} \times 0.0126 \text{ cm}} = 5.6 \times 10^{-5} \frac{\text{cm}}{\text{sec}}.$$

Example 5.6

Find the mass transfer coefficient (k_m) and the Sherwood number (Sh) for mass transfer from the surface of a sphere of radius R into an infinite quiescent fluid. Assume that the concentration of solute at the surface of a sphere is given by C_0.

Solution

In this case, the mass transfer occurs from the surface of the sphere into the surrounding quiescent fluid. We consider a thin shell volume in fluid of thickness Δr extending from r to $r + \Delta r$. We then perform a steady state solute balance (see Equation 1.8) on this shell volume using Fick's first law (see Equation 5.2):

$$-4\pi r^2 D \frac{dC}{dr}\bigg|_r - -4\pi r^2 D \frac{dC}{dr}\bigg|_{r+\Delta r} = 0.$$

The first term in this equation provides the rate at which solute enters the control volume by diffusion and the second term is the rate at which solute is leaving by diffusion. Dividing by Δr and taking the limit as $\Delta r \to 0$, the following differential equation is then obtained for the solute concentration (C) in the fluid surrounding the sphere:

$$\frac{d}{dr}\left(r^2 \frac{dC}{dr}\right) = 0.$$

The boundary conditions for this equation are

$$\text{BC1}: r = R, C = C_0$$

$$\text{BC2}: r = \infty, C = C_\infty.$$

If the previous differential equation is integrated twice, we then obtain

$$C(r) = -\frac{C_1}{r} + C_2.$$

Applying the boundary conditions to this equation allows us to find the integration constants C_1 and C_2. Hence, it is easily shown that $C_2 = C_\infty$ and $C_1 = R (C_\infty - C_0)$. With these constants, the concentration profile in the fluid surrounding the sphere is given by

$$C(r) = C_\infty - (C_\infty - C_0)\frac{R}{r}.$$

From Equation 5.13, we can next write the solute flux at the surface of the sphere as

$$j_S = k_m (C_0 - C_\infty) = -D \frac{dC}{dr}\bigg|_R.$$

However, from the previous equation for $C(r)$, it is easily shown that $-dC/dr|_R = (C_0 - C_R)/R$ and when this is substituted into the previous equation for the solute flux, we obtain the result that $k_m = D/R$. When this result is rearranged, we find that the Sherwood number ($Sh = k_m d_{sphere}/D$) based on the sphere diameter ($d_{sphere} = 2R$) as the characteristic length is equal to 2 for the case of diffusion of a solute from the surface of a sphere into a quiescent fluid.

Example 5.7

A drug has an equilibrium solubility in water of 0.0025 g cm^{-3}. The diffusivity of the drug in water is 0.9×10^{-5} cm^2 sec^{-1}. One gram of the drug is made into particles that are 0.1 cm in diameter and these particles have a density of 1.27 g cm^{-3}. These particles are then vigorously mixed in a stirred vessel containing 1 L of water. After mixing for 15 min, the drug concentration in the solution is 0.03 g L^{-1}. Estimate the value of the mass transfer coefficient (centimeters per second) for the drug under these conditions. Also, find the Sherwood number (Sh).

Solution

The concentration of the drug at the surface of the particles is equal to the equilibrium solubility of the drug in water, which is equal to 2.5 g L^{-1}, and this is C_{High} in Equation 5.13. The concentration of the drug dissolved in the solution starts at 0 g L^{-1} and increases after 15 min to 0.03 g L^{-1}. We can let C_{Low} in Equation 5.13 be given by the average of these values or 0.015 g L^{-1}. The total amount of drug that was dissolved over the 15 min period is equal to 0.03 g L$^{-1} \times 1$ L = 0.03 g. The average dissolution rate is therefore 0.03 g of drug in 15 min or 3.33×10^{-5} g sec^{-1}. The dissolution rate of the drug ($\dot{r}_{drug}$) is then given by the following adaptation of Equation 5.13, where the solute flux, j_S, is multiplied by the total surface area of all the drug particles to give the drug dissolution rate ($\dot{r}_{drug}$):

$$\dot{r}_{drug} = \bar{k}_m N_{particles} S_{particle} (C_{High} - C_{Low})$$

$$= \bar{k}_m N_{particles} S_{particle} \times 2.485 \text{ gL}^{-1} = 3.33 \times 10^{-5} \text{ gsec}^{-1}.$$

In this equation, the total area available for mass transfer is equal to the number of particles in the solution ($N_{particle}$) times the surface area of a single particle ($S_{particle}$). The area of a single drug particle is given by $4\pi R^2$, where R is the particle radius. Hence, we find that $S_{particle} = 0.0314$ cm^2. The number of particles can be found by dividing the mass of drug placed into the solution, by the drug density and then dividing this result by the volume of a given particle. When this is done, it is found that 1 g of drug is equivalent to 1504 particles. With these parameters found, the previous equation can now be solved for the mass transfer coefficient as follows:

$$\bar{k}_m = \frac{3.33 \times 10^{-5} \text{ gsec}^{-1}}{2.485 \frac{g}{L} \times \frac{1 \text{ L}}{1000 \text{ cm}^3}} \times \frac{1}{1504 \text{ particles} \times \frac{0.0314 \text{ cm}^2}{\text{particle}}} = 2.84 \times 10^{-4} \text{ cm sec}^{-1}.$$

The Sherwood for this example is calculated as

$$Sh = \frac{\bar{k}_m d_{particle}}{D} = \frac{2.84 \times 10^{-4} \text{ cmsec}^{-1} \times 0.1 \text{ cm}}{0.9 \times 10^{-5} \text{ cm}^2 \text{ sec}} = 3.15.$$

In this case, *Sh* is greater than 2 (see Example 5.6), indicating that solute transport from the drug particles in this example is a result of convection and diffusion.

5.4.6 MASS TRANSFER IN LAMINAR BOUNDARY LAYER FLOW OVER A FLAT PLATE

Figure 5.3 shows the laminar flow of a fluid across a semi-infinite flat plate of length L. The surface of the plate maintains a constant concentration of a solute (C_0) and the concentration of solute in the fluid is given by C. Unlike the situation shown in Figure 5.1, here the solute diffuses from the surface of the plate and is then swept away by the flowing fluid. Hence, in this problem, solute is transported away from the flat plate by a combination of diffusion and convection. We also assume that the bulk of the fluid is free of solute except in the region adjacent to the flat plate, which is defined as the concentration boundary layer. If the fluid approaching the plate has a solute concentration of C_∞, then the following analysis still applies; however, the concentration will need to be defined as $\bar{C} = C - C_\infty$.

In Section 4.15.2, we developed an approximate solution for the steady laminar flow of a fluid along a flat plate and determined the velocity profile within the boundary layer that is formed along the length of the plate. Here, we will extend this solution to determine the concentration profile of the solute in the concentration boundary layer that is also formed along the surface of the flat plate.

Consider the shell volume shown in Figure 5.2 of width W and located from x to $x + \Delta x$ and from $y = 0$ to $y = \delta_C(x)$, where δ_C is the concentration boundary layer thickness. We first perform a steady state solute balance on this shell volume, which is given by

$$W \int_0^{\delta_C} C v_x \, dy \Big|_x - W \int_0^{\delta_C} C v_x \, dy \Big|_{x+\Delta x} - C v_y \Big|_{\delta_C} W \Delta x + j_S \Big|_{y=0} W \Delta x = 0. \tag{5.15}$$

The first two terms in Equation 5.15 represent the net rate at which solute is being added to the shell volume by flow of the fluid in the x-direction. The third term represents the loss of solute from the top of the shell volume as a result of flow in the y-direction. The last term represents the rate at which solute is diffusing away from the surface of the flat plate. After eliminating the plate width, W, and dividing by Δx, taking the limit as $\Delta x \to 0$, we can write Equation 5.15 as

$$j_S \Big|_{y=0} = \frac{d}{dx} \int_0^{\delta_C} C v_x \, dy + C v_y \Big|_{y=\delta_C}. \tag{5.16}$$

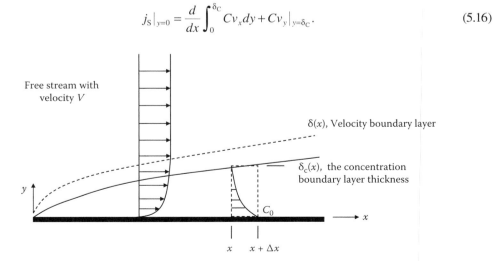

FIGURE 5.3 Laminar boundary layer flow in the vicinity of a flat plate of constant surface concentration.

Using Fick's first law, we can insert Equation 5.11 on the left-hand side of Equation 5.16 for $j_s|_{y=0}$ and obtain

$$-D\frac{dC}{dy}\bigg|_{y=0} = \frac{d}{dx}\int_0^{\delta_C} Cv_x\, dy + Cv_y\big|_{y=\delta_C}.$$ (5.17)

Solution of Equation 5.17 requires that we know how v_x and C depend on x and y in their respective boundary layers. Since we only want to obtain an approximate solution, we can use our previous result for $v_x(x, y)$, i.e., Equation 4.62:

$$\frac{v_x(x,y)}{V} = \frac{3}{2}\left(\frac{y}{\delta(x)}\right) - \frac{1}{2}\left(\frac{y}{\delta(x)}\right)^3,$$ (4.62)

and in a similar fashion propose that the concentration profile in the concentration boundary layer is approximately described by

$$C(y) = a_C + b_C y + c_C y^2 + d_C y^3.$$ (5.18)

The concentration profile described by Equation 5.18 also has to satisfy the following boundary conditions:

$$\text{BC1}:\ y=0,\, C = C_0,$$

$$\text{BC2}:\ y=\delta_C(x),\, C = 0,$$

$$\text{BC3}:\ y=\delta_C(x),\, \frac{\partial C}{\partial y} = 0,$$ (5.19)

$$\text{BC4}:\ y=0,\, \frac{\partial^2 C}{\partial y^2} = 0.$$

The first boundary condition expresses the fact that the concentration of the solute is constant along the surface of the flat plate. The second and third boundary conditions state that beyond the concentration boundary layer, the solute is not present. This allows us to eliminate the last term in Equation 5.17. Hence, Equation 5.17 becomes

$$-D\frac{dC}{dy}\bigg|_{y=0} = \frac{d}{dx}\int_0^{\delta_C} Cv_x\, dy.$$ (5.20)

The fourth boundary condition expresses the fact that the flux of solute along the surface of the plate is only a function of x.

After imposing the earlier above boundary conditions on Equation 5.18, the following expression is obtained for the concentration profile within the concentration boundary layer in terms of the concentration boundary layer thickness, $\delta_C(x)$, which at this time still needs to be determined:

$$\frac{C(x,y)}{C_0} = 1 - \frac{3}{2}\left(\frac{y}{\delta_C(x)}\right) - \frac{1}{2}\left(\frac{y}{\delta_C(x)}\right)^3.$$ (5.21)

We now can substitute Equations 4.62 and 5.21 into Equation 5.20. We also assume that the ratio of the boundary layer thicknesses, i.e., $\Delta = (\delta_C(x) / \delta(x))$, is a constant. The algebra is a bit overwhelming, but one can obtain the following differential equation for the thickness of the concentration boundary layer:

$$(0.15\Delta - 0.0107\Delta^3)\frac{d\delta_C}{dx} = \frac{3}{2}\frac{D}{V\delta_C}, \tag{5.22}$$

with the boundary condition that at $x = 0$, $\delta_C = 0$. Solution of Equation 5.22 is then given by

$$\delta_C(x) = \sqrt{\frac{3}{(0.15\Delta - 0.0107\Delta^3)}\frac{(Dx)}{V}}. \tag{5.23}$$

Recall from Equation 4.64 that the boundary layer thickness δ is given by

$$\delta(x) = 4.64\sqrt{\frac{\nu x}{V}} = 4.64\sqrt{\frac{\mu x}{\rho V}}. \tag{4.64}$$

Dividing Equation 5.23 by Equation 4.64 and simplifying results in the following equation for $\Delta = (\delta_C(x) / \delta(x))$:

$$1.0765\Delta^3 - 0.0768\Delta^5 = \frac{\rho D}{\mu} = \frac{1}{Sc}, \tag{5.24}$$

where $Sc = \mu / \rho D$ is a dimensionless number known as the *Schmidt number*. The Schmidt number is the ratio of momentum diffusion ($\nu = \mu/\rho$) to mass diffusion (D). For solutes diffusing through liquids, the Schmidt number is generally much greater than unity; hence, from Equation 5.24, we have that $\Delta < 1$ or $\delta > \delta_C$. This shows that for liquids, the concentration boundary layer lies within the velocity or momentum boundary layer. For solutes diffusing through gases, the Schmidt number is on the order of unity and Δ is approximately equal to unity or $\delta \approx \delta_C$. For solutes diffusing through materials like liquid metals, the Schmidt number is much less than unity and then $\Delta > 1$ and $\delta_C > \delta$.

In general, if one knows the value of the Schmidt number for the mass transfer problem being considered, then Equation 5.24 can be solved for the value of Δ. The velocity boundary layer thickness is then given by Equation 4.64 and the concentration boundary layer thickness is equal to $\Delta \delta(x)$.

For most mass transfer problems of interest to biomedical engineers, the Schmidt number is oftentimes much larger than unity. This makes the right-hand side of Equation 5.24 much smaller than unity. The only way this can happen is if Δ is also much smaller than unity. This means that the concentration boundary layer lies well within the boundary layer for the velocity. Hence, for $\Delta \ll 1$, Equation 5.24 simplifies to the following result:

$$\Delta = \frac{\delta_C(x)}{\delta(x)} \approx Sc^{-1/3}. \tag{5.25}$$

Combining this with Equation 4.64 results in Equation 5.26 for the thickness of the concentration boundary layer:

$$\frac{\delta_C(x)}{x} = 4.64\left(\frac{\mu}{\rho V x}\right)^{1/2}\left(\frac{\rho D}{\mu}\right)^{1/3} = 4.64\left(\frac{\nu}{V x}\right)^{1/2}\left(\frac{D}{\nu}\right)^{1/3} = 4.64\left(\frac{1}{\text{Re}_x}\right)^{1/2}\left(\frac{1}{Sc}\right)^{1/3}. \tag{5.26}$$

This equation has an extremely important result. Note that the concentration boundary layer thickness depends on two dimensionless parameters that describe the nature of the flow as given by the Reynolds number (Re_x) and the physical properties of the solute and the fluid as given by the Schmidt number (Sc).

We can also calculate the local value of the mass transfer coefficient defined by Equation 5.13. Here, $C_{High} = C_0$ and $C_{Low} = 0$. Using Equations 5.11 and 5.13, we find that the local mass transfer coefficient is given by the following expression:

$$k_m = -\frac{D}{(C_{High} - C_{Low})}\left.\frac{\partial C}{\partial y}\right|_{y=0} = -\frac{D}{C_0}\left.\frac{\partial C}{\partial y}\right|_{y=0}. \tag{5.27}$$

After substituting Equation 5.21 into Equation 5.27 and replacing $\delta_C(x)$ with Equation 5.26, we obtain the following expression for the local mass transfer coefficient:

$$\frac{k_m x}{D} = Sh_x = 0.323\left(\frac{\rho V x}{\mu}\right)^{1/2}\left(\frac{\mu}{\rho D}\right)^{1/3} = 0.323\,\text{Re}_x^{1/2}\,Sc^{1/3}, \tag{5.28}$$

where Sh_x is the local Sherwood number at location x. Recall that the Sherwood number is the ratio of the transport rate of solute by convection to that by diffusion. In Equation 5.28, k_m is the local value of the mass transfer coefficient. As the fluid progresses down the length of the plate, the concentration boundary layer thickness increases and, by Equation 5.28, we see that k_m decreases in inverse proportion to $x^{1/2}$.

For a plate of length L, the average mass transfer coefficient is given by

$$\bar{k}_m = \frac{1}{L}\int_0^L k_m(x)\,dx. \tag{5.29}$$

Substituting Equation 5.28 into Equation 5.29 results in Equation 5.30, which is also known as the *length averaged mass transfer coefficient*:

$$\frac{\bar{k}_m L}{D} = Sh = 0.646\left(\frac{\rho V L}{\mu}\right)^{1/2}\left(\frac{\mu}{\rho D}\right)^{1/3} = 0.646\,\text{Re}^{1/2}\,Sc^{1/3}. \tag{5.30}$$

Equations 5.28 and 5.30 describe the mass transfer of a solute from a flat plate for laminar boundary layer flow provided the local value of the Reynolds number is less than 300,000.

For flow over one side of a flat plate of length L and width W, the amount of solute transported can then be written as follows. In Equation 5.31, C_{High} is the concentration of the solute at the surface of the plate and C_{Low} is the concentration of the solute in the fluid outside the boundary layer:

$$J_S = L W \bar{k}_m (C_{High} - C_{Low}). \tag{5.31}$$

Example 5.8

Blood is flowing across the flat surface of a polymeric material coated with an anticoagulant drug. Equilibrium studies of the drug-coated polymer show that the drug has a solubility of 100 mg L^{-1} of blood. The fluid adjacent to the plate equilibrates quickly, hence the concentration of the drug at the surface of the plate is, for practical purposes, the same as this equilibrium value. Find the following:

a. The distance at which the laminar boundary layer flow over the surface of the polymeric material ends.
b. The thickness of the velocity and concentration boundary layers at the end of the polymeric material.
c. The local mass transfer coefficient at the end of the polymeric material and the average mass transfer coefficient of the drug.
d. The average elution flux of the drug from the polymeric material, assuming that the concentration of the drug in the blood is much smaller than at the surface.

Assume that the blood traveling across the polymeric material has a free stream velocity of 40 cm sec^{-1}. The polymeric material has a length in the direction of the flow of 30 cm. The diffusivity of the drug in blood is 4.3×10^{-6} cm^2 sec^{-1}.

Solution

For part (a) we use the fact that the transition to turbulence begins at $Re_x = \rho V x / \mu = 300{,}000$. Hence, we can solve this equation for the value of x at this transition as follows:

$$x = \frac{300{,}000 \times 3 \text{ cP} \times \dfrac{1 \dfrac{g}{\text{cm sec}}}{100 \text{ cP}}}{1.056 \dfrac{g}{\text{cm}^3} \times 40 \dfrac{\text{cm}}{\text{sec}}} = 213 \text{ cm}.$$

Note that this distance for the transition to turbulence is much greater than the actual length of the flat polymeric material, hence the flow is laminar over the region of interest. For part (b), we use Equations 4.64 and 5.22 to calculate the boundary layer thicknesses at the end of the polymeric material as shown in the following equation. Thus, the thickness of the velocity boundary layer is

$$\delta = 4.64 \sqrt{\frac{3 \text{ cP} \times \dfrac{1 \dfrac{g}{\text{cm sec}}}{100 \text{ cP}} \times 30 \text{ cm}}{1.056 \dfrac{g}{\text{cm}^3} \times 40 \dfrac{\text{cm}}{\text{sec}}}} = 0.68 \text{ cm}.$$

Next, we calculate the Schmidt number:

$$Sc = \frac{\mu}{\rho D} = \frac{3 \text{ cP} \times \dfrac{1 \dfrac{g}{\text{cm sec}}}{100 \text{ cP}}}{1.056 \dfrac{g}{\text{cm}^3} \times 4.3 \times 10^{-6} \dfrac{\text{cm}^2}{\text{sec}}} = 6607.$$

We see that $Sc \gg 1$ in this case and the concentration boundary layer will therefore lie well within the velocity boundary layer calculated previously:

$$\delta_C = \delta\, Sc^{-1/3} = 0.68 \text{ cm} \times 6607^{-1/3} = 0.036 \text{ cm}.$$

Note that both δ and δ_C are much larger than the plasma layer thickness (see Sections 4.13 and 4.14), hence we are justified in using the properties of whole blood in these calculations.

For part (c), we use Equations 5.25 and 5.27 to calculate the local and average mass transfer coefficients, respectively. The Reynolds number at the end of the polymeric surface is given by

$$\left(\frac{\rho V x}{\mu}\right) = \frac{1.056 \dfrac{g}{cm^3} \times 40 \dfrac{cm}{sec} \times 30\ cm}{3\ cP \times \dfrac{1 \dfrac{g}{cm\,sec}}{100\ cP}} = 42240$$

and

$$k_m = 0.323 \frac{4.3 \times 10^{-6}\ cm^2\ sec^{-1}}{30\ cm} \times 42{,}240^{1/2} \times 6607^{1/3} = 1.78 \times 10^{-4}\ cm\,sec^{-1}.$$

Since the average mass transfer coefficient at a given location is twice the local value, we then have that $\bar{k}_m = 3.57 \times 10^{-4}$ cm sec^{-1}. To calculate the average elution flux of the drug from the polymeric surface, we use Equation 5.10, which can be rearranged as

$$\bar{j}_S = \frac{1}{L} \int_0^L j_S\, dx = \bar{k}_m C_0 = 3.57 \times 10^{-4} \frac{cm}{sec} \times 100 \frac{mg}{L} \times \frac{1\,L}{1000\ cm^3}$$

$$= 3.57 \times 10^{-5}\ mg\,cm^{-2}\ sec^{-1}.$$

5.4.7 Mass Transfer from the Walls of a Tube Containing a Fluid in Laminar Flow

Next, we consider the situation shown in Figure 5.4. A fluid in laminar flow is flowing through a tube and comes into contact with a section of the tube that maintains a constant concentration of a solute that diffuses into the fluid. Much like we did with laminar flow over a flat plate, we want to determine the elution rate of the solute from the surface of the tube in terms of dimensionless quantities, such as the Reynolds number and the Schmidt number. In order to obtain an analytical solution to this rather complex problem, we will make some simplifying assumptions. First, we will only be interested in obtaining what is known as a *short contact time* solution. This means that the solute does not penetrate very far from the surface of the tube. Hence, the change in concentration of the solute occurs only near the surface of the tube wall.

From Chapter 4, we know that the velocity profile for laminar flow in a tube is given by the following expression, i.e., Equation 4.13:

$$v_z(r) = \frac{(P_0 - P_L)R^2}{4\mu L}\left[1 - \left(\frac{r}{R}\right)^2\right]. \tag{4.13}$$

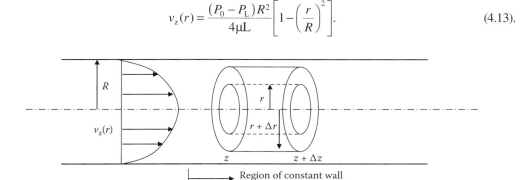

FIGURE 5.4 Laminar flow of a fluid in a tube with constant concentration of the solute along the surface.

Next, we perform a steady state solute balance on the cylindrical shell volume shown in Figure 5.4:

$$2\pi r\Delta r\left(v_z C\big|_z - v_z C\big|_{z+\Delta z}\right) + 2\pi r\Delta r\left(D\frac{\partial C}{\partial z}\bigg|_{z+\Delta z} - D\frac{\partial C}{\partial z}\bigg|_z\right)$$

$$+ 2\pi r\Delta z\left(D\frac{\partial C}{\partial r}\bigg|_{r+\Delta r} - D\frac{\partial C}{\partial r}\bigg|_r\right) = 0. \tag{5.32}$$

The first parenthetical term in Equation 5.32 represents the net rate at which solute is added to the shell volume by bulk flow of the fluid. The second term in parentheses represents the net rate at which solute is added to the shell volume by diffusion of solute in the z-direction. The final term in Equation 5.32 represents the net rate at which solute enters the shell volume by diffusion in the r-direction. Now, if we divide through by $2\pi r\Delta r\,\Delta z$ and then take the limit as Δr and $\Delta z \to 0$, we then obtain the following partial differential equation that describes the transport of solute at any value of r and z within the tube:

$$v_z\frac{\partial C}{\partial z} = D\left(\frac{1}{r}\frac{\partial}{\partial r}\left(r\frac{\partial C}{\partial r}\right) + \frac{\partial^2 C}{\partial z^2}\right). \tag{5.33}$$

The term on the left-hand side of Equation 5.33 represents the solute transport in the flowing fluid by axial (z) convection or bulk flow, of the fluid. The right-hand side represents the transport of solute by diffusion in the radial (r) and axial directions (z). Generally, more solute is transported in the z-direction by convection than is transported by axial diffusion, so at any given location in the fluid, we expect that $v_z\,(\partial C/\partial z) \gg D\,(\partial^2 C/\partial z^2)$.

Next, we let $s = R - r$ represent the distance from the wall of the tube. Since we are only interested in a solution near the wall, s will be small relative to R and we can neglect the effect of the curvature of the tube. With these assumptions, Equation 5.33 becomes

$$v_z\frac{\partial C}{\partial z} = D\frac{\partial^2 C}{\partial s^2}. \tag{5.34}$$

We also recognize that near the tube wall the velocity profile is nearly linear and $v_z \approx (\Delta PR / 2\mu L)s$. Using this for v_z, Equation 5.34 may then be written as

$$\left(\frac{\Delta PR}{2\mu L}\right)s\frac{\partial C}{\partial z} = D\frac{\partial^2 C}{\partial s^2}, \tag{5.35}$$

and we can state the following boundary conditions that can be used to solve the previous partial differential equation:

$$\text{BC1: } z = 0, \text{ and all } s,\ C = 0,$$

$$\text{BC2: } s = 0, \text{ and } z > 0,\ C = C_0,$$

$$\text{BC3: } s = \infty, \text{ and } z > 0,\ C = 0. \tag{5.36}$$

The last boundary condition results from the short contact time assumption, meaning that the solute does not penetrate very far into the fluid. We have also assumed that the bulk of the fluid is free of the solute except in the region adjacent to the tube wall. If the fluid enters the tube with a solute concentration of C_∞, then this analysis still applies; however, the concentration would need to be defined as $\bar{C} = C - C_\infty$.

Solution of Equations 5.35 and 5.36 is facilitated by rewriting these equations in terms of dimensionless variables defined as follows. We let $\theta \equiv C/C_0$, $\varepsilon \equiv z/R$, and $\sigma \equiv s/R$, where θ is defined as a dimensionless concentration, ε is a dimensionless axial position, and σ is a dimensionless distance from the tube wall. If these dimensionless variables replace C, z, and s in Equations 5.35 and 5.36, the following result is obtained:

$$N\sigma\frac{\partial \theta}{\partial \varepsilon} = \frac{\partial^2 \theta}{\partial \sigma^2}, \tag{5.37}$$

where $N \equiv (R^3\Delta P/2\mu DL) = (4\ RV_{average}/D)$, where $V_{average}$ is given by Equation 4.15. The boundary conditions become

$$\text{BC1: } \varepsilon = 0, \text{ and all } \sigma, \theta = 0,$$

$$\text{BC2: } \sigma = 0, \text{ and } \varepsilon > 0, \theta = 1,$$

$$\text{BC3: } \sigma = \infty, \text{ and } \varepsilon > 0, \theta = 0. \tag{5.38}$$

Now, to solve Equations 5.37 and 5.38, the *similarity transform technique* or *combination of variables* approach can be used. In this approach, we define a new independent variable, η, which is a suitable combination of ε and σ, that converts Equation 5.37 from a partial differential equation into an ordinary differential equation that only depends on η. If we let $\eta \equiv (N\sigma^3/9\varepsilon)^{1/3}$, then it can be shown with a little bit of effort that Equations 5.37 and 5.38 become

$$\frac{\partial^2 \theta}{\partial \eta^2} + 3\eta^2 \frac{\partial \theta}{\partial \eta} = 0, \tag{5.39}$$

and the boundary conditions become

$$\text{BC1': } \eta = \infty, \theta = 0,$$

$$\text{BC2': } \eta = 0, \theta = 1. \tag{5.40}$$

Integration of Equations 5.39 and 5.40 then gives the following result for the solute concentration profile as a function of $\eta \equiv (N\sigma^3/9\varepsilon)^{1/3}$:

$$C(\eta) = C_0 \frac{\displaystyle\int_{\eta}^{\infty} e^{-\eta^3} d\eta}{\Gamma\left(\dfrac{4}{3}\right)}. \tag{5.41}$$

Obtaining the previous solution involves the use of the *gamma function*, which is defined as $\Gamma(j) \equiv \int_0^{\infty} e^{-x} x^{j-1} dx$, along with the gamma function property that $j\ \Gamma(j) = \Gamma(j+1)$.

We can also use the earlier result to define the concentration boundary layer thickness at any position z as that distance away from the tube wall where $C(\eta) = 0.01\ C_0$. Hence, letting $(C(\eta)/C_0) = 0.01$, we can solve Equation 5.41 for η and obtain

$$\delta_C = 2.89\left(\frac{R^2 z}{N}\right)^{1/3} = 2.89\left(\frac{z R D}{4 V_{average}}\right)^{1/3}. \tag{5.42}$$

The solute flux from the tube surface at any value of z is given by

$$j_s = -D \frac{\partial C}{\partial r}\bigg|_{r=R} = -D \frac{\partial C_0}{\partial \sigma}\bigg|_{\sigma=0} = \frac{DC_0}{R\Gamma\left(\frac{4}{3}\right)} \left(\frac{N}{9\varepsilon}\right)^{1/3}, \tag{5.43}$$

where $\Gamma(4/3) = 0.893$. The local value of the mass transfer coefficient may then be written as

$$k_m = \frac{j_s}{C_0} = \frac{D}{\Gamma\left(\frac{4}{3}\right)} \left(\frac{4V_{\text{average}}}{9DRz}\right)^{1/3} = 1.077 D \left(\frac{V_{\text{average}}}{DD_{\text{tube}} z}\right)^{1/3}, \tag{5.44}$$

where $V_{\text{average}} = \Delta P \, R^2/8\mu L$ is the average velocity of the fluid in the tube (see Equation 4.15), which is also equal to the volumetric flow rate (Q) of the fluid divided by the tube cross-sectional area, i.e., $V_{\text{average}} = Q/\pi R^2$. In terms of the Sherwood number, Equation 5.44 becomes

$$Sh = \frac{k_m D_{\text{tube}}}{D} = 1.077 \left(\frac{\rho V D_{\text{tube}}}{\mu}\right)^{1/3} \left(\frac{\mu}{\rho D}\right)^{1/3} \left(\frac{D_{\text{tube}}}{z}\right)^{1/3}. \tag{5.45}$$

Using Equation 5.44, the total amount of solute eluted from the surface over a length of tube equal to L is given by

$$J_s = \int_0^L k_m C_0 2\pi R \, dz = \frac{2\pi R^{2/3} D^{2/3} C_0}{\Gamma\left(\frac{4}{3}\right)} \left(\frac{4V_{\text{average}}}{9}\right)^{1/3} \int_0^L z^{-1/3} dz. \tag{5.46}$$

Integration of Equation 5.46 provides the following result:

$$J_s = \frac{3\pi R D C_0}{\Gamma\left(\frac{4}{3}\right)} \left(\frac{2}{9}\right)^{1/3} \text{Re}^{1/3} Sc^{1/3} \left(\frac{L}{R}\right)^{2/3}, \tag{5.47}$$

where the Reynolds number in Equation 5.47 is defined as $Re \equiv \rho \, D_{\text{tube}} \, V_{\text{average}}/\mu$. J_s is also equal to the average mass transfer coefficient multiplied by the circumferential surface area of the tube ($2\pi RL$) and the concentration driving force, i.e., ($C_0 - 0$). Hence, the average mass transfer coefficient is given by

$$Sh = \frac{\bar{k}_m D_{\text{tube}}}{D} = \frac{3}{\Gamma\left(\frac{4}{3}\right)} \left(\frac{2}{9}\right)^{1/3} \text{Re}^{1/3} Sc^{1/3} \left(\frac{R}{L}\right)^{1/3} = 1.615 \, \text{Re}^{1/3} Sc^{1/3} \left(\frac{D_{\text{tube}}}{L}\right)^{1/3}. \tag{5.48}$$

This result is the short contact time solution, which means that the solute has not penetrated very far into the fluid from the surface of the tube. This solution is valid as long as $s \ll R$ or $\delta_C \ll R$. As in the case for laminar flow over a flat plate, we see that the mass transfer coefficient once again is a function of the Reynolds number and the Schmidt number.

Suppose the solute concentration in the fluid entering the tube is C_∞ and the concentration along the surface of the tube is C_0. The earlier results are still valid for the short contact time solution, since we can define a new concentration as $\bar{C} = C - C_\infty$ and because of the linearity of the differential equations, the same results will be obtained. The total amount of solute transported from the surface of the tube over length L is given by

$$J_s = \bar{k}_m \, 2\pi R L (C_0 - C_\infty). \tag{5.49}$$

If the fluid exiting the tube is then completely mixed, we can find the mixed solute concentration by performing the following overall steady state solute mass balance calculation between the entrance to the tube at $z = 0$ and the tube exit at $z = L$:

$$0 = \pi R^2 \, V_{average} \, C_\infty|_{z=0} - \pi R^2 \, V_{average} \, C_{mixed}|_{z=L} + \bar{k}_m \, 2\pi L (C_0 - C_\infty). \tag{5.50}$$

This equation can then be solved to give the mixed concentration of solute as

$$C_{mixed}|_{z=L} = C_\infty|_{z=0} + \left(\frac{2\bar{k}_m L}{R V_{average}} \right) (C_0 - C_\infty). \tag{5.51}$$

Example 5.9

Plasma is flowing at an average velocity of 10 cm sec^{-1} through a tube that is 1 cm in diameter. The walls of the tube are coated over a length of 10 cm with a drug that diffuses from the surface of the tube into the flowing fluid. The concentration of the drug in the plasma adjacent to the surface of the tube is 100 mg L^{-1}. Find the average mass transfer coefficient over the 10 cm length of the tube that is coated with the drug. Also, find the total transport rate of the solute and the mixed concentration of the drug in the plasma after the coated section of the tube. Show that the short contact time solution is a valid approach for solving this problem. The diffusivity of the drug in plasma is estimated to be 4×10^{-6} cm^2 sec^{-1}. The physical properties of plasma can be found in Table 4.1.

Solution
First, we calculate Re as

$$Re = \frac{\rho D_{tube} V_{average}}{\mu} = \frac{1.024 \, \frac{g}{cm^3} \times 1 \, cm \times 10 \, \frac{cm}{sec}}{1.2 \, cP \times \frac{1 \, \frac{g}{cm \, sec}}{100 \, cP}} = 853.3.$$

Therefore, based on Re, the flow is laminar. Sc is then calculated as

$$Sc = \frac{\mu}{\rho D} = \frac{1.2 \, cP \times \frac{1 \, \frac{g}{cm \, sec}}{100 \, cP}}{1.024 \, \frac{g}{cm^3} \times 4 \times 10^{-6} \, \frac{cm^2}{sec}} = 2929.7.$$

Using Equation 5.48, we can calculate Sh as shown in the following calculation:

$$Sh = 1.615 \, Re^{1/3} \, Sc^{1/3} \left(\frac{D_{tube}}{L} \right)^{1/3} = 1.615 \times 853.3^{1/3} \times 2929.7^{1/3} \times \left(\frac{1 \, cm}{10 \, cm} \right)^{1/3} = 101.7.$$

From this value of Sh, we can then calculate the average mass transfer coefficient over the 10 cm section of the tube as

$$\bar{k}_m = \frac{ShD}{D_{tube}} = \frac{101.7 \times 4 \times 10^{-6} \frac{cm^2}{sec}}{1\ cm} = 4.07 \times 10^{-4}\ cm\ sec^{-1}.$$

From Equation 5.49, we can calculate the average mass transfer rate of the solute from the surface of the tube:

$$J_S = \bar{k}_m\, 2\pi RL(C_0 - C_\infty) = 4.07 \times 10^{-4} \frac{cm}{sec} \times 2 \times \pi \times 0.5\ cm \times 10\ cm \times 100 \frac{mg}{L} \times \frac{1\ L}{1000\ cm^3}$$

$$= 1.279 \times 10^{-3}\ mg\ sec^{-1}.$$

The mixed concentration of the drug in the plasma after the coated section of the tube can then be found from Equation 5.51:

$$C_{mixed}\big|_{z=L} = \left(\frac{2\,\bar{k}_m L}{R V_{average}}\right) C_0 = \left(\frac{2 \times 4.07 \times 10^{-4} \frac{cm}{sec} \times 10\ cm}{0.5\ cm \times 10 \frac{cm}{sec}}\right) \times 100 \frac{mg}{L} = 0.163\ mg\,L^{-1}.$$

The thickness of the concentration boundary layer at the end of the coated section of the tube can be found from Equation 5.42:

$$\delta_C = 2.89\left(\frac{zRD}{4 V_{average}}\right)^{1/3} = 2.89 \left(\frac{10\ cm \times 0.5\ cm \times 4 \times 10^{-6} \frac{cm^2}{sec}}{4 \times 10 \frac{cm}{sec}}\right)^{1/3} = 0.023\ cm.$$

This value of δ_C is significantly less than the tube radius of 0.5 cm, hence the short contact time solution is valid for these conditions.

5.4.8 Mass Transfer Coefficient Correlations

The earlier mass transfer equations that we developed for specific situations also provide valuable insight as to how to develop correlations for mass transfer coefficients in more complicated flows and geometries. In many cases, we find that mass transfer coefficient data can be correlated with general functions that are very similar in form to these equations. For example, for problems involving forced convection or the flow of a fluid, one can propose a functional dependence for Sh on Re and Sc that is very similar to Equations 5.28 and 5.48, i.e., $Sh = A\,Re^m\,Sc^n$. The values of A, m, and n can then be found by fitting this equation to experimental data. Table 5.2 provides a summary (Cussler 1984) of some useful mass transfer coefficient correlations for a variety of flow situations.

For the laminar flow of a fluid (i.e., not turbulent) within a cylindrical tube, $\bar{k}_m$ may be estimated from (Thomas 1992):

$$Sh = 3.66 + \frac{0.104\, Re\, Sc\left(\dfrac{D_{tube}}{L}\right)}{1 + 0.016\left[Re\, Sc\left(\dfrac{D_{tube}}{L}\right)\right]^{0.8}}, \tag{5.52}$$

where $Sh\ (= \bar{k}_m D_{tube}/D)$ is the value averaged over the length L.

TABLE 5.2
A Selection of Useful Mass Transfer Coefficient Correlations

Physical Situation	Correlation
Sphere in a quiescent fluid	$\dfrac{k_{\mathrm{m}} D_{\mathrm{sphere}}}{D} = 2$
Forced convection around a sphere	$\dfrac{k_{\mathrm{m}} D_{\mathrm{sphere}}}{D} = 2.0 + 0.6\left(\dfrac{\rho D_{\mathrm{sphere}} V}{\mu}\right)^{1/2}\left(\dfrac{\mu}{\rho D}\right)^{1/3}$
Laminar flow over a flat plate	$\dfrac{k_{\mathrm{m}} z}{D} = 0.323\left(\dfrac{\rho V z}{\mu}\right)^{1/2}\left(\dfrac{\mu}{\rho D}\right)^{1/3}$
Laminar flow over a flat plate	$\dfrac{\bar{k}_{\mathrm{m}} L}{D} = 0.646\left(\dfrac{\rho V L}{\mu}\right)^{1/2}\left(\dfrac{\mu}{\rho D}\right)^{1/3}$
Laminar flow in a circular tube, short contact time solution	$\dfrac{\bar{k}_{\mathrm{m}} D_{\mathrm{tube}}}{D} = 1.615\left(\dfrac{\rho V D_{\mathrm{tube}}}{\mu}\right)^{1/3}\left(\dfrac{\mu}{\rho D}\right)^{1/3}\left(\dfrac{D_{\mathrm{tube}}}{L}\right)^{1/3}$
Laminar flow in a circular tube, short contact time solution	$\dfrac{k_{\mathrm{m}} D_{\mathrm{tube}}}{D} = 1.077\left(\dfrac{\rho V D_{\mathrm{tube}}}{\mu}\right)^{1/3}\left(\dfrac{\mu}{\rho D}\right)^{1/3}\left(\dfrac{D_{\mathrm{tube}}}{z}\right)^{1/3}$
Laminar flow in a circular tube, undeveloped flow and concentration profiles (average over a length L)	$\dfrac{\bar{k}_{\mathrm{m}} D_{\mathrm{tube}}}{D} = 3.66 + \dfrac{0.104\dfrac{\left(\dfrac{\rho V D_{\mathrm{tube}}}{\mu}\right)\left(\dfrac{\mu}{\rho D}\right)}{\dfrac{L}{D_{\mathrm{tube}}}}}{1 + 0.016\left(\dfrac{\left(\dfrac{\rho V D_{\mathrm{tube}}}{\mu}\right)\left(\dfrac{\mu}{\rho D}\right)}{\dfrac{L}{D_{\mathrm{tube}}}}\right)^{0.8}}$
Laminar flow in a circular tube, fully developed flow and concentration profiles	$\dfrac{\bar{k}_{\mathrm{m}} D_{\mathrm{tube}}}{D} = 3.66$
Turbulent flow within a horizontal slit	$\dfrac{k_{\mathrm{m}} D_{\mathrm{slit}}}{D} = 0.026\left(\dfrac{\rho D_{\mathrm{slit}} V}{\mu}\right)^{0.8}\left(\dfrac{\mu}{\rho D}\right)^{1/3}$, $\text{with } D_{\mathrm{slit}} = \dfrac{2}{\pi}\left(\text{slit width}\right)$
Turbulent flow through a circular tube	$\dfrac{k_{\mathrm{m}} D_{\mathrm{tube}}}{D} = 0.026\left(\dfrac{\rho D_{\mathrm{tube}} V}{\mu}\right)^{0.8}\left(\dfrac{\mu}{\rho D}\right)^{1/3}$
Laminar flow in a circular tube	$\dfrac{\bar{k}_{\mathrm{m}} D_{\mathrm{tube}}}{D} = 1.86\left(\dfrac{\rho D_{\mathrm{tube}} V}{\mu}\dfrac{\mu}{\rho D}\dfrac{D_{\mathrm{tube}}}{L}\right)^{1/3}$
Spinning disc	$\dfrac{\bar{k}_{\mathrm{m}} D_{\mathrm{disk}}}{D} = 0.62\left(\dfrac{\rho D_{\mathrm{disk}}^{2}\omega}{\mu}\right)^{1/2}\left(\dfrac{\mu}{\rho D}\right)^{1/3}$ where ω is the disc rotation rate in radians per second

(Continued)

TABLE 5.2 (Continued)
A Selection of Useful Mass Transfer Coefficient Correlations

Physical Situation	Correlation
Packed beds	$\dfrac{k_m}{V_0} = 1.17 \left(\dfrac{\rho D_{particle} V_0}{\mu} \right)^{-0.42} \left(\dfrac{\mu}{\rho D} \right)^{-.67}$ where V_0 is the superficial velocity defined as the volumetric flow rate divided by the unpacked tube cross-sectional area
Falling film	$\dfrac{k_m z}{D} = 0.69 \left(\dfrac{z V_{film}}{D} \right)^{1/2}$

Source: Cussler, E.L., *Diffusion: Mass Transfer in Fluid Systems*, Cambridge University Press, Cambridge, 1984.

Note: ρ is the fluid density; μ is the fluid viscosity; D is the solute diffusivity; D_{disk} is the disk diameter; $D_{particle}$ is the particle diameter; D_{sphere} is the sphere diameter; D_{tube} is the tube diameter; k_m is the local mass transfer coefficient; $\bar{k}_m$ is the length averaged mass transfer coefficient; L is the tube length; R is the tube radius; V is the average fluid velocity; V_{film} is the average film velocity; and z is the axial position.

Equation 5.52 is for the case where both the velocity and the concentration profiles are not fully developed. This means that the velocity and concentration profiles are changing with axial position. When the flow is fully developed in terms of the velocity and concentration profiles, the Sh number for laminar flow in a tube attains its asymptotic value of 3.66. For the velocity profile to be fully developed at any axial position z, experiments show that $(z/D_{tube}) > 0.05\, Re$, and for the concentration profile to be fully developed, we have that $(z/D_{tube}) > 0.05\, Re\, Sc$.

The length averaged mass transfer coefficient for undeveloped laminar flow in cylindrical tubes and channels can also be estimated from (Thomas 1992):

$$Sh = 1.86\, Re^{1/3}\, Sc^{1/3} \left(\frac{D_{tube}}{L} \right)^{1/3} \quad \text{for} \quad \frac{\dfrac{L}{D_{tube}}}{Re\, Sc} < 0.01. \tag{5.53}$$

For noncylindrical flow channels, the diameter (D_{tube}) in the mass transfer coefficient equations can be replaced with the *hydraulic diameter*, D_H, which is defined by (see Equation 4.78):

$$D_H = \frac{4 \times (\text{cross-sectional area})}{(\text{wetted perimeter})}. \tag{5.54}$$

Example 5.10

Show how the average mass transfer coefficient ($\bar{k}_m$) can be found for flow in a tube of arbitrary cross-sectional area A from the total solute mass transfer rate and the solute concentration entering and leaving the tube, i.e., $\bar{C}_0$ and $\bar{C}_L$, respectively. Assume that the diffusing solute concentration at the tube surface is C_S.

Solution

First, we perform a steady state shell balance on the solute in the direction of flow from x to $x + \Delta x$ as follows:

$$0 = V\, A\, \bar{C} \big|_x - V\, A\, \bar{C} \big|_{x + \Delta x} + W\, \Delta x\, k_m (C_S - \bar{C}),$$

where W represents the circumferential area of the tube per unit length of tube, V is the average velocity of the fluid flowing through the tube, and k_m is the local value of the mass transfer coefficient. In this equation, $\bar{C}$ is the mixing cup average concentration, which is defined as follows:

$$\bar{C} = \frac{\int_A v_x(y)C(x,y)dA}{\int_A v_x(y)dA}.$$

In this equation, $v_x(y)$ is the axial velocity in the tube and the integrations are over the cross-sectional area normal to this flow. The numerator of this equation reflects the fact that, in general, the flow is higher near the axis of the tube, so this flow will bring more solute into the mixing cup than the slower flow near the wall of the tube. The denominator is just the volumetric flow rate.

If the solute mass balance equation, given previously, is divided by Δx and the limit taken as $\Delta x \rightarrow 0$, then

$$\frac{d\bar{C}}{dx} = \frac{k_m}{V\,r_H}(C_S - \bar{C}),$$

with the boundary condition that at $x = 0$ then $\bar{C} = \bar{C}_0$. Note that the hydraulic radius (r_H) is the same as A/W, so the previous equation is valid for a tube of arbitrary cross section. Integration of this equation over the tube length (L) gives the following result:

$$\frac{C_S - \bar{C}_L}{C_S - \bar{C}_0} = \exp\left(-\frac{\bar{k}_m L}{V\,r_H}\right), \tag{A}$$

where the average mass transfer coefficient is defined in terms of the local mass transfer coefficient by $\bar{k}_m \equiv (1/L)\int_0^L k_m \, dx$. Note that if $V\,r_H \gg \bar{k}_m L$ then $\bar{C}_L \rightarrow \bar{C}_0$ and this situation is called *diffusion limited*, in the limit there is no transfer of solute into the fluid. If $\bar{k}_m L \gg V\,r_H$ then $\bar{C}_L \rightarrow C_S$ and the solute transport is *flow limited*, in the limit the solute concentration leaving the tube approaches that at the tube surface. We can also write a total mass balance on the solute over the tube length (L) and obtain the following result that expresses the total solute transport rate ($\dot{m}_{solute}$):

$$\dot{m}_{solute} = V\,A\left(\bar{C}_L - \bar{C}_0\right) = W\,L\,\bar{k}_m\,\Delta C_{LM}. \tag{B}$$

Here, ΔC_{LM} represents the concentration driving force that, when multiplied by the average mass transfer coefficient and the total surface area available for mass transfer, gives the solute transport rate. We can solve for ΔC_{LM} from Equation B:

$$\Delta C_{LM} = \frac{V\,A\left(\bar{C}_L - \bar{C}_0\right)}{W\,L\,\bar{k}_m} = \frac{r_H V\left(\bar{C}_L - \bar{C}_0\right)}{\bar{k}_m L}.$$

Using Equation A derived earlier, we have that

$$\frac{\bar{k}_m L}{r_H V} = \ln\left(\frac{C_S - \bar{C}_0}{C_S - \bar{C}_L}\right).$$

Therefore, the following equation defines how we determine the value of ΔC_{LM}:

$$\Delta C_{LM} = \frac{r_H V\left(\bar{C}_L - \bar{C}_0\right)}{\bar{k}_m L} = \frac{\left(\bar{C}_L - \bar{C}_0\right)}{\ln\left(\dfrac{C_S - \bar{C}_0}{C_S - \bar{C}_L}\right)}$$

$$\Delta C_{LM} = \frac{\left(C_S - \bar{C}_0\right) - \left(C_S - \bar{C}_L\right)}{\ln\left(\dfrac{C_S - \bar{C}_0}{C_S - \bar{C}_L}\right)}.$$

The previous result defines what is known as the *logarithmic mean concentration difference* (ΔC_{LM}) and this is the proper driving force to be used when calculating the total solute mass transfer rate using the average mass transfer coefficient as shown in Equation B. Equation B can also be used to determine the average mass transfer coefficient from measurements of the entering and exiting solute concentrations as follows:

$$\bar{k}_{\mathrm{m}} = \frac{\dot{m}_{\mathrm{solute}}}{W\,L\,\Delta C_{\mathrm{LM}}}.$$

5.5   SOLUTE TRANSPORT BY CAPILLARY FILTRATION

The perfusion of plasma across the capillary wall will carry with it a variety of solutes that are present in the blood. Because of the sizes of the pores in the capillary wall, this filtration of the plasma by the capillary wall will also tend to separate the solutes on the basis of their sizes. For example, smaller solutes like ions, glucose, and amino acids will readily pass through the capillary pores. However, larger proteins will be inhibited to varying degrees in their passage across the capillary wall by the sizes of the pores. This selective filtration by the capillary wall is also observed in a variety of membrane systems. The solute selectivity during this plasma filtration is described by the sieving coefficient (S_a), which is defined as the ratio of the solute concentration in the filtrate (C_f) to the solute concentration at the surface of the capillary on the blood side (C_{bs}). Theoretical expressions based on the motion of a spherical solute moving through a cylindrical pore have been developed to estimate the value of the sieving coefficient (Anderson and Quinn 1974; Deen 1987). The development of these expressions neglects secondary effects between the solute and the membrane pore, e.g., electrostatic, hydrophobic, and van der Waals interactions. The following expression can be used to estimate the sieving coefficient:

$$S_{\mathrm{a}} = \frac{C_{\mathrm{f}}}{C_{\mathrm{bs}}} = (1-\lambda)^2 \left[2 - (1-\lambda)^2\right]\left[1 - \frac{2}{3}\lambda^2 - 0.163\lambda^3\right], \tag{5.55}$$

where λ is defined as the ratio of the solute radius (a) to the capillary pore radius (r). If no other information is available, the solute radius can be estimated from the solute MW using Equation 5.5.

Under some conditions, the high filtration rate of plasma leads to the formation of a layer of retained proteins at a higher concentration near the nonfiltrate or feed side of the membrane. This effect, shown in Figure 5.5, is also called *concentration polarization* and the layer of retained proteins will affect the convective transport of the solutes. Concentration polarization is particularly important at the higher filtration rates used in commercial membrane-based plasmapheresis systems used for separating plasma from the cellular components of blood, as well as in membrane systems used to purify protein solutions.

The sieving coefficient for the case of concentration polarization (S_0) can be developed as follows. The flux of solute at any position y in the concentration polarization region shown in Figure 5.5 has to equal the rate at which the solute is diffusing back toward the bulk solution plus the amount of solute that is carried across the membrane by the filtration flow. This can be expressed by

$$qC = -D\frac{dC}{dy} + qC_{\mathrm{f}}, \tag{5.56}$$

where q is defined as the filtration flux (Q/S), Q is given by Equation 3.4, and S is the surface area normal to the direction of the filtration flow. Equation 5.56 can then be rearranged and integrated across the concentration polarization region of thickness δ as follows:

$$q\int_0^{\delta_C} dy = -D\int_{C_{bs}}^{C_b} \frac{dC}{C - C_f}. \tag{5.57}$$

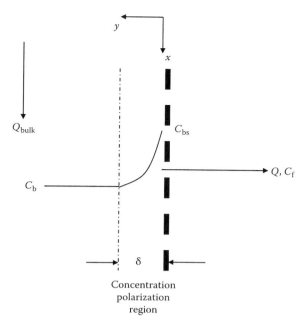

FIGURE 5.5   Sieving and concentration polarization.

After performing the integration, we obtain

$$q = -\frac{D}{\delta_C} \ln\left(\frac{C_b - C_f}{C_{bs} - C_f}\right) = -k_m \ln\left(\frac{C_b - C_f}{C_{bs} - C_f}\right). \qquad (5.58)$$

In Equations 5.57 and 5.58, δ_C is defined as the concentration boundary layer thickness, which also defines the thickness of the concentration polarization region. It should be apparent from our discussion of mass transfer in boundary layers that the mass transfer coefficient, k_m, is the same as D/δ_C.

The sieving coefficient for the case of concentration polarization (S_0) is defined as the ratio of the solute concentration in the filtrate (C_f) to that of the solute concentration in the bulk fluid (C_b). Equation 5.54 can then be rearranged and, after some algebra, we obtain the following expression for the sieving coefficient:

$$S_0 = \frac{C_f}{C_b} = \frac{S_a}{(1 - S_a)e^{-(q/k_m)} + S_a}, \qquad (5.59)$$

where S_a is given by Equation 5.55; k_m is the mass transfer coefficient and, as discussed in the previous sections, is dependent on the fluid properties and the nature of the flow field (see Table 5.2).

Example 5.11

Consider the filtration of a fluid flowing within a hollow fiber. Assume that the length of the hollow fiber is given by L and that the radius of the hollow fiber is R. Develop the differential equation that describes the change with axial position (z) of the volumetric flow rate of the fluid (F) as a result of filtration across the walls of the hollow fiber. Assume steady state flow.

Solution

Perform a shell balance across a finite axial shell volume of the hollow fiber equal to $\pi R^2 \Delta z$. The amount of fluid filtered across the wall of the hollow fiber within the shell volume is equal to

$2\pi R\,\Delta z\,q$, where $2\pi R\,\Delta z$ is the circumferential surface area of the shell volume. Our steady state shell balance may then be written as

$$0 = F\big|_z - F\big|_{z+\Delta z} - 2\pi R\Delta z q.$$

Dividing by Δz, and taking the limit as $\Delta z \to 0$, results in the following differential equation that describes the axial variation of F due to filtration:

$$\frac{dF(z)}{dz} = -2\pi R q(z).$$

If $q(z)$ is constant, then this equation can be integrated to give

$$F(z) = F_0 - 2\pi R q z,$$

where F_0 is the volumetric feed rate to the fiber.

Example 5.12

Consider a hollow fiber module that is being used to separate a protein solution at 37°C. The module contains 10,000 fibers and each fiber has a diameter of 400 μm and a length of 20 cm. The wall thickness of the hollow fibers is 75 μm. The nominal molecular weight cutoff (NMWCO) for the hollow fibers is 100,000 and the porosity of the hollow fibers (A_P/S) is equal to 0.40. The protein solution has a viscosity of 0.001 Pa sec. The protein solution enters the module at a flow rate of 250 mL min^{-1}. The composition of the protein solution is protein A (4 g L^{-1}, MW = 20,000), protein B (7 g L^{-1}, MW = 150,000), and protein C (6 g L^{-1}, MW = 300,000). The diffusivity of protein A in the solution is 6.42×10^{-7} cm^2 sec^{-1}. Determine the total filtration flow rate across the hollow fibers assuming that the pressure drop across the fibers is 750 mmHg. What is the composition of the fluid leaving on the filtrate side? What is the fractional removal of protein A from the feed solution to the hollow fiber module? What is the composition of the fluid leaving the hollow fibers?

Solution

We can calculate the filtration flux from Equation 3.4. Since proteins B and C are retained by the hollow fiber membranes, we find that their osmotic pressure contribution is 1.29 mmHg, according to Equation 3.3. In comparison to the pressure drop across the membranes of 750 mmHg, this osmotic pressure of the retained solutes is rather small; hence, we can assume for the most part that the filtration flux is constant along the length of the hollow fibers. Next, we calculate the hydraulic conductance of the membrane from Equation 3.7, assuming that the radius of the pores can be found from the NMWCO using Equation 5.5:

$$r = \left(\frac{3 \times 100,000 \dfrac{g}{\text{mole}}}{4\pi \times 1 \dfrac{g}{\text{cm}^3} \times 6.023 \times 10^{23} \dfrac{1}{\text{mole}}} \right)^{1/3} = 3.41 \times 10^{-7}\ \text{cm}$$

and

$$L_P = \frac{0.40 \times \left(3.41 \times 10^{-7}\ \text{cm}\right)^2}{8 \times 0.001\,\text{Pa sec} \times 0.0075\ \text{cm}} = 7.75 \times 10^{-10}\ \text{cmPa}^{-1}\text{sec}^{-1}.$$

Next, we can calculate the total filtration flow for the hollow fiber module using Equation 3.4:

$$Q = 7.75 \times 10^{-10} \frac{cm}{Pa\,sec} \times 10{,}000 \times 2\pi \times 0.02 \text{ cm} \times 20 \text{ cm} \times (750 - 1.29)\,mmHg$$

$$\times \frac{1 \text{ atm}}{760 \text{ mmHg}} \times \frac{101{,}325 \text{ Pa}}{1 \text{ atm}} = 1.94 \text{ cm}^3\,sec^{-1}.$$

The filtration flow rate is almost 50% of the total flow rate entering the hollow fiber module, which is $250 \text{ cm}^3 \text{ min}^{-1}$ or $4.17 \text{ cm}^3 \text{ sec}^{-1}$. Thus, the flow rate of fluid leaving the fibers of the hollow fiber module is $F(L) = (4.17 - 1.94) \text{ cm}^3 \text{ sec}^{-1} = 2.23 \text{ cm}^3 \text{ sec}^{-1}$. The filtration flux ($q$) equals Q divided by the total circumferential surface area of the fibers. Therefore, $q = (1.94(\text{cm}^3/\text{sec}))/(2\pi \times 0.02 \text{ cm} \times 20 \text{ cm} \times 10{,}000) = 7.72 \times 10^{-5} \text{ cm}\,sec^{-1}$. Next, we perform a solute balance on protein A over the length of a fiber from z to $z + \Delta z$. We assume because of the high filtration rate that the transport of solute A across the membrane is by convection (filtration) and not by diffusion:

$$FC_b\big|_z - FC_b\big|_{z+\Delta z} = 2\pi R\Delta z\,q\,C_f = 2\pi R\Delta z\,q\,S_0\,C_b.$$

After dividing by Δz and taking the limit as $\Delta z \to 0$, we obtain that

$$\frac{dFC_b}{dz} = F\frac{dC_b}{dz} + C_b\frac{dF}{dz} = -2\pi R q\,S_0\,C_b.$$

From Example 5.11, we also have that $(dF/dz) = -2\pi Rq$, and using this result the previous equation may then be written as follows, where we have also taken the filtration flux q to be constant along the length of the fibers because the pressure driving the filtration is also constant:

$$\frac{dC_b}{dz} = \frac{2\pi R(1 - S_0)q}{F}C_b = \frac{2\pi R(1 - S_0)q}{F_0 - 2\pi Rqz}C_b.$$

This equation can then be integrated as follows, to obtain an expression for how the bulk solute concentration for protein A changes with position in the hollow fiber module:

$$\int_{C_{b0}}^{C_b} \frac{dC_b}{C_b} = 2\pi R(1 - S_0)q\int_0^z \frac{dz}{F_0 - 2\pi Rqz},$$

$$C_b(z) = C_{b0}\left[1 - \left(\frac{2\pi Rq}{F_0}\right)z\right]^{-(1-S_0)}.$$

To find $C_b(z)$, we now need to calculate the sieving coefficient or S_0, which is also dependent on k_m according to Equation 5.59. Based on the average of the entrance and exit flow rates, i.e., 1/2 $(4.17 + 2.23)(\text{cm}^3/\text{sec}) = 3.2 \text{ cm}^3/\text{sec}$, the average velocity of the fluid in a fiber is given by

$$V = 3.2\frac{cm^3}{sec} \times \frac{1}{10{,}000} \times \frac{4}{\pi(0.04 \text{ cm})^2} = 0.25 \text{ cm}\,sec^{-1},$$

and the average Reynolds number of the fluid in a fiber is

$$Re = \frac{\rho V D_{tube}}{\mu} = \frac{1\dfrac{g}{cm^3} \times 0.04 \text{ cm} \times 0.25\dfrac{cm}{sec}}{0.01\dfrac{g}{cm\,sec}} = 1.0.$$

Therefore, the flow of fluid within the fibers is laminar. The diffusivity of protein A is given as 6.42×10^{-7} cm² sec⁻¹ and the Schmidt number is then calculated as follows:

$$Sc = \frac{\mu}{\rho D} = \frac{0.01\dfrac{g}{cm\,sec}}{1\dfrac{g}{cm^3} \times 6.42 \times 10^{-7}\dfrac{cm^2}{sec}} = 15,576.$$

Previously, we discussed that the flow is fully developed if $z > D_{tube} \times 0.05\,Re$. In this case, the flow is fully developed if $z > 0.002$ cm. For the concentration field to be fully developed, $z > D_{tube} \times 0.05\,Re\,Sc$, which in this case gives $z > 31$ cm, which is longer than the given length of the hollow fibers. Hence, the concentration profile is not fully developed and we should use Equation 5.52 to calculate the value of k_m as follows:

$$Sh = 3.66 + \frac{0.104 \times 1.0 \times 15,576 \times \dfrac{0.04\ cm}{20\ cm}}{1 + 0.016\left(1.0 \times 15,576 \times \dfrac{0.04\ cm}{20\ cm}\right)^{0.8}} = 6.25.$$

The mass transfer coefficient is then calculated as

$$\overline{k}_m = \frac{D\,Sh}{D_{tube}} = \frac{6.42 \times 10^{-7}\dfrac{cm^2}{sec} \times 6.25}{0.04\ cm} = 1.0 \times 10^{-4}\ cm\,sec^{-1}.$$

Next, we can use Equation 5.55 to find the value of S_a. The radius of protein A (a) can be estimated from the MW using Equation 5.5:

$$a = \left(\frac{3 \times 20,000\dfrac{g}{mole}}{4\pi \times 1\dfrac{g}{cm^3} \times 6.023 \times 10^{23}\dfrac{1}{mole}}\right)^{1/3} = 1.99 \times 10^{-7}\ cm = 1.994\ nm,$$

and the value of $\lambda = (a/r) = (1.994\ nm/3.41\ nm) = 0.585$. Hence, the sieving coefficient in the absence of concentration polarization is

$$S_a = (1 - 0.585)^2\left[2 - (1 - 0.585)^2\right]\left[1 - \tfrac{2}{3} \times 0.585^2 - 0.163 \times 0.585^3\right] = 0.233.$$

From Equation 5.59, we can then include the effect of concentration polarization on the sieving coefficient as shown in the following equation in the calculation of S_0:

$$S_0 = \frac{0.233}{(1 - 0.233)e^{-\frac{7.72 \times 10^{-5}\frac{cm}{sec}}{1.0 \times 10^{-4}\frac{cm}{sec}}} + 0.233} = 0.398.$$

Now we can calculate the concentration of protein A exiting ($z = L$) the hollow fiber module:

$$C_b\big|_{z=L} = 4\frac{g}{L}\left(1 - \left(\frac{2\pi \times 0.02\ cm \times 7.72 \times 10^{-5}\frac{cm}{sec}}{4.17\frac{cm^3}{sec} \times \frac{1}{10000}}\right) \times 20\ cm\right)^{-(1 - 0.398)} = 5.83\ g\,L^{-1}.$$

An overall solute balance can then be used to find the concentration of protein A in the filtrate:

$$QC_f\big|_{out} = FC_b\big|_{z=0} - FC_b\big|_{z=L},$$

$$C_f\big|_{out} = \frac{FC_b\big|_{z=0} - FC_b\big|_{z=L}}{Q} = \frac{4.17\frac{cm^3}{sec} \times 4\frac{g}{L} - 2.23\frac{cm^3}{sec} \times 5.83\frac{g}{L}}{1.94\frac{cm^3}{sec}}$$

$$C_f\big|_{out} = 1.90 \text{ gL}^{-1}.$$

The percent removal of protein A from the stream entering the hollow fiber module is then given by

$$\% \text{ removal of } A = \frac{FC_b\big|_{z=0} - FC_b\big|_{z=L}}{FC_b\big|_{z=0}}$$

$$= \frac{4.17\frac{cm^3}{sec} \times 4\frac{g}{L} - 2.23\frac{cm^3}{sec} \times 5.83\frac{g}{L}}{4.17\frac{cm^3}{sec} \times 4\frac{g}{L}} \times 100 = 22\%.$$

The concentrations of proteins B and C in the fluid exiting the hollow fibers may be calculated as

$$C_B\big|_{z=L} = \frac{FC_B\big|_{z=0}}{F\big|_{z=L}} = \frac{4.17\frac{cm^3}{sec} \times 7\frac{g}{L}}{2.23\frac{cm^3}{sec}} = 13.1 \text{ gL}^{-1},$$

$$C_C\big|_{z=L} = \frac{FC_C\big|_{z=0}}{F\big|_{z=L}} = \frac{4.17\frac{cm^3}{sec} \times 6\frac{g}{L}}{2.23\frac{cm^3}{sec}} = 11.2 \text{ gL}^{-1}.$$

Recall that the pressure drop across the hollow fiber membrane that drives the filtration flow was given as 750 mmHg. How does this value compare with the pressure drop in the direction of the fluid flowing within the hollow fibers? Recall that the total flow rate of the protein solution to the hollow fiber unit is 4.17 cm³ sec⁻¹ and the flow leaving the hollow fibers is 2.23 cm³ sec⁻¹. This gives an average flow rate within the hollow fiber unit of 3.2 cm³ sec⁻¹ or $Q_{fiber} = 3.2 \times 10^{-4}$ cm³ sec⁻¹ in each of the hollow fibers. Since the flow in the fibers is laminar, we can calculate the pressure drop using this average value of the flow rate according to Equation 4.11 as follows:

$$\Delta P = \frac{8Q_{fiber}\mu L}{\pi R^4} = \frac{8 \times 3.2 \times 10^{-4}\frac{cm^3}{sec} \times 0.001 \text{ Pa sec} \times 20 \text{ cm}}{\pi \times (0.02 \text{ cm})^4} \times \frac{760 \text{ mmHg}}{101{,}325 \text{ Pa}} = 0.76 \text{ mmHg}.$$

This pressure drop across the length of the hollow fiber is negligible in comparison to the pressure that is driving the filtration flow. Therefore, our assumption in this case of a constant filtration pressure of 750 mmHg is reasonable.

5.6 SOLUTE DIFFUSION WITHIN HETEROGENEOUS MEDIA

A unique aspect of biological systems is their heterogeneous nature. Therefore, we need to expand our understanding of solute diffusion to see how heterogeneous materials affect solute diffusion. Although the following discussion focuses a lot on the transport of a solute from blood across the

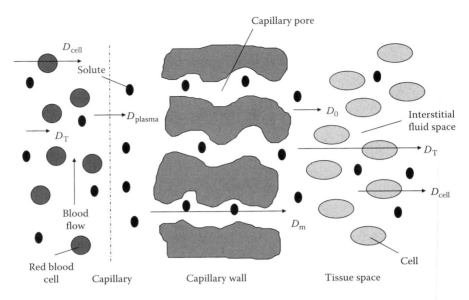

FIGURE 5.6   Solute transport from blood.

capillary wall, the concepts are general and may be applied to systems involving synthetic membranes as well.

Figure 5.6 illustrates the situation for diffusion of a solute from blood, through the cell-free plasma layer (see Chapter 4), across the capillary wall, and into the surrounding tissue space. The types of solute diffusivity that we need to consider are shown in Figure 5.6. These are the solute diffusivity in blood or tissue (D_T), the diffusivity in plasma (D_{plasma}), the diffusivity within the pores of the capillary wall (D_m), the diffusivity in the interstitial fluid (D_0), and the diffusivity through cells (D_{cell}). The following discussion will relate these various solute diffusivities to the diffusivity of solute in water (D).

We can use the Stokes–Einstein equation to estimate the solute diffusivity in plasma by adjusting for the increase of solution viscosity relative to that of water. The diffusivity of a solute in plasma at 37°C would therefore be given by

$$D_{plasma} = D_{water} \times \left(\frac{0.76 \text{ cP}}{1.2 \text{ cP}} \right) = 0.63 \ D_{water}. \tag{5.60}$$

For solute transport across the capillary wall, we must consider that the available surface area is not the total surface area S, but the pore area A_P. This assumes that the solute itself is not soluble in the continuous nonporous phase. Furthermore, the solute must follow a path through the pores that is tortuous, making the diffusion distance greater than the thickness of the capillary wall or the membrane (t_m).

In many cases in pore diffusion, the solute radius is also comparable to that of the pore radius. This leads to two additional effects, both reducing the diffusion rate of the solute. The first effect, called *steric exclusion*, is illustrated in Figure 5.7. Steric exclusion restricts the ability of the solute from entering the pore from the bulk solution. In this figure, we see that to a first approximation a solute can get no closer to the pore wall than its radius (a). Therefore, only a fraction of the pore cross-sectional area is available to the molecule. The fraction of pore cross-sectional area available to the solute is given by the ratio: $K = (\pi(r - a)^2)/(\pi r^2) = (1 - (a/r))^2$, which is also known as the solute *partition coefficient*. Because of steric exclusion, the equilibrium concentration of solute is less

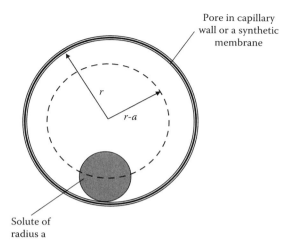

FIGURE 5.7   Steric exclusion.

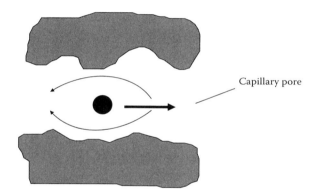

FIGURE 5.8   Solvent flow across solute in small pores increases drag.

within the pore mouth than in the bulk solution adjacent to the pore entrance. This simple expression for K ignores any secondary attractive or repulsive interactions between the solute and the pore and basically treats the molecule as a hard sphere.

As the solute molecule diffuses through the pore, it experiences hydrodynamic drag caused by the flow of solvent over the surface of the solute. This is shown in Figure 5.8. As the solute radius increases relative to that of the pore radius, this hydrodynamic drag increases, reducing or restricting the diffusion of the solute through the pore compared to its motion in the bulk fluid. This drag effect on the solute in the pore results in a decrease in the solute diffusivity relative to its value in the bulk solution and is called *restricted* or *hindered diffusion*.

Beck and Schultz (1970) studied the diffusion of a variety of solutes across track-etched mica membranes. The mica membranes were bombarded with U^{235} fission fragments. The etched particle tracks that were formed resulted in relatively straight pores with a narrow pore size distribution. Their data for a number of solutes are shown in Figure 5.9, where the ratio of the solute diffusivity in the pore (D_m) and the bulk diffusivity (D) is plotted as a function of the ratio of the solute radius to pore radius ($\lambda = a/r$). Note that when the solute size is only one-tenth of the pore size, the solute diffusivity is reduced by about 40%. When λ is equal to 0.40, the solute diffusivity has been reduced by nearly 90%.

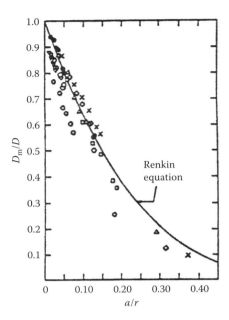

FIGURE 5.9 Apparent diffusivity in the pores of a membrane. (From Beck, R.E. and Schultz, J.S., *Science*, 170, 1302–5, 1970. With permission.)

The solid line in Figure 5.9 represents the fit of the Renkin equation (1954) to their data and is given by the following expression:

$$\frac{D_m}{D} = \left(1 - \frac{a}{r}\right)^2 = \left[1 - 2.1\left(\frac{a}{r}\right) + 2.09\left(\frac{a}{r}\right)^3 - 0.95\left(\frac{a}{r}\right)^5\right]. \tag{5.61}$$

We see that the Renkin equation describes their data rather well. The first term in parentheses on the right-hand side of Equation 5.61 is the partition coefficient mentioned earlier, i.e., $K(a/r) = (1 - a/r)^2$, and represents the effect of steric exclusion or the reduction in the solute concentration at the pore mouth compared to its value in the bulk solution. The bracketed term on the right-hand side of Equation 5.61, i.e., $\omega_r(a/r) = (1 - 2.1(a/r) + 2.09(a/r)^3 - 0.95(a/r)^5)$, accounts for the increase in hydrodynamic drag as the solute diffuses through the pore. In a general sense, we can then write Equation 5.61 as the product of a steric exclusion effect given by $K(a/r)$ and a hydrodynamic effect $\omega_r(a/r)$ or

$$\frac{D_m}{D} = K\left(\frac{a}{r}\right) \times \omega_r\left(\frac{a}{r}\right). \tag{5.62}$$

If the solute is already within the porous structure, then $K = 1$ and $D_m/D = \omega_r(a/r)$. Additional expressions for describing the hindered diffusion of large molecules in liquid-filled pores may be found in the review paper by Deen (1987).

With these corrections for the diffusion of a solute in the pores of the capillary wall or a synthetic membrane, we can now write Fick's first law as follows for a membrane of total surface area S with a total cross-sectional pore area of A_P:

$$J_s = DA_P\left(\frac{K\omega_r}{\tau}\right)\frac{dC}{dx}, \tag{5.63}$$

where τ is an experimentally defined parameter called the *tortuosity* and accounts for the fact that the solute diffusion distance is greater than that of the capillary wall or membrane thickness. If we divide Equation 5.63 by the total surface area (S) available for mass transfer of the solute, then we have that the solute flux, j_S, given by

$$j_s = -D \frac{A_P}{S} \left(\frac{K\omega_r}{\tau} \right) \frac{dC}{dx} = -D_e \frac{dC}{dx}, \tag{5.64}$$

where D_e is called the *effective diffusivity* of the solute in the heterogeneous material. Thus, D_e is given as follows:

$$D_e = D \frac{A_P}{S} \left(\frac{K\omega_r}{\tau} \right) = \frac{\varepsilon D}{\tau} K \omega_r = \frac{\varepsilon}{\tau} D_m, \tag{5.65}$$

where $\varepsilon = A_p/S$ is known as the *porosity*, or *void volume fraction* of the capillary wall or membrane.

Note the relationship and the distinction between D_m and D_e, as given in Equation 5.65. It is important to note that D_m describes the diffusion of the solute within a given pore of cross-sectional area πr^2, whereas D_e describes the diffusion of the solute through a membrane of total surface area S that contains a large number of pores of total surface area A_P. In both cases, however, the driving force for the diffusion process is the difference between the concentration of the solute in the bulk fluid adjacent to the entrance and exit of the pores.

Example 5.13

Estimate the diffusivity (D_m) of a spherically shaped molecule of MW equal to 36,000 through the cylindrical pores of a membrane that is 28 nm in diameter. Assume that the molecule enters the pores of the membrane from a well-stirred bulk solution.

Solution

In Example 5.2, we showed that this particular solute has an estimated diffusivity in water at 37°C of 8.12×10^{-7} cm^2 sec^{-1} and a solute radius of 2.43 nm. The pore radius is given as 14 nm, so $\lambda = 2.43$ nm/14 nm $= 0.174$. With these values, we can then use either Figure 5.9 or Equation 5.61 to estimate the diffusivity of the molecule within the pores of the membrane. From Figure 5.9, we find for this value of λ that $D_m/D = \approx 0.40$ and from Equation 5.61, we obtain $D_m/D = \approx 0.44$. From the latter value, we then obtain $D_m = 3.58 \times 10^{-7}$ cm^2 sec^{-1}.

Example 5.14

Suppose that the concentration of the solute in the previous example is 5 g L^{-1} in the bulk solution at the entrance to the pore and 0 g L^{-1} at the pore exit. Calculate the solute flux in grams per square centimeters per second of the solute through a 28 nm diameter pore assuming its length is 0.02 cm. Consider next a membrane with a total surface area of 10 cm^2 that consists of a multitude of these pores such that the porosity of the membrane (ε) is 0.40 and the tortuosity (τ) of the pores is 1.8. Calculate the transport rate in grams per second for the membrane.

Solution

First, we apply Fick's first law to the situation of solute diffusion through a single pore. Letting x denote the direction along the length (L) of the pore, we can write the following steady state solute balance over a control volume from x to $x + \Delta x$ as

$$0 = -\pi r^2 D_{pore} \left. \frac{dC}{dx} \right|_x -- \pi r^2 D_{pore} \left. \frac{dC}{dx} \right|_{x+\Delta x}.$$

In this equation, D_{pore} is the diffusivity of the solute within the pore, which is given by the product of its diffusivity in the bulk fluid, i.e., D, and the factor ω_r, which accounts for the reduction of diffusivity due to hydrodynamic drag in pores comparable in size to the solute. Taking the limit as $\Delta x \to 0$, the previous equation becomes

$$\frac{d^2 C}{d x^2} = 0.$$

This equation can then be integrated twice with the result that $C(x) = C_1 x + C_2$, where C_1 and C_2 are integration constants. This shows that the concentration profile within the pore is linear. These integration constants can be found from the boundary conditions written in terms of the bulk fluid adjacent to the entrance and exit of the pore, that is at $x = 0$, $C = C_{High}$, and at $x = L$, $C = C_{Low}$. Using these boundary conditions and accounting for solute partitioning at the pore entrance and exit, we can determine the integration constants and obtain the following equation that describes the concentration profile along the length of the pore:

$$\frac{C(x)}{K} = C_{High} - \left(C_{High} - C_{Low}\right)\frac{x}{L}.$$

Recall that K is the solute partition coefficient. From Fick's first law written for the solute inside the pore, the solute flux with $C(x)$ given as in the previous equation is then

$$j_s = -D_{pore}\frac{dC}{dx} = K D_{pore}\frac{\left(C_{High} - C_{Low}\right)}{L} = K D \omega_r \frac{\left(C_{High} - C_{Low}\right)}{L} = D_m \frac{\left(C_{High} - C_{Low}\right)}{L}.$$

Hence, the solute flux in a single pore can now be calculated as follows, where D_m was found for this solute in Example 5.13:

$$j_s = D_m \frac{\left(C_{High} - C_{Low}\right)}{L} = 3.58 \times 10^{-7}\,\frac{cm^2}{sec} \times \frac{(5-0)\frac{g}{L} \times \frac{1\,L}{1000\,cm^3}}{0.02\,cm} = 8.95 \times 10^{-8}\,g\,cm^{-2}\,sec^{-1}.$$

For a membrane containing a multitude of these pores, we use the effective diffusivity, i.e., D_e, which accounts for the effect of the membrane porosity ($\varepsilon = A_P/S$) and the tortuosity (τ) of the pores. If we take the expression for the flux in a single pore obtained previously, that is $j_s = D_m((C_{High} - C_{Low})/L)$, and correct the pore length by increasing L by the tortuosity (τ), and then multiply this equation by the area of all the pores, i.e., A_P, and then divide that result by S, we get the membrane solute flux, $j_s = D_e((C_{High} - C_{Low})/L)$, where by definition (see Equation 5.65) $D_e = (A_P/S)(1/\tau)$ $D_m = (\varepsilon/\tau)D_m$. Therefore, D_e is given by $D_e = (0.4/1.8) \times 3.58 \times 10^{-7}$ (cm^2/sec) $= 7.96 \times 10^{-8}$ cm^2/sec. We can then calculate the solute flux for the membrane as

$$j_s = D_e \frac{\left(C_{High} - C_{Low}\right)}{L} = 7.96 \times 10^{-8}\,\frac{cm^2}{sec} \times \frac{(5-0)\frac{g}{L} \times \frac{1\,L}{1000\,cm^3}}{0.02\,cm} = 1.99 \times 10^{-8}\,g\,cm^{-2}\,sec^{-1}.$$

Multiplying this result by the total membrane area of 10 cm^2 gives a solute transport rate of 1.99×10^{-7} g sec^{-1}.

5.6.1 DIFFUSION OF A SOLUTE FROM A POLYMERIC MATERIAL

Biocompatible polymers are widely used for the controlled release of a variety of drugs. Consider the situation shown in Figure 5.10. A drug is uniformly distributed within a microporous polymeric

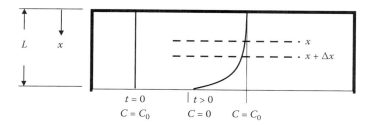

FIGURE 5.10   Diffusion of a solute within a microporous polymeric disk.

disc of radius R and thickness L. The disc is very thin, so $R \gg L$, making the diffusion process one dimensional in the x-direction. All of the surfaces of the disc are coated with an impermeable material except for the surface located at $x = L$. Since the drug release is assumed to occur over a long period of time, which can be many days or weeks, the rate limiting process for release of the drug from the polymeric support at $x = L$ is diffusion of the drug through the pores of the polymeric material. Using Fick's first law to describe the diffusion process, an unsteady solute mass balance over the region from x to $x + \Delta x$ shows that Fick's second law describes the diffusion of the solute within the polymeric material:

$$\frac{\partial C}{\partial t} = D_e \frac{\partial^2 C}{\partial x^2}. \tag{5.66}$$

Note that we now use the effective diffusivity (i.e., Equation 5.65) to describe the diffusion of the solute through the microporous structure of the polymeric material. The initial concentration of the drug in the material is C_0 and the drug concentration at $x = L$ is zero, since the drug is immediately taken up by the surroundings, a process that is much faster than the diffusion of the drug through the polymeric material. The initial and boundary conditions may then be written as

$$\text{IC: } t = 0, C(x,t) = C_0,$$

$$\text{BC1: } x = 0, \frac{dC}{dx} = 0, \tag{5.67}$$

$$\text{BC2: } x = L, C = 0.$$

Boundary condition one expresses the additional fact that the solute cannot diffuse out through the top surface of the polymeric disc, since this surface is coated with an impermeable material.

A solution to Equations 5.66 and 5.67 can be obtained using the *separation of variables* technique, which will be briefly described in the following discussion (check out any advanced calculus book for more information on this technique). We let $C(x, t)$ be given by the product of a function that depends only on x, i.e., $F(x)$, and we define another function that depends only on t, i.e., $T(t)$. Therefore, $C(x, t) = F(x) T(t)$. We substitute this equation into Equation 5.66 and obtain

$$\frac{1}{TD_e} \frac{dT}{dt} = \frac{1}{F} \frac{d^2 F}{dx^2} = -\lambda^2. \tag{5.68}$$

Note that the term on the left side of Equation 5.68 depends only on t and the term in the middle only depends on x. The only way that they can then be equal to each other is if they equal the same constant, i.e., λ^2. The minus sign is in front of λ^2 in order to obtain a solution that remains finite as time increases. Equation 5.68 can then be rearranged to give the following two ordinary differential equations:

$$\frac{dT}{dt} = -\lambda^2 D_e T \quad \text{and} \quad \frac{d^2F}{dx^2} + F\lambda^2 = 0. \tag{5.69}$$

These equations may then be integrated to provide the following results:

$$T = C_1 e^{-\lambda^2 D_e t} \quad \text{and} \quad F = C_2 \sin\lambda x + C_3 \cos\lambda x. \tag{5.70}$$

Applying boundary conditions one and two on the expression for F, we obtain $C_2 = 0$ and $C_3 \cos \lambda L = 0$. C_3 cannot be equal to zero since that provides a trivial solution. Hence, because of the periodic nature of the cosine function, we then find that there are an infinite number of λs that can satisfy the condition that $\cos \lambda L = 0$. It is easy to then show that these λs are given by

$$\lambda_n = \left(\frac{2n+1}{2}\right)\frac{\pi}{L} \quad \text{for } n=1,2,\dots,\infty. \tag{5.71}$$

λ_n are also known as *eigen values*, and $\cos\lambda_n x$ are known as *eigen functions*. Combining these results, our solution now is $C_n(x,t) = C_{3n} e^{-\lambda_n D_e t} \cos\lambda_n x$ for $n = 0, 1, 2,\dots, \infty$. Each value of n gives us a solution to the original partial differential equation, i.e., Equation 5.66, which satisfies boundary conditions one and two. We can also sum all of these n solutions and obtain the following result:

$$C(x,t) = \sum_{n=0}^{\infty} C_n e^{-\lambda_n^2 D_e t} \cos \lambda_n x. \tag{5.72}$$

C_n can then be obtained by requiring that Equation 5.72 satisfies the initial condition where $t = 0$, hence Equation 5.72 becomes

$$C_0 = \sum_{n=0}^{\infty} C_n \cos \lambda_n x. \tag{5.73}$$

Next, we multiply both sides of Equation 5.73 by $\cos \lambda_m x$ and integrate from $x = 0$ to $x = L$:

$$C_0 \int_0^L \cos\lambda_m x\, dx = \int_0^L \cos\lambda_m x \sum_{n=0}^{\infty} C_n \cos \lambda_n x\, dx. \tag{5.74}$$

Equation 5.74 can also be written as

$$\frac{C_0}{\lambda_n}\sin\lambda_n L = \sum_{n=0}^{\infty} C_n \int_0^L \cos\lambda_m x \cos\lambda_n x = C_n \int_0^L \cos^2 \lambda_n x\, dx, \tag{5.75}$$

where $\int_0^L \cos\lambda_m x \cos\lambda_n x\, dx = 0$ except when $m = n$. Hence, we then can solve Equation 5.75 for C_n, which is given by

$$C_n = \frac{4C_0(-1)^n}{(2n+1)\pi}. \tag{5.76}$$

Combining Equations 5.72 and 5.76 then provides the solution for the concentration distribution of the drug or solute within the polymeric material:

$$C(x,t) = \frac{4C_0}{\pi} \sum_{n=0}^{\infty} \frac{(-1)^n}{2n+1} e^{-\frac{(2n+1)^2 \pi^2 D_e t}{4L^2}} \cos\frac{(2n+1)\pi x}{2L}. \tag{5.77}$$

Of particular interest is the flux of drug leaving the polymeric disk, i.e., applying Equation 5.64 at $x = L$. After finding $dC/dx\big|_{x=L}$ from Equation 5.77, we can solve for the elution flux of the drug as

$$j_S\big|_{x=L} = \frac{2D_e C_0}{L} \sum_{n=0}^{\infty} e^{-\frac{(2n+1)^2 \pi^2 D_e t}{4L^2}}. \tag{5.78}$$

For thin drug patches that are applied to the skin, Equation 5.78 can also be combined with a pharmacokinetic model for drug distribution in the body to predict how the drug concentration in the body changes with time. This is discussed in Chapter 7.

The amount of drug or solute remaining in the polymeric material at any time t is also of interest. If we let S represent the total surface area of the polymeric material normal to the diffusion direction, then the total amount of drug (D_0) initially present in the polymeric material is equal to $D_0 = C_0 SL$. Note that C_0 is defined in terms of the total volume of the polymeric material, which includes the void space and the polymer. The amount of drug remaining in the polymeric material at time t, i.e., $D(t)$, is then given by integrating the concentration distribution at any time t, i.e., Equation 5.77, from $x = 0$ to $x = L$:

$$D(t) = S \int_0^L C(x,t)\,dx = \frac{8}{\pi^2} D_0 \sum_{n=0}^{\infty} \frac{1}{(2n+1)^2} e^{-\frac{(2n+1)^2 \pi^2 D_e t}{4L^2}}. \tag{5.79}$$

We can also define the cumulative fraction of the drug (f_R) that has been released as the amount of drug released (i.e., $D_0 - D(t)$) divided by the amount of drug originally present, i.e., D_0. Hence

$$f_R = \frac{D_0 - D(t)}{D_0} = 1 - \frac{8}{\pi^2} \sum_{n=0}^{\infty} \frac{1}{(2n+1)^2} e^{-\frac{(2n+1)^2 \pi^2 D_e t}{4L^2}}. \tag{5.80}$$

5.6.1.1 A Solution Valid for Short Contact Times

For small times the drug concentration change occurs only over a thin region near where $x = L$. In this case, Equation 5.66 still describes the diffusion of the drug within the polymeric material; however, we can replace the boundary conditions with those shown below, where now y represents the distance into the material from the exposed surface at $x = L$.

$$\text{IC:} \quad t = 0, C(y,t) = C_0,$$

$$\text{BC1:} \quad y = 0, C = 0, \tag{5.81}$$

$$\text{BC2:} \quad y = \infty, C = C_0.$$

Boundary condition two expresses the fact that at distances far from the exposed surface, the drug concentration within the polymeric material is still equal to its initial value. Equations 5.66

and 5.81 can be solved using the Laplace transform technique. If we let $C' \equiv C_0 - C$, then Equation 5.66 remains the same with the exception that C is replaced by C' and the initial and boundary conditions in Equation 5.18 become

$$\text{IC: } t = 0, C'(y,t) = 0,$$

$$\text{BC1: } y = 0, C' = C_0, \tag{5.82}$$

$$\text{BC2: } y = \infty, C' = 0.$$

The solution to this problem for C' using Laplace transforms is identical to the solution we obtained earlier for Equations 5.7 and 5.8, which is given by Equation 5.9, that is

$$C(y,t) = C_0 \text{erf}\left(\frac{y}{\sqrt{4D_e t}}\right). \tag{5.83}$$

The flux of drug (j_s) from the exposed surface is also given by Equation 5.12.

Example 5.15

Bawa et al. (1985) investigated the release of macromolecules from small slabs of a polymeric material made of ethylene-vinyl acetate copolymer. The polymeric material was formed into slabs that were 1×1 cm and 1 mm in thickness. All surfaces of the slab were coated with paraffin except one 1×1 cm face, which was exposed to the surroundings. During the formation of the polymeric slabs, bovine serum albumin (BSA) particles were incorporated within the polymer, forming a system of interconnected pores after the BSA became solubilized when the polymeric slabs were immersed in an aqueous solution. In one case, using BSA particles that ranged in size from 106 to 150 μm, the cumulative fraction of BSA released (i.e., f_R) as a function of the square root of time is shown in the following table.

Time$^{1/2}$ (h$^{1/2}$)	Cumulative Fraction of BSA Released (f_R)
2.5	0.05
5.0	0.1
7.0	0.12
11.0	0.20
13.0	0.22
15.5	0.28
17.5	0.30
20.0	0.32
22.0	0.35

If the initial concentration of BSA within the slab is 300 mg cm^{-3} and the porosity is 0.26, estimate the effective diffusivity of BSA within the slab.

Solution

Equation 5.80 applies and the only adjustable parameter in this equation is the effective diffusivity, D_e. Our goal is then to find the value of D_e that minimizes the error between the experimental values of f_R and those values of f_R predicted by Equation 5.80. For this type of problem, one typically uses the sum-of-the-square-of-the-error (SSE) as the objective function, defined by

$$\text{SSE} = \sum_{i=1}^{N} \left(f_{R\text{experimental}_i} - f_{R\text{theory}_i}\right)^2,$$

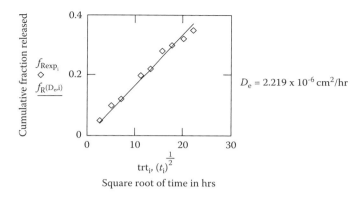

FIGURE 5.11 Cumulative fraction of BSA released as a function of time$^{1/2}$.

where $f_{R\,\text{experimental}}$ are the values in the table and $f_{R\,\text{theory}}$ are the values predicted by Equation 5.80. In the previous equation, N represents the total number of data points. The goal is then to find the value of D_e that minimizes SSE and this value then represents the best fit of Equation 5.80 to the experimental data. There are a variety of mathematical software programs that one may use to solve these types of problems, which are also known as *nonlinear regression problems*. Figure 5.11 shows the results of fitting Equation 5.80 to the data in the previous table. As shown in the figure, we obtain an excellent fit of Equation 5.80 to the experimental data. The value of D_e is found to be 2.22×10^{-6} cm^2 h^{-1}.

5.6.2 DIFFUSION IN BLOOD AND TISSUE

In heterogeneous regions like blood or the tissue space, solute can diffuse through both the continuous fluid space as well as through the cells themselves. The transport mechanism in blood or the tissue space is by diffusion and D_T represents the *effective diffusivity* of solute through the heterogeneous region. The simplest approach for estimating D_T is based on a model developed by Maxwell (1873):

$$\frac{D_T}{D_0} = \frac{2D_0 + D_{\text{cell}} - 2\phi(D_0 - D_{\text{cell}})}{2D_0 + D_{\text{cell}} + \phi(D_0 - D_{\text{cell}})}, \tag{5.84}$$

where D_0 is the diffusivity of the solute through the interstitial fluid space or in the case of blood, $D_0 = D_{\text{plasma}}$; D_{cell} is the diffusivity of the solute in the cells; and ϕ represents the volume fraction of the cells in the region of interest. Interestingly, Equation 5.84 does not depend on the size of the discrete particles, i.e., the cells, but only on their volume fraction.

Riley et al. (1994, 1995a, 1995b, 1995c, 1996) have developed an empirical relation based on Monte Carlo simulations for D_T/D_0 that shows good agreement with available data throughout a wide range of cell volume fractions, i.e., $0.04 < \phi < 0.95$:

$$\frac{D_T}{D_0} = 1 - \left(1 - \frac{D_{\text{cell}}}{D_0}\right)(1.727\phi - 0.8177\phi^2 + 0.09075\phi^3). \tag{5.85}$$

The interstitial fluid that lies between the cells has a gel-like consistency due to the presence of a variety of macromolecules. Therefore, the solute must diffuse around this random network of macromolecular obstacles. The reduction in the solute diffusivity relative to its value in pure water (D)

due to this network of macromolecules within the interstitial fluid gel can be described by Equation 5.86, developed by Brinkman (1947):

$$\frac{D_0}{D} = \omega_r = \frac{1}{1 + \kappa a + \frac{1}{3}(\kappa a)^2}, \tag{5.86}$$

where the ratio of D_0 to D depends on only one parameter, κ; and κ is a function of the gel's microstructure and can be found by fitting Equation 5.86 to experimental data for solute diffusion in the gel.

The steric exclusion, or equilibrium partitioning, of a solute entering from a bulk solution into a gel-like material, such as the interstitial fluid, may be described by Equation 5.87, developed by Ogston (1958):

$$K = \exp\left[-\phi\left(1 + \frac{a}{a_f}\right)^2\right], \tag{5.87}$$

where a_f represents the radius of the macromolecules, which are assumed to form very long cylindrical fibers. If the length of these fibers per volume of gel is given by L, then the volume fraction of these macromolecular fibers is given by $\phi = \pi \, a_f^2 \, L$ (Tong and Anderson 1996). Tong and Anderson (1996) showed that the product of Equations 5.86 and 5.87, and the solute diffusivity in water, i.e., $(KD_0/D) \times D = KD_0 = D_e$, provided excellent representation of the partitioning and diffusion of two representative globular proteins (albumin and ribonuclease-A) in a polyacrylamide gel.

Nugent and Jain (1984a, 1984b) examined the effective diffusion of various sized solutes through normal and tumor tissue. Their results are shown in Figure 5.12, where we see the ratio of the effective diffusivity of the solute in the tissue (D_T) to its diffusivity in water (D) plotted as a function of the solute radius. This figure may be used to provide reasonable estimates of the solute diffusivity within tissue. We see that normal tissue results in a greater reduction in solute diffusivity than that observed for tumor tissue. This reflects the smaller interstitial volume of normal tissue in comparison to tumor tissue.

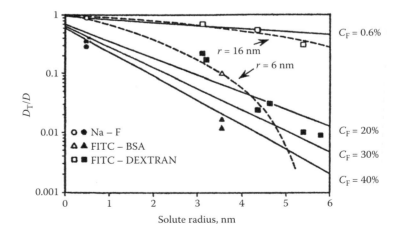

FIGURE 5.12 Ratios of the effective diffusivity in tissue to the aqueous diffusivity as a function of the solute radius. Open and closed symbols represent data for tumor and normal tissue, respectively. Solid lines represent fiber matrix model and dashed lines represent the Renkin pore model equation. (From Nugent, L.J. and Jain, R.K., *Cancer Res.*, 44, 238–44, 1984a. With permission.)

The solid and dashed lines in Figure 5.12 represent the corresponding fits of two diffusion models to their data. The solid lines represent a *fiber matrix model* (Curry and Michel 1980), wherein the diffusivity ratio is given by

$$\frac{D_T}{D} = \exp\left[-\left(1 + \frac{a}{a_f}\right)v^{1/2} C_F^{1/2}\right].$$

(5.88)

In Equation 5.88, the tissue interstitial space is modeled as a random matrix of straight fibers of radius a_f and the solute is considered to be a rigid sphere of radius a. The fiber concentration in the interstitial fluid is C_F, and v is the specific volume of the fibers. The fibers once again are assumed to be the macromolecules that form the gel-like consistency of the interstitial fluid. The quantity C_F v is related to the fiber volume fraction (ϕ) in the gel by the expression $\phi/(1-\phi)$.

The dashed lines in Figure 5.12 represent the predictions of the diffusivity ratio (D_T/D) based on the Renkin hydrodynamic pore model as given by the bracketed $\omega_r(a/r)$ term in Equation 5.61. In this case, the effect of solute partitioning is not included (i.e., $K = 1$) because the solute was already within the tissue region.

Example 5.16

The antibody IgG is diffusing from a bulk solution through a 5% agarose hydrogel at 37°C. The radius (a) of the IgG molecule based on Equation 5.5 and its MW of 150,000 is estimated to be 3.90 nm. The polymer fibers in the hydrogel have a radius (a_f) of 2.79 nm. The volume fraction of the polymer fibers in the hydrogel is $\phi = 0.05$ and the value of κ is 0.57. Estimate the effective diffusivity of IgG in this hydrogel.

Solution

Since IgG is diffusing into the gel from the bulk solution, we use the product of Equations 5.86 and 5.87 to find the diffusivity of this solute. Hence, we have

$$\frac{D_0 K}{D} = \frac{\exp\left[-\phi\left(1 + \frac{a}{a_f}\right)^2\right]}{1 + \kappa a + \frac{1}{3}(\kappa a)^2} = \frac{\exp\left[-0.05\left(1 + \frac{3.9\text{ nm}}{2.79\text{ nm}}\right)^2\right]}{1 + 0.57\frac{1}{\text{nm}}\times 3.9\text{ nm} + \frac{1}{3}\left(0.57\frac{1}{\text{nm}}\times 3.9\text{ nm}\right)^2} = 0.154.$$

The diffusivity of IgG in water at 37°C can be estimated from Figure 5.2 or by Equation 5.3, from which we find that $D = 4.21 \times 10^{-7}$ cm² sec⁻¹. Since $D_0 K$ is the same as D_e, we then have that $D_e = 0.154 \times 4.21 \times 10^{-7}$ cm² sec⁻¹ $= 6.48 \times 10^{-8}$ cm² sec⁻¹.

5.7   SOLUTE PERMEABILITY

Sometimes the term *permeability* (P_m) is used to describe solute transport across a membrane. Permeability lumps together all of the membrane properties that affect solute diffusivity and is defined by the following expression using the *total* membrane surface area S:

$$N_S = P_m S (C_{High} - C_{Low}),$$

(5.89)

where ($C_{High} - C_{Low}$) is the difference in the concentrations of the solute in the bulk fluid adjacent to either side of the membrane. In the absence of external diffusion effects, i.e., a well-stirred solution, this concentration difference is also equal to the difference in the bulk concentration of the solution on either side of the membrane.

Recall Example 5.13, where we showed that the flux across a membrane of thickness L is given by $j_s = D_e((C_{High} - C_{Low})/L)$. If this equation for the flux is multiplied by the total membrane surface area (i.e., S), then by comparing this result to Equation 5.89 we find that the membrane permeability is equal to the *effective diffusivity* of the solute in the membrane divided by the thickness of the membrane, i.e., D_e/L. Permeability may then be expressed in terms of the membrane properties as given by the following relationship:

$$P_m = \frac{D_e}{L} = \frac{D}{L}\left(\frac{A_P}{S}\right)\left(\frac{K\omega_r}{\tau}\right). \tag{5.90}$$

For a membrane with cylindrically shaped pores, the quantity, $K\omega_r$, can be estimated from Equation 5.61. The tortuosity (τ) can be found from experimental permeability data. On the other hand, for diffusion through a gel-like membrane, (A_p/S) $(K\omega_r/\tau)$ is given by the product of Equations 5.86 and 5.87. It is important to note that in both cases, if the solute is already within the membrane, that $K = 1$.

In Equation 5.89, we see the product of the solute permeability and the membrane surface area, i.e., $P_m S$. Although we just developed a framework for describing P_m in terms of the membrane properties, i.e., Equation 5.90, there is still a lot of uncertainty in the values that go into estimating the permeability for a given solute by this equation. In some cases, e.g., for solute diffusion through the capillaries in a large region of tissue, it is also difficult to estimate the surface areas of the capillaries involved. In cases like this, it is more convenient to work with the product of P_m and S, rather than specific values of P_m or S. This is because of the uncertainty in accurately defining either the permeability or the capillary surface area in biological systems.

Renkin and Curry (1979) summarized the results of a variety of experiments that defined continuous capillary $P_m S$ values for solutes of various sizes. These data are shown in Figure 5.13. Note that in some cases for a given size solute, the value of $P_m S$ can vary by almost an order of magnitude. This arises from molecular shape effects, hydration of the solute that increases the effective diffusion size of the solute, and other interactions between the solute and the membrane material. The $P_m S$ value is also highly dependent on the nature of the capillaries that exist in the tissue of interest.

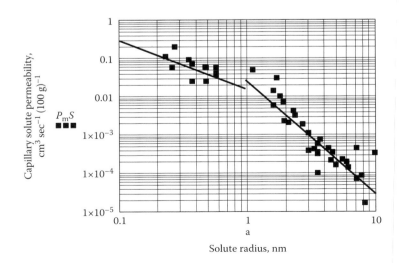

FIGURE 5.13 Solute permeability of capillaries. (Based on data from Renkin, E.M. and Curry, F.E., *Membrane Transport in Biology*, Springer-Verlag, New York, 1979.)

In addition, fenestrated and discontinuous capillaries can have P_mS values that are several order of magnitudes higher than those given in Figure 5.13.

The solid lines in Figure 5.13 represent a linear least square regression of the data. The following empirical equations may be used to estimate capillary P_mS values for a given solute of radius (a). Note that the units on P_mS are cubic centimeters per second per 100 g of tissue:

$$P_mS = 0.0184\, a - 1.223\,, a < 1 \text{ nm,} \tag{5.91a}$$

$$P_mS = 0.0287\, a - 2.92\,, a > 1 \text{ nm.} \tag{5.91b}$$

5.8 IRREVERSIBLE THERMODYNAMICS OF MEMBRANE TRANSPORT

The transport of solute across membranes can occur, as discussed in the previous sections, by a combination of bulk flow or convection, and by diffusion. In some cases, the relative magnitudes of these two transport rates for a given solute are comparable and we must consider the fact that these processes are therefore interdependent. This interaction makes the description of the solute transport rate much more complicated and to arrive at the proper result requires an understanding of the thermodynamics of irreversible processes.

The application of irreversible thermodynamics to membrane processes was developed by Staverman (1948) and Kedem and Katchalsky (1958). For dilute solutions, the theory of irreversible thermodynamics states that the filtration rate of the solvent (Q), and the filtration rate due to the solute relative to that of the solvent (J_v), both depend in a linear manner on the driving forces ΔP (pressure for flow) and $RT\Delta C$ (concentration for diffusion). This linear combination of the driving forces is given by the following equations:

$$Q = SL_p\Delta P + SL_{pS}RT\Delta C, \tag{5.92a}$$

$$J_V = SL_{SP}\Delta P + SL_S RT\Delta C. \tag{5.92b}$$

Q and J_V are vectors and must have the correct sign sense for proper interpretation. This is easily accomplished if we simply assume a vertical orientation for the membrane. Flow of solvent and solute from left to right is considered positive. The Δ sign in this case also represents the value of the property (P or C) on the left side of the membrane minus the corresponding value on the right side of the membrane.

The cross coefficients (L_{pS} and L_{SP}) represent secondary effects that are caused by the primary driving forces. For example, the primary ΔP driving force generates the filtration flow (Q) and also induces a relative flow (J_V) between solute and solvent represented by $L_{SP}\Delta P$. This relative flow between solute and solvent is capable of producing a separation of solute and solvent, by the sieving mechanism discussed earlier, and is referred to as *ultrafiltration*. Along the same line of reasoning, the $RT\Delta C$ primary driving force is responsible for solute diffusion. This diffusive transport produces an additional contribution to the filtration flow that is given by $L_{pS}RT\Delta C$. From Equation 3.1, Van't Hoff's equation, we recognize this as an osmotic flow.

The coefficient L_p is the hydraulic conductance of the membrane defined by Equation 3.7. We will soon show that L_S is related to the permeability of the membrane.

Another basic theorem from the work of Onsager is that the cross coefficients L_{SP} and L_{pS} are equal. We can then simplify Equation 5.92 as follows:

$$Q = L_p S(\Delta P - \sigma RT\Delta C), \tag{5.93a}$$

$$J_V = L_p S\left(-\sigma\Delta P + \frac{L_S}{L_p}RT\Delta C\right). \tag{5.93b}$$

The parameter σ, defined as $-(L_{SP}/L_P)$ or $-(L_{PS}/L_P)$, is called the Staverman *reflection coefficient*.

Note the similarity between the filtration flow rate (Q) in Equation 5.93a and that given by Equation 3.4, recognizing that $RT\Delta C$ is equivalent to $(\pi_C - \pi_{IF})$. These equations are identical when $\sigma = 1$. This implies that the pores of the membrane are impermeable to the solute as required by the derivation of Equation 3.4, the solute is therefore completely "reflected" by the membrane and the $RT\Delta C$ term in Equation 5.93a is, in fact, the osmotic pressure difference. We then recover Starling's equation. When $\sigma = 0$ in Equation 5.93a, then from our previous discussion we have no secondary effect of osmotic flow caused by the primary concentration difference driving force. This means that the solute flows through the pores completely unimpeded, just as easily as the solvent, hence the solute is not "reflected." The concept of the reflection coefficient is illustrated in Figure 5.14. Most solutes have a reflection coefficient that lies between these two extremes.

The total rate of transfer of solute (N_S) through the pores of the capillary wall is given by the product of the solute concentration (C) and the combined flow rate of the solution due to the applied pressure and concentration differences, i.e., $N_S = C(Q + J_V)$. Combining Equations 5.93a and 5.93b in this fashion results in Equation 5.94 for the transport rate of the solute:

$$N_S = CL_P S\left[(1-\sigma)\Delta P + \left(\frac{L_S}{L_P} - \sigma\right)RT\Delta C\right]. \tag{5.94}$$

This equation can be rewritten by using Equation 5.93a to express ΔP as a function of Q and ΔC. Hence, we then obtain

$$N_S = C(1-\sigma)J + CSL_P\left(\frac{L_S}{L_P} - \sigma\right)RT\Delta C. \tag{5.95}$$

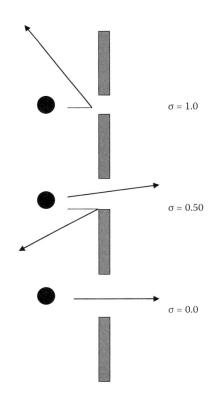

FIGURE 5.14   The reflection coefficient.

The second term in Equation 5.95 contains the following factor, i.e., $((L_S/L_P) - \sigma^2)$, that must be determined. Recognizing for the case of no filtration flow that $Q = 0$ and diffusion is the only solute transport mechanism, then by Equation 5.93a we have

$$\Delta P|_{Q=0} = \sigma RT \Delta C|_{Q=0}. \tag{5.96}$$

Substituting this expression into Equation 5.94 for the case where $Q = 0$ gives Equation 5.97, where we must recognize that in the absence of Q, the transport rate of the solute across the membrane must be equal to our previous result given by Equation 5.89:

$$N_S|_{Q=0} = C L_P S (RT \Delta C)|_{Q=0} \left(\frac{L_S}{L_P} - \sigma^2 \right) = P_m S \Delta C|_{Q=0}. \tag{5.97}$$

We can then solve Equation 5.97 for $((L_S/L_P) - \sigma^2)$, which is given by

$$\left(\frac{L_S}{L_P} - \sigma^2 \right) = \frac{P_m}{L_P RTC}, \tag{5.98}$$

where P_m is the membrane permeability defined earlier by Equation 5.89. Equation 5.95 can now be simplified using Equation 5.98 to give our final result for the solute transport rate through the pores of the capillary wall, recognizing that Q is the filtration flow rate defined earlier by Equation 5.93a:

$$N = C(1 - \sigma)Q + P_m S \Delta C, \tag{5.99}$$

The first term in Equation 5.99 represents the amount of solute transported by bulk flow (convection) of the solvent and the second term represents the contribution of diffusion. The solute concentration on the side of the membrane is represented by C, where the filtration flow (Q) originates.

5.8.1 FINDING L_P, P_M, AND σ

We must know the three parameters, L_P, P_m, and σ, in Equations 5.93a and 5.99 to calculate the transport rate of a solute. Recall that L_P is the hydraulic conductance and it can be estimated using Equation 3.7, or it can be measured using a pure solvent whose value of $\sigma = 0$, and measuring the flow rate Q across the membrane for a given ΔP. Then, by Equation 5.93a, $L_P = J/S\Delta P$.

Recall that P_m is the permeability of the solute in the membrane and may be estimated by Equation 5.90. Alternatively, in the absence of any bulk flow across the membrane (i.e., $Q = 0$), the permeability can be obtained by measuring the solute transport rate (N_S) for a given concentration difference. Then, from Equation 5.99, the permeability is given by $P_m = N_S/S\Delta C$.

The reflection coefficient, σ, can be measured in one of two ways. In the first method, one employs a pressure drop across the membrane and measures the filtration flow rate, Q, as well as the transport rate of the solute, N_S, across the membrane in the absence of a solute concentration difference across the membrane. Then, from Equation 5.99, we obtain the fact that $(1 - \sigma) = N_S/CJ$. Alternatively, we can measure the filtration flow rate, Q, in the absence of a pressure drop across the membrane, but under the control of a concentration difference across the membrane (osmotic flow). Then, from Equation 5.93a, the reflection coefficient is given by $\sigma = (-Q/(SL_P RT\Delta C))$.

Durban (1960) has obtained the data shown in Figure 5.15 for the reflection coefficient of a variety of solutes in dialysis tubing, cellophane, and in a wet gel. One can use this figure to directly estimate the reflection coefficient for a solute of known (a/r).

The reflection coefficient can also be estimated using theories developed to describe the motion of particles in small pores. Anderson and Quinn (1974) reexamined the basic hydrodynamic equations

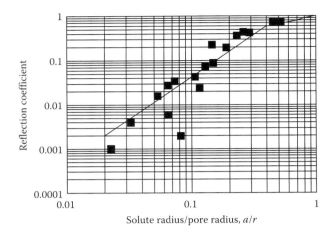

FIGURE 5.15   Reflection coefficient. (Based on data from Durbin, R.P., *J. Gen. Physiol.*, 44, 315–26, 1960.)

that describe hindered particle motion in small pores. Under the assumptions of a rigid and spherical solute, and no electrostatic interactions with the pore wall, which is assumed to be inert, they showed that the sieving coefficient, $S_a \equiv N_S/CQ$, is the same as $(1 - \sigma)$. In this expression, C is the concentration of the solute on the nonfiltrate side of the membrane. Recall that their expression for S_a is given by Equation 5.55. This equation can be rewritten to give the following predictive equation for the reflection coefficient:

$$\sigma = 1 - \left(1 - \frac{a}{r}\right)^2 \left[2 - \left(1 - \frac{a}{r}\right)^2\right] \times \left[1 - \frac{2}{3}\left(\frac{a}{r}\right)^2 - 0.163\left(\frac{a}{r}\right)^3\right]. \tag{5.100}$$

The solid line in Figure 5.15 represents a comparison of Equation 5.100 to the experimental data obtained by Durbin (1960) for the reflection coefficient. Equation 5.100 provides a reasonable comparison to Durbin's data at higher values of (a/r), where the (a/r) ratio is expected to be the dominant factor controlling the value of the reflection coefficient. At lower values of (a/r), other effects not accounted for in the derivation of Equation 5.100 dominate. In the absence of experimental data, Equation 5.100 can be used to provide an estimate of the reflection coefficient.

5.8.2 Multicomponent Membrane Transport

Equations 5.93a and 5.99 describe the filtration flow rate and the solute transport rate in a system consisting of only a single solute and a solvent. However, similar equations can be derived for a multicomponent solute and solvent system by recognizing that now Equations 5.92a and b become

$$Q = SL_P\Delta P + SRT \sum_i L_{PS_i}\Delta C_i, \tag{5.101a}$$

$$J_V = S\Delta P \sum_i L_{SP_i} + SRT \sum_i L_{S_i}\Delta C_i. \tag{5.101b}$$

The summations are over each solute in the system. Using the approach outlined earlier for the derivation of Equations 5.93a and 5.99, the following equations are obtained for the filtration rate (Q) and the transport rate of solute i (N_i) across the membrane in multicomponent systems:

$$Q = SL_P \left[\Delta P - RT \sum_i \sigma_i \Delta C_i \right],$$ (5.102a)

$$N_i = C_i (1 - \sigma_i) Q + P_{mi} S \Delta C_i.$$ (5.102b)

5.9 TRANSPORT OF SOLUTES ACROSS THE CAPILLARY WALL

In this section, we will use the capillary wall as a representative membrane to illustrate in the following examples the calculation of solute transport rates using the relationships developed in the previous sections. In our discussion, we will find it convenient to classify solutes into three categories: the first category concerns the transport of small water soluble but lipid insoluble substances such as Na$^+$ and K$^+$ ions, glucose, amino acids, and other metabolites. The second category of solutes are lipid soluble substances such as oxygen and carbon dioxide, and the third category we will discuss is the transport of large lipid insoluble substances such as proteins (macromolecules).

Example 5.17

Calculate the rate of transport by filtration and diffusion of a small water soluble but lipid insoluble solute, such as glucose. The *diffusivity of glucose (D)* is 0.91×10^{-5} cm^2 sec^{-1} and its molecular radius is 0.36 nm. The average concentration of glucose in plasma is about 5 μmol cm^{-3} and it is assumed that all the glucose transported to the extracapillary space is consumed instantly by the cells. Assume that all of the plasma proteins are retained by the capillary wall. Also, assume that the tortuosity of the capillary wall is equal to 2.

Solution

The filtration flow rate of plasma (Q) across the capillary wall is given by Equation 5.93a with $\sigma = 1.0$ and was shown earlier in Example 5.1 to be equal to 5.75×10^{-9} cm^3 h^{-1}. Equation 5.99 may be used to calculate the solute transport rate of glucose across the capillary wall. We first need to estimate the reflection coefficient (σ) for glucose. Equation 5.100 can be used to obtain an estimate assuming that the diameter of the capillary pores is 7 nm (see Table 5.1):

$$\sigma = 1 - \left(1 - \frac{0.36}{3.5}\right)^2 \left[2 - \left(1 - \frac{0.36}{3.5}\right)^2\right] \times \left[1 - \frac{2}{3}\left(\frac{0.36}{3.5}\right)^2 - 0.163\left(\frac{0.36}{3.5}\right)^3\right].$$

$$\sigma = 0.045$$

The permeability of glucose in the capillary wall can be estimated using Equation 5.90. The quantity $K \omega_r$ can be obtained from Equation 5.61 assuming that the pores in the capillary wall are cylindrical in shape:

$$P_m = \frac{0.91 \times 10^{-5} \frac{cm^2}{sec}}{2 \times 0.5 \times 10^{-6} \ m} \times \frac{1 \ m}{100 \ cm} \times 0.001 \times \left(1 - \frac{0.36 \ nm}{3.5 \ nm}\right)^2$$

$$\times \left[1 - 2.1\left(\frac{0.36 \ nm}{3.5 \ nm}\right) + 2.09\left(\frac{0.36 \ nm}{3.5 \ nm}\right)^3 - 0.95\left(\frac{0.36 \ nm}{3.5 \ nm}\right)^5\right] = 5.76 \times 10^{-5} \ cm \, sec^{-1}.$$

Renkin (1977) reports values of the glucose permeability ranging from 1.6×10^{-6} cm sec^{-1} for capillaries in the dog brain to as high as 9.6×10^{-5} cm sec^{-1} for cat leg capillaries. On average, the capillary glucose permeability is about 2.65×10^{-5} cm sec^{-1}, thus the value calculated previously is of the correct order of magnitude. We may now calculate, by Equation 5.99, the transport rate of glucose across the capillary wall:

$$N_S = 5\frac{\mu mole}{cm^3}(1-0.05) \times 5.75 \times 10^{-9}\frac{cm^3}{h} + 5.76 \times 10^{-5}\frac{cm}{sec} \times \pi$$

$$\times 10 \times 10^{-6}\ m \times 0.001\ m \times (5-0)\frac{\mu mol}{cm^3} \times \frac{(100\ cm)^2}{m^2} \times \frac{3600\ sec}{h},$$

$$N_S = 2.73 \times 10^{-8}\frac{\mu mole}{h} + 3.26 \times 10^{-4}\frac{\mu mole}{h} = 3.26 \times 10^{-4}\frac{\mu mole}{h}.$$

The first term in the calculation of N_S by Equation 5.99 represents the transport rate of glucose by filtration. The second term represents the transport rate by diffusion. Comparing the previous result, we find that the transport rate of small water soluble but lipid insoluble solutes such as glucose across the capillary wall is several thousand times higher by diffusion than by convection! We may therefore neglect the convective or filtration transport of these substances across the capillary wall.

Example 5.18

Now consider the transport rate of lipid soluble solutes such as oxygen. In this case, the entire surface area of the capillary wall is available for transport. Oxygen has considerable solubility in the cell membrane and will readily diffuse through the endothelial cells of the capillary wall. Oxygen can also diffuse through the pores in the capillary wall. The *diffusivity of oxygen* (D) is 2.11×10^{-5} cm^2 sec^{-1} and its molecular radius is 0.16 nm. The average concentration of oxygen in the capillary is 0.09 µmol cm^{-3}. Once again, assume that the tortuosity of the capillary wall is 2.

Solution

The filtration flow rate of plasma (Q) across the capillary wall was shown earlier in Example 5.1 to be equal to 5.75×10^{-9} cm^3 h^{-1}. Equation 5.99 can be used to calculate the solute transport rate of oxygen through the pores in the capillary wall. We first need to estimate the reflection coefficient for oxygen. Equation 5.100 can be used to obtain an estimate assuming that the diameter of the capillary pores is 7 nm:

$$\sigma = 1 - \left(1 - \frac{0.16}{3.5}\right)^2\left[2 - \left(1 - \frac{0.16}{3.5}\right)^2\right] \times \left[1 - \frac{2}{3}\left(\frac{0.16}{3.5}\right)^2 - 0.163\left(\frac{0.16}{3.5}\right)^3\right]$$

$$\sigma = 0.0094$$

The permeability of oxygen in the capillary pores may now be determined using Equations 5.61 and 5.90:

$$P_m = \frac{2.11 \times 10^{-5}\frac{cm^2}{sec}}{2 \times 0.5 \times 10^{-6}\ m} \times \frac{1\ m}{100\ cm} \times 0.001 \times \left(1 - \frac{0.16\ nm}{3.5\ nm}\right)^2$$

$$\times \left[1 - 2.1\left(\frac{0.16\ nm}{3.5\ nm}\right) + 2.09\left(\frac{0.16\ nm}{3.5\ nm}\right)^3 - 0.95\left(\frac{0.16\ nm}{3.5\ nm}\right)^5\right] = 1.74 \times 10^{-4}\ cm\,sec^{-1}.$$

We can now calculate, by Equation 5.99, the transport rate of oxygen through the pores of the capillary wall:

$$N_S = 0.09\frac{\mu mole}{cm^3}(1-0.01) \times 5.75 \times 10^{-9}\frac{cm^3}{h} + 1.74 \times 10^{-4}\frac{cm}{sec}$$

$$\times \pi \times 10 \times 10^{-6}\ m \times 0.001\ m \times (0.09-0)\frac{\mu mol}{cm^3} \times \frac{(100\ cm)^2}{m^2} \times \frac{3600\ sec}{h},$$

$$N_S = 5.12 \times 10^{-10} \frac{\mu\text{mole}}{\text{hr}} + 1.77 \times 10^{-5} \frac{\mu\text{mole}}{\text{hr}} = 1.77 \times 10^{-5} \frac{\mu\text{mole}}{\text{hr}}.$$

Comparing the contributions of filtration and diffusion in the earlier result, we again find that the transport rate of small molecules like oxygen through the capillary pores is several thousand times higher by diffusion than by convection. The permeability of oxygen through the capillary wall itself is given by

$$P_{\text{oxygen}} = \frac{K_{\text{oxygen}}D}{t_m} = \frac{0.25 \times 2.11 \times 10^{-5} \frac{\text{cm}^2}{\text{sec}}}{0.5 \times 10^{-6}\text{m}} \times \frac{1\,\text{m}}{100\,\text{cm}} = 0.11 \frac{\text{cm}}{\text{sec}}$$

K_{oxygen} is the oxygen partition coefficient and represents the ratio of the oxygen concentration in the capillary wall to that in the bulk solution. The value of 0.25 was used in order to match the oxygen permeability of the capillary wall with the accepted value of about 0.1 cm sec^{-1}. The oxygen transport rate from the capillary is given by Equation 5.89 and is equal to

$$N_{\text{oxygen}} = 0.11 \frac{\text{cm}}{\text{sec}} \times \pi \times 10 \times 10^{-6}\,\text{m} \times 0.001\,\text{m} \times (0.09 - 0)\frac{\mu\text{mol}}{\text{cm}^3} \times \frac{(100\,\text{cm})^2}{\text{m}^2} \times \frac{3600\,\text{sec}}{\text{hr}}$$

$$= 0.011 \frac{\mu\text{mol}}{\text{hr}}$$

This is essentially the total oxygen transport rate from the capillary since the contribution due to the capillary pores found earlier is negligible in comparison. The smaller value of the oxygen transport rate through the pores (N_S) in comparison to the value given previously for the entire capillary wall (N_{oxygen}) represents the fact that the pore surface area is 1/1000th that of the capillary wall. Since the permeability of oxygen through the pores of the capillary and the capillary wall itself need to be multiplied by the same values of S and ΔC to calculate the transport rate of oxygen, it is seen that regardless of these values, the diffusive transport of oxygen is 1000-fold higher through the endothelial cells lining the capillary walls than what is transported by diffusion through the capillary pores. Thus, we can conclude that the dominant mode for transport of oxygen is simple diffusion across the entire surface area of the capillary walls.

Example 5.19

Our final category of solute to consider is lipid insoluble macromolecules such as proteins. These solutes are characterized by a molecular radius that is comparable to the pores in the capillary wall. Calculate the rate of transport by filtration and diffusion through the capillary wall of a protein with an MW of 50,000. Assume that the concentration of the protein in the plasma is 0.05 µmol cm^{-3}.

Solution

The filtration flow rate of plasma (Q) across the capillary wall, as shown in Examples 5.17 and 5.18, is equal to 5.75×10^{-9} cm^3 h^{-1}. Since we only know the MW of the protein, we need to first estimate its radius and diffusivity. From Equation 5.5, we find that the radius of the solute is equal to 2.71 nm. We can also use Equation 5.3 to estimate the diffusivity. Therefore, we find that $D = 7 \times 10^{-7}$ cm^2 sec^{-1}. Equation 5.99 may be used to calculate the solute transport rate of the protein. We first need to estimate the reflection coefficient for the protein. Equation 5.100 can be used to obtain an estimate assuming that the diameter of the capillary pores is 7 nm:

$$\sigma = 1 - \left(1 - \frac{2.71}{3.5}\right)^2 \left[2 - \left(1 - \frac{2.71}{3.5}\right)^2\right] \times \left[1 - \frac{2}{3}\left(\frac{2.71}{3.5}\right)^2 - 0.163\left(\frac{2.71}{3.5}\right)^3\right]$$

$$\sigma = 0.95.$$

The permeability of the protein in the capillary wall can then be found using Equation 5.90. The quantity, $K\omega_r$, can be estimated from Equation 5.61 assuming that the pores in the capillary wall are cylindrical in shape:

$$P_m = \frac{7.0 \times 10^{-7} \frac{cm^2}{sec}}{2 \times 0.5 \times 10^{-6} m} \times \frac{1\,m}{100\,cm} \times 0.001 \times \left(1 - \frac{2.71\,nm}{3.5\,nm}\right)^2$$

$$\times \left[1 - 2.1\left(\frac{2.71\,nm}{3.5\,nm}\right) + 2.09\left(\frac{2.71\,nm}{3.5\,nm}\right)^3 - 0.95\left(\frac{2.71\,nm}{3.5\,nm}\right)^5\right] = 2.85 \times 10^{-8} \; cm\,sec^{-1}.$$

This value of the permeability is representative of the experimental values for macromolecules of this size reported by Renkin (1977). We can now calculate, by Equation 5.99, the transport rate of the protein across the capillary wall:

$$N_s = 0.05 \frac{\mu mol}{cm^3}(1-0.95) \times 5.75 \times 10^{-9} \frac{cm^3}{hr} + 2.85 \times 10^{-8} \frac{cm}{sec}$$

$$\times \pi \times 10 \times 10^{-6} \; m \times 0.001\,m \times (0.05-0)\frac{\mu mol}{cm^3} \times \frac{(100\;cm)^2}{m^2} \times \frac{3600\;sec}{hr}$$

$$N_s = 1.44 \times 10^{-11} \frac{\mu mole}{hr} + 1.61 \times 10^{-9} \frac{\mu mole}{hr} = 1.61 \times 10^{-9} \frac{\mu mole}{hr}.$$

Once again, we observe that even for a macromolecular solute with an MW of 50,000, the diffusive transport across the capillary wall is still significantly higher than the amount of solute transported by convection.

Example 5.20

Repeat Example 5.19 assuming that the mean capillary pressure is increased from 17.3 to 37.3 mmHg.

Solution

The filtration flow rate of plasma (Q) across the capillary wall will be significantly increased as a result of the increase in mean capillary pressure. The net filtration pressure will increase from 0.3 to 20.3 mmHg. We can calculate the filtration rate of the plasma following the procedure outlined in Example 5.1. We can then show that the filtration rate equals 3.89×10^{-7} cm³ h⁻¹. This represents a nearly 68-fold increase in the plasma filtration rate and leads to edema in the interstitium. The reflection coefficient and solute permeability still have the same values as calculated in Example 5.19, i.e., $\sigma = 0.95$ and $P_m = 2.85 \times 10^{-8}$ cm sec⁻¹. We can now calculate, by Equation 5.99, the transport rate of the protein across the capillary wall for this special case of increased mean capillary pressure:

$$N_s = 0.05 \frac{\mu mol}{cm^3}(1-0.95) \times 3.89 \times 10^{-7} \frac{cm^3}{hr} + 2.85 \times 10^{-8} \frac{cm}{sec}$$

$$\times \pi \times 10 \times 10^{-6} \; m \times 0.001\,m \times (0.05-0)\frac{\mu mol}{cm^3} \times \frac{(100\;cm)^2}{m^2} \times \frac{3600\;sec}{hr}$$

$$N_s = 9.73 \times 10^{-10} \frac{\mu mole}{hr} + 1.61 \times 10^{-9} \frac{\mu mole}{hr} = 2.58 \times 10^{-9} \frac{\mu mole}{hr}.$$

The first term in the calculation of N_s by Equation 5.99 represents the transport rate of the protein by filtration. The second term represents the transport rate by diffusion. Unlike the calculations for solute transport in the previous examples, we now observe that the increased filtration pressure has significantly increased the transport rate of the solute by convection. Convection of the solute now accounts for 38% of the total solute transport rate and is responsible for the 60% increase in the solute transport rate as a result of the increased mean capillary pressure.

5.10   TRANSPORT OF A SOLUTE BETWEEN A CAPILLARY AND THE SURROUNDING TISSUE SPACE

Now that we have developed an understanding of solute transport across a semipermeable membrane like the capillary wall, it is possible to develop a model for the spatial distribution and consumption or production of a particular species within the tissue space surrounding a given capillary. The following equations also describe the transport of a solute from within a hollow fiber to the cylindrical space surrounding the fiber. In the case of oxygen, the situation is somewhat more complex since the red blood cell provides an additional source or sink of oxygen through interactions with hemoglobin. We will discuss oxygen transport later in Chapter 6.

5.10.1   THE KROGH TISSUE CYLINDER

Microscopic studies support, somewhat, the notion that a bed of capillaries in tissue can be represented as a repetitive arrangement of capillaries surrounded by a cylindrical layer of tissue. Figure 5.16 shows an idealized sketch of the capillary bed and the corresponding cylindrical layer of tissue of thickness equal to r_T. As mentioned at the beginning of this chapter, the residence time for blood in the capillary is on the order of 1 sec. The diffusivity of a solute in tissue is on the order of 10^{-5} cm^2 sec^{-1}. The characteristic time for diffusion in the Krogh tissue cylinder can be expressed by the following relation: $t_{\text{diffusion}} \approx \text{order}(r_T^2/D_T)$. Since the blood residence time and the diffusion time must be of the same order of magnitude, we can use this simple relationship to estimate that the Krogh tissue cylinder radius is on the order of 30 μm. Krogh (1919) used this cylindrical capillary-tissue model to study the supply of oxygen to muscle. This approach is now known as the *Krogh tissue model*.

A steady state mathematical model for the transport of a solute can be developed using the Krogh tissue model. Although this model is developed for a single capillary surrounded by tissue, it can also be extended to a single hollow fiber membrane found in a bioartificial organ or in a bioreactor. In the case of the hollow fiber, the radius and length of the fiber is much larger than the dimensions of a single capillary.

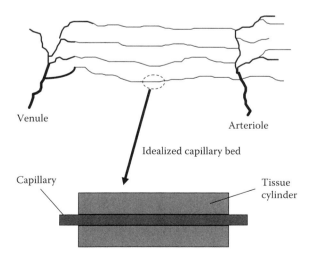

FIGURE 5.16   The Krogh tissue cylinder.

We will treat the tissue space surrounding the capillary as a continuous phase and ignore the fact that it is composed of discrete cells. Therefore, the diffusion of the solute in the tissue space may be described by an effective diffusivity (D_T) as defined earlier by Equations 5.84, 5.85, and 5.88. The driving force for diffusion of the solute is the consumption (or production) of the solute by the cells within the tissue space.

We also need to have an expression that relates the consumption or production of the solute to the concentration of the solute in the tissue space ($\bar{C}$). The *Michaelis–Menten equation* can be used to describe the metabolic consumption (or production) of the solute in the tissue space. The Michaelis–Menten equation can be written as follows:

$$R\bar{C} = \frac{V_{max}\bar{C}}{K_m + \bar{C}}. \tag{5.103}$$

For consumption of the solute, $R(\bar{C})$ will be assumed to have a positive value, whereas if the solute is produced, then $R(\bar{C})$ will have a negative value. In this equation, V_{max} represents the maximum reaction rate. The maximum reaction rate occurs when $\bar{C} \gg K_m$, and the reaction rate is then said to be zero-order in the solute concentration (i.e., $\bar{C}^0$). K_m is called the Michaelis constant and represents the value of $\bar{C}$ for which the reaction rate is one-half the maximal value. When $\bar{C} \ll K_m$, then $R(\bar{C}) = (V_{max} / K_m)\bar{C}$, and the reaction is first-order in the solute concentration (i.e., $\bar{C}^1$). In many cases for biological reactions, $\bar{C} \gg K_m$, and we may reasonably assume that the reaction is zero order and that $R(\bar{C}) = V_{max} = R_0$. This will considerably simplify the following mathematical analysis and allow us to find an analytical solution for solute diffusion in the tissue space that surrounds a given capillary.

5.10.2 A MODEL OF THE KROGH TISSUE CYLINDER

Figure 5.17 illustrates a finite volume section of the Krogh tissue cylinder extending from z to $z + \Delta z$. Within the capillary, the solute is transported primarily by convection in the axial direction and by diffusion in the radial direction. Diffusion of the solute from the blood and through the capillary wall is proportional to the concentration difference of the solute between that in the bulk fluid and that at the interface between the outer capillary wall and the surrounding tissue. This concentration difference is multiplied by an overall mass transfer coefficient (K_0) to give the solute flux. The overall mass transfer coefficient represents the combined resistance of the fluid flowing through the capillary (i.e., k_m) and the permeability of the solute in the capillary wall (i.e., P_m). The overall mass transfer coefficient, K_0, can be evaluated by recognizing that the solute flux at any value of z is constant (neglecting the curvature of the capillary wall) from the blood to the capillary wall at r_c, and across the capillary wall to the beginning of the tissue space at $r_c + t_m$. The transport rate across the capillary wall is given by Equation 5.99, recognizing that in most cases, as shown in the previous examples, the convective term, $C(1-\sigma)Q$, is negligible in comparison to the amount of

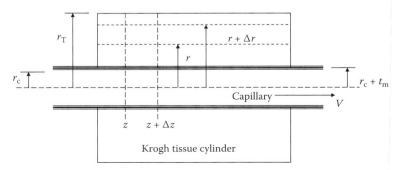

FIGURE 5.17 Geometry of the Krogh tissue cylinder.

solute transported by diffusion. We can then write at a particular axial location z along the length of the capillary that

$$N = 2\pi r_C \Delta z k_m \left(C - \hat{C}\big|_{r_C}\right) = 2\pi r_C \Delta z P_m \left(\hat{C}\big|_{r_C} - \bar{C}\big|_{r_C + t_m}\right) = 2\pi r_C \Delta z K_0 \left(C - \bar{C}\big|_{r_C + t_m}\right), \quad (5.104)$$

where $\hat{C}$ now denotes the solute concentration at the surface of the capillary wall on the blood side. Equation 5.104 can then be rearranged to obtain the following expression for the overall mass transfer coefficient:

$$K_0 = \frac{1}{\dfrac{1}{k_m} + \dfrac{1}{P_m}} \quad (5.105)$$

P_m is the solute permeability across the capillary and k_m is the blood side film mass transfer coefficient defined earlier in this chapter.

When calculating the value of k_m, one needs to determine whether the transport properties, as represented by the viscosity and the solute diffusivity, should be based on their respective plasma values or on those of blood. Recall from Chapter 4 that the red blood cells have a tendency to accumulate along their axis of flow. This creates a cell-free plasma layer (δ_{plasma}) adjacent to the capillary wall that is on the order of a few microns in thickness (see Table 4.4). Also, the thickness of the concentration boundary layer can be approximated from the fact that $k_m \approx (D/\delta_C)$. Hence, one can first determine the value of k_m using the properties of blood. Then the concentration boundary layer thickness is approximately given by, $\delta_C \approx (D/k_m)$. If $\delta_C > \delta_{plasma}$, then the value of k_m should be based on the properties of blood, since the concentration boundary layer is expected to be composed of blood. Otherwise, if $\delta_C < \delta_{plasma}$, then the concentration boundary layer lies within the plasma layer and the physical properties should be those of plasma.

We also treat the blood as a continuous phase and ignore the finite size of the red blood cells. We will assume that the blood flows through the capillary with an average velocity of V. With these assumptions, a steady state shell balance on the solute in the blood from z to $z + \Delta z$ (see Figure 5.17) provides the following equation:

$$V\pi r_c^2 C\big|_z - V\pi r_c^2 C\big|_{z+\Delta z} = 2\pi r_c \Delta z K_0 \left(C - \bar{C}\big|_{r_c + t_m}\right). \quad (5.106)$$

Dividing by Δz, and taking the limit as $\Delta z \to 0$, results in the following differential equation for the solute concentration in the blood:

$$-V\frac{dC}{dz} = \frac{2}{r_c} K_0 \left(C - \bar{C}\big|_{r_c + t_m}\right). \quad (5.107)$$

A steady state shell balance at a given value of z from r to $r + \Delta r$ may also be written for the solute concentration in the tissue space.

$$-D_T \frac{d\bar{C}}{dr} 2\pi r \Delta z\big|_r + D_T \frac{d\bar{C}}{dr} 2\pi r \Delta z\big|_{r+\Delta r} = R(\bar{C}) 2\pi r \Delta r \Delta z. \quad (5.108)$$

This equation neglects axial diffusion in comparison to radial diffusion within the tissue space. After dividing by $2\pi r \Delta r$, and taking the limit as $\Delta r \to 0$, the following differential equation results for the solute concentration in the tissue space:

$$\frac{D_T}{r} \frac{d}{dr}\left(r\frac{d\bar{C}}{dr}\right) - R(\bar{C}) = 0. \quad (5.109)$$

The boundary conditions for differential Equations 5.107 and 5.109 are

$$\text{BC1}: z = 0, \quad C = C_0,$$

$$\text{BC2}: r = r_c + t_m, \quad \bar{C} = \bar{C}\big|_{r_c+t_m},$$

$$\text{BC3}: r = r_T, \quad \frac{d\bar{C}}{dr} = 0$$

(5.110)

It should be noted that even though we neglect axial diffusion in the tissue space in comparison to radial diffusion, the solute concentration in the tissue space will still have an axial dependence. This axial dependence of the tissue space solute concentration results from the boundary condition at $(r_c + t_m)$ and the solute flux term, $K_0\left(C - \bar{C}\big|_{r_c+t_m}\right)$, which couples the solutions for the solute concentration within the capillary and the tissue space. In this way, the axial variation of the solute concentration in the capillary is impressed on that of the tissue space.

The solute concentration in the tissue space is then easily found by solving Equation 5.109 subject to boundary conditions two and three in Equation 5.110. Here, we assume that the consumption of solute is zero-order as discussed earlier, hence $R(\bar{C}) = R_0$ is a constant. The following equation is then obtained:

$$\bar{C}(r,z) = \bar{C}(z)\big|_{r_c+t_m} + \frac{R_0}{4D_T}\left[\left(r^2 - (r_c + t_m)^2\right)\right] - \frac{R_0 r_T^2}{4D_T}\ln\left(\frac{r}{r_c + t_m}\right).$$

(5.111)

This equation provides no information on the value of $\bar{C}(z)\big|_{r_c+t_m}$. However, at any axial position z from the capillary entrance, we must have at steady state that the change in solute concentration within the blood must equal the consumption of solute in the tissue space. Therefore, we can express this requirement by writing the following equation:

$$V\pi r_c^2 C_0 - V\pi r_c^2 C\big|_z = \pi\left[r_T^2 - (r_c + t_m)^2\right]z R_0$$

(5.112)

This equation can be rearranged to give Equation 5.113, which provides for the axial variation of the solute concentration in the capillary:

$$C(z) = C_0 - \frac{R_0}{V r_c^2}\left[r_T^2 - (r_c + t_m)^2\right]z.$$

(5.113)

Equation 5.113 can now be used to find dC/dz in Equation 5.107 with the result that we can solve for $\bar{C}(z)\big|_{r_c+t_m}$, which is given by

$$\bar{C}(z)\big|_{r_c+t_m} = C(z) - \frac{R_0}{2r_c K_0}\left(r_T^2 - (r_c + t_m)^2\right).$$

(5.114)

Note that $C(z)$ in Equation 5.114 makes $\bar{C}(z)\big|_{r_c+t_m}$ depend on z, and by Equation 5.111, the tissue space solute concentration then depends on both r and z as discussed earlier. Equations 5.111, 5.113, and 5.114 can now be combined to give Equation 5.115 for the solute concentration in the tissue space:

$$\bar{C}(r,z) = C_0 - \frac{R_0}{V r_c^2}\left[r_T^2 - (r_c + t_m)^2\right]z - \frac{R_0}{2r_c K_0}\left[r_T^2 - (r_c + t_m)^2\right]$$

$$+ \frac{R_0}{4D_T}\left[r^2 - (r_c + t_m)^2\right] - \frac{R_0 r_T^2}{2D_T}\ln\left(\frac{r}{r_c + t_m}\right).$$

(5.115)

Under some conditions, either the delivery of the solute to the capillary is limited by the capillary flow rate, or the transport rate of the solute across the capillary wall is limited, or the consumption of the solute by the tissue is very rapid. Any one of these conditions can lead to regions of tissue that have no solute. We can then define a *critical radius* in the tissue, $r_{\text{critical}}(z)$, defined as the distance beyond which no solute is present in the tissue. For this situation, we need to modify boundary condition three in Equation 5.110 to the following:

$$BC3': r = r_{\text{critical}}(z), \frac{d\bar{C}}{dr} = 0 \quad \text{and} \quad \bar{C} = 0. \tag{5.116}$$

Under these conditions, the solute concentrations in the capillary, i.e., $C(z)$, at the interface between the capillary and the tissue space, i.e., $\bar{C}(z)|_{r_C+t_m}$, and in the tissue space itself, i.e., $\bar{C}(r,z)$, are still given, respectively, by Equations 5.113 through 5.115; however, the Krogh tissue cylinder radius in these equations, i.e., r_T, is replaced with $r_{\text{critical}}(z)$ once the solute concentration in the tissue at a particular location has reached zero. The critical radius is found by recognizing that at $r_{\text{critical}}(z)$, $\bar{C}(r,z) = 0$. Thus, we can use Equation 5.108, with $r_T = r_{\text{critical}}(z)$ and $\bar{C}(r,z) = 0$, to obtain the following expression for the critical radius:

$$\left(\frac{r_{\text{critical}}(z)}{r_C + t_m}\right)^2 \ln\left(\frac{r_{\text{critical}}(z)}{r_C + t_m}\right)^2 - \left(\frac{r_{\text{critical}}(z)}{r_C + t_m}\right)^2 + 1 = \left(\frac{4D_TC_0}{R_0(r_C + t_m)^2}\right)$$

$$- \frac{4D_T}{Vr_C^2}\left[\left(\frac{r_{\text{critical}}(z)}{r_C + t_m}\right)^2 - 1\right]z - \frac{4D_T}{r_CK_0}\left[\left(\frac{r_{\text{critical}}(z)}{r_C + t_m}\right)^2 - 1\right]. \tag{5.117}$$

This is a nonlinear algebraic equation that can be solved for the critical radius at a given axial position (i.e., z) along the length of the capillary. This is illustrated in Example 5.21. Of particular interest are the conditions under which the solute just begins to be depleted within the tissue space. The depletion of the solute will start at the downstream corner of the Krogh tissue cylinder represented by the coordinates $z = L$ and $r = r_T$. Equation 5.115 can be used to explore the conditions when the solute depletion just begins by recognizing that $\bar{C}(r_T, z) = 0$.

Example 5.21

Calculate the glucose concentration profiles for an exercising muscle capillary. Assume that the capillary properties summarized in Table 5.1 apply and that the glucose consumption rate, R_0, is equal to 0.01 μmol sec^{-1} cm^{-3}. Also assume that the Krogh tissue cylinder radius, r_T, is 40 μm.

Solution
Using Equations 5.113 and 5.115, we obtain the solution shown in Figure 5.18. Shown in the figure are the axial solute concentrations within the capillary and in the tissue space at the Krogh tissue cylinder radius (r_T). As shown in the figure, we observe that the axial solute concentration within the capillary and in the tissue space decreases in a linear fashion. We also observe very little difference between the glucose concentration at the outer surface of the capillary wall and at the Krogh tissue cylinder radius. This indicates that there is little mass transfer resistance to glucose within the tissue space. It is also important to note for glucose transport, that the bulk of the mass transfer resistance is the capillary wall, as reflected by the value of the glucose permeability, P_m. We see that the glucose permeability found in Example 5.16 (5.76 × 10^{-5} cm sec^{-1}) and used here is about three orders of magnitude smaller than the film mass transfer coefficient k_m (0.033 cm sec^{-1}), and for practical purposes we can ignore the blood side mass transfer effect since it offers little resistance in comparison to that of the capillary wall. We also see that because of the large

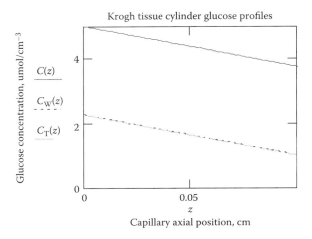

FIGURE 5.18   Glucose concentration as a function of axial position in a capillary.

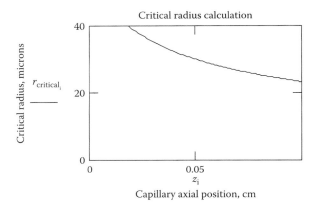

FIGURE 5.19   Calculation of critical radius.

mass transfer resistance of the capillary wall, a significant concentration difference must exist between the capillary and the tissue space to obtain the required mass transfer rate. Transport of glucose for these conditions is therefore said to be diffusion limited.

Example 5.22

Calculate the critical radius for the capillary conditions used in Example 5.21. Assume that the flow rate in the capillary has been reduced by a factor of 10 to $V = 0.005$ cm sec^{-1}.

Solution

Figure 5.19 shows the solution for the critical radius as a function of the axial position in the capillary. The calculation shows that the glucose concentration at the Krogh tissue cylinder radius is nonzero until 0.018 cm into the capillary of total length 0.1 cm. The region above the curve in Figure 5.19 represents the region of the tissue where the glucose concentration is zero. It is also found that the exiting glucose concentration in the blood for the reduced flow conditions in this example, i.e., 0.887 µmol cm^{-3}, is significantly smaller in comparison to the result obtained in Example 5.21, i.e., 3.8 µmol cm^{-3}. Although the exiting glucose concentration in the capillary is lower, this does not necessarily mean that the total amount of glucose transferred is higher. In fact, the amount of solute transferred in this example is about one-third lower than the previous example because of the reduced velocity of the blood. For solutes with reasonably high capillary permeability, i.e., lipophilic solutes like oxygen or small lipophobic solutes like glucose, the solute transport rate as in this example can be flow limited. The transport rate of the solute can therefore be increased to the tissue by increasing the blood flow rate as in Example 5.21. However, for larger lipid insoluble solutes, the capillary solute permeability is significantly reduced and the solute transport rate is diffusion limited. In this case, the blood flow rate has little effect on the solute transport rate.

5.10.2.1 Comparison of Convection and Diffusion Effects

Recall that in developing Equations 5.113 and 5.115, we ignored axial diffusion of solute within the capillary in comparison to axial convection of solute, which is the solute carried along by the flowing blood. We can easily check the validity of this assumption through the following arguments. The amount of solute carried by convection is on the order of:

$$\text{solute transport by axial convection} = \frac{\pi}{4} D_C^2 V C_0, \tag{5.118}$$

and the amount of solute carried by axial diffusion is on the order of:

$$\text{solute transport by axial diffusion} = \frac{\pi}{4} D_C^2 \left(\frac{D C_0}{L} \right). \tag{5.119}$$

By taking the ratio of Equations 5.118 and 5.119, we obtain what is called the *Peclet (Pe) number*, whose magnitude represents the importance of axial convection in comparison to axial diffusion. The criterion for ignoring axial diffusion is then given by the fact that $Pe = VL/D \gg 1$. For the example problems presented previously: $Pe = (V L / D) = (0.05(\text{cm/sec}) \times 0.1 \text{ cm}) / (0.91 \times 10^{-5} (\text{cm}^2/\text{sec}) = 550 > 1$, and we conclude that we can ignore axial diffusion in comparison to axial convection.

A similar line of reasoning can be applied to the tissue space to support the neglect of axial diffusion in comparison to radial diffusion. The solute transport by radial diffusion is on the order of:

$$\text{solute transport by radial diffusion} = \left(\frac{D_T C_0}{r_T - r_C} \right) 2\pi r_C L, \tag{5.120}$$

and the amount of solute transport by axial diffusion is on the order of:

$$\text{solute transport by axial diffusion} = D_T \left(\frac{C_0}{L} \right) \pi (r_T^2 - r_C^2). \tag{5.121}$$

The ratio of Equation 5.120 and 5.121 is a measure of the relative importance of radial diffusion in comparison to axial diffusion. If this ratio, that is $(2 r_C L^2)/((r_T - r_C)(r_T^2 - r_C^2)) = 2 r_C L^2 / ((r_T - r_C)(r_T^2 - r_C^2))$ $\gg 1$, then axial diffusion can be neglected. For Example 5.20, we then have

$$\frac{2 r_c L^2}{(r_T - r_c)(r_T^2 - r_c^2)} = \frac{2 \times 0.5 \times 10^{-3} \text{ cm} \times (0.1 \text{ cm})^2}{(4 \times 10^{-3} - 0.5 \times 10^{-3})\left[(4 \times 10^{-3})^2 - (0.5 \times 10^{-3})^2 \right] \text{cm}^3} = 181 \gg 1,$$

and we conclude for this example that we can ignore axial diffusion in the tissue space.

5.10.3 The Renkin–Crone Equation

The previous analysis can also be simplified considerably for the special case where the solute concentration in the tissue space is zero or much smaller than the concentration in the blood. We can also assume that the blood offers little resistance to the solute transport rate in comparison to that of the capillary wall, hence $K_0 \approx P_m$. For these conditions, Equation 5.107 simplifies to

$$\frac{dC}{dz} = -\frac{2 P_m}{V r_C} C = -\frac{2 \pi r_C}{Q} P_m C. \tag{5.122}$$

Here, we have replaced the capillary blood velocity (V) with the volumetric flow rate of blood in the capillary (Q). Equation 5.122 is easily integrated to obtain the solute concentration in the capillary at any axial position z:

$$E = \frac{[C_0 - C(z)]}{C_0} = 1 - \exp\left(-\frac{2\pi r_C P_m z}{Q}\right). \tag{5.123}$$

This equation is known as the *Renkin–Crone equation* and describes how the solute concentration varies along the length of the capillary for the special case where the solute concentration in the tissue space is very small in comparison to its value in the blood. The solute extraction is represented by E. Since the quantity, $Q\,C_0$, represents the maximum amount of solute that can be transported across the capillary, then $E \times Q C_0$ represents the actual solute transport rate for the given conditions. The solute extraction is therefore a measure of the removal efficiency of the solute from the capillary. This equation readily shows that solute transport from the capillary is strongly dependent on the ratio of the permeability of the capillary wall ($2\pi r_C P_m z$) to the blood flow rate (Q). This ratio can be used to define conditions under which the solute transport is either flow limited or diffusion limited as follows:

$$\text{if } \frac{2\pi r_C P_m z}{Q} \gg 1, \text{ then flow limited and } E \to 1$$

$$\text{if } \frac{2\pi r_C P_m z}{Q} \ll 1, \text{ then diffusion limited and } E \to 0 \tag{5.124}$$

Flow limited means that the mass transport rate of the solute out of the capillary is much faster than the rate at which the solute is entering the capillary by the flowing blood. In this situation, solute transport to the tissue surrounding a given capillary can only be increased by increasing the blood flow rate. In the diffusion limited case, blood supply is more than adequate and the limitation is the difficulty of transporting the solute across the capillary wall. For regions of tissue containing many capillaries, we can replace the quantity $2\pi r_C P_m z$ with the group $P_m S$, where S represents the total surface area of the capillaries within the tissue region of interest.

5.10.3.1   Determining the Value of $P_m S$

The multiple tracer indicator diffusion technique can be used to obtain a test solute's permeability ($P_m S$) across the capillary wall in organs and large tissue regions. This technique can also be used to evaluate the solute permeability in medical devices that employ membranes, e.g., in hemodialyzers, bioreactors, and bioartificial organs.

In this technique, a solution containing equal concentrations of a permeable test solute and a nonpermeable reference solute (typically labeled albumin) is rapidly injected into a main artery leading to the region of interest (Levick 1991). Venous blood samples are then collected over the next few seconds and analyzed for concentrations of the test and reference solutes. Because the test solute is permeable, its concentration in the venous blood will initially fall below that of the reference solute due to its diffusion into the tissue space that surrounds the capillaries. This is shown in Figure 5.20. Because of the transient nature of this technique and the mass transfer resistance of the capillary wall, the concentration of the test solute in the tissue space surrounding the capillaries can be assumed to be close to zero or much smaller than the value within the blood. The concentration of the reference solute in the venous blood provides an estimate of what the test solute concentration would have been at a particular time if no transport of the test solute out of the capillaries had occurred. After several seconds, the reference solute concentration in the blood will fall below that of the test solute, since the reference solute has nearly been washed out of the capillary and the test

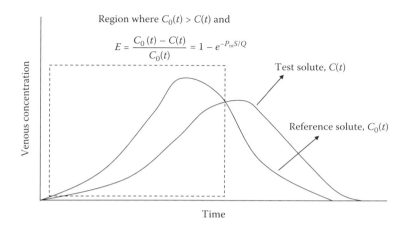

Region where $C_0(t) > C(t)$ and

$$E = \frac{C_0(t) - C(t)}{C_0(t)} = 1 - e^{-P_m S/Q}$$

Test solute, $C(t)$

Reference solute, $C_0(t)$

Venous concentration

Time

FIGURE 5.20 Multiple tracer indicator diffusion technique.

solute concentration in the tissue space is now larger than that in the blood. Hence, the test solute starts to diffuse back into the blood.

The concentrations of the test (i.e., C) and reference solutes (i.e., C_0) in the venous blood at a particular time may then be used to calculate, using Equation 5.123 (with $2\pi r_C P_m z = P_m S$), the test solute extraction (E) and the value of $P_m S$ for a given blood flow rate (Q). It is important to note that accurate results require that the test be repeated at increasing blood flow rates (Q) to ensure that the solute transport is not flow limited.

To show that the solute transport is not flow limited, it is convenient to express Equation 5.123 in terms of the *clearance*. The term clearance is often used to describe the removal of a solute from a flowing fluid such as in a capillary. Clearance, given the symbol CL, is the volumetric flow rate of the fluid that has been totally cleared of the solute. Clearance is then defined by the following expression where we see that $CL = Q\,E$:

$$CL(z) \equiv \frac{Q(C_0 - C(z))}{C_0} = QE = Q\big(1 - e^{-(P_m S/Q)}\big). \tag{5.125}$$

This expression shows that capillary solute clearance is linearly dependent on the flow rate (i.e., $CL = Q$) under flow limited conditions where $P_m S \gg Q$ and $E \approx 1$. However, under diffusion limited conditions where $P_m S \ll Q$, the parenthetical term in Equation 5.125 becomes equal to $P_m S/Q$ and we obtain the result that the clearance becomes independent of the flow rate Q and is equal to the capillary wall permeability (i.e., $CL = P_m S$). Hence, in experiments to determine the value of $P_m S$ using the multiple tracer indicator diffusion technique, the value of $P_m S$ should not be dependent on the flow rate (Q).

Example 5.23

Blood is flowing through a hollow fiber-based bioartificial organ at a flow rate (Q) of 8 cm³ min⁻¹ for each fiber. Surrounding the fibers in the shell space are living cells obtained from a pig's liver. Each fiber has a diameter of 800 μm and a length of 30 cm. At time equal to zero, equal concentrations of a permeable (relative to the fiber wall) test solute and labeled albumin (impermeable) are injected into the blood entering the device. Exiting blood samples are collected over the next few seconds and analyzed for concentrations of the test solute (C) and the reference solute, albumin (C_0). At a particular time, the value of C/C_0 was found to equal 0.60. Estimate the solute permeability (P_m) of the hollow fiber membrane (centimeters per second).

Solution

We can use Equation 5.125 to first calculate the clearance of a given hollow fiber for the test solute as follows:

$$CL = Q\left(1 - \frac{C}{C_0}\right) = 8\frac{\text{cm}^3}{\text{min}} \times (1 - 0.6) = 3.2 \text{ cm}^3 \text{min}^{-1}.$$

Next, we calculate from Equation 5.125 the value of P_m:

$$P_m = -\frac{Q}{S}\ln\left(1 - \frac{CL}{Q}\right) = -\frac{8\dfrac{\text{cm}^3}{\text{min}} \times \dfrac{1 \text{ min}}{60 \text{ sec}}}{2\pi \times 0.04 \text{ cm} \times 30 \text{ cm}}\ln\left(1 - \frac{3.2\dfrac{\text{cm}^3}{\text{min}}}{8\dfrac{\text{cm}^3}{\text{min}}}\right) = 0.009 \text{ cm sec}^{-1}.$$

5.10.4 SOLUTE TRANSPORT IN VASCULAR BEDS, THE WELL-MIXED ASSUMPTION

The Krogh tissue cylinder model provides a description of solute transport within a single capillary. In many cases, a model is needed to describe the solute transport in a much larger region containing numerous capillaries, as illustrated for the capillary bed shown in Figure 5.16. Equation 5.123 with $2\pi\, r_C P_m z$ replaced with $P_m S$ applies to vascular beds, but assumes that the solute concentration is zero in the tissue space.

One simple approach to describe solute transport in this situation is to treat the blood in the capillary bed and the tissue space as separate well-mixed regions. To facilitate the analysis, we introduce two parameters that characterize the degree of vascularization in the region of interest. The first parameter is the *surface area capillary density* (*s*), which is the capillary surface area per unit volume of tissue. Assuming the geometry of the Krogh tissue cylinder shown in Figure 5.17, Equation 5.126 provides an estimate of (*s*) based on given values of the capillary radius, r_C, and the Krogh tissue cylinder radius, r_T:

$$s = \frac{2r_C}{r_T^2}. \tag{5.126}$$

The second parameter is the *volume of capillaries per unit volume of tissue* (*v*). In terms of the Krogh tissue model, this may also be expressed in terms of r_C and r_T.

$$v = \left(\frac{r_C}{r_T}\right)^2. \tag{5.127}$$

A steady state solute balance, assuming only diffusion is relevant for solute transport across the capillary wall, can then be written for the capillary bed illustrated in Figure 5.16. We can write that the solute carried into the capillary bed by blood flow must equal the amount removed by diffusion across the capillary walls and that which leaves with the blood flow:

$$QC_0|_{\text{in}} = P_m S(C - \bar{C}) + QC|_{\text{out}}. \tag{5.128}$$

In this equation, Q is the volumetric flow rate of blood to the region of interest. If V_T is the volume of the tissue region considered, then $q_b = Q/V_T$ is defined as the *tissue blood perfusion rate*. Assuming a capillary of length L, and blood velocity within the capillary of V, it is easily shown that

$$q_b = \left(\frac{r_C}{r_T}\right)^2\left(\frac{V}{L}\right) = \frac{v}{\tau}, \tag{5.129}$$

where τ is the blood residence time, i.e.,

$$\tau = \frac{L}{V} = \frac{0.1 \text{ cm}}{0.05 \dfrac{\text{cm}}{\text{sec}}} = 2 \text{ sec}.$$

At steady state, the amount of solute diffusing out of the capillaries must equal the consumption rate of the solute in the tissue space. Hence, the diffusion term, $P_m S(C - \bar{C})$ must equal $V_T R_0$ for a zero-order reaction occurring within the tissue space. After equating these two expressions, we can write the solute concentration in the tissue space as

$$\bar{C} = C - \frac{V_T R_0}{P_m S}. \tag{5.130}$$

Replacing the diffusion term in Equation 5.128 with $V_T R_0$ allows for the calculation of the solute concentration in the blood exiting the tissue region of interest:

$$C = C_0 - \frac{V_T R_0}{Q} = C_0 - \frac{R_0}{q_b}. \tag{5.131}$$

When this equation is substituted into Equation 5.130 for C, the solute concentration within the tissue space is then given by

$$\bar{C} = C_0 - R_0 \left[\frac{1}{q_b} + \frac{V_T}{P_m S} \right]. \tag{5.132}$$

The following example illustrates the use of these equations.

Example 5.24

Consider the transport of glucose from capillary blood to exercising muscle tissue. As a basis, consider 1 g of tissue. The glucose consumption of the tissue is 60 µmol min^{-1} (100 g)$^{-1}$ or 0.01 µmol sec^{-1} g^{-1}. Blood flow to the region is 60 cm^3 min^{-1} (100 g)$^{-1}$ or 0.01 cm^3 sec^{-1} g^{-1}. The arterial glucose concentration is 5 mM (5 µmol cm^{-3}). The value of $P_m S$ based on capillary recruitment during exercise is 0.33 cm^3 sec^{-1} (100 g)$^{-1}$ or for the 1 g of tissue considered here, 0.0033 cm^3 sec^{-1}. Calculate the glucose concentration in the tissue space and in the exiting blood (based on data provided in Levick 1991).

Solution
From Equation 5.130, we have

$$C = 5 \frac{\text{µmol}}{\text{cm}^3} - \frac{0.01 \dfrac{\text{µmol}}{\text{g sec}}}{0.01 \dfrac{\text{cm}^3}{\text{g sec}}} = 4.0 \frac{\text{µmol}}{\text{cm}^3} = 4 \text{ mM},$$

and by Equation 5.132:

$$\bar{C} = 5 \frac{\text{µmol}}{\text{cm}^3} - 0.01 \frac{\text{µmol}}{\text{g sec}} \left(\frac{1}{0.01 \dfrac{\text{cm}^3}{\text{g sec}}} + \frac{1}{0.0033 \dfrac{\text{cm}^3}{\text{g sec}}} \right) = 0.97 \frac{\text{µmol}}{\text{cm}^3} = 0.97 \text{ mM}.$$

PROBLEMS

1. Consider a polymeric membrane that is 30 μm thick, with a total surface area of 1 cm². The membrane has pores equivalent in size to a spherical molecule with an MW of 100,000, a porosity of 80%, and a tortuosity of 2.5. On one side of the membrane, we have a protein solution that is to be concentrated (protein properties, $a = 3$ nm and $D = 6.0 \times 10^{-7}$ cm² sec⁻¹) at a concentration of 80 g L⁻¹. Assume that the solution viscosity is 1 cP. On the other side of the membrane is pure water. The pressure on the protein side of the membrane is 20 pounds per square inch (psi) higher than on the filtrate side of the membrane. Determine the convective flow rate of the solution and the rate at which the protein crosses the membrane.

2. Hemoglobin has an MW of 64,460. Estimate its molecular radius and diffusivity in water at 37°C using Equations 5.3 (or Figure 5.2) and 5.4. Explain why these results differ.

3. Show graphically that the Stokes–Einstein equation (Equation 5.4) provides a reasonable estimation of the data shown in Figure 5.2.

4. Tong and Anderson (1996) obtained for BSA the following data in a polyacrylamide gel for the partition coefficient (K) as a function of the gel volume fraction (ϕ). The BSA they used had an MW of 67,000, a molecular radius of 3.6 nm, and a diffusivity of 6×10^{-7} cm² sec⁻¹. Compare the Ogston equation (Equation 5.87) to their data and obtain an estimate for the radius of the cylindrical fibers (a_f) that comprise the gel.

Gel Volume Fraction (ϕ)	K_{BSA}
0.00	1.0
0.025	0.35
0.05	0.09
0.06	0.05
0.075	0.017
0.085	0.02
0.105	0.03

5. Tong and Anderson (1996) also obtained the following data for the diffusive permeability of BSA (KD_0) in polyacrylamide gels as a function of the gel volume fraction (ϕ). Show that the product of the Brinkman and Ogston equations (Equations 5.86 and 5.87) may be used to represent these data.

Gel Volume Fraction (ϕ)	Diffusive Permeability (KD_0), cm² sec⁻¹
0.00	5.5×10^{-7}
0.02	1.0×10^{-7}
0.025	6.5×10^{-8}
0.05	8.0×10^{-9}
0.065	4.5×10^{-9}
0.075	1.0×10^{-9}

6. Show for a protein with an MW of 50,000 that $S_0 = S_a$ for blood flowing through a capillary.

7. Investigate the fit of Equations 5.84 and 5.85 to the data presented in Figure 5.12.

8. The following data were obtained by Iwata et al. (1996) for the effective diffusivity of several solutes through a 5% agarose hydrogel. Show how well the Brinkman (1947) and Ogston (1958) equations and the Renkin (1954) equation represents these data, i.e.,

Equations 5.86, 5.87, and 5.61. What is the effective pore size of the gel based on the Renkin equation?

Solute	Molecular Weight	Gel Diffusivity (KD_0), cm^2 sec^{-1}
Glucose	180	4.5×10^{-6}
Vitamin B$_{12}$	1,200	1.7×10^{-6}
Myoglobin	17,000	4.0×10^{-7}
BSA	69,000	1.0×10^{-7}
IgG	150,000	1.3×0^{-7}

9. Iwata et al. (1996) also measured the sieving coefficient for a variety of solutes using an XM-50 ultrafilter reported to have an NMWC of 50,000. Compare their measured sieving coefficients to predictions for the sieving coefficient calculated by Equation 5.55. What pore diameter is needed to get a reasonable representation of the data?

Solute Molecular Weight	Sieving Coefficient
180	0.98
1,200	0.94
14,300	0.55
17,000	0.12
69,000	0.01

10. Starting with Equations 5.107 and 5.109, derive Equations 5.113 and 5.115. The boundary conditions are given by Equation 5.110.

11. Membrane plasmapheresis (Zydney 1995; Zydney and Colton 1986) is a technique used to separate plasma from the cellular components of blood. The plasma that is collected may be processed further to yield a variety of substances that are useful for the treatment of a variety of blood disorders. Membrane-based systems employ blood flow that is parallel to the membrane. The filtrate of plasma is therefore perpendicular to the membrane. A cell-free filtrate with minimal retention of plasma proteins is obtained, since the membrane pores are typically 0.2–1.0 μm in diameter. The filtrate flux is limited, however, by the accumulation (concentration polarization) of cells along the surface of the membrane that provides an additional hydraulic resistance.

At any axial position (z) from the device inlet, this layer of cells at steady state represents a balance between the flux of cells carried to the membrane surface by the filtration flux, i.e., $q(z) C(z)$, and the diffusive flux of cells carried away from the membrane, i.e., $D(dC(z)/dy)$. Hence, we can write

$$q(z)C(z) = -D\frac{dC(z)}{dy}.$$

Here, $q(z)$ is the filtration flux ($Q(z)/S$), $C(z)$ is the concentration of the cells in the blood, D is the diffusivity of the red blood cells, and y is the distance from the membrane surface on the blood side. Show that after equating these flux expressions as mentioned previously and integrating over the thickness of the cell boundary layer (δ), that the following expression is obtained for the filtrate flux:

$$q(z) = \frac{D}{\delta}\ln\left(\frac{C_w}{C_b}\right) = k_m \ln\left(\frac{C_w}{C_b}\right),$$

where k_m represents the local film mass transfer coefficient (i.e., D/δ) and C_w and C_b represent the cell concentration at the membrane surface and in the bulk blood, respectively. Also, the thickness of the cell boundary layer is much thinner than that of the flow channel. For these conditions, the axial velocity profile will vary linearly across the cell boundary layer and the mass transfer coefficient for this situation is given by an expression similar to Equation 5.53 with the lead constant of 1.86 replaced with the value of 1.03 and the length (L) is replaced with the local position z. The shear-induced diffusivity of the red blood cells is given by (Zydney and Colton 1986):

$$D = 0.03a^2\gamma w.$$

Here, a represents the radius of the red blood cells, about 4.2 μm, and γ_w is the wall shear rate given by the following equation for the flow of blood in a single cylindrical hollow fiber of radius equal to R (discussed in Chapter 4):

$$\gamma_w = \frac{4 F(z)}{\pi R^3}.$$

Show that Equation 5.53 (with 1.86 replaced with 1.03) may be rewritten as follows for the flow of blood in the hollow fiber membrane:

$$k_m = 0.05\left(\frac{a^4}{z}\right)^{1/3} \gamma_w.$$

The previous expressions may be substituted into the following equation, which was obtained in Example 5.10 for the axial change of the volumetric flow rate of the blood (F_B) as a result of plasma filtration:

$$\frac{d F_B(z)}{dz} = -2\pi R q(z).$$

The plasma filtration rate (Q_f) is defined as the difference between the flow rate of the entering blood ($F_B(0)$) and that leaving the device ($F_B(L)$). With $Q_f = F_B(0) - F_B(L)$, the previous equation can be integrated to find the fractional filtrate yield, defined as the ratio of the total plasma filtrate formed (Q_f) and the inlet blood flow rate $F_B(0)$. However, to obtain an analytical expression useful for design calculations, several assumptions are needed.

To obtain an analytical expression, Zydney and Colton (1986) assumed that $\ln(C_w/C_b) \approx \ln(C_w/C_b(0))$, hence the bulk cell concentration remained constant at its inlet value ($C_b(0)$) and they also assumed that the wall shear rate (γ_w) varied linearly with $F(z)$. Using these assumptions, show that the following equation is obtained for the fractional filtration yield, $Q_f/F_B(0)$:

$$\frac{Q_f}{Q_B(0)} = 1 - \exp\left[-0.90\beta\ln\left(\frac{C_w}{C_b(0)}\right)\right] \quad \text{where } \beta = \frac{2}{3}\left(\frac{a^2 L}{R^3}\right)^{2/3}.$$

In this equation, β is a dimensionless length. Zydney and Colton (1986) provided the following performance data for a variety of membrane plasmapheresis units. Show that the previous equation provides a remarkable prediction of the fractional filtration yield for these units. Use 0.35 and 0.95 for the concentration of the red blood cells entering the blood and at the wall, respectively.

β	Fractional Filtrate Yield
0.20	0.12
0.27	0.20
0.27	0.23
0.32	0.22
0.32	0.25
0.35	0.25
0.40	0.30

β	Fractional Filtrate Yield
0.42	0.32
0.43	0.30
0.44	0.39
0.45	0.38
0.49	0.36
0.51	0.35
0.58	0.40
0.60	0.35
0.62	0.37
0.70	0.41

12. If the surface area of capillaries is 7000 cm² per 100 g of tissue, estimate the Krogh tissue cylinder radius.

13. In a discussion of glucose transport through exercising muscle capillaries, Renkin (1977) reports the capillary glucose permeability (P_m) to be on the order of 12×10^{-6} cm sec⁻¹ with a capillary surface area (S) of 7000 cm² per 100 g of tissue. How does the value of $P_m S$ compare to the date shown in Figure 5.13?

14. Using the Renkin–Crone equation, make a graph of capillary solute extraction (E) as a function of capillary position (z) for various values of the ratio ($2\pi r_C P_m z/Q$). Identify conditions on your plot for which the solute transport is flow limited and diffusion limited.

15. Levick (1991) presented the following data obtained by Renkin (1977) for the capillary clearance of antipyrene and urea as a function of blood flow in skeletal muscle. Based on these data, which of these solutes' transport rate is flow limited and which is diffusion limited? Calculate the value of $P_m S$ for urea. How does this value of $P_m S$ for urea compare to the results presented in Figure 5.13? Explain why the $P_m S$ value for antipyrene cannot be found from these data.

Solute	Blood Flow (Q, mL min⁻¹ (100 g)⁻¹)	Clearance (CL, cm³ min⁻¹ (100 g)⁻¹)
Antipyrene	4.5	4
	4.5	4.5
	6.5	6.5
	10	10
Urea	5	3.5
	5.5	3
	7	4
	8	2.5
	8.5	2.6
	8.5	4
	10	2.6
	10.5	2.5
	12	4.5
	13.5	2.6

16. Mann et al. (1979) obtained the following concentration vs. time data for several solutes in cat fenestrated salivary glands using the multiple tracer indicator diffusion technique. The test solutes were cyanocobalamin (vitamin B_{12}, MW = 1353) and insulin (MW = 5807). Albumin (MW = 69000) served as the nonpermeable reference solute. The perfusion rate for the gland was 8 mL min⁻¹ g⁻¹. The following table summarizes the concentration of these

solutes (percent dose per 0.2 mL of perfusate sample) for various periods of time following their injection into the gland. From these data, calculate the value of P_mS for B_{12} and insulin. How do these P_mS values for B_{12} and insulin compare to the results shown in Figure 5.13?

Time (sec)	C (B_{12})	C (Insulin)	C (Albumin)
0.3	1.5	2.2	2.3
0.66	4.4	6.3	7.3
1.0	7.3	10.0	11.9
1.5	8.6	11.0	12.9
1.9	7.6	10.2	10.7
2.1	6.0	7.3	7.8
2.5	5.0	5.9	5.7
2.8	4.1	4.2	3.9
3.1	3.8	3.7	3.0
3.5	3.0	2.6	2.2
3.8	2.7	2.2	2.0

17. Estimate the diffusivity of a spherically shaped molecule with an MW equal to 35,000 through a membrane containing cylindrical pores that are 8 nm in diameter. Assume that the molecule enters the pores of the membrane from a well-stirred bulk solution.

18. Some folks suggest that the cell walls of living cells have pores that are 30 angstroms in diameter (1 angstrom = 10^{-8} cm). Estimate the diffusivity (square centimeters per second) at 37°C for a solute 5 angstroms in diameter through such pores. Assume that the pores are filled with water having a viscosity of 0.76 cP (1 cP = 0.01 g cm^{-1} sec^{-1}).

19. A molecule with a radius of 2.5 nm enters, from a bulk solution, a membrane having pores with a radius of 7.5 nm. For a temperature of 37°C, and assuming that the fluid phase is water, what is the apparent diffusivity of the molecule in the pores of the membrane? Express your answer in square centimeters per second.

20. Rework Example 5.9 assuming that the concentration and velocity profiles are undeveloped as they enter the coated section of the tube.

21. Your group has developed a new drug for the treatment of hilariosis. This drug is lipid insoluble and has an MW of 1200. As part of the design of a controlled release system for this drug, you have been asked to estimate the permeability, cubic centimeters per second per 100 g, of this drug through the capillary wall.

22. Estimate the reflection coefficient for a molecule that has a radius of 0.75 nm, assuming that the pores in the membrane have a radius of 2.5 nm.

23. The following data for the mass transfer coefficient of water flowing within hollow fiber tubes were reported by Yang and Cussler (1986). Compare their results to the correlation given by Equation 5.53. Pe is a dimensionless number, known as the Peclet number, and in this case is defined as shown in the following table. Pe represents the ratio of solute transport by convection to that by diffusion.

$Pe = Re\ Sc\ \dfrac{D_{tube}}{L}$	$Sh = \dfrac{k_m D_{hollow\ fiber}}{D}$
30	4.8
50	6.0
100	7.6
300	9.8
500	13.0
1000	18.0
3000	30.0

24. A new photosensitizer drug has an MW of 1400. Estimate the diffusivity of this drug in water at 37°C using Equations 5.3 and 5.4. Express your answer in square centimeters per second.

25. A single hollow fiber is placed within a larger diameter glass tube forming a shell space that surrounds the hollow fiber. The hollow fiber is 20 cm in length and has a diameter of 400 μm. The flow rate of a liquid through the hollow fiber is 1 cm^3 min^{-1} and in the shell space, another liquid also flows through at a flow rate of 100 cm^3 min^{-1}. The liquid entering the hollow fiber also contains a permeable solute. It is found that the concentration of the permeable solute exiting the hollow fiber is 10% of the concentration of this solute when entering the hollow fiber in the liquid. Estimate the permeability for this solute in centimeters per second.

26. Blood is flowing through a hollow fiber that is 800 μm in diameter and 30 cm in length. The average velocity of the blood within the hollow fiber is 25 cm sec^{-1}. The concentration of a drug is maintained at 10 mg L^{-1} along the inside surface of the hollow fiber. The diffusivity of the drug in blood is 4×10^{-6} cm^2 sec^{-1}. Estimate the mass transfer coefficient, k_m, for the drug. Assuming no drug enters the hollow fiber with the blood, estimate the exiting concentration of the drug in the blood.

27. Derive Equation 5.21.

28. Derive Equation 5.22.

29. Derive Equation 5.39.

30. Derive Equation 5.41.

31. Derive Equation 5.59.

32. Bawa et al. measured the drug distribution inside a polymer matrix material. These polymeric slabs of thickness 1 mm were loaded with albumin at an initial concentration of 484 mg cm^{-3} of slab. The following table shows two distributions of the drug concentration as a function of x/L after 6.5 and 172 h. Find the value of D_e that best fits these data using Equation 5.77. How well does Equation 5.83 fit the 6.5 h concentrations?

$\dfrac{x}{L}$	C @ 6.5 h	C @ 172 h
0.039	484 mg cm^{-3}	175 mg cm^{-3}
0.115	484	200
0.20	484	213
0.32	484	188
0.38	484	175
0.46	484	150
0.58	475	113
0.66	450	100
0.73	275	75
0.82	200	38
0.89	112	25
0.93	80	12.5

33. Bawa et al. obtained the following data for the cumulative fraction released (f_R) for lysozyme from 1 mm thick polymeric slabs. Estimate the value of D_e that best fits these data using Equation 5.80. The initial concentration of lysozyme is about 400 mg cm^{-3} and the diffusivity of lysozyme in water is 3.74×10^{-3} cm^2 h^{-1}.

Time$^{1/2}$ (h$^{1/2}$)	f_R
3	0.03
5	0.12
7	0.18
8	0.21

Time$^{1/2}$ (h$^{1/2}$)	f_R
9.5	0.24
13	0.32
15	0.40
16.5	0.45
18.5	0.50
21	0.52

34. Investigate strategies for improving the removal of protein A by membrane filtration in Example 5.12.

35. For Example 5.12, show that diffusion of solute A across the hollow fiber membrane is significantly less than that transported by the filtration flow.

36. Show the calculations behind the results presented in Examples 5.21 and 5.22.

37. A stainless steel disc is coated on one side with a drug called *Rocketinfusiola* having an MW of 341. The radius of the disc is 0.9 cm. The coated disc is placed in a solution of ethanol and the disc is then rotated at an angular speed of 80 rpm (rpm = revolutions per minute, where 2 π radians = 1 revolution). The density of ethanol is 0.79 g cm^{-3} and its viscosity is 0.68 cP. Estimate the dissolution rate (percent per minute, i.e., relative to the initial amount of the drug on the disc) of *Rocketinfusiola* assuming that the initial mass coated on the disc is 41.2 mg and that the drug's equilibrium solubility in ethanol is equal to 5.86 mg cm^{-3}. Assume that the concentration of *Rocketinfusiola* in the bulk solution surrounding the spinning disc is zero.

38. The filtration flux across a membrane is 5×10^{-5} cm sec^{-1}. The mass transfer coefficient of a particular solute for this situation is 10^{-4} cm sec^{-1}. The ratio of the solute radius to the radius of the membrane pores is $\lambda = 0.4$. If the concentration of this solute is 5 g L^{-1} in the bulk solution, estimate the concentration of the solute in the filtrate.

39. A well-mixed protein solution is being filtered across a semipermeable membrane. The NMWC of the membrane is 100,000 and the protein has an MW of 20,000. The protein concentration on the feed side of the membrane in the filtration unit is 1 g L^{-1}. The protein's mass transfer coefficient is 3×10^{-4} cm sec^{-1} and the filtration flux of the protein solution across the membrane is 10^{-4} cm sec^{-1}. Assuming that the protein has a molecular radius of 2.0 nm, what is the concentration of the protein in the filtrate solution?

40. The antibody IgG has an MW of 150,000. Estimate its diffusivity in water at a temperature of 20°C.

41. Consider the flow of an aqueous solution within a 400 μm diameter hollow fiber that is 20 cm in length. A drug is impregnated into the wall of the fiber and the equilibrium solubility of the drug in water is 3 mg L^{-1}. The Reynolds number for the flow conditions is 16,000 and the Schmidt number is 10,000. If the diffusivity of the solute of interest in the water is 0.6×10^{-6} cm^2 sec^{-1}, what is the maximum total elution rate (milligrams per hour) of the drug from the fiber surface?

42. A particular drug is a solid at 300 K and is soluble in water. The equilibrium solubility of the drug in water is 0.005 g cm^{-3}. The MW of the drug is 600. One hundred grams of the drug is made into particles that are 0.2 cm in diameter. The density of the drug particles is 1.05 g cm^{-3}. These particles are then very gently mixed in a stirred vessel containing 100 L of water. After 10 min of mixing, estimate how many grams of the drug have dissolved into the water.

43. A well-mixed protein solution is being filtered across a semipermeable membrane. The pores in the membrane have an average radius of 3.6 nm and the radius of the protein molecule is 2.2 nm. The protein concentration in the filtration unit is 1.5 g L^{-1}. The protein's mass transfer coefficient is 3.5×10^{-4} cm sec^{-1} and the filtration flux of the protein solution across the membrane is 1.5×10^{-4} cm sec^{-1}. What is the concentration of the protein in the filtrate solution?

44. Estimate the diffusivity of a drug in a solvent at 315 K. The MW of the drug is 526. The viscosity of the solvent at this temperature is 0.295 cP and the viscosity of water at 310 K is 0.76 cP.

45. A flat plate coated with a layer of a volatile organic material is exposed to a parallel flow of air at 20°C with a free stream air velocity of 5 m sec^{-1}. The length of the flat plate in the direction of flow is 0.5 m. Estimate how long it will take for the average thickness of the volatile organic layer to decrease by 1 mm. The saturation concentration of the organic material in air at these conditions is 1.85×10^{-9} mol cm^{-3} and its diffusivity in air is 0.068 cm^2 sec^{-1}. The density of the volatile organic material as a solid is 1.21 g cm^{-3} and its MW is 140.2. Also, the kinematic viscosity of air at these conditions is 0.155 cm^2 sec^{-1}.

46. A cylindrical tube, 1 cm in diameter and 10 cm in length, has a drug immobilized on its surface. The equilibrium solubility of the drug in the fluid flowing through the tube is 100 mg L^{-1}. The fluid flows through the tube at an average velocity of 10 cm sec^{-1}. If the exiting concentration $(\bar{C}_L)$ of the drug in the fluid is 0.163 mg L^{-1}, estimate the average value of the mass transfer coefficient $(\bar{k}_m)$ assuming that there is no drug in the fluid that enters the tube.

47. A liquid extraction process is proposed for removing a drug from the solution leaving a fermentation system. As part of the design process, you need to provide an estimate of the diffusivity of this drug at 17°C in the extraction solvent cyclohexane. The MW of the drug is 765 g mol^{-1}. The viscosity of cyclohexane at this temperature is 1.02 cP as compared to the viscosity of water, which is 0.76 cP at 37°C.

48. A spinning disk system is being proposed as a room deodorizer. The deodorizer is basically a small cylindrical disc 5 cm in diameter. One side of the disk is coated with a layer of a volatile substance that smells really good. The disk will rotate at 300 rpm and the average room temperature is 20°C. Estimate how long it will take (in hours) for the average thickness of the volatile substance layer to decrease by 1 mm. The saturation concentration of the volatile substance in air at these conditions is 3×10^{-8} mol cm^{-3} and the diffusivity of the volatile substance in air is given by $D = 0.06$ cm^2 sec^{-1}. Also, the kinematic viscosity of air at these conditions is 0.151 cm^2 sec^{-1}. The density of the volatile substance layer is 1.1 g cm^{-3} and its MW is 175 g mol^{-1}.

49. A rotating disc 1 cm in diameter is coated on one side with a drug that bears a fluorescent tag that allows for determining its concentration in solution. The spinning disc system is placed in 100 cm^3 of pure water at 37°C and the disc is spun at 25 rpm. After 60 min, the concentration of the drug in the water was found to equal 0.075 g L^{-1}. After a long period of time, the drug concentration in the solution was equal to 3.0 g L^{-1}. From this information, estimate the diffusivity of the drug in water in square centimeters per second. Assume that the viscosity and density of the water at these conditions are 0.76 cP and 1 g cm^{-3}, respectively.

50. Water containing a solute is flowing through a 1 cm diameter tube, which is 50 cm in length, at an average velocity of 1 cm sec^{-1}. The temperature of the water is 37°C. The walls of the tube are coated with an enzyme that makes the solute disappear instantly. The solute enters the tube at a concentration of 0.1 M and the solute has an MW of 300 g mol^{-1}. Estimate the fraction of the solute that leaves the tube. Assume that the viscosity and density of the water at these conditions are 0.76 cP and 1 g cm^{-3}, respectively.

51. Estimate the permeability of a spherically shaped molecule that has an MW of 42,000 g mol^{-1} through a microporous polycarbonate membrane. Assume that the molecule enters the pores of the membrane from a well-stirred bulk solution that is at 37°C. The membrane pores are cylindrical and have a diameter of 32 nm and an overall length of 325 μm. The membrane itself is 250 μm in thickness. The pores comprise 42% of the membrane volume.

6 Oxygen Transport in Biological Systems

6.1 DIFFUSION OF OXYGEN IN MULTICELLULAR SYSTEMS

The growth of simple multicellular systems beyond a diameter of 100 μm or so is limited by the availability of oxygen. The metabolic needs of the cells near the center of the cell aggregate exceed the supply of oxygen available by simple diffusion. This is illustrated in Figure 6.1 for isolated islet organs (Brockman bodies) of the fish, *Osphronemus gorami*. The islet organs, typically 800 μm in diameter, were placed in culture, and the oxygen partial pressure in the surrounding media and within the cells was measured radially using an oxygen microelectrode.

For the case with no convection in the media (closed symbols), we see that the oxygen partial pressure drops sharply and has been reduced to nearly zero within a few hundred microns of the islet organ's surface (electrode position of 0 μm). Those cells that experience reduced oxygen levels suffer from *hypoxia*. If the oxygen level is reduced beyond some critical value, the cells may die, which is called *necrosis*. Note that with increased convection (fluid mixing) in the surrounding media (open symbols), the partial pressure of oxygen (pO_2) at the surface and core regions of the islets is increased significantly in comparison to the case without convection. This is a result of the decrease in the resistance to oxygen transport of the fluid surrounding the islets. However, we still observe a large decrease in the oxygen partial pressure within the islet body.

Clearly, the development of such complex cellular systems as vertebrates required the development of oxygen delivery systems that are more efficient than diffusion alone. This has been accomplished by the development of two specialized systems that enhance the delivery of oxygen to cells. The first is the circulatory system, which carries oxygen in blood by convection to tiny blood vessels called *capillaries* in the neighborhood of the cells. Here, oxygen is released and then diffuses over much shorter distances to the cells. Secondly, a specialized oxygen carrier protein, called *hemoglobin*, is contained within the red blood cells (RBCs) that are suspended within the blood plasma. The presence of hemoglobin overcomes the very low solubility of oxygen in water, which is the major component of plasma. In this chapter, we will investigate the factors that affect oxygen delivery to cells.

6.2 HEMOGLOBIN

Figure 6.2 shows the basic structural features of the hemoglobin molecule. Hemoglobin found in adult humans is called *hemoglobin A*. Hemoglobin consists of four polypeptide chains (a polymer of amino acids), two of one kind called the α *chain* and two of another kind called the β *chain*. These polypeptide chains of hemoglobin A are held together by noncovalent attractions. The hemoglobin molecule is nearly spherical with a diameter of about 5.5 nm and a molecular weight of 68,000.

The oxygen-binding capacity of hemoglobin depends on the presence of a nonpolypeptide unit called the *heme group*. The iron atom within heme gives blood its distinctive red color. The heme group is also called a *prosthetic group* because it gives the hemoglobin protein its overall functional

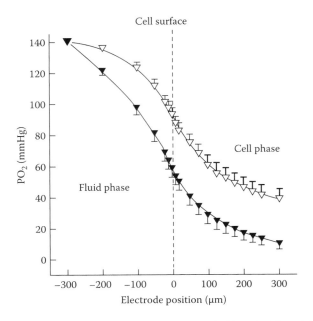

FIGURE 6.1 pO$_2$ profiles of 12 Brockman bodies with (open symbols) and without (closed symbols) convection in the medium. Measurements were started 300 μm outside the tissue (–300). At the surface of the body, the electrode position is 0 μm. (From Schrezenmeir, J., Kirchgessner, J., Gero, L., Kunz, L.A., Beyer, J., and Mueller-Klieser, W., *Transplantation*, 57, 1308–14, 1994. With permission.)

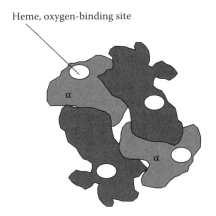

FIGURE 6.2 Structure of hemoglobin.

activity. The heme group consists of an organic part that binds through four nitrogen atoms with the iron atom within the center. The organic part is called the *protoporphyrin* and consists of four pyrole rings linked by methene bridges. The iron atom can form up to six bonds, four of which are committed to the pyrole rings. So, we have two additional bonds available, one on each side of the heme plane. A heme group can therefore bind one molecule of oxygen.

The four heme groups in a hemoglobin molecule are located near the surface of the molecule and the oxygen-binding sites are about 2.5 nm apart. Each of the four chains contains a single

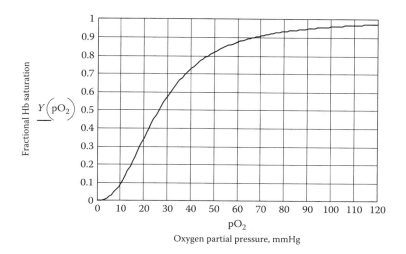

FIGURE 6.3   Oxygen–hemoglobin dissociation curve.

heme group and therefore a single oxygen-binding site. A molecule of hemoglobin, therefore, has the potential to bind four molecules of oxygen. Hemoglobin is an *allosteric protein* and its oxygen-binding properties are affected by interactions between these separate and nonadjacent oxygen-binding sites. The binding of oxygen to hemoglobin enhances the binding of additional oxygen molecules and is therefore said to be *cooperative.**

6.3   OXYGEN–HEMOGLOBIN DISSOCIATION CURVE

The binding of oxygen with hemoglobin is described by the oxygen dissociation curve shown in Figure 6.3. This is an equilibrium curve that expresses, for a given oxygen partial pressure (or pO_2), the fractional occupancy of the hemoglobin–oxygen binding sites (Y). The equilibrium pO_2 above a solution of blood is related to the dissolved oxygen concentration by *Henry's law*, which is expressed by

$$pO_2 \equiv Py_{oxygen} = H_{oxygen}C_{oxygen} = HC. \qquad (6.1)$$

In this equation, H_{oxygen} or H is Henry's constant with a value for normal blood at 37°C of 0.74 mmHg μm⁻¹. C_{oxygen} is the dissolved oxygen concentration (microns) and y_{oxygen} is the mole fraction of oxygen in the gas phase. It is important to realize that pO_2 is exerted only by the dissolved oxygen. Oxygen bound to hemoglobin does not affect pO_2, but serves only as a source or sink for oxygen.

6.4   OXYGEN LEVELS IN BLOOD

Table 6.1 summarizes some typical properties of arterial and venous blood. The concentration of hemoglobin in blood (see Table 4.1) is about 150 g L⁻¹ or 2200 μM. Since each molecule of

---

* An analogy for cooperative binding would be several people entering a life raft from the water. Because of the high sides
  of the life raft, it is very difficult for the first person to get aboard. Once one person is aboard, he/she can help the next
  person. Two people on board make it even easier to load the next person and so on.

TABLE 6.1
Oxygen Levels in Blood

Oxygen Property	Arterial	Venous
Partial pressure (tension) (pO_2, mmHg)	95	40
Dissolved O_2 (μM)	130	54
As oxyhemoglobin (μM)	8500	6424
Total effective (dissolved + oxy-Hb) (μM)	8630	6478

Note: $H_{oxygen} = 0.74$ mmHg μM^{-1}; saturated oxyhemoglobin, $C'_{SAT} = 8800$ μM; $P_{50} \sim 26$ mmHg; $n = 2.34$.

hemoglobin can bind up to four molecules of oxygen, the saturated ($Y = 1$) concentration of oxygen bound to hemoglobin is equal to 8800 μM. Comparing this saturated value, or even the arterial or venous values shown in Table 6.1 (i.e., as oxyhemoglobin), to the arterial dissolved oxygen value of 130 μm clearly shows that the bulk of the oxygen that is available in the blood is carried by hemoglobin. The P_{50} value represents the value of pO_2 at which 50% of the oxygen-binding sites are filled, that is $Y = 0.5$. From Figure 6.3, we see that this occurs when $pO_2 \approx 26$ mmHg.

6.5 THE HILL EQUATION

In 1913, Archibald Hill proposed that the sigmoidal shape of the oxygen dissociation curve, shown in Figure 6.3, could be described by the following equilibrium reaction:

$$Hb(O_2)_n \xrightleftharpoons[k_1]{k_{-1}} Hb + nO_2. \tag{6.2}$$

In this equation, n represents the number of oxygen molecules that bind with hemoglobin (Hb) to form oxyhemoglobin, i.e., $[Hb(O_2)_n]$. Writing an elementary rate expression for the appearance of hemoglobin, with k_1 and k_{-1} representing the reaction rate constants, we obtain

$$\frac{d[Hb]}{dt} = k_1[Hb(O_2)_n] - k_{-1}[Hb][O_2]^n = 0, \tag{6.3}$$

where the brackets imply the concentration of the respective chemical species. Assuming that the reaction is at equilibrium, we set Equation 6.3 equal to zero and solve for the concentration of oxygenated hemoglobin in terms of free hemoglobin and dissolved oxygen concentrations:

$$[Hb(O_2)_n] = \kappa[Hb][O_2]^n. \tag{6.4}$$

In Equation 6.4, κ is defined as (k_{-1}/k_1). The fraction of hemoglobin that is saturated (Y) is defined by the following relationship:

$$Y \equiv \frac{[Hb(O_2)_n]}{[Hb(O_2)_n]+[Hb]} = \frac{\dfrac{[Hb(O_2)_n]}{[Hb]}}{1+\dfrac{[Hb(O_2)_n]}{[Hb]}}. \tag{6.5}$$

This equation can be simplified using Equations 6.1 and 6.4 to obtain the following result:

$$Y = \frac{\kappa[O_2]^n}{1+\kappa[O_2]^n} = \frac{(pO_2)^n}{P_{50}^n + (pO_2)^n}. \tag{6.6}$$

The P_{50}^n value was substituted for (H_{oxygen}^n/κ). This equation is called the *Hill equation* and can be used to provide a mathematical relationship between Y and pO_2 that is convenient to use in design calculations.

Regression analysis of Y and pO_2 data is facilitated by the following rearrangement of the Hill equation:

$$\ln \frac{Y}{1-Y} = n \ln(pO_2) - n \ln P_{50}. \tag{6.7}$$

A plot of $\ln(Y/1-Y)$ vs. $\ln(pO_2)$ is linear with a slope equal to n and a y-intercept equal to $-n \ln P_{50}$. This is called the *Hill plot*. Example 6.1 illustrates the use of the Hill equation to represent the oxygen–hemoglobin dissociation curve.

Example 6.1

The following table presents data representing the oxygen–hemoglobin dissociation. Show that the Hill equation provides excellent representation of this data.

Oxygen Partial Pressure (pO$_2$, mmHg)	Fractional Hemoglobin Saturation
10	0.12
20	0.28
30	0.56
40	0.72
50	0.82
60	0.88
70	0.91
80	0.93
90	0.95
100	0.96

Source: Adapted from Guyton, A.C., *Textbook of Medical Physiology*, W.B. Saunders Co., Philadelphia, PA, 1991.

Solution

Performing a linear regression analysis on the previous data using Equation 6.7, we find that the P_{50} value is 26 mmHg and the value of n is 2.34. We also see in Figure 6.4 (dashed line) that the Hill equation provides for an excellent representation of the oxygen–hemoglobin dissociation curve.

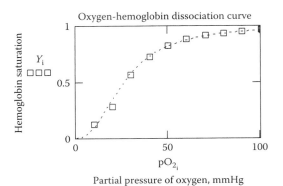

FIGURE 6.4   The Hill equation.

6.6   OTHER FACTORS THAT CAN AFFECT THE OXYGEN DISSOCIATION CURVE

Other molecules, such as H^+, CO_2, and organic phosphates (such as 2,3 diphosphoglycerate, DPG), also bind on specific sites on the hemoglobin molecule and greatly affect its oxygen-binding ability through *allosteric interactions*. A lowering of the pH shifts the oxygen dissociation curve to the right, as shown in Figure 6.5. This decreases for a given pO_2, the oxygen affinity of the hemoglobin molecule. CO_2 at increased levels and constant pH also lowers the oxygen affinity of hemoglobin. In metabolically active tissues such as muscle, the higher levels of CO_2 and H^+ in the capillaries have a beneficial effect by promoting the release of oxygen. This is called the *Bohr effect*.

As shown in Figure 6.5, we see that the oxygen–hemoglobin dissociation curve shifts to the right when the partial pressure of carbon dioxide (pCO_2) is increased. Under these conditions, CO_2 can reversibly bind to hemoglobin, displacing oxygen to form a compound known as *carbaminohemoglobin*. The reverse process also occurs, that is, oxygen binding to hemoglobin has the ability to displace CO_2 that is bound to hemoglobin. This is important both in tissues and in the lungs. In the tissue capillaries, the higher pCO_2 in the tissue space results in the formation of carbaminohemoglobin, which increases the release of oxygen from hemoglobin. Within the lungs, where pO_2 is much higher within the gas space of the alveoli, there is displacement of CO_2 from hemoglobin by oxygen. This is called the *Haldane effect*.

The affinity of hemoglobin for oxygen within the RBC is significantly lower than that of hemoglobin in free solution. Within the RBC, there exists the organic phosphate, DPG, at about the same concentration as hemoglobin. DPG binds to hemoglobin and reduces its oxygen-binding capacity, as shown in Figure 6.5. Without DPG, the value of P_{50} is reduced to about 1 mmHg. This makes it very difficult for hemoglobin to release the bound oxygen at low values of pO_2. Note that without DPG, there would essentially be no release of oxygen in the physiological range of pO_2 values from 40 to 95 mmHg. Therefore, the presence of DPG within the RBC is vital for hemoglobin to perform its oxygen-carrying role.

6.7   TISSUE OXYGENATION

With this understanding of the oxygen-binding ability of hemoglobin, we can now focus on tissue oxygenation. Hemoglobin-bound oxygen within the RBCs is transported by the circulatory system to the capillaries where a change in pO_2 levels along the length of the capillary causes a release of oxygen. The released oxygen then diffuses across the capillary wall and into the surrounding tissue space where it is consumed by the cells.

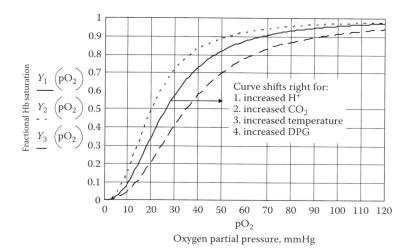

FIGURE 6.5 Shifts in the oxygen–hemoglobin dissociation curve.

Consider the capillary bed shown in Figure 5.16. What is the nominal oxygen consumption rate of the tissue lying within this capillary bed? To answer this question, we can develop the following simple mathematical model to describe oxygen consumption by tissue. We assume that the capillary bed and the surrounding tissue space are each well mixed with respect to oxygen. This assumption allows us to write the following mass balance equations for oxygen. We will let V_T represent the volume of the tissue space (including the capillaries), $\Gamma_{metabolic}$ represents the tissue metabolic volumetric oxygen consumption rate (μM per second); q is the tissue blood perfusion rate, i.e., in units of cubic centimeters of blood (cm^3 tissue)$^{-1}$ min^{-1}; C represents the dissolved oxygen concentration in the blood (μM) and C' represents the concentration of oxygen in the blood that is bound to hemoglobin (μM); C_T represents the concentration of oxygen dissolved within the interstitial space of the tissue (μM); P_C (centimeters per second) is the permeability of oxygen through the capillary wall; and S_C represents the total surface area of the capillaries (square centimeters). With these definitions, we can then write the following oxygen mass balances for the blood and tissue regions:

$$\text{Blood: } 0 = qV_T(C+C')_A - qV_T(C+C')_V - P_C S_C(C_V - C_T), \tag{6.8}$$

$$\text{Tissue: } 0 = P_C S_C(C_V - C_T) - V_T\Gamma_{metabolic}. \tag{6.9}$$

Note that the convective terms (containing q) in Equation 6.8 include oxygen transported in the dissolved state as well as that bound to hemoglobin. However, the mass transfer of oxygen across the capillary wall is only based on the difference between the dissolved oxygen concentration in the blood and that within the tissue interstitial space. Equations 6.8 and 6.9 can then be added together and solved to give Equation 6.10 for the value of the metabolic oxygen consumption rate:

$$\Gamma_{metabolic} = q[(C+C')_A - (C+C')_V]. \tag{6.10}$$

To solve Equation 6.10 for the metabolic oxygen consumption rate of tissue requires a value for the tissue blood perfusion rate, q. Table 6.2 provides representative values for the blood flow to various organs and tissues in the body. The blood perfusion in Table 6.2 is expressed on the

TABLE 6.2
Blood Flow to Different Organs and Tissues under Basal Conditions

Organ	Percent	mL min⁻¹	mL min⁻¹(100 g of tissue)⁻¹
Brain	14	700	50
Heart	4	200	70
Bronchial	2	100	25
Kidneys	22	1100	360
Liver	27	1350	95
Portal	(21)	(1050)	
Arterial	(6)	(300)	
Muscle (inactive state)	15	750	4
Bone	5	250	3
Skin (cool weather)	6	300	3
Thyroid gland	1	50	160
Adrenal glands	0.5	25	300
Other tissues	35.5	175	1.3
Total	100.0	5000	–

Source: Adapted from Guyton, A.C., *Textbook of Medical Physiology*, W.B. Saunders Co., Philadelphia, PA, 1991. With permission.

basis of flow in milliliters per minute per 100 g of tissue or organ. The kidneys are highly perfused because of their role in the filtration of blood. A nominal tissue perfusion rate is on the order of 0.5 mL cm⁻³ min⁻¹ (assuming the density of tissue, $\rho_{tissue} \approx 1$ g cm⁻³). The nominal arterial and venous pO_2 levels are 95 and 40 mmHg, respectively. We can then use Equation 6.10 to obtain an estimate of the metabolic oxygen consumption rate, $\Gamma_{metabolic}$, as shown in the following example.

Example 6.2

Calculate the metabolic oxygen consumption rate in microns per second using the previous nominal values for the blood perfusion rate and the arterial and venous pO_2 levels.

Solution

$$\Gamma_{metabolic} = \frac{0.5 \text{ mL}}{cm^3 \, min} \times (8630 - 6478) \mu m \times \frac{1 \, min}{60 \, sec}$$

$$\times \frac{1000 \text{ cm}^3}{L_{tissue}} \times \frac{\mu \, mol}{L_{blood} \, \mu m} \times \frac{1 \, L_{blood}}{1000 \, mL} \times \frac{\mu m \, L_{tissue}}{\mu \, mol} = 17.93 \frac{\mu M}{sec}.$$

This calculation shows that the tissue metabolic oxygen consumption rate is on the order of 20 μM sec⁻¹. The actual value for any given tissue will depend on the cells specific requirement for oxygen.

If the value of $\Gamma_{metabolic}$ were known for a given tissue, then Equation 6.10 can be rearranged to solve for the change in blood oxygenation in order to provide the oxygen demands of the tissue:

$$(C + C')_V = (C + C')_A - \frac{\Gamma_{metabolic}}{q}. \tag{6.11}$$

To illustrate this type of calculation, where $\Gamma_{metabolic}$ is known, consider the *islets of Langerhans*. The islets of Langerhans are a specialized cluster of cells located within the pancreas. The islets amount to about 1%–2% of the pancreatic tissue mass. The cells that lie within the islet are responsible for the control of glucose metabolism through secretion of the hormones insulin (β cells) and glucagon (α cells). Loss of β cells as a result of a defect in the immune system results in diabetes.

An islet of Langerhans is only about 150 μm in diameter. The *single islet blood flow* has been determined to be about 7 nL min^{-1} for an islet tissue perfusion rate of about 4 mL cm^{-3} min^{-1} (Lifson et al. 1980). This value is about four times the blood perfusion rate of the pancreas itself, and is probably related to the hormonal function of the islet.

The oxygen consumption rate for tissue like islets may be described by Michaelis–Menten type kinetics.* Here we see that the oxygen consumption rate of the tissue is dependent on the pO$_2$ in the tissue:

$$\Gamma_{metabolic} = \frac{V_{max}pO_2}{K_m + pO_2}. \tag{6.12}$$

For islets, the value of K_m is 0.44 mmHg and the value of V_{max} is 26 μM sec^{-1} when the islets are exposed to basal levels of glucose (100 mg dL^{-1}) and 46 μM sec^{-1} under stimulated glucose levels (300 mg dL^{-1}) (Dionne et al. 1989, 1991).

Because of the small value of K_m, the tissue oxygen consumption rate is generally independent of the tissue pO$_2$ until the pO$_2$ in the tissue reaches a value of just a few mmHg. Cellular metabolic processes are therefore relatively insensitive to the local pO$_2$ level until it reaches a value of 1–2 mmHg. Therefore, we can approximate the value of $\Gamma_{metabolic}$ as simply the value of V_{max}. Only at very low pO$_2$ levels would this approximation no longer be valid.

From Equation 6.11, we know the value of $\Gamma_{metabolic}$, and the value of $(C + C')_A$ is the nominal arterial value of 8630 μM (from Table 6.1). We now must solve for the value of $(C + C')_V$, recognizing that C' (the amount of oxygen bound to hemoglobin) depends on C (the dissolved oxygen concentration) through the Hill equation (Equation 6.6), where we make use of the fact that pO$_2 = HC$ (i.e., Henry's law). Hence, C' is given by

$$C' = \frac{C'_{SAT}(HC)^n}{P_{50}^n + (HC)^n}. \tag{6.13}$$

Substituting this equation into Equation 6.11 for the venous value of C', we obtain Equation 6.14 for the dissolved oxygen concentration in venous blood:

$$\left[C + \frac{C'_{SAT}(HC)^n}{P_{50}^n + (HC)^n} \right] = (C + C')_A - \frac{\Gamma_{metabolic}}{q}. \tag{6.14}$$

For given values of q, $\Gamma_{metabolic}$, and $(C + C')_A$, Equation 6.14 can be solved for the dissolved oxygen concentration (C) in the blood leaving the tissue, which in this case is an islet of Langerhans. Once we have obtained the value of C, then the concentration of oxygen bound to hemoglobin, C', can be found from Equation 6.13. The pO$_2$ of the exiting blood is then given by Henry's law (Equation 6.1). The following examples illustrate these calculations.

---

* A derivation of the Michaelis–Menten equation can be found in Section 8.8.3.

Example 6.3

Calculate the change in blood oxygenation within an islet of Langerhans. Perform the calculations under conditions of normal blood perfusion for basal and stimulated levels of glucose. Also, find the critical blood perfusion rate to maintain the exiting pO_2 at 20 mmHg under conditions of basal and stimulated levels of glucose.

Solution

Equation 6.14 provides the solution. Since this equation is nonlinear in C, a Newton root-finding method is needed to solve for C in the venous blood. For the case of basal glucose stimulation, the venous dissolved oxygen concentration is found to be 102.7 μM. The corresponding value of the exiting pO_2 from Equation 6.1 is 76 mmHg and the fractional saturation of the hemoglobin from Equation 6.6 or Figure 6.3 is 0.925. The exiting oxygen levels under stimulated glucose conditions are found to be 86.7 μM with a pO_2 of 64 mmHg and a fractional saturation of 0.89. In both these cases, we observe a modest decrease in oxygen levels in the blood exiting the islet. This is because the islets are highly perfused with blood. The critical perfusion rate is defined as that blood flow rate to the tissue that just maintains a critical pO_2 level, e.g., 20 mmHg. Hence, we solve Equation 6.14 for the value of the blood perfusion rate, i.e., q. We then find that the critical blood perfusion rate for an islet is 0.28 mL cm^{-3} min^{-1} under basal glucose conditions. A similar calculation shows that the critical blood perfusion rate is 0.50 mL cm^{-3} min^{-1} under stimulated glucose conditions.

These concepts can also be used to determine the amount of oxygen transported to blood in medical devices such as blood oxygenators. This is illustrated in the following example.

Example 6.4

Blood travels through a membrane oxygenator at a flow rate of 5000 mL min^{-1}. The entering blood pO_2 is 30 mmHg and the exiting blood pO_2 is 100 mmHg. Calculate the amount of oxygen transported into the blood in micromoles per second.

Solution

The difference between the amount of oxygen in the blood leaving and entering the oxygenator is the amount of oxygen transported into the blood. We can modify Equation 6.10 and replace $\Gamma_{metabolic}$ with the oxygen transport to the blood, i.e., Q_{oxygen}, and replace q with the total blood flow through the device, i.e., Q_{blood}. For this situation, we can then write the oxygen material balance between the entrance and the exit of the membrane blood oxygenator as

$$Q_{oxygen} = Q_{blood} \left[(C + C')_{out} - (C + C')_{in} \right].$$

In this equation, Q_{oxygen} is the amount of oxygen transported to the blood and Q_{blood} is the blood flow rate through the oxygenator. Next, we need to find the amount of oxygen that is dissolved in the blood at 30 and 100 mmHg. This can be found from Henry's law given by Equation 6.1:

$$C_{in} = \left(\frac{pO_2}{H} \right)_{in} = \frac{30 \text{ mmHg}}{0.74 \frac{\text{mmHg}}{\mu m}} = 40.54 \ \mu M,$$

$$C_{out} = \left(\frac{pO_2}{H} \right)_{out} = \frac{100 \text{ mmHg}}{0.74 \frac{\text{mmHg}}{\mu m}} = 135.14 \ \mu M.$$

The amount of oxygen bound to hemoglobin can be found from Figure 6.3. At 30 mmHg, the hemoglobin saturation is $Y = 0.58$ and at 100 mmHg it is $Y = 0.97$. Therefore, we have

$$C'_{in} = 0.58 \times 8800 \ \mu m = 5104 \ \mu M,$$

$$C'_{out} = 0.97 \times 8800 \ \mu m = 8536 \ \mu M.$$

Using these values, we can calculate the amount of oxygen transported to the blood:

$$Q_{oxygen} = Q_{blood} \left[(C + C')_{out} - (C + C')_{in} \right] = 5000 \frac{mL}{min} \times \left[(135.14 + 8536) - (40.54 + 5104) \right] \mu M$$

$$\times \frac{1 \ \mu mol}{L \ \mu M} \times \frac{1 \ min}{60 \ sec} \times \frac{1 \ L}{1000 \ mL}$$

$$Q_{oxygen} = 293.9 \frac{\mu mol}{sec}.$$

6.8 OXYGEN TRANSPORT IN BIOARTIFICIAL ORGANS AND TISSUE-ENGINEERED CONSTRUCTS

This discussion on oxygen transport is also of critical importance in the design of bioartificial organs (Colton 1995) and constructs used for tissue engineering (Radisic et al. 2006). In one type of bioartificial organ that is proposed for the treatment of diabetes, therapeutic cells, such as the islets of Langerhans or other insulin-secreting cells, are sandwiched between two permselective membranes, which are adjacent to a polymer matrix material that has been vascularized by the host. The permselective membrane has a porous structure at the molecular level such that small molecules like oxygen, key nutrients, and the therapeutic agent are readily permeable; however, the components of the host immune system cannot cross the membrane. The cells are therefore said to be *immunoprotected*.* In tissue engineering applications,† cells are placed within porous polymeric constructs that, in many cases, can be represented as thin disks. In these applications, key nutrients are brought to the cells primarily by diffusion and, in most cases, the critical species that affect the design is the transport of oxygen. The following discussion will focus on calculating the oxygen gradient in thin planar disks that contain cells with the goal of providing sufficient oxygen to the cells. In general, the oxygen gradient will depend on the cell density in the region of interest, the viability of the cells, and the metabolic activity of the cells.

Consider the bioartificial organ or tissue-engineered construct shown in Figure 6.6. As a first approximation, we will assume that the major resistance to oxygen transport lies in the membrane and the layer of cells. The membrane encloses the layer of cells and can also immunoprotect the layer of cells. We will prescribe the pO_2 level at the outer surface of the membrane to be some average of the blood pO_2 found in the capillaries adjacent to the membrane. A steady state shell balance for oxygen over a thickness Δx of the cell layer of cross-sectional area A may be written as

$$0 = -A D_e \left. \frac{dC}{dx} \right|_x + A D_e \left. \frac{dC}{dx} \right|_{x+\Delta x} - \Gamma_{metabolic} A \Delta x (1 - \varepsilon). \tag{6.15}$$

---

* We will discuss immunoprotection in greater detail in Chapter 10.
† Tissue engineering is discussed in Chapter 9.

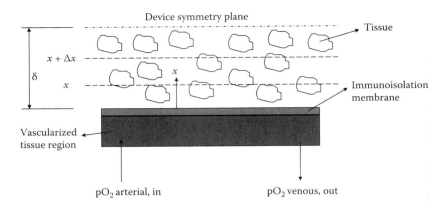

FIGURE 6.6 Conceptual model for a bioartificial organ.

Note that the volumetric metabolic oxygen consumption rate of the tissue, $\Gamma_{\text{metabolic}}$ (assumed constant), is multiplied by the cell volume, i.e., $A \, \Delta x \, (1-\varepsilon)$, to get the amount of oxygen consumed by the tissue in the shell of thickness Δx. The void volume within the cell layer is given by ε. The effective diffusivity of oxygen within the combined cell layer and void space is D_e, which was defined earlier in Chapter 5, e.g., by Equations 5.84 and 5.85, where in this case $D_e = D_T$. After dividing Equation 6.15 by Δx, and taking the limit as $\Delta x \to 0$, Equation 6.16 is obtained. Henry's law was used to express the dissolved oxygen concentration in terms of the oxygen partial pressure, i.e., $pO_2 = HC$:

$$D_e \frac{d^2 pO_2}{dx^2} = \Gamma_{\text{metabolic}} H (1-\varepsilon). \tag{6.16}$$

Since this equation is a second order differential equation, we require the following two boundary conditions in order to obtain the solution for the pO_2 profile within the tissue layer:

$$\text{BC1:} \ x = 0, pO_2 = pO_2^{x=0},$$

$$\text{BC2:} \ x = \delta, \frac{dpO_2}{dx} = 0. \tag{6.17}$$

The first boundary condition expresses the pO_2 level at the interface between the membrane and the layer of cells. The second boundary condition results from device symmetry and expresses the fact that there is no net flow of oxygen across the symmetry plane. Equation 6.16 can now be integrated twice and Equation 6.17 can be used to find the integration constants. Equation 6.18 is then obtained for the oxygen profile within the tissue layer:

$$pO_2(x) = pO_2^{x=0} + \left[\frac{\Gamma_{\text{metabolic}} H (1-\varepsilon) \delta^2}{2 D_e} \right] \left[\left(\frac{x}{\delta} \right)^2 - 2 \left(\frac{x}{\delta} \right) \right]. \tag{6.18}$$

At this point, the value of $pO_2^{x=0}$ in Equation 6.18 is unknown. We need to develop an additional equation that relates the transport rate of oxygen across the membrane to the total consumption of oxygen by the cells. This is given by

$$\frac{A P_m}{H} \left(pO_2^B - pO_2^{x=0} \right) = A \delta (1-\varepsilon) \Gamma_{\text{metabolic}}, \tag{6.19}$$

where P_m is the permeability of the immunoisolation membrane and pO_2^B is the average pO_2 in the blood in the vascularized region adjacent to the membrane, which is assumed to be known. Rearranging Equation 6.19 provides the value of $pO_2^{x=0}$:

$$pO_2^{x=0} = pO_2^B - \left[\frac{\Gamma_{metabolic} H (1-\varepsilon) \delta}{P_m} \right]. \tag{6.20}$$

Using this result for $pO_2^{x=0}$, we can now write the pO_2 gradient within the cell layer as

$$pO_2(x) = pO_2^B - \left[\frac{\Gamma_{metabolic} H (1-\varepsilon) \delta}{P_m} \right] + \left[\frac{\Gamma_{metabolic} H (1-\varepsilon) \delta^2}{2 D_e} \right] \left[\left(\frac{x}{\delta} \right)^2 - 2 \left(\frac{x}{\delta} \right) \right]. \tag{6.21}$$

The previous equations can now be used for basic calculations on the effect of oxygen transport on the design of a bioartificial organ or in a tissue-engineered construct. These calculations are illustrated in Examples 6.5 and 6.6.

Example 6.5

Consider a bioartificial organ such as that shown in Figure 6.6. Assume that the value of pO_2^B is the mean of the arterial and venous oxygen values, i.e., 68 mmHg. Let the cells be the islets of Langerhans, assumed spheres with a diameter of 150 µm, with the basal metabolic oxygen consumption rate of 25.9 µM sec^{-1}. The half-thickness (δ) of the islet layer is 200 µm and the void volume fraction (ε) is equal to 0.85. The oxygen permeability of the membrane (P_m) is equal to 4×10^{-3} cm sec^{-1}. Calculate the pO_2 profile in the islet layer of the bioartificial organ.

Solution

Using a value for the bulk diffusivity of oxygen as 2.11×10^{-5} cm^2 sec^{-1}, the effective diffusivity by Equation 5.84, for the given void fraction of 0.85, is equal to 1.67×10^{-5} cm^2 sec^{-1}. Equation 6.21 can then be solved for the oxygen profile in the layer of islets. Figure 6.7 shows a plot of the pO_2 profile within the islet layer of the bioartificial organ. The graph shows that for these conditions, the tissue layer is fully oxygenated.

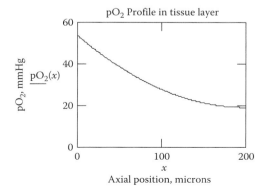

FIGURE 6.7 Oxygen profile in a bioartificial organ.

Example 6.6

The only life that is present on the planet *Biophilus* is a small, slimy flat worm that is about 200 mm in length and about 20 mm in width. The slimy gel-like material that covers the flesh of the worm has a thickness of 200 μm. The worm survives through metabolic processes that are limited by the transport of methane present in the planet's atmosphere. Based on the slimy coverings found on earth-based worms, it is estimated that the diffusivity of methane through the worm's slimy covering is 75% of the water diffusivity of methane (MW = 16) at 37°C. The effective diffusivity of methane in the flesh of the worm is estimated to be about 1.4×10^{-5} cm^2 sec^{-1}. The concentration of methane in the planet's atmosphere was found to be 0.20 mol L^{-1} and the molar equilibrium solubility ratio of methane in the slimy covering of this worm relative to atmospheric methane is 0.10 and is given by

$$\text{Solubility ratio} = \frac{C_{\text{methane}}\left(\text{slimy covering}\right)}{C_{\text{methane}}\left(\text{atmosphere}\right)} = 0.1.$$

If the fleshy part of the worm is 2 mm thick, estimate the metabolic rate of methane consumption for the worm based on the total volume of the worm, i.e., $\Gamma'_{\text{metabolic}} = \Gamma_{\text{metabolic}}\,(1-\varepsilon)$, in μM per second.

Solution

This problem is similar to the discussion and mathematical development for the bioartificial organ shown in Figure 6.6. In this case, the slimy covering encases the fleshy part of the worm where methane is consumed as part of the worm's metabolism. By analogy to the situation shown in Figure 6.6, the slimy covering is like the immunoisolation membrane and the cell layer is the fleshy part of the worm. We also make the analogy that methane is like oxygen to the worm on this planet, so we can use Equation 6.21. However, since we do not have any information on the Henry's constant for the solubility of methane in the worm's interstitial fluid, we will solve the problem in terms of the methane concentration. Therefore, we can replace the pO_2s in Equation 6.21 with HC and in doing this, the Henry's constant cancels itself out and is no longer needed. Hence, Equation 6.21 can be rewritten in terms of the concentration of methane in the worm tissue as

$$C_{\text{methane}}(x) = C_{\text{methane}}\big|_{\text{slime surface}} - \left[\frac{\Gamma_{\text{metabolic}}(1-\varepsilon)\delta}{P_m}\right] + \left[\frac{\Gamma_{\text{metabolic}}\,(1-\varepsilon)\delta^2}{2D_e}\right]\left[\left(\frac{x}{\delta}\right)^2 - 2\left(\frac{x}{\delta}\right)\right].$$

In this equation, the permeability of the slimy layer (P_m) is equal to the diffusivity of methane through the slime divided by the thickness of the slime layer. The diffusivity of methane in water at 37°C can be found from Equation 5.3 and is equal to 2.83×10^{-5} cm^2 sec^{-1}. Since it is given that the diffusivity of methane in the slimy layer is 75% of this value, the slime diffusivity is then equal to 2.12×10^{-5} cm^2 sec^{-1}. Since the slime layer is 200 μm thick (0.02 cm), we can show that

$$P_m = \frac{2.12 \times 10^{-5}\,\dfrac{\text{cm}^2}{\text{sec}}}{0.02\,\text{cm}} = 0.0011\,\frac{\text{cm}}{\text{sec}}.$$

Note that we have also replaced pO_2^B with the concentration of methane at the surface of the slime layer that is in equilibrium with the methane in the atmosphere. From the equilibrium solubility ratio of 0.1, we can then find the methane concentration at the surface of the slime layer as

$$C_{\text{methane}}\big|_{\text{slime surface}} = 0.1 \times 0.20\,\frac{\text{mol}}{\text{L}} = 0.02\,\frac{\text{mol}}{\text{L}}.$$

To find the metabolic consumption rate of methane by the worm, we assume that $C_{methane} = 0$ at $x = 0$ along the central longitudinal axis of the worm. With $C_{methane} = 0$ at $x = 0$, we can then solve the $C_{methane}(x)$ equation shown earlier for $\Gamma'_{metabolic}$ as

$$\Gamma'_{metabolic} = \frac{C_{methane}|_{slime\,surface}}{\left(\dfrac{\delta}{P_m} + \dfrac{\delta^2}{2D_e}\right)} = \frac{0.02\dfrac{mol}{L}}{\left(\dfrac{0.1\,cm}{0.0011\dfrac{cm}{sec}} + \dfrac{(0.1\,cm)^2}{2\times1.4\times10^{-5}\dfrac{cm^2}{sec}}\right)}$$

$$= 4.46\times10^{-5}\frac{mol}{L\,sec}\times\frac{10^6\,\mu M}{\dfrac{mol}{L}sec},$$

$$\Gamma'_{metabolic} = 44.6\frac{\mu M}{sec}.$$

Although Equation 6.21 was derived for the case of the bioartificial organ shown in Figure 6.6, it can be generalized to other situations, as shown in Example 6.6. In many cases, the cells may be incorporated into a polymeric construct such that there is no membrane surrounding the cell layer or there is no other diffusional barrier that offers mass transfer resistance. In this case, $pO_2^{x=0} = pO_2^B$ and Equation 6.18 becomes

$$pO_2(x) = pO_2^B + \left[\frac{\Gamma_{metabolic}H(1-\varepsilon)\delta^2}{2D_e}\right]\left[\left(\frac{x}{\delta}\right)^2 - 2\left(\frac{x}{\delta}\right)\right]. \tag{6.22}$$

In this case, pO_2^B represents the oxygen partial pressure in the blood adjacent to the cell layer or pO_2^B can represent the oxygen partial pressure in a well-stirred nutrient medium in which the cell layer construct is immersed. If this nutrient medium is not sufficiently agitated, then there will also exist a concentration boundary layer in the fluid surrounding the cell layer construct (see Figure 6.1) and this external mass transfer resistance can be accounted for by a mass transfer coefficient, k_m. In this case, Equation 6.19 can be modified to describe the transport rate of oxygen from the bulk solution to the surface of the cell layer construct and at steady state this transport rate of oxygen must equal the total consumption rate of oxygen by the cells:

$$\frac{Ak_m}{H}(pO_2^B - pO_2^{x=0}) = A\delta(1-\varepsilon)\Gamma_{metabolic}. \tag{6.23}$$

In this case, we can therefore replace in Equation 6.21 the immunoisolation membrane permeability (P_m) with k_m:

$$pO_2(x) = pO_2^B - \left[\frac{\Gamma_{metabolic}H(1-\varepsilon)\delta}{k_m}\right] + \left[\frac{\Gamma_{metabolic}H(1-\varepsilon)\delta^2}{2D_e}\right]\left[\left(\frac{x}{\delta}\right)^2 - 2\left(\frac{x}{\delta}\right)\right]. \tag{6.24}$$

Example 6.7

Malda et al. (2004) measured oxygen gradients in tissue-engineered cartilaginous constructs. Using a microelectrode, oxygen concentrations were measured as a function of distance into the tissue-engineered constructs containing the chondrocytes. In addition, they also determined the chondrocyte distribution as a function of position within the construct. The average number of chondrocytes per cubic centimeter of total construct (i.e., the cell density) 28 days after seeding is about 1.5×10^8 cells cm^{-3}. The following table summarizes the pO$_2$ levels that were measured within the construct as a function of the distance from the exposed surface of the construct.

Depth (µm)	pO$_2$ (mmHg)
0	160
250	114
500	95
750	76
1000	61
1250	53
1500	42
1750	38
2000	30
2500	19

They also estimated the diffusivity of oxygen in the construct to be 3.8×10^{-6} cm^2 sec^{-1}. From these data, estimate the metabolic oxygen consumption rate of the chondrocytes in µM per second, assuming that the diameter of a chondrocyte is 20 µm.

Solution

Since the chondrocyte construct is directly exposed to the surrounding nutrient medium, we can use Equation 6.22 with pO$_2^B$ = 160 mmHg. The volume of one cell is equal to 4.19×10^{-9} cm^3 cell^{-1}. Multiplying the cell volume by the cell density provides the value of $1 - \varepsilon = 0.63$. A regression analysis can then be performed to find the best value of $\Gamma_{\text{metabolic}}$ in Equation 6.22 that fits the previous data. Figure 6.8 shows the results of these calculations. $\Gamma_{\text{metabolic}}$ is found to equal 0.037 µM sec^{-1} and we see that Equation 6.22 provides a good representation of the data. It is also important to note that chondrocytes have very low respiratory activity, which helps them survive in the relatively avascular regions of cartilage tissue. Heywood et al. (2006), in a study involving bovine articular chondrocytes, reported oxygen consumption rates ranging from 9.6×10^{-16} moles cell^{-1} h^{-1} at basal levels of glucose to 18.4×10^{-16} moles cell^{-1} h^{-1} at reduced levels of glucose. On the basis of a cell diameter of 20 µm, this equates to a metabolic oxygen consumption rate for chondrocytes of 0.06–0.12 µM sec^{-1}, which is of the same order as found here for the data obtained by Malda et al. (2004).

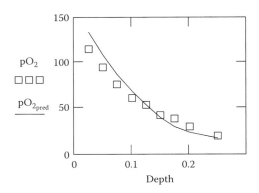

FIGURE 6.8 Measured and predicted oxygen gradients in engineered cartilaginous constructs, from Example 6.5.

6.9 STEADY STATE OXYGEN TRANSPORT IN A PERFUSION BIOREACTOR

Tilles et al. (2001) and Allen and Bhatia (2003) developed steady state microchannel perfusion bioreactor systems for studying the effect of oxygen on a variety of cellular functions. The microchannel bioreactor also serves as a model system for the flow channels that are formed in perfusion bioreactors that are used for the production of tissue-engineered constructs. The oxygen concentration within the bioreactor can be controlled by adjusting the flow rate of media through the device.

Figure 6.9 illustrates their parallel-plate microchannel bioreactor. Nutrient media flows through the bioreactor with an average velocity equal to V. On the lower surface there is a monolayer of attached cells that consumes the oxygen that is in the flowing stream. The entering oxygen concentration is C_{in}. A steady state shell balance for oxygen on a shell volume of the fluid of width W from x to $x + \Delta x$ and from y to $y + \Delta y$ can be written as

$$V \Delta y W C|_x - V \Delta y W C|_{x+\Delta x} + D \Delta x W \frac{\partial C}{\partial y}\bigg|_{y+\Delta y} - D \Delta x W \frac{\partial C}{\partial y}\bigg|_y = 0. \tag{6.25}$$

After dividing by $\Delta x \Delta y$, and taking the limit as Δx and $\Delta y \to 0$, the following differential equation is obtained:

$$V \frac{\partial C}{\partial x} = D \frac{\partial^2 C}{\partial y^2}. \tag{6.26}$$

This is a convective-diffusion equation and describes the transport of oxygen within the parallel-plate microchannel bioreactor. The term on the left-hand side of Equation 6.26 represents the transport of oxygen by convection, or the bulk flow of the fluid flowing through the bioreactor, and the term on the right-hand side represents oxygen transport by diffusion in the y direction. We have assumed that the transport of oxygen in the axial direction (x) by convection is much larger than that transported by axial diffusion. Recall from our discussion in Section 5.10.2.1 that this means that the Peclet number is much larger than unity, or $Pe = (VL/D) \gg 1$. The boundary conditions for Equation 6.26 are

$$\text{BC1: } x = 0, C(0,y) = C_{in},$$

$$\text{BC2: } y = h, -D\frac{\partial C}{\partial y}\bigg|_{y=h} = \Gamma_{metabolic}\,\delta_{cell}(1-\varepsilon), \tag{6.27}$$

$$\text{BC3: } y = 0, \frac{\partial C}{\partial y}\bigg|_{y=0} = 0.$$

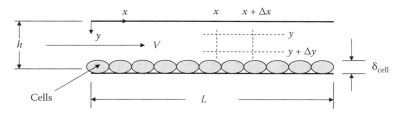

FIGURE 6.9 Parallel-plate microchannel bioreactor.

Boundary condition one expresses the fact that fluid flowing through the bioreactor enters at a uniform oxygen concentration, i.e., C_{in}. Boundary condition two expresses the fact that the flux of oxygen at the cellular surface is equal to the rate of oxygen consumption by the cells. The thickness of the cellular layer along the lower surface is δ_{cell} and $(1 - \varepsilon)$ represents the volume fraction of the cells along the lower surface. Boundary condition three represents the fact that oxygen is not consumed or lost from the upper surface of the bioreactor.

After introducing the following dimensionless variables into Equations 6.26 and 6.27 (i.e., $\hat{x} = x/L$, $\hat{y} = y/h$, $\hat{C} = C/C_{in}$), the following dimensionless equations are obtained:

$$\frac{\partial \hat{C}}{\partial \hat{x}} = \frac{\alpha}{Pe} \frac{\partial^2 \hat{C}}{\partial \hat{y}^2}, \tag{6.28}$$

$$\text{BC1: } \hat{x} = 0, \hat{C} = 1,$$

$$\text{BC2: } \hat{y} = 1, \frac{\partial \hat{C}}{\partial y}\bigg|_{\hat{y}=0} = -Da, \tag{6.29}$$

$$\text{BC3: } \hat{y} = 0, \frac{\partial \hat{C}}{\partial \hat{y}}\bigg|_{\hat{y}=1} = 0.$$

The Peclet number (Pe) in this case is based on the channel thickness (h), defined as $(VL/D) \times (h/L) \times (Vh/D)$, and represents the ratio of oxygen transport by axial convection to that by axial diffusion. The dimensionless oxygen flux is also known as the *Damkohler number*, defined as $Da = \Gamma_{metabolic}(1-\varepsilon)h\delta_{cell}/DC_{in}$. Equations 6.28 and 6.29 can be solved analytically using separation of variables as discussed in Chapter 5. The resulting dimensionless oxygen concentration profile can be shown to be given by Equation 6.30 (Allen and Bhatia 2003):

$$\hat{C}(\hat{x}, \hat{y}) = 1 + Da\left[\frac{1 - 3\hat{y}^2}{6} - \frac{\alpha}{Pe}\hat{x} + \frac{2}{\pi^2} \sum_{n=1}^{\infty} \frac{(-1)^n}{n^2} \exp\left(-\frac{\alpha n^2 \pi^2 \hat{x}}{Pe}\right) \cos(n\pi\hat{y})\right]. \tag{6.30}$$

Equation 6.30 provides the oxygen concentration within the bioreactor at any position $(\hat{x}, \hat{y})$. For large Pe numbers (where the convective flow dominates axial diffusion), the summation term in Equation 6.30 becomes insignificant at a short distance into the bioreactor and this solution can then be approximated by the following result:

$$\hat{C}(\hat{x}, \hat{y}) \approx 1 + Da\left(\frac{1 - 3\hat{y}^2}{6} - \frac{\alpha}{Pe}\hat{x}\right). \tag{6.31}$$

Of particular interest is the location along the cell surface where the oxygen concentration becomes equal to zero, i.e., $\hat{C}(\hat{x}_{critical}, \hat{y} = 1) = 0$. We can then solve Equation 6.31 for this critical value of x as given by

$$\hat{x}_{critical} \approx \frac{Pe}{\alpha}\left(\frac{1}{Da} - \frac{1}{3}\right). \tag{6.32}$$

We can also calculate the average oxygen concentration at any value of $\hat{x}$ by integrating Equation 6.30 from $\hat{y} = 0$ to $\hat{y} = 1$. The following result is then obtained for the average oxygen concentration in the bioreactor as a function of $\hat{x}$:

$$\hat{C}_{\text{average}}(\hat{x}) = 1 - \left(\frac{\alpha Da}{Pe}\right)\hat{x}. \tag{6.33}$$

Example 6.8

Calculate the distance into the bioreactor where the cell surface pO_2 reaches 0 mmHg. The flow rate through the bioreactor is 0.5 cm^3 min^{-1} and the inlet pO_2 is 75 mmHg. Use the bioreactor parameters in the following table from Allen and Bhatia (2003).

Parameter	Value	Units
D, oxygen diffusivity	2×10^{-5}	cm^2 sec^{-1}
V_{max}, max O_2 uptake	0.38	nmol sec^{-1} $(10^6$ cells$)^{-1}$
ρ, surface cell density	1.7×10^5	cells cm^{-2}
$pO_{2\,in}$, inlet pO_2	75	mmHg
Q, flow rate	0.5	cm^3 min^{-1}
h, height of flow channel	100	μm
W, width of flow channel	2.8	cm
L, length of flow channel	5.5	cm

Solution

First we need to calculate α, Da, and Pe: $\alpha = L/H = 550$. For the Da number, we also recognize that $\rho V_{max} = \Gamma_{\text{metabolic}} (1 - \varepsilon) \delta_{\text{cell}}$. Therefore, we can calculate the Da number as shown in the following equation. Note that we have divided the inlet pO_2 by the Henry's constant for oxygen in media ($\sim$1.04 mmHg μM^{-1}) to estimate the inlet oxygen concentration as 72 nmol cm^{-3}:

$$Da = \frac{\rho V_{max} h}{D C_{in}} = \frac{1.7 \times 10^5 \text{ cells cm}^{-2} \times 0.38 \text{ nmol sec}^{-1} 10^{-6} \text{ cell}^{-1} \times 0.01 \text{ cm}}{2 \times 10^{-5} \text{ cm}^2 \text{ sec}^{-1} \times 72 \text{ nmol cm}^{-3}} = 0.45.$$

The Pe number is then calculated as

$$Pe = \frac{Vh}{D} = \frac{\left(\frac{Q}{h \times W}\right)h}{D} = \frac{0.5 \frac{cm^3}{min} \times \frac{1 \text{ min}}{60 \text{ sec}} \times \frac{1}{(0.01 \text{ cm} \times 2.8 \text{ cm})} \times 0.01 \text{ cm}}{2 \times 10^{-5} \text{ cm}^2 \text{ sec}^{-1}} = 149.$$

Using the previous values of α, Da, and Pe, we find that Equations 6.30 and 6.32 give an $\hat{x}_{\text{critical}} = 0.511$ or an x_{critical} of 0.511×5.5 cm $= 2.81$ cm.

Example 6.9

Allen and Bhatia (2003) obtained the following outlet pO_2s (equivalent to the average pO_2 at $x = L$) for their bioreactor, discussed in Example 6.8, at various flow rates for an inlet $pO_2 = 158$ mmHg. Compare these results to that predicted by Equation 6.33.

Flow rate (cm^3 min^{-1})	pO$_{2\,\text{out}}$ (mmHg)
0.5	42
0.75	73
1.0	92
1.5	112
2.0	122
3.0	133

Solution

Once again, we first need to calculate α, Da, and Pe. From before, $\alpha = L/H = 550$. We can calculate the Da number as shown in the following equation. Note that we have divided the inlet pO$_2$ by the Henry's constant for oxygen (~1.04 mmHg μM^{-1}) to estimate the inlet oxygen concentration as 152 nmol cm^{-3}:

$$Da = \frac{\rho V_{\max}\, h}{D C_{\text{in}}} = \frac{1.7 \times 10^5\ \text{cells cm}^{-2} \times 0.38\ \text{nmol sec}^{-1} 10^{-6}\ \text{cell}^{-1} \times 0.01\ \text{cm}}{2 \times 10^{-5}\ \text{cm}^2\ \text{sec}^{-1} \times 152\ \text{nmol cm}^{-3}} = 0.21.$$

For the Peclet number, we have to calculate a value for each of the flow rates:

$$Pe = \frac{Vh}{D} = \frac{\left(\dfrac{Q}{h \times W}\right)h}{D} = \frac{Q\,\dfrac{\text{cm}^3}{\text{min}} \times \dfrac{1\ \text{min}}{60\ \text{sec}} \times \dfrac{1}{(0.01\ \text{cm} \times 2.8\ \text{cm})} \times 0.01\ \text{cm}}{2 \times 10^{-5}\ \text{cm}^2\ \text{sec}^{-1}} = 298Q.$$

We can then rewrite Equation 6.33 at $x = L$ as shown in the following equation, replacing the concentrations with pO$_2$s and recognizing that the units on Q are cubic centimeters per minute:

$$pO_{2\text{out}} = pO_{2\text{in}}\left(1 - \frac{\alpha Da}{Pe}\right) = 158\ \text{mmHg}\left(1 - \frac{550 \times 0.21}{298Q}\right) = 158\ \text{mmHg}\left(1 - \frac{0.388}{Q}\right).$$

Figure 6.10 shows the comparison between the measured outlet pO$_2$ values and those predicted by Equation 6.33 as given by the previous equation for this example. The agreement between the data and the model is quite good.

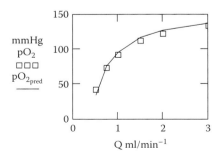

FIGURE 6.10 Comparison of actual and predicted exiting pO$_2$ levels from the parallel-plate microchannel bioreactor, See Example 6.8.

6.10 OXYGEN TRANSPORT IN THE KROGH TISSUE CYLINDER

Our previous analysis for tissue oxygenation (Equation 6.10) treats the tissue and blood as well-mixed regions, i.e., each region is spatially lumped or averaged. This type of analysis provides no information about the oxygen diffusion and consumption within the tissue surrounding a given capillary. Oxygen transport by diffusion to the tissue surrounding a given capillary is usually the rate-limiting step as far as maintaining viable tissue is concerned. The effective oxygenation distance within the tissue outside the capillary is quite small, typically only several capillary diameters, thereby requiring an extensive network of capillaries to meet the tissue oxygen demand.

We can use the Krogh tissue cylinder approach (Krogh 1919) developed earlier in Chapter 5 to analyze oxygen transport to the tissue surrounding a given capillary. The Krogh tissue model is a good place to start before considering more advanced models that are described in the literature (Mirhashemi et al. 1987; Tsai et al. 1990; Lagerlund and Low 1993; Secomb et al. 1993; Intaglietta 1997; Li et al. 1997).

Figure 5.17 illustrates a concentric shell of the Krogh tissue cylinder. Considering blood first, the shell balance equation for oxygenated hemoglobin may be written as*

$$(2\pi r \Delta r \Delta x)\frac{\partial C'}{\partial t} = 2\pi r \Delta r V\, C'|_{z} - 2\pi r \Delta r V\, C'|_{z+\Delta z} + R_{\text{HbO}}\, 2\pi r \Delta r \Delta z. \tag{6.34}$$

In this equation, we ignore the particulate nature of blood as well as the mass transfer resistance of the RBC. The blood is assumed to be in plug flow with an average velocity represented by V. Also note that the hemoglobin is carried along by the RBC at the average blood velocity (V) and, accordingly, there is no diffusive transport of hemoglobin. The production rate of oxygenated hemoglobin per unit volume of blood is represented by R_{HbO}. After dividing by $2\pi\, r\, \Delta r\, \Delta z$, and taking the limit as $\Delta z \to 0$, we obtain the following differential equation that describes the mass balance for oxygenated hemoglobin within the blood flowing through the capillary:

$$\frac{\partial C'}{\partial t} + V\frac{\partial C'}{\partial z} = R_{\text{HbO}}. \tag{6.35}$$

A shell balance for the dissolved oxygen in the blood is given by

$$(2\pi r \Delta r \Delta z)\frac{\partial C}{\partial t} = 2\pi r \Delta r V C\,|_{z} - 2\pi r \Delta r V C\,|_{z+\Delta z} + (-2\pi r \Delta z)D\frac{\partial C}{\partial r}\bigg|_{r}$$

$$- (-2\pi r \Delta z)D\frac{\partial C}{\partial r}\bigg|_{r+\Delta r} + (-2\pi r \Delta r)D\frac{\partial C}{\partial z}\bigg|_{z} - (-2\pi r \Delta r)D\frac{\partial C}{\partial z}\bigg|_{z+\Delta z} \tag{6.36}$$

$$+ R\, 2\pi r \Delta r \Delta z.$$

---

* Unlike the shell balances developed in Chapter 5, we will now leave in the accumulation term so that we can consider, if necessary, unsteady state problems.

In addition to the convective flow of the dissolved oxygen (VC terms), we also have axial and radial diffusion of the dissolved oxygen ($D(\partial C/\partial z)$ and $D(\partial C/\partial r)$ terms). In Equation 6.36, R represents the production rate of dissolved oxygen per unit volume of blood. Once again, we divide by $2\pi r \, \Delta r \, \Delta z$ and take the limit as Δr and $\Delta z \to 0$. We then obtain the following differential equation that describes the mass balance on dissolved oxygen within the capillary:

$$\frac{\partial C}{\partial t} + V\frac{\partial C}{\partial t} = D\left[\frac{1}{r}\frac{\partial}{\partial r}\left(r\frac{\partial C}{\partial r}\right) + \frac{\partial^2 C}{\partial z^2}\right] + R. \tag{6.37}$$

The two reaction rate terms in Equations 6.35 and 6.37 are simply the negative of each other, therefore $R_{HbO} = -R$. This is true since the rate of disappearance of dissolved oxygen must equal the rate of appearance of oxygenated hemoglobin.

We can also make use of the oxygen dissociation curve to relate C' to C as follows:

$$\frac{\partial C'}{\partial t} = \frac{\partial C}{\partial t}\left(\frac{\partial C'}{\partial C}\right) = m\frac{\partial C}{\partial t},$$

$$\frac{\partial C'}{\partial z} = \frac{\partial C}{\partial z}\left(\frac{\partial C'}{\partial C}\right) = m\frac{\partial C}{\partial z}, \tag{6.38}$$

where $m = dC'/dC$ is simply related to the slope of the oxygen–hemoglobin dissociation curve.

Equations 6.35 and 6.37 can now be added together, using Equation 6.38 and Henry's law (Equation 6.1), to express the dissolved oxygen concentration in terms of pO_2. We then obtain Equation 6.39, which expresses the mass balance for dissolved oxygen within the capillary in terms of the oxygen partial pressure:

$$(1+m)\frac{\partial pO_2}{\partial t} + V(1+m)\frac{\partial pO_2}{\partial z} = D\left[\frac{1}{r}\frac{\partial}{\partial r}\left(r\frac{\partial pO_2}{\partial r}\right) + \frac{\partial^2 pO_2}{\partial z^2}\right]. \tag{6.39}$$

Before we go any further, let us take a closer look at how to evaluate the parameter, m, which can be related to the blood pO_2 using Henry's law and recognizing that $Y = C'/C'_{Sat}$. Equation 6.39 then provides a way to calculate m, which depends on the local value of pO_2. Note that m is dimensionless:

$$m = HC'_{SAT}\frac{dY}{dpO_2}. \tag{6.40}$$

Using Hill's equation (Equation 6.6), we can then evaluate dY/dpO_2 and obtain the following expression for the dependence of m on the local pO_2:

$$m = nP_{50}^n HC'_{SAT}\frac{pO_2^{n-1}}{(P_{50}^n + pO_2^n)^2}. \tag{6.41}$$

A shell balance on the dissolved oxygen within the tissue interstitial space of void volume ε_T surrounding the capillary may be written as

$$2\pi r \Delta r \Delta z \varepsilon_T \frac{\partial C^T}{\partial t} = (-2\pi r \Delta z) D_T \frac{\partial C^T}{\partial r}\bigg|_r$$

$$-(-2\pi r \Delta z) D_T \frac{\partial C^T}{\partial r}\bigg|_{r+\Delta r} + (-2\pi r \Delta r) D_T \frac{\partial C^T}{\partial z}\bigg|_z \qquad (6.42)$$

$$-(-2\pi r \Delta r) D_T \frac{\partial C^T}{\partial z}\bigg|_{z+\Delta z} - 2\pi r \Delta r \Delta z \Gamma_{\text{metabolic}}.$$

Within the tissue, D_T for oxygen can once again be estimated using the methods discussed in Chapter 5. After dividing by $2\pi\, r\, \Delta r\, \Delta z$ and taking the limit as Δr and $\Delta z \to 0$, the following differential equation is obtained for the oxygen mass balance within the tissue region. Henry's law was used to express the dissolved oxygen concentration in the tissue, C^T, in terms of the tissue region oxygen partial pressure, pO_2^T:

$$\varepsilon^T \frac{\partial \text{pO}_2^T}{\partial t} = D_T \left[\frac{1}{r} \frac{\partial}{\partial r} \left(r \frac{\partial \text{pO}_2^T}{\partial r} \right) + \frac{\partial^2 \text{pO}_2^T}{\partial z^2} \right] - \Gamma_{\text{metabolic}} H^T. \qquad (6.43)$$

The boundary conditions needed to solve Equations 6.39 and 6.43 are

$$\text{Blood region}\ \ 0 \le z \le L\ \ \text{and}\ \ 0 \le r \le r_C. \qquad (6.44)$$

$$\text{BC1:}\ \ z = 0, \text{pO}_2 = \text{pO}_2(t),$$

$$\text{BC2:}\ \ z = L, \frac{\partial \text{pO}_2}{\partial z} = 0,$$

$$\text{BC3:}\ \ r = 0, \frac{\partial \text{pO}_2}{\partial r} = 0,$$

$$\text{BC4:}\ \ r = r_C, \text{pO}_2 = \text{pO}_2^T \ \ \text{and}\ \ D \frac{\partial \text{pO}_2}{\partial r} = D \frac{\partial \text{pO}_2^T}{\partial r}.$$

Boundary condition one expresses the fact that the pO_2 of the blood entering the capillary is assumed to be known and may be a function of time. Boundary condition two simply states that oxygen cannot leave the capillary in the axial direction by axial diffusion. Boundary condition three assumes that the oxygen profile is symmetric with respect to the capillary centerline. The last boundary condition assumes that the capillary wall has negligible mass transfer resistance* and expresses the requirement that the dissolved oxygen concentrations and the oxygen flux are continuous at the interface between the capillary and tissue regions.

The tissue region boundary conditions may be written as follows, assuming that there are no anoxic regions:

$$\text{Tissue region}\ \ 0 \le z \le L\ \ \text{and}\ \ r_C \le r \le r_T \qquad (6.45)$$

---

* Recall that oxygen is lipid soluble and readily permeates the entire surface of the capillary wall.

$$\text{BC5: } z = 0, \frac{\partial pO_2^T}{\partial z} = 0,$$

$$\text{BC6: } z = L, \frac{\partial pO_2^T}{\partial z} = 0,$$

$$\text{BC7: } r = r_T, \frac{\partial pO_2^T}{\partial r} = 0.$$

Boundary conditions five and six state that oxygen cannot leave the tissue region at either end by axial diffusion. Boundary condition seven states that the oxygen profile in the tissue region between capillaries spaced a distance $2\, r_T$ (the Krogh tissue cylinder diameter) apart is symmetric.

Under certain conditions, anoxic regions may develop within the tissue region. These regions will be defined by a *critical radius*, $r_{anoxic}\,(z)$, a distance beyond which there is no oxygen in the tissue. If an anoxic region exists, then boundary condition seven becomes

$$\text{Tissue region} - \text{anoxic: } \text{BC8: } r = r_{anoxic}(z), \frac{\partial pO_2}{\partial r} = 0 \text{ and } pO_2 = 0. \tag{6.46}$$

6.11 APPROXIMATE SOLUTION FOR OXYGEN TRANSPORT IN THE KROGH TISSUE CYLINDER

Solution of the previous equations for the oxygen concentrations within the blood and tissue regions is a formidable problem and requires a numerical solution (Lagerlund and Low 1993). However, with some simplifications, a reasonable analytical solution can be obtained, which is a good starting point for exploring the key factors that govern oxygenation of the tissue surrounding a given capillary. We can also limit ourselves to a steady state solution, thereby eliminating the time derivatives in these equations. Another useful approximation is to treat m as a constant. Recall that m is related to the slope of the oxygen dissociation curve and is given by Equation 6.41. An average value of m can then be used in the range of pO_2 levels of interest. Since the capillary is much longer in length than the Krogh tissue cylinder radius, we can also ignore axial diffusion within the tissue region. This greatly simplifies the equation for the tissue region and Equation 6.43 becomes

$$\frac{d}{dr}\left(r\frac{\partial pO_2^T}{dr}\right) = \frac{r\Gamma_{metabolic}H^T}{D_T}. \tag{6.47}$$

Within the capillary, we can ignore axial diffusion in comparison to axial convection and from Equation 6.39 we obtain

$$(1+m)V\frac{\partial pO_2}{\partial z} = D\left(\frac{1}{r}\frac{\partial}{\partial r}\left(r\frac{\partial pO_2}{\partial r}\right)\right). \tag{6.48}$$

This a partial differential equation and is still tough to solve. One approximate approach is to eliminate the radial diffusion term by lumping (i.e., integrating the equation) over the r-direction, as illustrated in the following equation. Radial averaging of the capillary oxygen levels is appropriate here since the bulk of the oxygen mass transfer resistance is not within the capillary. Hence, we do not expect steep gradients in the oxygen concentration in the radial direction within the blood:

$$2\pi \int_0^{r_C} (1+m)V \frac{\partial pO_2}{\partial z} r\,dr = 2\pi D \int_0^{r_C} \frac{1}{r}\left(\frac{\partial}{\partial r}\left(r\frac{\partial pO_2}{\partial r}\right)\right) r\,dr.$$ (6.49)

This equation can then be integrated and written as

$$2\pi(1+m)V \frac{d}{dz}\int_0^{r_C} pO_2 r\,dr = 2\pi D r_C \frac{\partial pO_2}{\partial r}\bigg|_{r_C}.$$ (6.50)

We next recognize that the radially averaged pO_2 level in the blood, i.e., $\langle pO_2\rangle$, at a given axial location z, is defined as

$$\langle pO_2\rangle = \frac{2\pi \int_0^{r_C} pO_2 r\,dr}{\pi r_C^2}.$$ (6.51)

This allows Equation 6.50 to be rewritten in terms of the average pO_2 level in the blood as given by

$$(1+m)\frac{d\langle pO_2\rangle}{dz} = \frac{2D}{r_C V}\frac{dpO_2}{dr}\bigg|_{r_C} = \frac{2D_T}{r_C V}\frac{dpO_2^T}{dr}\bigg|_{r_C}.$$ (6.52)

Note that the solutions for the average oxygen level in the blood, i.e., $\langle pO_2\rangle$, and tissue regions, i.e., pO_2^T, are connected through the oxygen flux terms on the right-hand side of Equation 6.52, which arises from the use of boundary condition four in Equation 6.44.

We can now proceed to obtain an analytical solution for the oxygen levels within the capillary and the tissue region. First, we integrate Equation 6.47 twice and use boundary conditions four and seven in Equations 6.44 and 6.45 to obtain

$$pO_2^T(r,z) = \langle pO_2(z)\rangle - \frac{r_C^2 \Gamma_{\text{metabolic}} H^T}{4D_T}\left[1-\left(\frac{r}{r_C}\right)^2\right]$$
$$- \frac{r_T^2 \Gamma_{\text{metabolic}} H^T}{2D_T}\ln\left(\frac{r}{r_C}\right).$$ (6.53)

Although we ignored axial diffusion within the tissue region, note that the axial average capillary pO_2 level, $\langle pO_2(z)\rangle$, impresses an axial dependence on the tissue region pO_2 level. Thus, Equation 6.53 depends on the local capillary oxygen pO_2 and is valid so long as $pO_2^T \geq 0$ throughout the tissue region.

In some cases, part of the tissue region may become anoxic and Equation 6.47 must be solved using the anoxic boundary condition given by Equation 6.46. In this case, the tissue pO_2 is given by

$$pO_2^T(r,z) = \langle pO_2(z)\rangle - \frac{r_C^2 \Gamma_{\text{metabolic}} H^T}{4D_T}\left[1-\left(\frac{r}{r_C}\right)^2\right]$$
$$- \frac{r_{\text{anoxic}}(z)^2 \Gamma_{\text{metabolic}} H^T}{2D_T}\ln\left(\frac{r}{r_C}\right).$$ (6.54)

We can find $r_{\text{anoxic}}(z)$ from the additional condition in Equation 6.46 that requires $pO_2^T = 0$ at $r_{\text{anoxic}}(z)$. By setting the left-hand side of Equation 6.54 equal to zero and letting $r = r_{\text{anoxic}}(z)$, we then obtain the following nonlinear equation that can be solved for $r_{\text{anoxic}}(z)$:

$$\left(\frac{r_{\text{anoxic}}(z)}{r_C}\right)^2 \ln\left(\frac{r_{\text{anoxic}}(z)}{r_C}\right)^2 - \left(\frac{r_{\text{anoxic}}(z)}{r_C}\right)^2 + 1 = \frac{4D_T\langle pO_2(z)\rangle}{r_C^2 \Gamma_{\text{metabolic}} H^T}. \tag{6.55}$$

Equations 6.53 through 6.55 provide the pO_2 level within the tissue region under both non-anoxic and anoxic conditions for a given average capillary pO_2 level.

The capillary pO_2 level changes with axial position and can now be found by solving Equation 6.52. Solution of this equation requires that we know the value of $D_T(dpO_2^T/dr)|_{r_C}$. We can get this by differentiating Equation 6.52 or Equation 6.53 and evaluating the derivative at r_C. On substitution of this result for the non-anoxic case, we obtain

$$\frac{d\langle pO_2(z)\rangle}{dz} = -\frac{\Gamma_{\text{metabolic}} H^T}{(1+m)V}\left[\left(\frac{r_T}{r_C}\right)^2 - 1\right]. \tag{6.56}$$

This equation can be integrated to give the following result for the axial change in the capillary oxygen partial pressure. Note that this equation predicts that the capillary pO_2 level decreases linearly with axial position in the capillary:

$$\langle pO_2(z)\rangle = \langle pO_2\rangle_{\text{in}} - \frac{\Gamma_{\text{metabolic}} H^T}{(1+m)V}\left[\left(\frac{r_T}{r_C}\right)^2 - 1\right]z. \tag{6.57}$$

For anoxic conditions, we can simply replace r_T by $r_{\text{anoxic}}(z)$ in Equation 6.57.

At a given value of z, we can substitute the anoxic version of Equation 6.57 into Equation 6.55 to obtain the axial dependence of $r_{\text{anoxic}}(z)$. This would be the radial position for a given z at which the tissue pO_2 is equal to zero:

$$\left(\frac{r_{\text{anoxic}}(z)}{r_C}\right)^2 \ln\left(\frac{r_{\text{anoxic}}(z)}{r_C}\right)^2 - \left(\frac{r_{\text{anoxic}}(z)}{r_C}\right)^2 + 1$$

$$= \left[\frac{4D_T\langle pO_2\rangle_{\text{in}}}{r_C^2 \Gamma_{\text{metabolic}} H^T}\right] - \frac{4D_T}{(1+m)r_C^2 V}\left[\left(\frac{r_{\text{anoxic}}(z)}{r_C}\right)^2 - 1\right]z. \tag{6.58}$$

Of particular interest are the conditions under which anoxia first begins. Anoxia will first start at the corner of the Krogh tissue cylinder represented by the coordinates $z = L$ and $r = r_T$. This critical tissue pO_2 level, equal to zero at the *anoxic corner*, can be found by first solving Equation 6.57 for the pO_2 level in the blood exiting the capillary. Equation 6.53 can then be solved with $r = r_T$, to find the value of the pO_2^T at the anoxic corner. One can then adjust various parameters, such as V and $\Gamma_{\text{metabolic}}$, until pO_2^T equals zero at the anoxic corner.

The Krogh tissue cylinder approach assumes that we know the Krogh tissue cylinder radius (r_T). In Chapter 5, we showed that the blood perfusion rate can be related to the Krogh tissue cylinder radius, the capillary radius and length, and the average blood velocity. This relationship is given by Equation 5.129, which when solved for r_T, gives the following equation:

$$r_T = r_C \sqrt{\frac{V}{q_b L}}. \tag{6.59}$$

This completes our analysis of tissue oxygenation in the region surrounding a given capillary. Example 6.10 illustrates how these equations can be solved to obtain the pO_2 levels within the capillary and the tissue region that surrounds the capillary.

Example 6.10

Calculate the oxygen profiles within a capillary and its surrounding tissue for an islet of Langerhans. Assume that blood enters the capillary at $pO_2 = 95$ mmHg. Also, calculate the critical blood perfusion rate. Determine the anoxic boundary for a blood perfusion rate of 0.25 mL cm^{-3} min^{-1}.

Solution

We use Equations 6.53, 6.54, 6.57, and 6.58 to obtain the solution for the islet tissue perfusion rate of 4 mL cm^{-3} min^{-1}. Note that the value of m is based on the average pO_2 of the blood, which must be found by trial and error. This means that one must first assume an exiting pO_2 in the blood to start the calculation. The average of the entering and exiting blood pO_2s is then used to calculate the value of m by Equation 6.41. The calculations are then performed and the calculated exiting pO_2 is compared to the value previously assumed. This process is repeated until convergence is obtained. Figure 6.11 shows the resulting solution. The solid line represents the change in the capillary pO_2 and the dashed line represents the change in the tissue pO_2 at the Krogh tissue cylinder radius. If the tissue perfusion rate is decreased to about 0.62 mL cm^{-3} min^{-1}, the pO_2 level becomes zero at the anoxic corner. This result is shown as the dashed line in Figure 6.12. If the tissue blood perfusion rate is decreased further, we will develop an anoxic layer of tissue whose

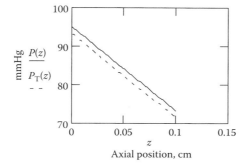

FIGURE 6.11 Oxygen profile in the Krogh tissue cylinder.

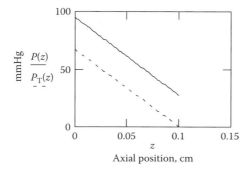

FIGURE 6.12 Oxygen profile in the Krogh tissue cylinder with an anoxic corner.

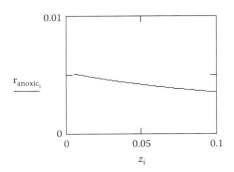

FIGURE 6.13   Anoxic profile in the Krogh tissue cylinder.

axial boundary is defined by $r_{anoxic}(z)$. This boundary can be found by solving Equation 6.58. Figure 6.13 shows the anoxic boundary for the case where the tissue perfusion rate has been decreased to 0.25 mL cm^{-3} min^{-1}.

6.12 ARTIFICIAL BLOOD

Blood is crucial for survival and is frequently needed for treatment of life-threatening injuries, to replace blood loss during surgery, and to treat a variety of blood disorders. The world demand for blood amounts to over 100 million units* per year. Within the United States, a blood transfusion occurs about every 4 sec (Lewis 1997). Blood banking provides, in most cases, an immediate source of blood to meet these needs. However, blood banking requires an extensive infrastructure for collection, testing for disease, cross matching, and storage. In addition to potential contamination of the blood supply with infectious agents, such as HIV and hepatitis, blood itself can cause hemolytic transfusion reactions. The risks associated with blood transfusions range from 0.0004%[†] for HIV infection, 0.001% for a fatal hemolytic reaction, 0.002%–0.03% for contracting the hepatitis virus, and 1% for a minor reaction (Intaglietta and Winslow 1995).

Concern in recent years about the safety of the blood supply has resulted in major efforts to develop a substitute for blood. A blood substitute can replace human donor blood and potentially be used for trauma and surgery, as well as for treatment of chronic blood disorders that require frequent transfusions. Blood substitutes are also of interest to the military because of the potential for a significant decrease in the special handling and logistical requirements of human blood on the modern battlefield.

An artificial blood must meet the two most important functions of blood given by transfusion. These are the replacement of lost plasma and the ability to transport sufficient amounts of oxygen. Hence, blood substitutes can be categorized as either *volume expanders* and/or *oxygen carriers* or *oxygen therapeutics*. Blood substitutes, known as volume expanders, are inert aqueous materials that carry no more than dissolved oxygen and these are used to increase blood volume to replace lost plasma. Examples of volume expanders include donated plasma, Ringer's solution, and 5% dextrose. Blood substitutes, known as oxygen therapeutics, also serve as volume expanders and have the capability to transport significant amounts of oxygen. These materials are designed to mimic the oxygen transport properties of human blood. The most promising oxygen therapeutics tend to be based on the use of compounds with enhanced oxygen solubility, like perfluorocarbons (PFCs) (Castro and Briceno 2010), or solutions that contain hemoglobin or hemoglobin-like compounds. The solubility of oxygen in solutions of PFCs can be a hundred times higher than that in plasma. Oxygen therapeutics that use hemoglobin are either using solutions containing

---

* A unit of blood is 1 pint equal to about 500 mL.
[†] 1% is equal to 1 transfusion incident per 100 units transfused.

stroma-free* hemoglobin or camouflaged RBCs. Many products based on these approaches are in clinical trials and utilize either some form of hemoglobin, PFCs, or camouflaged RBCs (Winslow 1997; Scott et al. 1997; Goorha et al. 2003).

Both of these general approaches have their advantages and disadvantages and, regardless, a successful oxygen therapeutic must be as readily available as donated blood. Furthermore, these materials need to be universal or compatible across all blood types. They also must have minimal side effects, not cause allergic reactions, and be free of infectious agents. In order to be competitive in the marketplace, an oxygen therapeutic should also be relatively inexpensive and have a long storage life.

Several products are based on the use of stroma-free hemoglobin and are also known as *hemoglobin-based oxygen carriers* (HBOCs). The hemoglobin can be derived from a variety of sources. For example, hemoglobin can be of human or bovine origin, it may be derived from bacteria via recombinant DNA techniques, or transgenic methods in animals and plants. Hemoglobin, as discussed earlier, consists of four polypeptide chains, two α chains and two β chains. The hemoglobin molecule outside the RBC is not stable. For example, if hemoglobin is diluted in comparison to its concentration in RBCs, it will spontaneously dissociate into these smaller chains. These lower molecular weight chains are then rapidly removed from the circulation and excreted by the kidneys. Also recall that hemoglobin by itself has a very high affinity for oxygen. Therefore, to be effective as a blood substitute, artificial bloods based on hemoglobin must chemically modify the hemoglobin in order to improve its stability and to decrease its affinity for oxygen. Therefore, the P_{50} value obtained for artificial blood based on hemoglobin needs to be comparable to that of normal human blood.

PFCs are hydrocarbons in which the bonds between C and H are replaced by the much stronger C and F bonds. These strong C–F bonds give PFCs their chemical inertness (Shah and Mehra 1996). PFCs also exhibit high oxygen solubility, they are for the most part biocompatible, and have a low cost. However, a disadvantage of PFCs is that they are immiscible with water. In order to be used as a blood substitute, they must first be emulsified into tiny droplets, typically in the range of 0.1–0.2 μm in diameter. The droplets are also frequently coated with phospholipids derived from egg yolk to stabilize the emulsion. Oxygenated blood carries about 20 mL of oxygen (based on 37°C and 760 mmHg) per 100 mL of blood. A pure PFC solution in equilibrium with oxygen at 760 mmHg has an average oxygen solubility of about 50 mL of oxygen per 100 mL of PFC (Gabriel et al. 1996). PFCs also exhibit a linear equilibrium relationship between oxygen solubility and pO_2. Unlike blood containing hemoglobin, PFCs do not saturate and they can therefore transport more oxygen by simply increasing the pO_2 level in the gas breathed by the patient. Because of their relatively high vapor pressure, a portion of PFCs are removed from the circulation by evaporation via respiration.

Example 6.11

Show that oxygenated blood carries about 20 mL of oxygen (based on 37°C and 760 mmHg) per 100 mL of blood.

Solution

First, we need to calculate the gas density of oxygen at these conditions, as shown by the following calculation:

$$\rho_{oxygen} = \frac{P}{RT} = \frac{760 \text{ mmHg} \times \dfrac{1 \text{ atm}}{760 \text{ mmHg}}}{0.0826 \dfrac{\text{atmL}}{\text{molK}} \times 310 \text{ K}} = 0.0391 \frac{\text{mol}}{\text{L}}.$$

---

* Stroma-free in this case means without the RBC and its cellular framework.

From Table 6.1, oxygenated blood has a total oxygen (dissolved and bound to hemoglobin) concentration of 8620 μM. We can then recast this result in terms of the volume of oxygen as a gas that is dissolved in 100 mL of blood as

$$V_{oxygen} = 8630 \ \mu M \times \frac{\mu mol \ oxygen}{\mu M \times L \ blood} \times \frac{1 \ L \ blood}{1000 \ mL \ blood} \times \frac{100 \ mL \ blood}{100 \ mL \ blood} \times \frac{1 \ mol \ oxygen}{10^6 \ \mu mol}$$

$$\times \frac{1 \ L \ oxygen}{0.0391 \ mol \ oxygen} \times \frac{1000 \ mL \ oxygen}{L \ oxygen}$$

$$= 22.1 \frac{mL \ oxygen}{100 \ mL \ blood}.$$

Example 6.12

Based on the solubility of oxygen in a pure solution of PFC, show that the Henry constant is equal to about 0.04 mmHg μM⁻¹. Also show that a 40 vol% PFC emulsion when saturated with pure oxygen at 760 mmHg has the same oxygen-carrying capacity as saturated blood.

Solution

As previously discussed, the oxygen solubility in 100 mL of a pure solution of PFC is 50 mL at 37°C and 760 mmHg. Using the density of oxygen at the conditions found in Example 6.11, we can calculate the Henry's constant as

$$H = \frac{760 \ mmHg}{\dfrac{50 \ mL \ oxygen}{100 \ mL \ blood} \times \dfrac{1000 \ mL \ blood}{1 \ L \ blood} \times \dfrac{1 \ L \ oxygen}{1000 \ mL \ oxygen} \times \dfrac{0.0391 \ mol \ oxygen}{L \ oxygen} \times \dfrac{10^6 \ \mu mol \ oxygen}{1 \ mol \ oxygen}}{L}$$

$$H = 0.039 \frac{mmHg}{\mu M}$$

Since we are given that there are 50 mL of oxygen in 100 mL of a pure PFC solution, then if we have a solution that is 40% by volume PFC, the oxygen-carrying capacity is reduced to 20 mL of oxygen in 100 mL of this solution, which is comparable to the oxygen-carrying capacity of blood.

Example 6.13

An artificial blood-like fluid is flowing at the rate of 250 mL min⁻¹ through the lumens of a hollow fiber bioreactor that contains hepatoma cells on the shell side. The Henry's constant for dissolved oxygen in the aqueous portion of the artificial blood is 0.85 mmHg μM⁻¹. The artificial blood also contains an insoluble PFC material at a volume fraction of 0.60, which forms an emulsion that has an enhanced solubility for oxygen. The Henry's constant for this material is equal to 0.04 mmHg μM⁻¹, where μM refers to the volume of the PFC oxygen-carrying material only. The pO_2 of the entering artificial blood is 160 mmHg. The exiting pO_2 of the artificial blood is 35 mmHg. Estimate the oxygen transport rate to the hepatoma cells in micromoles per second and in milliliters of oxygen (37°C and 760 mmHg) per minute.

Solution

The artificial blood can be considered a fluid where oxygen is dissolved in both the water phase and in the emulsion phase that has enhanced oxygen solubility. The difference between the amount of oxygen in the artificial blood entering the bioreactor and leaving is equal to the amount of oxygen transported to the hepatoma cells. Once again, we can modify Equation 6.10 and replace $\Gamma_{metabolic}$ with the oxygen transport to the hepatoma cells, i.e., Q_{oxygen}, and replace q with the total flow of artificial blood through the device, i.e., $Q_{artificial\,blood}$. For this situation, we can then write the oxygen material balance between the entrance and the exit of the membrane blood oxygenator as follows, where only the oxygen dissolved in the aqueous and emulsion phases needs to be considered:

$$Q_{oxygen} = Q_{artificial\,blood}\left[\left(0.6C_{aqueous} + 0.4C_{emulsion}\right)_{in} - \left(0.6\,C_{aqueous} + 0.4C_{emulsion}\right)_{out}\right].$$

In this equation, Q_{oxygen} is the amount of oxygen transported from the blood to the cells and $Q_{artificial\,blood}$ is the blood flow rate through the bioreactor. Note that in the parenthetical terms, the flow rate of the artificial blood is multiplied by the respective volume fraction of the artificial blood that is the aqueous phase or the emulsion phase. We can then express the dissolved oxygen concentrations in each of these phases using Henry's law as given by Equation 6.1:

$$Q_{oxygen} = 250\frac{mL}{min} \times \frac{1\,min}{60\,sec} \times \frac{1\,L}{1000\,mL} \times \left[\left(0.40 \times \frac{160\,mmHg}{0.85\frac{mmHg}{\mu M}} + 0.60 \times \frac{160\,mmHg}{0.04\frac{mmHg}{\mu M}}\right)_{in} - \left(0.40 \times \frac{35\,mmHg}{0.85\frac{mmHg}{\mu M}} + 0.60 \times \frac{35\,mmHg}{0.04\frac{mmHg}{\mu M}}\right)_{out}\right]$$

$$\times \frac{1\,\mu\,mol}{\frac{L}{\mu M}},$$

$$Q_{oxygen} = 8.06\frac{\mu mol}{sec}.$$

Using the density of oxygen at 37°C and 1 atm from Example 6.11, we can find the volumetric flow rate of oxygen that was transported to the hepatoma cells, as shown by the following calculation:

$$\dot{V}_{oxygen} = 8.06\frac{\mu mol}{sec} \times \frac{60\,sec}{min} \times \frac{1\,L}{0.0391\,mol} \times \frac{1\,mol}{10^6\,\mu\,mol} \times \frac{1000\,mL}{L},$$

$$\dot{V}_{oxygen} = 12.36\frac{mL}{min}.$$

Artificial bloods based on modified hemoglobin or PFCs still face some significant development problems. Key among these problems are gastrointestinal complaints and vasoconstriction (reduction in blood vessel diameter) when using hemoglobin products, and flu-like symptoms and thrombocytopenia (very low quantity of platelets) when using PFCs. These artificial bloods also exhibit short lifetimes in the circulation. Modified hemoglobins last on the order of 12 h to 2 days, whereas PFCs are removed in about 12 h. By comparison, normal RBCs have lifetimes of about 120 days. Until the lifetimes of these products can be significantly improved, they will only be of use in acute situations such as trauma and surgery (Winslow 1997).

Another promising approach that may overcome some of the problems of artificial blood based on modified hemoglobin or PFCs (Scott et al. 1997) is to covalently bind substances like polyethylene glycol (PEG) to the surface of intact RBCs. The RBCs do not appear to be affected by the presence of the PEG coat. The PEG molecules have the effect of masking or camouflaging the surface antigens on the RBC that lead to the various blood types in the case of human blood or that trigger rejection in the case of animal RBCs. The result is a universal blood type that could allow the use of unmatched human or animal RBCs. The camouflaged RBCs should have a lifetime that is comparable to that of a normal RBC.

PROBLEMS

1. The only life that is present on the planet *Biophilus* is a small, slimy flat worm that is about 200 mm in length and about 20 mm in width. The worm survives through metabolic processes that are limited by the transport of methane present in the planet's atmosphere. The metabolic rate of methane consumption is estimated to be 15 μm sec^{-1} based on total flesh volume. The effective diffusivity of methane in the flesh of the worm is estimated to be about 4×10^{-6} cm^2 sec^{-1}. The concentration of methane in the planet's atmosphere was found to be 0.20 mol L^{-1} and the molar equilibrium solubility ratio of methane in the flesh of this worm relative to atmospheric methane is 0.10. Estimate the maximum thickness of this flat worm in millimeters.

2. Determine the volumetric tissue oxygen consumption rate ($\Gamma_{metabolic}$, μM sec^{-1}) from the data shown in Figure 6.1 for Brockman bodies. Carefully state all assumptions.

3. Calculate the external mass transfer coefficient (k_m, cm sec^{-1}) for oxygen from the data shown in Figure 6.1. Recall that the flux of oxygen into the spherical Brockman body by diffusion must equal the oxygen transport rate from the bulk fluid as described by the following equation:

$$-D_{effective} \left. \frac{dC_{tissue}}{dr} \right|_{surface} = k_m (C_{bulk} - C_{tissue}|_{surface}) \approx \frac{1}{3} \pi R_{Brockman\,body} \Gamma_{metabolic}.$$

Show that for the case without convection, the Sherwood number (Sh) is equal to 2.

4. Derive Equation 6.18.

5. For the situation described in Example 6.5, determine the critical loading of islets, i.e., the void volume of the tissue space for which the pO$_2$ becomes equal to zero at the centerline of the symmetric islet layer.

6. For the results obtained in Example 6.5, determine an estimate of the blood flow rate needed to sustain a total of 750,000 islets. Each islet may be assumed to be 150 μm in diameter. What would be the diameter of such a device?

7. For the situation described in Example 6.5, determine the critical oxygen permeability of the membrane, i.e., the value for which pO$_2$ becomes equal to zero at the centerline of the symmetric islet layer.

8. Write a short paper that discusses blood types.

9. Derive Equation 6.52.

10. Derive Equation 6.56.

11. Develop a Krogh tissue cylinder model for blood oxygenation using PFCs. Using this model, determine the tissue oxygen profile for an artificial blood perfusion rate of 0.75 mL cm^{-3} min^{-1}.

12. Consider a long and wide planar aggregate or slab of hepatocytes (liver cells) used in a bioartificial liver. The slab is suspended in a growth medium at 37°C with no mass transfer limitations between the slab and the medium. The medium is saturated with air, giving a pO$_2$ of 160 mmHg or 190 μM. The cell density in the slab is 1.25×10^8 cells cm^{-3}. Each cell is spherical in shape and 20 μm in diameter. The oxygen consumption rate of the cells may be described by the Michaelis–Menten relationship (Equation 6.12) with $V_{max} = 0.4$ nmol 10^{-6} cells sec^{-1} and $K_m = 0.5$ mmHg. Estimate the maximum half-thickness of the slab of hepatocytes.

13. Photodynamic therapy (PDT) involves the localized photoirradiation of dye-sensitized tissue toward the end goal of causing cell death in tumors. The tumor containing the photosensitizer is irradiated by a laser with light of the proper wavelength (around

630 nm) to generate excited singlet molecules of the sensitizer. In the presence of molecular oxygen, the singlet sensitizer forms a very reactive form of oxygen called *singlet oxygen*, which destroys the tumor. Therefore, sufficient levels of molecular oxygen are needed in the tissue to kill the tumor by PDT. The action of PDT on tissue oxygen levels is like having an additional sink for oxygen that is dependent on the laser fluence rate, ϕ_0. Foster et al. (1993) states that this PDT-induced oxygen consumption rate may be estimated by

$$\Gamma_{PDT} = 0.14(\mu M \; cm^2 \; sec^{-1} \; mW^{-1}) \times \phi_0 (mW \; cm^{-2}),$$

where ϕ_0 is the fluence rate.

Consider a typical capillary within the tumor undergoing PDT treatment. Assume a capillary diameter of 8 μm and a capillary length of 300 μm. The Krogh tissue cylinder radius is assumed to be 40 μm. The metabolic oxygen consumption rate is 11.5 μm sec^{-1}. The average velocity of blood in the capillary is 0.04 cm sec^{-1}. The entering blood pO$_2$ is 95 mmHg. Answer the following questions:

a. What is the exiting pO$_2$ of the blood from the capillary with the laser off?
b. What is the pO$_2$ in the lethal corner ($z = L$ and $r = r_T$) with the laser off?
c. At what fluence rate is the pO$_2$ in the lethal corner equal to zero? Note, this is the point at which PDT starts to be ineffective for those cancer cells at the outer periphery of the Krogh tissue cylinder.

14. Consider the design of a hollow fiber unit for an extracorporeal bioartificial liver. Blood flows through the lumens of the hollow fibers contained within the device at a flow rate of 400 mL min^{-1}. The unit consists of 10,000 fibers and the lumen diameter of a fiber is 400 μm. The blood enters the fiber with a pO$_2$ of 95 mmHg and must exit the device with a pO$_2$ no lower than 40 mmHg. Surrounding each fiber is a multicellular layer of cloned human liver cells. These cells have a void fraction of 80% and they consume oxygen at the rate of 17.5 μM sec^{-1} (based on cellular volume). The hollow fibers are 25 cm in length and the fiber wall provides negligible resistance to the transport of oxygen. Determine the maximum thickness of the layer of cells that can surround each fiber. Carefully state your assumptions.

15. A nonwoven mesh of polyglycolic acid (PGA) is proposed to serve as the support for growing chondrocytes in vitro for the regeneration of cartilage. The PGA mesh is a square pad, 1×1 cm and 0.5 mm thick. The PGA mesh has a porosity of 97% and the edges of the mesh are clamped within a support. The chondrocytes were found to consume oxygen at a rate of 2.14×10^{-13} mol cell^{-1} h^{-1} (based on cellular volume). Assuming that the cells are cultured in a well-perfused growth media that is saturated with air, estimate the maximum number of cells that can be grown within each PGA mesh. Assume that the diameter of a cell is about 20 μm. Carefully state your assumptions.

16. Blood perfuses a region of tissue at a flow rate of 0.35 mL min^{-1} cm^{-3} of tissue. The pO$_2$ of the entering blood is 95 mmHg and the exiting pO$_2$ of the blood is 20 mmHg. Calculate the metabolic oxygen consumption rate of the tissue in μM per second.

17. Consider a slab layer of cells being grown within an artificial support structure. The layer of cells is immersed in a well-mixed nutrient medium maintained at an oxygen pO$_2$ of 150 mmHg. The cells are known to consume oxygen at the rate of 40 μM sec^{-1}. Estimate the maximum half-thickness of the cell layer (centimeters), assuming that the void volume fraction in the tissue layer is 0.20.

18. A tumor spheroid, 400 μm in diameter, is suspended in an infinite and quiescent media at 37°C with a pO$_2$ of 120 mmHg. The pO$_2$ at the surface of the spheroid was measured to be 100 mmHg. Estimate the oxygen consumption rate of the cells in the spheroid in microns per second. Assume that $H = 0.74$ mmHg μM^{-1}.

19. Blood flows through a membrane oxygenator at a flow rate of 5000 mL min^{-1}. The entering pO$_2$ of the blood is 40 mmHg and the exiting blood pO$_2$ is 95 mmHg. Calculate the amount of oxygen transported into the blood in micromoles per second and in milliliters of oxygen per minute at 37°C and 1 atm.

20. Consider a slab layer of cells being grown between two microporous support membranes. The half-thickness of the cell layer is 125 μm and the void volume fraction in the cell layer is 0.90. The permeability of oxygen through the support membrane is estimated to be equal to 1.5×10^{-3} cm sec^{-1}. The layer of cells is immersed in a well-mixed nutrient medium maintained at an oxygen pO$_2$ of 150 mmHg. An oxygen microelectrode placed at the centerline of the layer of cells gives a pO$_2$ reading of 15 mmHg. Estimate the rate at which these cells are consuming oxygen in μM per second.

21. Blood is flowing at the rate of 200 mL min^{-1} through the lumens of a hollow fiber unit containing hepatocytes on the shell side. The pO$_2$ of the entering blood is 95 mmHg and the exiting pO$_2$ of the blood is 20 mmHg. The volume of hepatocytes on the shell side of the device is estimated to be about 600 mL. Estimate the metabolic oxygen consumption rate of the hepatocytes in μM per second.

22. Consider a slab layer of cells being grown within an artificial support structure. The layer of cells is immersed in a well-mixed nutrient medium maintained at an oxygen pO$_2$ of 130 mmHg. The cells are known to consume oxygen at the rate of 10 μM sec^{-1} and the half-thickness of the slab of cells is 35 μm. Estimate the void volume fraction in the tissue.

23. Blood perfuses a region of tissue at a flow rate of 0.50 mL min^{-1} cm^{-3} of tissue. The pO$_2$ of the entering blood is 95 mmHg and the exiting pO$_2$ of the blood is 30 mmHg. Calculate the metabolic oxygen consumption rate of the tissue in μM per second.

24. Consider a slab layer of cells being grown within an artificial support structure. The layer of cells is immersed in a well-mixed nutrient medium maintained at an oxygen pO$_2$ of 140 mmHg. The cells are known to consume oxygen at the rate of 30 μM sec^{-1}. Estimate the maximum half-thickness of the cell layer (microns), assuming that the void volume fraction in the tissue layer is 0.20.

25. In a tissue-engineered vascularized tissue construct, pO$_2$ measurements were taken in vivo in the region equidistant from the capillaries, using luminescent oxygen-sensitive dyes. This pO$_2$ value, which is basically at the Krogh tissue cylinder radius, was found to be 10 mmHg and the average concentration of oxygen in the blood in the capillaries was 100 μM. Histological analysis of the tissue samples indicated that the capillaries were pretty much parallel to each other. The average distance between the capillaries, measured center to center, was found to be 130 μm. The capillaries themselves are 7 μm in diameter and the rate of oxygen uptake for the cells surrounding the capillaries is estimated to be 30 μM sec^{-1}. From these data, estimate the diffusivity of oxygen through the tissue surrounding the capillaries.

26. Islets of Langerhans are sequestered from the immune system in a device similar to that shown in Figure 6.6. The pO$_2$ of the blood in the capillaries adjacent to the immunoisolation membrane is 40 mmHg. The membrane oxygen permeability is 9.51×10^{-4} cm sec^{-1}. If the islets consume oxygen at the rate of 25.9 μM sec^{-1}, estimate the maximum half-thickness of the islet tissue in centimeters, assuming a void volume in the islet layer of ε = 0.95.

27. A laboratory scale bioartificial liver consists of a single hollow fiber that is 500 μm in diameter and 25 cm in length. The flow rate of blood through the fiber is 0.2 mL min^{-1}. The blood enters the fiber with a pO$_2$ of 95 mmHg and exits the fiber at a pO$_2$ of 35 mmHg. Surrounding the hollow fiber is a confluent layer of cloned human liver cells that consume oxygen at the rate of 25 μM sec^{-1} (based on cell volume). Assume that the hollow fiber wall provides negligible resistance to the transport of oxygen. Determine the maximum thickness of the layer of cells that can surround the hollow fiber.

28. A laboratory scale bioartificial liver consists of 10,000 hollow fibers that are 500 μm in diameter and 25 cm in length. The blood enters the fibers with a pO$_2$ of 95 mmHg and exits the fibers at a pO$_2$ of 35 mmHg. Surrounding each of the hollow fibers is a single confluent layer of cloned human liver cells that consume oxygen at the rate of 25 μM sec^{-1}. The thickness of the layer of cells is 25 μm. Assume that the hollow fiber wall provides negligible resistance to the transport of oxygen. Determine the total flow rate of the blood in milliliters per minute needed to provide these conditions.

29. Design a planar disk bioartificial pancreas for the treatment of diabetes (what are the dimensions, thickness, and radius?). Assume that the device will contain a total of 50 million genetically modified human β cells. Each of these cells has a diameter of 15 μm and when packed into the device, the void volume (ε) must be 0.15. The immunoisolation membrane has a thickness of 50 μM and a porosity of 0.80. Assume that the cells consume oxygen at the rate of 25 μM sec^{-1} and that the oxygen pO_2 in the blood adjacent to the immunoisolation membrane is 40 mmHg.

30. Rework Problem 29 assuming that there are 500 million cells.

31. Generate the cell surface oxygen concentration profiles in the bioreactor studied by Allen and Bhatia (2003) for an inlet flow rate of 0.5 cm^3 min^{-1} and inlet oxygen concentrations from 90 to 190 nmol cm^{-3}. Other data may be found in Example 6.8.

32. Starting with Equation 6.3, derive Equation 6.6 where Y is given in terms of P_{50}, n, and pO_2. Also show how Equation 6.6 can be rearranged to give Equation 6.7.

33. Show the calculations behind the results presented in Example 6.10.

34. Write a short paper that reviews the current state of artificial blood development.

35. Blood is flowing at the rate of 250 mL min^{-1} through the lumens of a hollow fiber unit that contains hepatoma cells on the shell side. The pO_2 of the entering blood is 95 mmHg. The volume of the hepatoma cells is 500 mL. The hepatoma cells consume oxygen at the rate of 20 μM sec^{-1} (based on the cell volume). Show that the exiting pO_2 of the blood is about 37.5 mmHg.

36. Consider the oxygen transport within a flat, disk-shaped tumor of total thickness 250 μm. The surfaces of the tumor are covered with capillaries such that the oxygen concentration at the surface of the tumor is equal to 120 μM. A microelectrode placed at the center of the tumor showed that the pO_2 was equal to 15 mmHg and a tissue sample indicated that the volume fraction of the tumor cells was equal to 0.45. From these data, estimate the oxygen consumption rate of the tumor cells in μM per second.

37. A blood oxygenator for a small child needs to deliver 0.5 mol h^{-1} of oxygen to the blood. Assuming that the blood enters the oxygenator at a pO_2 of 40 mmHg and leaves at 95 mmHg, estimate the required flow rate of blood (liters per minute) through this oxygenator.

7 Pharmacokinetic Analysis

7.1 TERMINOLOGY

In this chapter, we will focus our attention on some useful techniques in the field of pharmacokinetics. *Pharmacokinetics* is the study of the processes that affect drug distribution and the rate of change of drug concentrations within various regions of the body. These processes are also collectively referred to as ADMET for drug *a*dsorption, *d*istribution, *m*etabolism, *e*xcretion, and *t*oxicity. Although pharmacokinetics is of utmost importance in the treatment of diseases using drugs, our major focus in the remainder of this book will be on using this technique as a tool for understanding the transport processes in drug delivery systems and between an artificial device and the body.

We also need to distinguish pharmacokinetics from similar terms. Therefore, *pharmaceutics* concerns the formulation and preparation of the drug to achieve a desired drug availability within the body; *pharmacodynamics* is concerned with the time course of the treatment response that results from a given drug. For example, how is the cholesterol concentration in the plasma affected by daily doses of a particular drug? The actual physiological reason for the response that results from a drug is the subject of *pharmacology*.

7.2 ENTRY ROUTES FOR DRUGS

There are two routes through which a drug can enter the body. The *enteral* route refers to drugs that are given via the gastrointestinal tract (GI tract). All other routes are called *parenteral*. The enteral route includes drugs that are absorbed via one or more of the following components of the GI tract: the buccal cavity (mouth) and sublingually (beneath the tongue), gastrically (stomach), intestinally (small and large intestines), and rectally. Once the drug is absorbed from the GI tract, the drug enters the blood and is distributed throughout the body. It is important, however, to point out that only drugs absorbed from the buccal cavity and the lower rectum enter the *systemic circulation* directly. Drugs absorbed from the stomach, intestines, colon, and upper rectum enter the *splanchnic circulation*. The splanchnic circulation then takes the drug to the liver via the portal vein and after leaving the liver, the drug enters the systemic circulation. Since the liver contains many enzymes capable of degrading the drug (metabolism), a significant portion of the drug may be removed during this first pass through the liver before the drug is available to the general circulation. Table 7.1 summarizes all of the other routes for drug administration that are called parenteral. For the most part, drugs given parenterally enter the systemic circulation directly.

As the drug is being absorbed into the body, its presence will be noticed within the circulation as its concentration in the plasma portion of the blood changes. Recall that plasma refers to the clear supernatant fluid that results from blood after the cellular components have been removed. If the blood sample is simply allowed to clot, then the resulting clear fluid is referred to as serum, since the clotting proteins have been removed by the clotting process. For the most part, the concentration of drug in plasma or serum is identical and no distinction is needed. It is usually only the plasma drug concentration in the body that is known, and this is what is meant when we speak of the concentration of the drug. For the most part, it has been found that the physiological activity of most drugs can be related to the plasma concentration of the drug. Pharmacokinetic modeling is then used to predict and/or adjust the dosing strategies for a particular drug.

TABLE 7.1
Parenteral Routes for Drug Administration

Route	Result
Intravenous (IV)	Introduced directly into the venous circulation
Intramuscular	Within the muscle
Subcutaneous	Beneath the epidermal and dermal skin layers
Intradermal	Within the dermis, usually a local effect
Percutaneous	Topical treatment applied to the skin
Inhalation	Mouth/nose, pharynx, trachea, bronchi, bronchioles, alveolar sacs, alveoli
Intraarterial	Introduced directly into an artery, regional drug delivery
Intrathecal	Cerebrospinal fluid, subarachnoid space
Intraperitoneal (IP)	Within the peritoneal cavity
Vaginal	Within the vagina

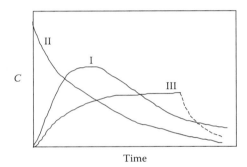

FIGURE 7.1   Plasma drug concentrations following I: absorption, II: rapid intravenous, and III: continuous infusion.

Plasma drug concentration as measured is usually the total plasma concentration (C_{total}), which includes the concentration of the drug that is bound to plasma proteins (C_P) and that which is unbound (C_U). However, the effect of a particular drug and its transport out of the circulatory system is due to the unbound drug concentration. The drug fraction unbound or f_U is defined as ($C_U/(C_P + C_U)$) = (C_U/C_{total}) and f_U for a particular drug is usually constant, so either C_U or C_{total} can be used in a pharmacokinetic model; however, it is important to make sure you know and understand which concentration you are using.

Figure 7.1 illustrates typical plasma drug concentrations as a function of time after their introduction into the body. Curve I illustrates the case where the drug is first slowly absorbed, resulting in increasing plasma concentrations, followed by a plateau, and then the plasma concentration decreases as the drug is eliminated by a variety of body processes. Curve II represents a very rapid injection (bolus) of the drug, usually intravenously (IV injection), where we see that the drug quickly reaches its peak plasma concentration near the time of injection, and then it is slowly eliminated. Curve III results from a continuous infusion, usually intravenously or through a controlled release formulation. In this case, we see that after a short period of time the drug reaches a steady state plasma concentration, since the infusion rate is equal to the elimination rate of the drug. If the infusion is stopped, then the drug plasma concentration falls due to its elimination, as shown by the dashed portion of the curve.

7.3 MODELING APPROACHES

Our goal is then to develop mathematical models that can be used to describe profiles of plasma drug concentration such as those shown in Figure 7.1. This will allow for the description of the drug concentration in the body as a function of time. Through pharmacology and pharmacodynamics, the effects of the drug on the body can then be related to the plasma drug concentration. For example, the physiological response to a particular drug often saturates at high drug concentrations (C) and can be described as follows:

$$R = \frac{R_{max} \, C^n}{C_{50}^n + C^n},\tag{7.1}$$

where R represents the physiological response to the drug, often expressed in terms of a percent change in some key parameter. Common examples include body temperature, heart rate, glucose levels, cholesterol levels, and blood pressure. R_{max} is the saturation response at high drug concentrations, where $C \gg C_{50}$. C_{50} is the drug concentration where $R = 0.50 \, R_{max}$. The values of C result from measurements of the drug concentration or from the solution to a pharmacokinetic model.

Pharmacokinetic models will incorporate several unknown parameters that we find by fitting our models to experimental drug concentration data. The development of pharmacokinetic models has generally followed three approaches—the *compartmental approach*, the *physiological approach*, and the *model independent approach*.

The compartmental approach assumes that the drug distributes into one or more "compartments" in the body, which usually represent a particular region of the body, an organ, a group of tissues, or body fluids. The compartments are assumed to be well-mixed, so that the concentration of the drug within the compartment is spatially uniform. Spatial distribution of the drug within the body is accounted for by the use of multiple compartments. The assumption of well-mixed compartments is usually pretty good, since the cardiac output in humans is nominally 5 L min⁻¹ and the blood volume is about 5 L, giving an effective residence time for one passage through the circulatory system of about 1 min. Also, recall that there is a net filtration of several milliliters per minute of the blood plasma forming the interstitial fluid that bathes and circulates around all the cells in the body. Hence, the mixing time of the body fluids occurs much faster than the observed temporal changes of the drug concentration, which can occur over periods of hours or even days. Since drug distribution occurs over periods of hours, and the body's fluids are moving over periods of minutes, and mass transfer occurs on the order of seconds, the well-mixed assumption is appropriate. Movement of the drug between the compartments is usually described by simple irreversible or reversible first order rate processes, i.e., the rate of transport is proportional to the difference in the drug concentration between compartments.

In physiological models, the movement of the drug is based on the blood flow rate through a particular organ or tissue and includes consideration of the rates of mass transport processes within the region of interest. Experimental blood and tissue drug concentrations are needed to define the model parameters. The Krogh tissue cylinder model, developed in Chapter 6 to describe the changes in oxygen levels within the capillary and the surrounding tissue region, is an example of a physiological model. In this case, however, the "drug" is oxygen. Since the liver plays a major role in the metabolism and excretion of drugs from the body, much attention has been focused on the development of physiological models to describe liver function (Bass and Keiding 1988; Niro et al. 2003).

The model independent approach does not try to make any physiological connection like the compartmental approach or the physiological approach. Instead, one simply finds the best set of mathematical equations that describe the situation of interest (Notari 1987). These models are considered to be linear if the plasma drug concentration can be represented by a simple weighted summation of exponential decays. For example, Equation 7.2 may be used to describe the time course of the plasma drug concentration:

$$C = \sum_{i=1}^{n} C_i e^{-\lambda_i t}, \tag{7.2}$$

where the constants λ_i and C_i are adjusted to provide the best fit to the drug concentration as a function of time.

Our focus in this chapter will be on the use of compartmental models to describe pharmacokinetic data like that shown in Figure 7.1. Compartmental models are commonly used and provide a convenient framework for building a pharmacokinetic model and are conceptually simple to use (Cooney 1976; Gibaldi and Perrier 1982; Welling 1986; Notari 1987; Rowland and Tozer 1995; Tozer and Rowland 2006; Hacker et al. 2009).

7.4 FACTORS AFFECTING DRUG DISTRIBUTION

Before we can investigate some simple compartmental models, we first must discuss the following factors that influence how a particular drug is distributed throughout the body. Table 7.2 summarizes these factors and they are discussed in greater detail in the following discussion.

7.4.1 Drug Distribution Volumes

Recall that Table 3.1 summarized the types of fluids found within the body. Since a drug can only distribute within these fluid volumes, we call these the *true distribution volumes*. If a drug only distributes within the plasma volume of the circulatory system, then the true distribution volume for this drug is about 3 L. If a drug readily penetrates the capillary walls, then it can also distribute throughout the extracellular fluid volume for a total true distribution volume of about 15 L. If a drug can also permeate the cell wall, then it will also be found within the intracellular fluid spaces as well, giving a total true distribution volume of about 40 L.

The rate at which a drug is distributed throughout these fluid volumes is controlled by the rate at which the drug is delivered to a region of interest by the blood, i.e., the tissue blood perfusion rate, and by the rate at which the drug diffuses from within the vascular system into the extravascular spaces. If a drug is lipid soluble, then its mass transfer across the capillary wall is usually not rate limiting and equilibrium is quickly reached between the amount of drug in the tissue region and that found in the blood. In this case, the distribution of the drug in a particular tissue region is *perfusion rate limited*. For lipid insoluble drugs, the capillary membrane permeability controls the rate at which the drug distributes between the blood and tissue regions. The distribution of the drug is then said to be *diffusion rate limited*.

Drugs are also capable of binding to proteins found in the plasma and in the extravascular spaces as well. Albumin is perhaps the most common of the blood proteins that will bind with drugs. Recall

---

TABLE 7.2
Factors Influencing Drug Distribution

Blood perfusion rate
Capillary permeability
Drug biological affinity
Metabolism of drug
Renal excretion

---

that the major difference between plasma and interstitial fluid is the amount of protein that is present. For the most part, plasma proteins have a difficult time diffusing through the capillary wall to enter the extravascular space. Therefore, a drug that binds strongly with plasma proteins will, for the most part, be limited to just the volume of the plasma, i.e., about 3 L. On the other hand, extravascular proteins or cell surface receptors that bind strongly to a drug will tend to accumulate the drug in the extravascular space at the expense of the plasma. In this case, the distribution volume of the drug may appear to be larger than the physical fluid volumes. Thus, protein binding of the drug can have a significant effect on the calculated or *apparent distribution volumes* as the following discussion will show.

As mentioned earlier, it is important to point out that most analytical methods measure the total drug concentration (bound and unbound) that is in the plasma. Since this total plasma concentration (C_{total}) is readily measured, the *apparent distribution volume* ($V_{apparent}$) for the drug is simply given by

$$V_{apparent} = \frac{A}{C_{total}}, \tag{7.3}$$

where A is the total amount of drug within the body. The amount of drug in the body is typically expressed in units of milligrams (mg), and the concentration of drug is in micrograms per milliliter of plasma ($\mu g\ mL^{-1}$), which is the same as milligrams per liter of plasma ($mg\ L^{-1}$), resulting in the apparent distribution volume having units of liters (L).

It is also important to note that the concentration of drug within any particular tissue region may vary significantly from the concentration of drug in plasma. Therefore, although the apparent distribution volume seems to have some physical significance, it really is just a factor having units of volume that, when multiplied by the plasma drug concentration, provides the total amount of drug in the body at a particular time.

The apparent distribution volume calculated by Equation 7.3 can also be vastly different from the true distribution volume (V), related to the volume of the various body fluids, because of the effect of protein binding. To examine the effect of protein binding on the apparent distribution volume, consider the conceptual model shown in Figure 7.2 (Oie and Tozer 1979). In their model, unbound drug is free to distribute between three compartments: the plasma (P), the extracellular fluid outside the plasma (E), and the remainder of the body (R) where the drug is bound to the surface of the cell by cell–surface receptors or internalized by the cell by a variety of cellular transport processes. A mass balance on the total amount of drug in the body (A) at any time may then be written as

$$A = A_P + A_E + A_R, \tag{7.4}$$

where A_P, A_E, and A_R represent the total amount of drug in the plasma, the extracellular fluid, and the remainder of the body as represented by the intracellular fluid volume, respectively.

The amount of drug in any of these three compartments is also related to the corresponding concentration of total drug in a given compartment, i.e., $A_P = C\ V_P$, $A_E = C_E\ V_E$, and $A_R = C_R\ V_R$, where C, C_E, and C_R are the respective total drug concentration in each compartment and V_P, V_E, and V_R are the corresponding compartment volumes. Using these relationships, we can rewrite Equation 7.4 as

$$V = V_P + V_E\left(\frac{C_E}{C}\right) + V_R\left(\frac{C_R}{C}\right), \tag{7.5}$$

where we also recognize that by definition $A = V\ C$ for a particular compartment.

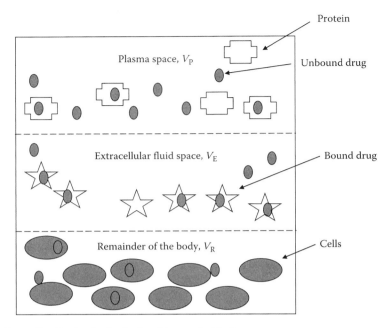

FIGURE 7.2   Drug distribution in the body with protein binding.

Ignoring active transport type processes, it is assumed that at distribution equilibrium the unbound concentration of the drug, i.e., C_U, is the same in all of the compartments. In addition, in the extracellular fluid compartment, we have that $C_E = C_U + C_{EB}$, where C_{EB} is the bound drug concentration in the extracellular fluid. We now define f_U as (C_U/C), which is the fraction of unbound drug in the plasma and we also define f_{UT} as (C_U/C_R), which is the fraction of unbound drug in the extracellular space. Both f_U and f_{UT} are usually determined experimentally by equilibrating a known amount of drug with plasma or extracellular fluid and determining from a drug mass balance the fraction of drug that is unbound. With these relationships, Equation 7.5 becomes

$$V = V_P + V_E\, f_U \left(\frac{C_U + C_{EB}}{C_U} \right) + V_R\, \frac{f_U}{f_{UT}}. \tag{7.6}$$

Unbound drug in the plasma and extracellular spaces can bind with proteins in the plasma (P_P) and extracellular spaces (P_E), as described by the following chemical reactions:

$$C_U + P_P \Leftrightarrow C_{PB} \quad \text{and} \quad C_U + P_E \Leftrightarrow C_{EB}, \tag{7.7}$$

where C_{PB} and C_{EB} represent the concentrations of drug-bound protein in the plasma and extracellular spaces. Assuming that the binding sites on these proteins have the same affinity for the unbound drug, then at equilibrium we can write

$$K_a = \frac{C_{PB}}{C_U\, P_P} = \frac{C_{EB}}{C_U\, P_E}, \tag{7.8}$$

where K_a is the equilibrium or affinity constant. The total protein concentration in the plasma (P_{PT}) and extracellular spaces (P_{ET}) must be equal to the concentration of the respective proteins that are bound to drug (C_{PB} and C_{EB}) plus the concentration of these proteins that are not bound to drug (P_P and P_E). Hence, we can write

$$P_P + C_{PB} = P_{PT} \quad \text{and} \quad P_E + C_{EB} = P_{ET}. \tag{7.9}$$

Solving the previous equations for P_P and P_E and substituting this result into Equation 7.8, taking the inverse of this equation, and solving for C_{EB}, we have

$$C_{EB} = C_{PB} \frac{P_{ET}}{P_{PT}}. \tag{7.10}$$

Recall that P_{ET} and P_{PT} represent the total protein concentrations in the extracellular and plasma spaces, respectively. The total amount of drug-binding protein in the plasma and extracellular spaces equals $P_{PT} V_P$ and $P_{ET} V_E$. $R_{E/I}$ is then defined as the ratio of the total amount of drug-binding protein in the extracellular space to that in the plasma space, which is equal to $(P_{PT}V_P/(P_{ET}V_E))$. Therefore, Equation 7.10 can be written as

$$C_{EB} = C_{PB} R_{E/I} \frac{V_P}{V_E}. \tag{7.11}$$

Now we can substitute Equation 7.11 into Equation 7.6 for C_{EB} and obtain

$$V = V_p + f_U \left(\frac{V_E C_U + C_{PB} V_P R_{E/I}}{C_U} \right) + V_R \frac{f_U}{f_{UT}}. \tag{7.12}$$

Recognizing that $(C_{PB}/C_U) = (1 - f_U)/f_U$, we then obtain what is known as the *Oie–Tozer equation*:

$$V = V_P (1 + R_{E/I}) + f_U V_P \left(\frac{V_E}{V_P} - R_{E/I} \right) + \frac{V_R f_U}{f_{UT}}. \tag{7.13}$$

From Table 3.1, we have for a 70 kg man that the plasma volume, V_P, is ~ 3 L and that the extracellular fluid volume outside the plasma, V_E, is ~ 12 L. The intracellular fluid volume represented by V_R is in the range of 25–27 L. Also, the amount of drug-binding protein in the extracellular fluid space to that in the plasma, $R_{E/I}$, is ~ 1.4. Using these values, Equation 7.13 simplifies to approximately

$$V = 7 + 8 f_U + V_R \frac{f_U}{f_{UT}}. \tag{7.14}$$

Equation 7.14 shows that if a drug is only distributed within the extracellular fluid space and cannot enter the cells, i.e., $V_R = 0$, the apparent distribution volume of the drug then attains its smallest value and is equal to $V_{min} = 7 + 8 f_U$. If, in addition, the drug is not bound to plasma proteins, i.e., $f_U = 1$, then the apparent distribution volume of the drug is limited to 15 L, which is the extracellular fluid volume. On the other hand, if the drug is completely bound ($f_U = 0$), then the distribution volume cannot be less than 7 L and this corresponds to the distribution volume of albumin, which is found in both the plasma and the extracellular compartments. If the drug enters the cells but is not significantly bound ($f_U = 1$ and $f_{UT} = 1$), then from Equation 7.14 the apparent volume of distribution is that of the total body water, i.e., 40–42 L. An example of a drug that distributes throughout the total body water is alcohol. Also note that as $f_{UT} \rightarrow 0$, Equation 7.14 predicts that the apparent distribution volume will increase and exceed that of the total body water volume. It is also important to note that for a high molecular weight drug (>60,000 gmol⁻¹), it is very difficult for the drug to diffuse across the capillary walls and leave the vascular system, and hence the distribution volume becomes that of the plasma volume, i.e., 3 L.

Equations 7.13 and 7.14 are therefore useful for investigating a variety of factors that affect the apparent distribution volume of a drug. Many of these changes can occur as a result of age, disease, injury, and interactions with other drugs.

Example 7.1

A drug has an apparent distribution volume of 35 L and is found to be 95% bound to plasma proteins. Unbound drug is also found to distribute freely throughout the total volume of water found in the body. Estimate the unbound fraction of the drug with the extracellular fluids, i.e., f_{UT}.

Solution

The value of f_U is $1-0.95 = 0.05$. From Equation 7.14, we can solve for the value of f_{UT} as shown by the following calculation:

$$f_{UT} = \frac{f_U V_R}{V - 7 - 8 f_U} = \frac{0.05 \times 27 \text{ L}}{35 \text{ L} - 7 - 8 \times 0.05} = 0.0489.$$

7.4.2 Drug Metabolism

Once the drug is absorbed and distributed throughout the body, a variety of biochemical reactions will begin to degrade the drug. This breakdown of the drug is part of the body's natural defense against foreign materials. Metabolism of the drug is beneficial in the sense that it limits the time of drug action and, in some cases, it produces the active form of the drug. These biochemical reactions, driven by existing enzymes, occur in a variety of organs and tissues. However, the major site of drug metabolism is within the liver; other important sites include the kidneys, lungs, blood, and the GI wall. The enzymatic destruction of the drug reduces its pharmacological activity because the active site related to the drug's molecular structure is destroyed. Also, the *metabolites* that result tend to have increased water solubility, which decreases their capillary permeability and enhances their removal from the body via the kidneys.

Enzymatic reactions that degrade the drug in the tissues tend to follow the Michaelis–Menten rate model,* where the reaction rate is given by the following expression:

$$r_{\text{metabolic}} = -\frac{V_{\text{max}} C}{K_{\text{m}} + C}. \tag{7.15}$$

Usually, the drug concentration is much smaller than the value of K_{m}. Therefore, the rate at which the drug is degraded by metabolic reactions can be represented by an irreversible first order rate law, where $k_{\text{metabolic}} = V_{\text{max}}/K_{\text{m}}$, i.e:

$$r_{\text{metabolic}} = -k_{\text{metabolic}} C. \tag{7.16}$$

In this equation, C represents the total plasma concentration of the drug, $k_{\text{metabolic}}$ represents the first order rate constant for the metabolic reactions, and $r_{\text{metabolic}}$ represents the rate at which the drug is

---

* A derivation of the Michaelis–Menten model can be found in Section 8.3.3.

degraded on a per unit volume basis. We will use this expression later when we write mass balances for the drug using compartmental models.

7.4.3 RENAL EXCRETION OF THE DRUG

The kidneys also have a major role in the elimination of the drug from the body. Part of this elimination process is enhanced by the enzymatic degradation of the drug by the liver and the formation of more water-soluble products. The kidneys receive about 1100 mL min^{-1} of blood or about 22% of the cardiac output. On a per mass basis, they are the highest perfused organ in the body. Their primary purpose is to remove unwanted metabolic end products, such as urea, creatinine, uric acid, and urates from the blood, and to control the concentrations of such ions as sodium, potassium, chloride, and hydrogen. The concentrated urine that is formed by the kidneys amounts to about 1.5 L of urine per day.

Figure 7.3 illustrates the basic functional unit of the kidney called the *nephron*. Each kidney contains about one million nephrons and each nephron contributes to the formation of urine. The operation of the kidney can be explained by considering how an individual nephron functions. The nephron consists of the following major components: the *glomerulus* contained within *Bowman's capsule*, the *convoluted proximal tubule*, *Henle's loop*, the *convoluted distal tubule*, and, finally, the *collecting tubule* that branches into the *collecting duct*. The glomerulus, proximal tubule, and distal tubule have a major role in drug elimination from the kidney.

Blood enters the glomerulus via the *afferent arteriole* and leaves by the *efferent arteriole*. The glomerulus consists of a tuft of highly permeable fenestrated capillaries, several hundred times more permeable to solutes than the typical capillary. *Glomerular filtrate* is formed across these capillaries within Bowman's capsule due to hydraulic and osmotic pressure differences that exist

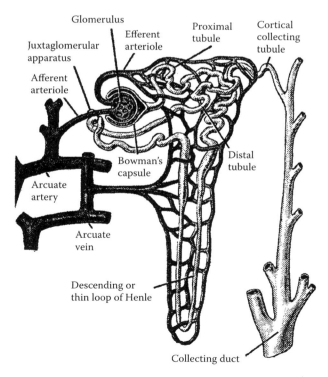

FIGURE 7.3 The nephron. (From Guyton, A.C., *Textbook of Medical Physiology*, W.B. Saunders, Philadelphia, PA, 1991. With permission.)

between the fluids within the glomerular capillaries and the Bowman space. The glomerular capillary membrane for the most part is impermeable to the plasma proteins, retaining those molecules with a molecular weight larger than about 69,000. Therefore, drugs that are bound to plasma proteins cannot be removed from the bloodstream by filtration through the glomerulus. However, unbound drugs with a molecular weight less than 69,000 are readily filtered out of the bloodstream with the glomerular filtrate. The *glomerular filtration rate* (GFR) is the sum total of all the glomerular filtrate formed per unit time by the nephrons in the two kidneys. This is about 125 mL min^{-1} or about 180 L each day, which is more than twice the weight of the body. Clearly, this volume of fluid must be recovered and we find that over 99% of the glomerular filtrate is reabsorbed in the tubules. The filtration rate of unbound drug in the glomerulus is given by

$$\text{Unbound drug removal rate}\big|_{\text{glomerulus}} = \text{GFR} \times C_U = f_U \times \text{GFR} \times C. \qquad (7.17)$$

After the glomerular filtrate leaves Bowman's capsule, it enters the proximal tubule that is primarily concerned with the active reabsorption of sodium ions and water. Other substances, such as glucose and amino acids, are also reabsorbed. The proximal tubule is important in drug elimination in that it can also actively secrete drugs from the peritubular capillaries surrounding the proximal tubule. This active transport of the drug is sufficiently fast that even drugs that are bound to plasma proteins will dissociate, freeing up the drug for active secretion across the tubule wall. Within the descending portion of Henle's loop, continued reabsorption of water and other ions occurs primarily by passive diffusion. The ascending loop of Henle is considerably less permeable to water and urea, resulting in a very dilute tubular fluid that is rich in urea. Henle's loop does not play a significant role in regard to drug elimination. The first portion of the distal tubule is similar to that of the ascending loop of Henle in that it continues to absorb ions, but is impermeable to water and urea. Within the latter portions of the distal tubule and the collecting tubule, ions and water continue to be absorbed along with acidification of the urine, thereby controlling the acid-base balance of the body fluids.

The reabsorption rate of sodium ions is controlled by the hormone *aldosterone*. The water permeability of these distal segments is controlled by *antidiuretic hormone*, providing a means for controlling the final volume of urine that is formed. Drugs can also be reabsorbed within the distal tubule, thereby affecting the rate of drug elimination from the body. The collecting duct continues to reabsorb water under the control of antidiuretic hormone and does not, for the most part, affect drug elimination. We thus find that the kidney affects drug elimination through the following mechanisms: filtration through the glomerulus, secretion from the peritubular capillaries into the proximal tubule, and reabsorption within the distal tubule.

7.5 DRUG CLEARANCE

7.5.1 RENAL CLEARANCE

The elimination of a drug by the kidney is described by the term *renal clearance*. The renal clearance is simply the volume of plasma that is totally "cleared" of the drug per unit time as a result of the drug's elimination by the kidneys. Figure 7.4 illustrates the concept of renal clearance and allows for a mathematical definition for the case where the only elimination pathway for the drug is through the kidneys. As we shall see, however, this figure can be generalized to represent multiple elimination pathways.

In this figure, a drug is uniformly distributed within an apparent distribution volume, V_{apparent}, and has a total plasma concentration represented by C. CL_{renal} represents the renal plasma flow rate (typically milliliters per minute) that is totally cleared of the drug, that is $C = 0$ in the blood leaving the kidneys. Therefore, by definition, the renal clearance is given by CL_{renal}, and the rate at which the drug is removed from the body by the kidneys is given by $CL_{\text{renal}} \times C$. We can then write an

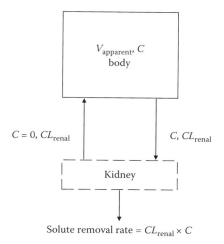

FIGURE 7.4   Concept of renal clearance.

unsteady mass balance on the drug in the body's apparent distribution volume assuming that the only elimination pathway for the drug is through the kidneys:

$$V_{apparent} \frac{dC}{dt} = -CL_{renal}C. \tag{7.18}$$

This equation may be integrated to provide an equation for how the drug concentration within the body changes with time, starting from an initial plasma drug concentration C_0:

$$C(t) = C_0 e^{-(CL_{renal}/V_{apparent})t} = C_0 e^{-k_{renal}t}. \tag{7.19}$$

We see that drug elimination by the kidneys can be described as a first order process exhibiting an exponential decay in plasma drug concentration with time. The renal elimination rate constant, k_{renal}, is related to the renal clearance and the apparent distribution volume by

$$k_{renal} = \frac{CL_{renal}}{V_{apparent}}. \tag{7.20}$$

At any particular time, if Q_{urine} represents the urine volumetric flow rate, and C_{urine} is the concentration of drug in the urine, then another drug mass balance requires the following relationship:

$$CL_{renal}C = Q_{urine}C_{urine} \quad \text{or} \quad CL_{renal} = \frac{Q_{urine}C_{urine}}{C}. \tag{7.21}$$

This equation then provides a connection between the renal clearance and the flow rate of the urine that contains the excreted drug. Knowing the urine flow rate and the plasma and urine drug concentrations, Equation 7.21 can be used to calculate the renal clearance.

The polysaccharide *inulin*, with a molecular weight of about 5000, is particularly important to mention at this point because its renal clearance can be used to measure kidney function. Inulin does not bind with plasma proteins, is not metabolized by the body, readily passes through the glomerular capillaries, and is neither reabsorbed nor secreted within the tubules. From Equation 7.17 and Figure 7.4, we have that the mass removal rate of unbound inulin across the glomerular capillaries is given by $C_U \times CL_{renal} = GFR \times C_U = f_U \times GFR \times C$. Since f_U is equal to unity and $C_U = C$, this result shows that the rate at which plasma is being cleared of inulin in the kidneys (i.e., renal clearance, CL_{renal}) is the same as the GFR.

Creatinine, another substance that is naturally found in the body, can be used to determine the GFR. Creatinine is found in plasma and is a product of endogenous protein degradation, specifically the breakdown of creatine phosphate in muscles. Creatinine, for the most part, is produced at a fairly constant rate by the body, although its production rate can depend on muscle mass. So, for example, there will be a difference between males and females in the creatinine production rate and a body builder is expected to have a higher production rate of creatinine than a normal person.

The production rate of creatinine ($\bar{M}$) in the body is about 1.2 mg min^{-1} in healthy young men and about 1 mg min^{-1} in healthy young women. Creatinine does not bind with the plasma proteins and, unlike inulin, is secreted by the tubules, which will affect the accuracy of GFR estimation from creatinine concentrations in the plasma. At steady state, the production rate of creatinine in the body has to be equal to the amount of creatinine that shows up in the urine. This is expressed by Equation 7.22, where C is the plasma concentration of creatinine:

$$\bar{M} = GFR \times C = Q_{urine} \times C_{urine}. \tag{7.22}$$

Knowing the urine flow rate and the plasma and urine creatinine concentrations, Equation 7.22 can be used to calculate the GFR. Since the production rate of creatinine is relatively constant, Equation 7.22 also provides a means for calculating the GFR based on a measured concentration of creatinine in the plasma, i.e.:

$$GFR\big|_{creatinine} = \frac{\bar{M}_{creatinine}}{C_{creatinine}}. \tag{7.23}$$

Example 7.2

The normal concentration of creatinine in blood is about 1 mg per 100 mL or 1 mg dL^{-1}. Estimate the GFR for this value of the creatinine concentration. The normal GFR is about 125 mL min^{-1}. Suppose the creatinine concentration in the blood was found to be 8 mg dL^{-1}, what is the GFR now?

Solution

We can use Equation 7.23 with $\bar{M}$ equal to 1.2 mg min^{-1}. For a creatinine concentration of 1 mg dL^{-1}, the GFR is 120 mL min^{-1}, which compares favorably with the normal GFR of 125 mL min^{-1}. If the creatinine concentration in the body rises to 8 mg dL^{-1}, then the GFR decreases to a value of 15 mL min^{-1}, which indicates significant impairment of kidney function and the possible need for hemodialysis.

Most drugs, unlike inulin and creatinine, are secreted and reabsorbed in the tubules of the nephron, making the calculation of renal clearance for these drugs more complicated. Secretion of the drug may be inferred when the rate of excretion ($CL_{renal} \times C$) exceeds the rate of drug filtration by the glomeruli ($f_U \times GFR \times C$) or when $CL_{renal} > f_U \times GFR$. Tubular reabsorption of the drug is important whenever the rate of drug excretion is less than the rate of drug filtration or when $CL_{renal} < f_U \times GFR$.

7.5.2 Plasma Clearance

Plasma clearance (CL_{plasma}) is the term used to represent the sum of all the drug elimination processes of the body. This includes the two primary ones already discussed, metabolism and the kidneys, and other secondary processes such as sweating, bile, respiration, and feces. Since these are all assumed to be first order processes that depend on the plasma concentration of the drug, we can write

$$V_{\text{apparent}} \frac{dC}{dt} = -(CL_{\text{metabolism}} + CL_{\text{renal}} + CL_{\text{sweat}} + CL_{\text{bile}} + CL_{\text{respiration}} + CL_{\text{feces}})C. \qquad (7.24)$$

Each elimination process has its own characteristic clearance value and a first order elimination rate constant* can be defined for each as follows:

$$k_i = \frac{CL_i}{V_{\text{apparent}}}, \qquad (7.25)$$

where i indicates metabolic, renal, sweat, bile, respiration, and feces.

The plasma clearance will then be related to the sum of all the first order drug elimination rate constants as given by the following relationship. The total first order elimination rate constant is given by k_{te} and is equal to the sum of the elimination rate constants for all the elimination processes:

$$k_{\text{te}} = \sum_i k_i = \frac{1}{V_{\text{apparent}}} \sum_i CL_i = \frac{CL_{\text{plasma}}}{V_{\text{apparent}}}. \qquad (7.26)$$

7.5.3 BIOLOGICAL HALF-LIFE

The *biological half-life* $(t_{1/2})$ of a drug is the time needed for the drug concentration in the plasma to be reduced by one-half. To obtain the half-life, we can replace the renal clearance in Equations 7.18 and 7.19 with the plasma clearance, which accounts for all elimination pathways. When this is done, we can rewrite Equation 7.19 in terms of the total first order elimination rate constant, i.e., k_{te}. The change in drug concentration with time, taking into account all elimination processes, is then given by

$$C(t) = C_0 e^{-k_{\text{te}}t}. \qquad (7.27)$$

For first order processes, the half-life time can be related to the total first order elimination rate constant by simply letting $C/C_0 = 0.5$ in Equation 7.27. When solved for $t_{1/2}$, the following result is obtained:

$$t_{1/2} = \frac{0.693}{k_{\text{te}}} = \frac{0.693 V_{\text{apparent}}}{CL_{\text{plasma}}}. \qquad (7.28)$$

7.6 MODEL FOR INTRAVENOUS INJECTION OF DRUG

Equation 7.27 represents our first pharmacokinetic model. It is based on a single drug distribution compartment with all the elimination processes lumped into a single first order elimination rate constant, i.e., k_{te}. Equation 7.27 predicts that the drug concentration decreases at an exponential rate with a general shape like that of curve II in Figure 7.1. The initial drug concentration is C_0 and this is the result of a rapid bolus intravenous or intraarterial injection. The initial drug concentration is

---

* First order means that the rate of elimination depends on the concentration to the first power.

related to the dose (D) and the apparent distribution volume, as given by Equation 7.29, assuming no drug is eliminated during the rapid injection phase of the drug:

$$C_0 = \frac{D}{V_{apparent}}. \tag{7.29}$$

If the dose and the initial concentration of the drug are known, then Equation 7.29 can be used to solve for the apparent distribution volume of the drug.

The total area under a C vs. t curve, such as those shown in Figure 7.1, is given the special symbol $\mathrm{AUC}^{0 \to \infty}$ and is defined mathematically as

$$\mathrm{AUC}^{0 \to \infty} \equiv \int_0^\infty C(t)\,dt. \tag{7.30}$$

Equation 7.27 provides a mathematical relationship for the change of the drug concentration with time for the special case of a rapid bolus intravenous injection of a drug that distributes throughout a single compartment. Substituting Equation 7.27 into Equation 7.30, we obtain the following result for the value of $\mathrm{AUC}^{0 \to \infty}$:

$$\mathrm{AUC}^{0 \to \infty} = \frac{C_0}{k_{te}} = \frac{D}{V_{apparent} k_{te}} = \frac{D}{CL_{plasma}}. \tag{7.31}$$

This equation provides a simple relationship between the total area under the curve, the drug dose, and the plasma clearance. Note that $\mathrm{AUC}^{0 \to \infty}$ is directly proportional to the dose and inversely proportional to the plasma clearance and is a measure of the exposure to the drug.

7.7   ACCUMULATION OF DRUG IN THE URINE

As we discussed earlier, some of the drug will be eliminated by the kidneys and show up in the urine. In some cases, it may be convenient to use additional data on the amount of drug (not its metabolites) found in the urine to supplement a pharmacokinetic analysis. The rate at which the drug accumulates in the urine is given by Equation 7.32, where M_{urine} is the mass of drug (unchanged) in the urine at any given time:

$$\frac{dM_{urine}}{dt} = k_{renal} V_{apparent} C = CL_{renal} C. \tag{7.32}$$

Continuing our discussion with the bolus intravenous injection of drug, we can substitute Equation 7.27 for the value of the drug concentration in the plasma at any time and integrate to obtain the following expression for the mass of drug in the urine at any time:

$$M_{urine}(t) = \left(\frac{k_{renal} C_0 V_{apparent}}{k_{te}} \right)(1 - e^{-k_{te}t}). \tag{7.33}$$

After a sufficiently long period of time ($t \to \infty$), Equation 7.33 simplifies to provide the total amount of drug collected in the urine, M_{urine}:

$$M_{\text{urine}} = \frac{k_{\text{renal}} C_0 V_{\text{apparent}}}{k_{\text{te}}} = D \frac{k_{\text{renal}}}{k_{\text{te}}} = \text{AUC}^{0 \to \infty} CL_{\text{renal}}. \tag{7.34}$$

Thus, we find that a relationship exists between the total amount of drug collected in the urine, the renal clearance, and the total area under the curve. Note that if the only drug elimination pathway is through the kidneys ($k_{\text{renal}} = k_{\text{te}}$), then the total amount of drug collected in the urine is equal to the total drug dose, as it should be.

7.8 CONSTANT INFUSION OF DRUG

So far, we have discussed the case of first order drug elimination following a rapid bolus intravenous injection of a drug. The drug was assumed to be uniformly distributed throughout a single compartment. In this case, the drug concentration vs. time is expected to follow curve II of Figure 7.1. In many cases, a constant infusion of the drug is given. Here, the concentration of the drug increases and reaches a steady state level once a balance is reached between the drug infusion rate and the drug elimination rate. The concentration vs. time curve for constant infusion follows something like curve III of Figure 7.1. This result is similar to the so-called *controlled release* drug delivery systems that are commonly used, for example, to treat motion sickness, for birth control, for extended chemotherapy, and to help people quit smoking by delivering nicotine. The constant infusion of a drug can also be used to provide a steady starting concentration of the drug for pharmacokinetic studies.

Figure 7.5 illustrates a single compartment model for continuous infusion of drug. I_0 represents the drug infusion rate (milligram per minute) assumed here to be constant. However, one could generalize this infusion rate to be any arbitrary time-varying function. For example, in Section 5.6.1, we developed an analytical solution describing the diffusion of solute from a polymeric material. In this case, the solute is the drug and the polymeric material is a controlled release device. The elution flux of the drug ($j_S|_{x=L}$) is then given by Equation 5.78 (assuming that the concentration of drug in the polymeric material is much larger than that in the body so that BC2 in Equation 5.67 is satisfied) and I_0 in Equation 7.35 is $I_0(t) = j_S|_{x=L} S$, where S is the exposed surface area of the device. An unsteady mass balance can then be written for the drug as

$$V_{\text{apparent}} \frac{dC}{dt} = I_0 - CL_{\text{plasma}} C = I_0 - V_{\text{apparent}} k_{\text{te}} C. \tag{7.35}$$

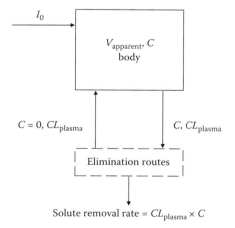

FIGURE 7.5 Single compartment model for continuous infusion with first order elimination.

Equation 7.35 may be integrated under the assumption of a constant infusion rate and that no drug is initially present in the body, to give

$$C(t) = \left(\frac{I_0}{k_{te}V_{apparent}} \right)(1 - e^{-k_{te}t}). \tag{7.36}$$

The general shape of the curve described by this equation is that of curve III in Figure 7.1. For long periods of time, this equation predicts a steady state plasma drug concentration given by

$$C_{SS} = \frac{I_0}{k_{te}V_{apparent}} = \frac{I_0}{CL_{plasma}}. \tag{7.37}$$

When the infusion is stopped, then the drug is eliminated according to Equation 7.27, where the initial drug concentration, C_0, is equal to the steady state drug concentration, C_{SS}, arising from the constant infusion. Now the drug concentration decreases, as shown by the dotted line of curve III in Figure 7.1.

Example 7.3

The following table provides some data for the elimination of radioactive inulin from a 392 g laboratory rat. The animal was given a bolus injection at $t = 0$ equivalent to 1.01×10^5 cpm (counts per minute, radioactivity directly proportional to the inulin concentration). This was done to hasten the development of a steady state plasma inulin level. Next, a continuous infusion of inulin was started at the rate of 2.76×10^3 cpm min^{-1}. The inulin infusion was then stopped after a total of 80 min. From these data, determine the GFR, the renal elimination rate constant, and the apparent distribution volume for inulin.

Inulin Elimination from a Laboratory Rat

Time (min)	Inulin Concentration (cpm mL^{-1})
0	0
30	849
40	845
60	903
70	888
75	873
80	882 (infusion stopped)
90	565
100	412
110	271

Source: From Sarver, J.G., PhD dissertation, The University of Toledo, 1994.

Solution

We see that after about 30 min a steady state plasma inulin concentration of about 873 cpm mL^{-1} is reached. Once the infusion is stopped, the plasma inulin concentration decreases rapidly. As discussed earlier, inulin is only removed by the kidneys and its renal clearance is the same as the GFR. Therefore, for the case of this experiment with inulin, Equation 7.37 can be rearranged to solve for the GFR in terms of the known values of the infusion rate and the steady state plasma concentration:

$$k_{te}V_{apparent} = GFR = \frac{I_0}{C_{SS}}, \text{ for inulin only.}$$

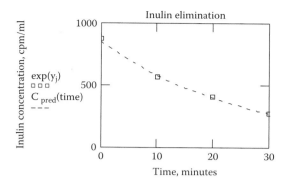

FIGURE 7.6   Measured and predicted inulin concentrations.

For the experimental conditions leading to the data shown in the previous table, the GFR is then calculated to be 3.16 mL min^{-1}. Since the GFR is proportional to body weight, this can be expressed as 0.48 mL h^{-1} gBW^{-1}. The renal elimination rate constant can be found by performing a regression analysis of the data obtained after the infusion is stopped using Equation 7.27. Note that the time in this equation represents the time since the infusion was stopped. The initial concentration is equal to 873 cpm mL^{-1} and k_{te} is equal to k_{renal}, since the only elimination process is that of the kidneys. Note that Equation 7.27 can be linearized by taking the natural logarithm of each side. The regression equation is then given by

$$\ln C(t) = \ln C_0 - k_{te}t.$$

The intercept is the natural logarithm of the initial concentration and the slope is the negative of the elimination rate constant. After performing the linear regression, we find that $C_0 = 860.1$ cpm mL^{-1} and that $k_{te} = k_{renal} = 0.038$ min^{-1}. Figure 7.6 shows a comparison between the data and the inulin concentrations predicted by Equation 7.27. The apparent distribution volume for inulin is given by

$$V_{apparent} = \frac{GFR}{k_{renal}}.$$

For these data, the apparent distribution volume is about 83 mL. Since the distribution volume is also proportional to body weight, the apparent distribution volume for a rat can be expressed as 0.21 mL gBW^{-1}.

7.8.1   Application to Controlled Release of Drugs by Osmotic Pumps

There is now considerable interest focused on the development of novel drug delivery systems (Verma et al. 2002). Novel drug delivery systems have several advantages, including elimination of first pass metabolism by the liver, minimal discomfort, lower risk of infections, sustained release of the drug, and better patient compliance. For example, osmotic pumps are small implantable devices that can be used to deliver a drug at a constant infusion rate to a specific site in the body for very long periods of time. A recent example is the DUROS osmotic pump (Wright et al. 2001), which is being used to deliver the gonadotropin-releasing hormone agonist leuprolide for the palliative treatment of advanced prostate cancer. Serum leuprolide levels of about 1 ng mL^{-1} for long periods of time will reduce serum testosterone levels in humans to below castration levels, which is important in controlling the growth of prostate cancer cells. The device is inserted subcutaneously and is designed to deliver leuprolide at a constant rate for up to 1 year.

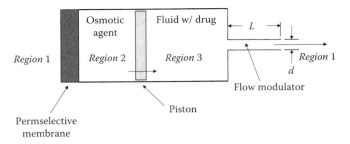

FIGURE 7.7 Osmotic pump drug delivery system.

The DUROS osmotic pump is made from a cylindrically shaped piece of titanium alloy and measures 4 mm in diameter and 45 mm in length. As shown in Figure 7.7, the device is basically a cylindrical reservoir that is capped at one end by a permselective membrane and at the other end by a flow modulator through which the drug is released to the surroundings.

Within the reservoir are two chambers separated by a piston. In the chamber between the permselective membrane and the piston, there is an osmotic agent that increases the osmotic pressure of the fluid within the chamber relative to the osmotic pressure of the body fluid that surrounds the device. The osmotic agent in the DUROS osmotic pump is NaCl at amounts much higher than the NaCl solubility. As water from the surroundings flows across the permselective membrane by osmosis, the NaCl dissolves, and because of being in excess, maintains saturation levels and hence a constant osmotic pressure throughout the lifetime of the implant. As water enters the device by osmosis, the fluid volume containing the osmotic agent increases and exerts pressure on the piston, causing the piston to move. The movement of the piston then acts on the second chamber that contains a fluid with the drug. The piston movement then forces the fluid containing the drug out of the device through the flow modulator at a rate equal to the rate at which water enters the device by osmosis.

The rate at which water enters the device, shown in Figure 7.7, from the surroundings is given by Equation 3.4, i.e.:

$$Q_{in} = L_P S[(\pi_2 - \pi_1) - (P_2 - P_1)], \tag{7.38}$$

where the numerical subscripts refer to the regions shown in Figure 7.7 and S refers to the surface area of the permselective membrane. Since the piston moves very slowly, we can also assume that the pressures in the osmotic agent chamber and the drug chamber are the same, i.e., $P_2 = P_3$.

If the flow modulator is closed or plugged, then water continues to enter the device by osmosis until the hydrodynamic pressure difference, i.e., $(P_2 - P_1)$, balances the osmotic pressure difference $(\pi_2 - \pi_1)$, stopping the influx of water by osmosis, or

$$P_2 = P_1 + (\pi_2 - \pi_1) \text{ equilibrium pressure.} \tag{7.39}$$

With the flow modulator open, there must be at steady state some value of the internal pressure, i.e., P_2 or P_3, which gives a flow rate out of the device that is equal to the rate at which water enters the device by osmosis. The value of P_2 can be found by applying Bernoulli's equation, i.e., Equation 4.72, across the flow modulator as shown by

$$\frac{P_2}{\rho} = \frac{P_1}{\rho} + \frac{\alpha V_1^2}{2} + h_{friction}. \tag{7.40}$$

In this equation, h_{friction} represents the frictional losses due to the contraction and expansion of the fluid as it enters and leaves the flow modulator, as well as the friction loss as it flows through the flow modulator of length L and diameter d. Recall that h_{friction} is given by Equation 4.75. V_1 is the velocity of the drug-containing fluid as it exits the flow modulator. The volumetric flow rate of the drug-containing fluid is given by

$$Q_{\text{out}} = \frac{\pi d^2}{4} V_1. \tag{7.41}$$

At steady state, one needs to find the value of P_2 in Equation 7.40 that makes the Q_{out}, calculated by Equation 7.41, equal the osmotic flow into the device, as calculated by Equation 7.38. The drug delivery rate, or the drug infusion rate I_0 in Equation 7.37, is then given by the product of Q_{out} and the drug concentration within the osmotic pump, i.e., C_{drug}:

$$I_0 = Q_{\text{out}} \, C_{\text{drug}}. \tag{7.42}$$

Because the flow rate of the drug from the device is usually so small and occurs over such a long period of time, the pressures within the chambers of the osmotic pump, i.e., P_2 and P_3, are very close to the pressure of the surroundings, i.e., P_1. In addition, the osmotic pressure of the osmotic agent chamber, i.e., π_2, is much larger than the osmotic pressure of the body fluids surrounding the device, i.e., π_1. Hence, for these practical reasons, the value of Q_{in} and Q_{out} is given by Equation 7.43, which results from a simplification of Equation 7.38:

$$Q_{\text{in}} = Q_{\text{out}} = L_{\text{P}} \, S \, \pi_2. \tag{7.43}$$

Combining Equations 7.42 and 7.43 results in Equation 7.44 for the infusion rate of a drug by an osmotic pump:

$$I_0 = C_{\text{drug}} \, L_{\text{P}} \, S \, \pi_2. \tag{7.44}$$

This equation shows that the infusion rate of the osmotic pump is directly proportional to the drug concentration within the device and the osmotic pressure of the osmotic agent that is used. The proportionality constant is the product of the surface area of the permselective membrane and its hydraulic conductance.

7.8.2 Application to the Transdermal Delivery of Drugs

In transdermal delivery of a drug (i.e., a skin patch) (Panchagnula 1997), the drug molecules diffuse across the skin and enter the systemic circulation. However, the passive transport of the drug across the skin can be a major problem. Skin acts as a major defense barrier for the body, and most drugs have very low skin permeability, which can limit their ability to achieve therapeutic drug levels in the body. Therefore, transdermal delivery of a drug is best suited for drugs that have high pharmacological activity (potency) and good skin permeability (lipophilic). Examples where transdermal delivery of drugs is being used include the treatment of hypertension, estrogen replacement, birth control, pain, overactive bladder, motion sickness, testosterone replacement therapy, and for cessation of smoking.

The bulk of the resistance for the transport of a drug across skin, as shown in Figure 7.8, results from the *stratus corneum* (SC) (Mitragotri 2003). The SC is the outermost layer of skin consisting of about 15 layers of keratin-filled cells (keratinocytes). The spaces between these cells are filled

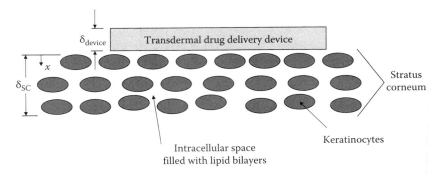

FIGURE 7.8   Transdermal delivery of drugs.

with lipid bilayers. The rate of transdermal transport of a drug across the SC, i.e., the drug infusion rate, can be described by an equation similar to Equation 5.89, i.e.:

$$I_0 = P_{SC}\, S\, (C_{drug} - 0) = P_{SC}\, S\, C_{drug}, \tag{7.45}$$

where P_{SC} is the permeability of the drug in the SC, S is the total surface area of the transdermal device, and C_{drug} is the average concentration of the drug within the transdermal device. The permeability of a drug across the SC can be measured experimentally using samples of skin or it can be estimated as discussed in the next section. It is also assumed that the concentration of the drug at $x = \delta_{SC}$ is zero, since the drug is immediately taken up by the blood supply. The SC permeability, i.e., P_{SC}, can be described by an equation similar to Equation 5.90:

$$P_{SC} = \frac{\varepsilon_{SC} D_{SC}\, K}{\delta_{SC}}, \tag{7.46}$$

where ε_{SC} represents the volume fraction of the lipid bilayers through which the drug diffuses, D_{SC} is the diffusivity of the drug in the lipid bilayer material, and $K = (C_{SC}/C_{drug})|_{x=0}$ represents the equilibrium solubility of the drug in the SC.

In Equation 7.45, the value of C_{drug} will change with time as the drug is depleted from the device. In the most rigorous case, one could use Equation 5.77 to calculate the average concentration of the drug within the device at a given time. One can also assume that the drug concentration within the device is many times larger than the concentration of the drug within the SC, thereby satisfying BC2 in Equation 5.67. However, within the transdermal device, we can also make the reasonable assumption that the concentration profile of the drug at any time is flat, since the device is very thin, and diffusion of the drug out of the device is a very slow process because the bulk of the mass transfer resistance for the drug lies within the SC. Therefore, an unsteady mass balance on the amount of drug within the device at any time may be written as

$$V_{device} \frac{dC_{drug}}{dt} = -S\, P_{SC}\, C_{drug}. \tag{7.47}$$

With the initial condition that $C_{drug} = C_{drug0}$, Equation 7.47 can be integrated to provide the amount of drug within the transdermal delivery device at any time:

$$C_{drug}(t) = C_{drug0}\, e^{-(P_{SC}/\delta_{device})t}, \tag{7.48}$$

where $\delta_{device} = V_{device}/S$ is the thickness of the transdermal patch. Equations 7.45 and 7.48 can then be combined with Equation 7.35 to obtain the following differential equation that describes the concentration within the apparent distribution volume during continuous infusion of a drug by a transdermal delivery system:

$$V_{apparent} \frac{dC}{dt} + V_{apparent} k_{te} C = P_{SC} S C_{drug0} e^{-(P_{SC}/\delta_{device})t}, \tag{7.49}$$

with the initial condition that at $t = 0$, $C = C_0$, where C_0 is the initial concentration of the drug in the body. Equation 7.49 may be easily solved using Laplace transforms (Table 4.5) to give Equation 7.50 for the plasma drug concentration as a function of time:

$$C(t) = C_0 e^{-k_{te} t} + \frac{P_{SC} S C_{drug0}}{V_{apparent}} \left(\frac{e^{-(P_{SC}/\delta_{device})t} - e^{-k_{te} t}}{k_{te} - \dfrac{P_{SC}}{\delta_{device}}} \right). \tag{7.50}$$

Note that if the device is very large, i.e., $\delta_{device} \gg 0$, or the SC permeability is very small, then Equation 7.50 simplifies for $C_0 = 0$ to the previous result given by Equation 7.36 with $I_0 = P_{SC} S C_{drug0}$.

7.8.2.1 Predicting the Permeability of Skin

Oftentimes for preliminary design calculations using Equation 7.50, it may be necessary in the absence of experimental values to have an estimate of the permeability of a drug in the SC. Potts and Guy (1992) developed a relatively simple model for the SC permeability based on the size of the drug molecule (i.e., its molecular weight, MW) and its octanol/water partition coefficient ($K_{O/W}$). The octanol/water partition coefficient is a commonly used measure of the lipid solubility of a drug. They assumed that the diffusivity of a drug in the SC depends on the molecular weight of the drug as given by

$$\varepsilon_{SC} D_{SC} = (\varepsilon D)^0_{SC} e^{-\beta MW}. \tag{7.51}$$

Substituting Equation 7.51 into Equation 7.46 and replacing the equilibrium solubility of the drug in the lipid bilayers (K) with the octanol/water partition coefficient ($K_{O/W}$), we obtain

$$P_{SC} = K_{O/W} \frac{(\varepsilon D)^0_{SC}}{\delta_{SC}} e^{-\beta MW}. \tag{7.52}$$

After taking the $\log_{10}$ of each side, Equation 7.52 becomes

$$\log P_{SC} = \log \left[\frac{(\varepsilon D)^0_{SC}}{\delta_{SC}} \right] + \alpha \log K_{O/W} - \beta MW, \tag{7.53}$$

where α is an adjustable constant added to improve the regression analysis and is expected to be on the order of unity. Potts and Guy (1992) then performed a regression analysis using Equation 7.53 on a data set of more than 90 drugs for which the SC permeability is known. The drugs considered in their data set ranged in molecular weight from 18 to 750 and had octanol/water partition coefficients, i.e., log $K_{O/W}$, from −3 to +6. The regression analysis found the values of $[(\varepsilon D)_{SC}^0 / \delta_{SC}]$, α, and β that best represented the data set. Their regression analysis resulted in

$$\log P_{SC}\,(\text{cm sec}^{-1}) = -6.3 + 0.71 \log K_{O/W} - 0.0061\,\text{MW}. \tag{7.54}$$

This equation can be expected to yield predicted values of P_{SC} that are within several fold of the actual values of the SC permeability for a given drug.

Example 7.4

Estimate the SC permeability for caffeine. The molecular weight of caffeine is 194 and its octanol/water partition coefficient, $K_{O/W}$, is equal to 1 (Joshi and Raje 2002).

Solution

We use the Potts and Guy equation to estimate the SC permeability for caffeine as

$$\log P_{SC} = -6.3 + 0.71 \log 1 - 0.0061 \times 194 = -7.483,$$

$$P_{SC} = 3.29 \times 10^{-8}\ \text{cm sec}^{-1} = 1.18 \times 10^{-4}\ \text{cm h}^{-1}.$$

The value reported by Joshi and Raje for the SC permeability of caffeine is 1×10^{-4} cm h^{-1}, which compares quite well to the value predicted by the Potts and Guy equation.

7.9 FIRST ORDER DRUG ABSORPTION AND ELIMINATION

Many drugs are given orally or by injection. Other routes may include the nasal cavity or by inhalation. In all of these non-intravenous cases, the drug is first absorbed and makes its way into the plasma by diffusion, where it is distributed throughout the body. The plasma concentration of the drug gradually increases and then finally decreases as the elimination processes begin to overcome the rate of drug absorption. Curve I of Figure 7.1 illustrates the general trend of the plasma drug concentration with time during the absorption and elimination phases. Although the drug absorption rate is strongly affected by the dosage form, i.e., solution, capsule, tablet, and suspension, it has been found that, for the most part, the drug absorption process in many cases can be adequately described by simple first order kinetics.

Figure 7.9 illustrates the simplest compartmental model for the case of first order drug absorption and elimination. Note, a separate mass balance equation is needed for the drug whose amount decreases with time due to its absorption into the body. The factor (f) accounts for the fraction of the dose (D) that is actually absorbable. An amount of the dose equal to $(1 - f)\,D$ is assumed to be removed from the body without ever entering the apparent distribution volume.

A mass balance on the amount of drug within the apparent distribution volume can be written as

$$V_{\text{apparent}}\,\frac{dC}{dt} = k_a\,A - CL_{\text{plasma}}\,C = k_a\,A - V_{\text{apparent}}\,k_{te}\,C, \tag{7.55}$$

where A represents the mass of drug remaining to be absorbed, and k_a represents the drug's first order absorption rate constant.

A mass balance on the drug yet to be absorbed is given by

$$\frac{dA}{dt} = -k_a A. \tag{7.56}$$

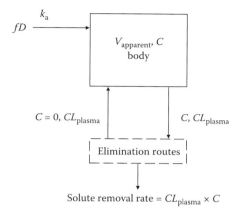

FIGURE 7.9 Single compartment model with first order absorption and elimination.

When this equation is integrated, recognizing that initially $A_0 = fD$, then the amount of drug at anytime waiting to be absorbed is given by

$$A(t) = fDe^{-k_a t}. \tag{7.57}$$

This equation can then be substituted into Equation 7.55 to obtain an expression for the total amount of drug in the plasma as a function of time, assuming that the initial concentration of drug is zero:

$$C(t) = \frac{fD}{V_{apparent}}\left(\frac{k_a}{k_a - k_{te}}\right)(e^{-k_{te}t} - e^{-k_a t}). \tag{7.58}$$

We see that the concentration of drug in the plasma is dependent on the first order rate constants for absorption and elimination, i.e., k_a and k_{te}. The concentration is directly proportional to the fraction of the drug dose that is absorbable and inversely proportional to the apparent distribution volume.

An expression for the time at which the concentration peaks can be obtained by finding the derivative of Equation 7.58 with respect to time, setting this result equal to zero, and solving for the time. This time, called τ_{max}, is given by

$$\tau_{max} = \frac{1}{k_a - k_{te}}\ln\left(\frac{k_a}{k_{te}}\right). \tag{7.59}$$

Note that the time the maximum concentration is reached is independent of the dose. The corresponding concentration at the time of the peak may be found by substituting τ_{max} into Equation 7.58 for the time and simplifying to obtain

$$C_{max} = \left(\frac{fD}{V_{apparent}}\right)\left(\frac{k_a}{k_{te}}\right)^{k_{te}/(k_{te}-k_a)}. \tag{7.60}$$

We can also find the value of $\text{AUC}^{0\rightarrow\infty}$ by substituting Equation 7.58 into Equation 7.30. On integration, the following result is obtained for the value of $\text{AUC}^{0\rightarrow\infty}$:

$$\text{AUC}^{0\rightarrow\infty} = \frac{fD}{V_{apparent}k_{te}} = \frac{fD}{CL_{plasma}} \tag{7.61}$$

It is interesting to note that the value of $AUC^{0 \to \infty}$ does not depend on the absorption rate constant. However, $AUC^{0 \to \infty}$ is directly proportional to the absorbed dose, fD, and inversely proportional to the plasma clearance. $AUC^{0 \to \infty}$ is a measure of the body's exposure to the drug.

To complete our mathematical description of the process shown in Figure 7.9, the rate at which the drug enters the urine can be found from Equation 7.32. Substituting Equation 7.58 for the total plasma drug concentration into this equation gives the following result for the amount of drug in the urine at any given time:

$$M_{\text{urine}}(t) = \left(\frac{fD k_{\text{renal}}}{k_{\text{te}}} \right) \left[1 - \frac{1}{k_{\text{a}} - k_{\text{te}}} \left(k_{\text{a}} e^{-k_{\text{te}}t} - k_{\text{te}} e^{-k_{\text{a}}t} \right) \right]. \tag{7.62}$$

After all of the drug has cleared the body ($t \to \infty$), the amount of drug in the urine is given by

$$M_{\text{urine}}^{\infty} = fD \frac{k_{\text{renal}}}{k_{\text{te}}}. \tag{7.63}$$

Note that if the only elimination pathway for the drug is through the kidneys, i.e., $k_{\text{renal}} = k_{\text{te}}$, then $M_{\text{urine}}^{\infty} = fD$.

If Equation 7.63 is subtracted from Equation 7.62, then an equation is obtained for the quantity of drug remaining to be excreted by the kidneys. This equation is of a convenient form for determining the absorption and elimination rate constants from urinary excretion data:

$$M_{\text{urine}}^{\infty} - M_{\text{urine}}(t) = \frac{M_{\text{urine}}^{\infty}}{(k_{\text{a}} - k_{\text{te}})} \left(k_{\text{a}} e^{-k_{\text{te}}t} - k_{\text{te}} e^{-k_{\text{a}}t} \right). \tag{7.64}$$

These parameters (k_{a} and k_{te}) may be determined by performing a nonlinear regression on urinary excretion data in a manner similar to that shown in Example 7.5. Since $M_{\text{urine}}^{\infty}$ is known from the urinary excretion data, one can then determine the renal elimination rate constant (k_{renal}) from Equation 7.63, given the dose actually absorbed (fD) and the value of k_{te}.

Example 7.5 illustrates the use of the single compartment model with first order absorption and elimination to describe the plasma concentration of an experimental drug after an oral dose of 500 mg. This example requires that a nonlinear regression be performed to find the three unknown parameters in Equation 7.58, i.e., fD/V_{apparent}, k_{a}, and k_{te}. Once again, we choose as an objective function the sum-of-the-square-of-the-error, defined as $SSE = \sum_{i=1}^{N} (C_{\text{experimental}_i} - C_{\text{theory}_i})^2$, where $C_{\text{experimental}}$ are the measured plasma drug concentrations and C_{theory} are the values predicted by the pharmacokinetic model, e.g., from Equation 7.58. N represents the total number of data points that are considered. The goal is then to find the set of parameters in the pharmacokinetic model, i.e., fD/V_{apparent}, k_{a}, and k_{te}, that minimizes SSE. These values of the parameters that minimize SSE represent the best fit of the pharmacokinetic model to the data. There are a variety of mathematical software programs that have routines for solving multiple parameter nonlinear regression problems. It is important to note that many times when performing a nonlinear regression analysis, a unique solution may not be found. In other words, there can be different values for the regression parameters that can fit the data to the same level of accuracy.

To obtain the values of f and V_{apparent} from the value of fD/V_{apparent} requires additional data, such as that obtained from urinary excretion or intravenous injection studies. Other techniques based on graphical methods may also be used to determine pharmacokinetic parameters, as discussed by Cooney (1976), Gibaldi and Perrier (1982), Welling (1986), and Notari (1987) and in the problems at the end of this chapter.

Example 7.5

The following table summarizes the plasma concentration levels of an experimental drug following a 500 mg oral dose. From the data provided in the table, determine the pharmacokinetic parameters that describe the absorption and elimination of this drug. The drug was also found to have an absorption factor (*f*) of 0.8.

Time (min)	Plasma Drug Concentration(mg L^{-1})
0	0
1	1.5
4	4.5
8	8.2
12	10.5
16	11.5
20	12.5
24	13.2
28	13.1
32	13.0
40	12.5
70	8.5
100	5.9
150	2.5
200	1.4
300	0.3

Solution

One of the issues with the nonlinear regression analysis of pharmacokinetic data is obtaining suitable initial guesses of the parameters. In this example, we need to find the values of $(fD/V_{apparent}) \equiv I_0$, k_a, and k_{te}. One approach for guessing I_0 is to extrapolate the data after the concentration peak back to the value at $t = 0$, since this represents in a sense what the initial drug concentration would be if the absorption process was very fast. This is usually facilitated by plotting the log of the concentration data vs. time. For the data mentioned in the table, it is then found that $I_0 \sim 25$ mg L^{-1}. Since the absorption process, in most cases, is many times faster than the elimination process, for times after the peak concentration, the elimination processes dominate. Hence, Equation 7.58 becomes

$$\ln C\big|_{t>>0} \approx \ln\left[I_0\left(\frac{k_a}{k_a - k_{te}}\right)\right] - k_{te}\,t.$$

Therefore, if ln *C* is plotted as a function of time for the data after the peak, the slope of these data is approximately equal to k_{te}. In this case, an estimate of $k_{te} = 0.014$ min^{-1} is found. Since the absorption process is much faster than the elimination process, a value of $k_a = 10 \times k_{te}$ can be assumed as a starting point. Figure 7.10 shows the results of the nonlinear regression. The open symbols are the data points from the previous table and the solid line is the resulting best fit of Equation 7.58 to these data. The value of $(fD/V_{apparent}) = 19.9$ mg L^{-1}, $k_a = 0.07$ min^{-1}, and $k_{te} = 0.015$ min^{-1}. From Equation 7.59, the predicted maximum in the plasma drug concentration occurs after 27.9 min and the peak plasma concentration from Equation 7.60 is 13.1 mg L^{-1}. The fit during the initial absorption phase and the elimination phase is excellent. Since the dose is given as 500 mg and the fraction of the dose absorbed is 0.8, we find that the apparent distribution volume for this drug is

$$V_{apparent} = \frac{fD}{I_0} = \frac{0.8 \times 500 \text{ mg}}{19.9 \dfrac{\text{mg}}{\text{L}}} = 20.1 \text{ L}.$$

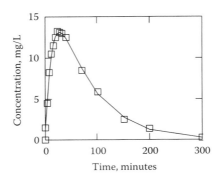

FIGURE 7.10 A comparison of actual and predicted plasma drug concentrations, Example 7.5.

7.10 TWO COMPARTMENT MODEL

In many cases, the single compartment model is not sufficient to represent the time course of the plasma drug concentration. The single compartment model assumes that the drug rapidly distributes throughout a single homogeneous apparent distribution volume. In fact, the drug may distribute at different rates in different tissues of the body and it may require an extended period of time before the drug concentration equilibrates between the different tissue and fluid regions of the body. Clearly, this a complex process and one could envision building a model in which a central compartment representing the plasma is in direct communication with a multitude of other compartments representing other tissue and fluid regions of the body.

A reasonable simplification of this type of model is the two compartment model shown in Figure 7.11. Drug is introduced into the central compartment that generally represents the plasma and other fluids in which the drug rapidly equilibrates. The drug then slowly distributes into the remaining tissue and fluid spaces represented by a single tissue compartment. The constants k_{pt} and k_{tp} represent the rate constants describing the first order transport between the plasma and tissue compartments. Since most or all of the drug is eliminated by the highly perfused kidneys and liver, the elimination processes (metabolic and renal) are assumed to be associated with the plasma compartment.

The two compartment model has a characteristic biphasic temporal plasma concentration profile that distinguishes it from the single compartment model. During the early period after drug absorption, the drug is eliminated at a more rapid pace from the plasma volume due to the combined effects of the usual elimination processes as well as drug absorption by the tissues and fluids of the secondary compartment. After a period of time, the two compartments equilibrate and the drug elimination rate falls off.

7.10.1 TWO COMPARTMENT MODEL FOR AN INTRAVENOUS INJECTION

First, let us develop a pharmacokinetic model for the case of a bolus intravenous injection. A separate unsteady mass balance equation for the drug is then written for the plasma and the tissue compartments:

$$\text{Plasma compartment: } V_p \frac{dC_{\text{plasma}}}{dt} = k_{\text{tp}}V_{\text{tissue}}C_{\text{tissue}} - (k_{\text{te}} + k_{\text{pt}})V_{\text{plasma}}C_{\text{plasma}}, \tag{7.65}$$

$$\text{Tissue compartment: } V_{\text{tissue}} \frac{dC_{\text{tissue}}}{dt} = k_{\text{pt}}V_{\text{plasma}}C_{\text{plasma}} - k_{\text{tp}}V_{\text{tissue}}C_{\text{tissue}}, \tag{7.66}$$

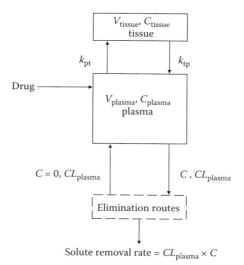

FIGURE 7.11   A two compartment model.

where V_{plasma} and V_{tissue} represent the apparent distribution volumes of the plasma and tissue compartments and C_{plasma} and C_{tissue} represent the concentrations of the drug within these respective compartments. These equations may be solved analytically using Laplace transforms with the following initial condition: $t = 0$, $C_{\text{plasma}} = D/V_{\text{plasma}}$, and $C_{\text{tissue}} = 0$. The equation for the concentration of drug within the plasma compartment can be given by

$$C_{\text{plasma}}(t) = \frac{D}{V_{\text{plasma}}(A - B)}\left[(k_{\text{tp}} - B)e^{-Bt} - (k_{\text{tp}} - A)e^{-At}\right], \tag{7.67}$$

and that for the concentration of drug in the tissue compartment is given by

$$C_{\text{tissue}}(t) = \frac{Dk_{\text{pt}}}{V_{\text{tissue}}(A - B)}\left[e^{-Bt} - e^{-At}\right]. \tag{7.68}$$

The constants A and B are given by the following equations in terms of the first order rate constants:

$$\begin{aligned}
A &= \tfrac{1}{2}\{(k_{\text{pt}} + k_{\text{tp}} + k_{\text{te}}) + [(k_{\text{pt}} + k_{\text{tp}} + k_{\text{te}})^2 - 4k_{\text{tp}}k_{\text{te}}]^{1/2}\} \\
B &= \tfrac{1}{2}\{(k_{\text{pt}} + k_{\text{tp}} + k_{\text{te}}) - [(k_{\text{pt}} + k_{\text{tp}} + k_{\text{te}})^2 - 4k_{\text{tp}}k_{\text{te}}]^{1/2}\}
\end{aligned} \tag{7.69}$$

A nonlinear regression of plasma drug concentration data using Equation 7.67 can be performed to obtain the constants in this equation, i.e., V_{plasma}, A, B, and k_{tp}. The other first order rate constants, k_{pt} and k_{te}, can then be found through the following process. We first multiply A and B in Equation 7.69 to obtain the following result:

$$k_{\text{te}} = \frac{AB}{k_{\text{tp}}}. \tag{7.70}$$

Then, A and B in Equation 7.69 are added together to give

$$k_{pt} = (A + B) - (k_{tp} + k_{te}).$$ (7.71)

Equations 7.70 and 7.71 may then be solved to obtain the values of k_{te} and k_{pt} from the known values of A, B, and k_{tp}.

The apparent distribution volume of the tissue compartment can be found by assuming that once the two compartments are at equilibrium, then the derivative in Equation 7.66 is equal to zero and the concentrations in the two compartments are the same. Equation 7.66 then simplifies to give the following result for the volume of the tissue space:

$$V_{tissue} = V_{plasma} \left[\frac{k_{pt}}{k_{tp}} \right].$$ (7.72)

The overall apparent distribution volume is then equal to $V_{plasma} + V_{tissue}$ and is given by

$$V_{apparent} = V_{plasma} \left[1 + \frac{k_{pt}}{k_{tp}} \right].$$ (7.73)

The data in Example 7.6 illustrate how the pharmacokinetic parameters of the two compartment model for an intravenous dose can be determined. Other techniques based on graphical methods may also be used to determine pharmacokinetic parameters (Welling 1986; Notari 1987). Oftentimes, the graphical approaches can provide reasonable starting values for the parameters in a nonlinear regression.

Example 7.6

The data shown in the following table were given in Koushanpour (1976) and represent plasma inulin concentrations following a rapid intravenous injection of 4.5 g of inulin in an 80 kg human. Determine the pharmacokinetic parameters of the two compartment model given by Equations 7.67 and 7.68.

Time (min)	Plasma Inulin Concentration ($\mu g\ mL^{-1}$)
10	440
20	320
40	200
60	150
90	110
120	80
150	60
175	48
210	35
240	25

Solution

The log-normal graph of these data is given in Figure 7.12 and shows that the elimination phase in this case is not linear but biphasic, indicating that a two compartment model is needed to provide an adequate representation of the data. Figure 7.13 shows the results of the nonlinear regression. The open symbols are the data points from the previous table and the solid line is the resulting best fit of Equation 7.67 to these data. We see that the two compartment model fits these data extremely well. The following table summarizes the results of the regression analysis and these parameter values provide the best fit of Equation 7.67 to the data in the previous table.

Parameter	Value
V_{plasma}	7.8 L
A	0.036 min^{-1}
B	0.0049 min^{-1}
k_{tp}	0.012 min^{-1}
k_{pt}	0.014 min^{-1}
k_{te}	0.016 min^{-1}
V_{tissue}	9.5 L
V_{total}	17.3 L

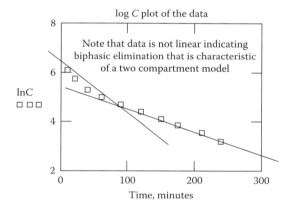

FIGURE 7.12 Biphasic elimination of inulin, Example 7.6.

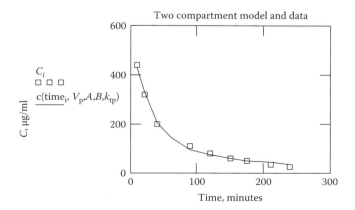

FIGURE 7.13 A comparison of actual and predicted plasma inulin levels, Example 7.6.

Since for inulin the only elimination pathway is through the kidneys, our earlier discussion showed that the renal clearance is the same as the GFR. Therefore, for this example, the renal clearance (CL_{renal}), or GFR, is equal to $V_{plasma} \times k_{te} = 125$ mL min^{-1}. Inulin is known to only distribute throughout the extracellular fluid volume. Hence, inulin can be used to measure the total extracellular fluid. The value of 17 L compares well with the reported value of 15 L for the extracellular fluid volume (Table 3.1).

7.10.2 Two Compartment Model for First Order Absorption

Continuing our discussion of the two compartment model, consider the case now of first order absorption of the drug. Unsteady mass balance equations may once again be written for each of the compartments. The equation for the tissue compartment is still given by Equation 7.66 and the equation that describes the amount of drug yet to be absorbed (A) is still given by Equations 7.56 and 7.57. The plasma compartment equation is then given by

$$V_{plasma} \frac{dC_{plasma}}{dt} = k_a A + k_{tp} V_{tissue} C_{tissue} - (k_{te} + k_{pt}) V_{plasma} C_{plasma}. \quad (7.74)$$

The initial conditions are that at time equal to zero, no drug is present in either compartment. These equations can then be solved using Laplace transforms to obtain the following expressions for the concentration of drug within the plasma and tissue compartments. For the plasma concentration we have

$$C_{plasma}(t) = \frac{fDk_a}{V_{plasma}} \left[\frac{(k_{tp} - A)}{(k_a - A)(B - A)} e^{-At} + \frac{(k_{tp} - B)}{(k_a - B)(A - B)} e^{-Bt} + \frac{(k_{tp} - k_a)}{(A - k_a)(B - k_a)} e^{-k_a t} \right], \quad (7.75)$$

and for the tissue concentration:

$$C_{tissue}(t) = \frac{fDk_a k_{tp}}{V_t} \left[\frac{e^{-At}}{(k_a - A)(B - A)} + \frac{e^{-Bt}}{(k_a - B)(A - B)} + \frac{e^{-k_a t}}{(A - k_a)(B - k_a)} \right]. \quad (7.76)$$

The constants A and B have the same dependence on k_{pt}, k_{tp}, and k_{te} as the two compartment IV case and are also given by Equation 7.69. Depending on the type of data one has, Equation 7.75 can be fitted to plasma concentration data to obtain the unknown parameters: fD, V_{plasma}, k_a, k_{tp}, A, and B. In some cases, it may not be possible to resolve the group fD/V_{plasma} without additional data, perhaps from an IV bolus injection and/or urinary excretion data. Once these parameters have been determined, Equations 7.70 through 7.73 may be used to find the remaining parameters. Other techniques based on graphical methods may also be used to determine pharmacokinetic parameters, as discussed by Cooney (1976), Gibaldi and Perrier (1982), Welling (1986), and Notari (1987).

Example 7.7 illustrates the use of the two compartment model with first order absorption to describe the plasma nicotine levels in a 49-year-old man after smoking one cigarette yielding 0.8 mg of nicotine.

Example 7.7

The following table (McNabb et al. 1982) presents plasma nicotine levels in a 49-year-old man during and after smoking one cigarette yielding 0.8 mg of nicotine. Using the two compartment model given by Equations 7.75 and 7.76, determine the pharmacokinetic parameters.

Time (min)	Plasma Nicotine Concentration (ng mL^{-1})
0	0
2	5
3	7
4	11
5	10
6	14
8	12
9	13
11	13
12	12
15	12
21	9
25	9
35	8
41	7.5
49	7
59	7
71	6

Solution

Figure 7.14 illustrates the solution. We see that the two compartment model fits these data rather well. The following table summarizes the results of the regression analysis.

Parameter	Value
fD/V_{plasma}	15.4 ng mL^{-1}
A	0.035 min^{-1}
B	0.0013 min^{-1}
k_{tp}	0.013 min^{-1}
k_{pt}	0.02 min^{-1}
k_{te}	0.0037 min^{-1}

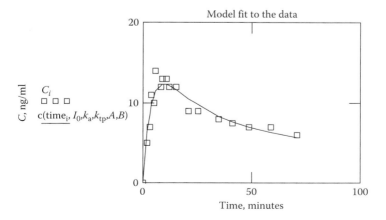

FIGURE 7.14   A comparison of actual and predicted plasma nicotine levels, Example 7.7.

PROBLEMS

1. Derive Equation 7.36.
2. Derive Equation 7.58.
3. Derive Equations 7.59 and 7.60.
4. Derive Equation 7.62.
5. The rate constants in Equation 7.58 (k_a and k_{te}) can also be found using a graphical technique called "feathering" (Notari 1987). Note that the plasma concentration in Equation 7.58 is related to the sum of two exponential terms. In most cases, k_a is larger than k_{te}, hence the second exponential in Equation 7.58 vanishes for long periods of time. Therefore, for long time periods, Equation 7.58 may be approximated as

$$C(t)_{TL} \approx \frac{fD}{V_{apparent}} \left(\frac{k_a}{k_a - k_{te}} \right) e^{-k_{te} t}.$$

   On a log-normal plot (ln C vs. t), this equation predicts for long periods of time, that the concentration data should form a straight line. This region is referred to as the terminal elimination phase. The slope of the terminal line (TL) that fits the data during this period of time is then equal to $-k_{te}$ and the intercept is equal to ln $(fD/V_{apparent})(k_a/(k_a-k_{te}))$. We can also use the equation for $C(t)_{TL}$ to form the difference or residual, $C(t)_{TL} - C(t)$, with the following result:

$$C(t)_{TL} - C(t) = \frac{fD}{V_{apparent}} \left(\frac{k_a}{k_a - k_{te}} \right) e^{-k_a t}.$$

   This equation is valid for all values of time starting from time equal to zero. Note that for short periods of time, $C(t)_{TL}$ is obtained either by extrapolation of the terminal line or by estimating $C(t)_{TL}$ by the first equation above. For long periods of time, this difference is equal to zero. Now, if a plot is made of ln $[C(t)_{TL} - C(t)]$ vs. time, the slope is equal $-k_a$ and the intercept is once again equal to ln $(fD/V_{apparent})(k_a/(k_a-k_{te}))$. With k_a and k_{te} now known, $fD/V_{apparent}$ may be obtained from the value of the intercept.
   Use this feathering technique to obtain the parameters for the data given in Example 7.5. How do these parameters obtained using the feathering technique compare with those obtained from a nonlinear regression? Under what conditions do you expect the feathering technique to be difficult to use?
6. Rework Example 7.6 assuming a single compartment model.
7. You are taking a final examination, and it is unlike any examination you have taken before, plus you had a short night of sleep, you had car problems yesterday, you had too much to drink last night, and the telephone kept ringing early this morning. The result—you have a terrible headache, unlike any headache you have ever had before. You reach for your bottle of a new headache medicine that claims maximum relief within 30 min and contains a well-known pain reliever. An amazing coincidence is that the examination you are taking includes a problem related to the pharmacokinetics of this pain reliever. You are given the following information:

Elimination rate constant	0.277 h^{-1}
Apparent distribution volume	35 L
Fraction absorbed	80%
Therapeutic range	10–20 µg mL^{-1}

   You plan to take this pain reliever. What dose in milligrams should you take? At what time should you think about taking another dose? How much should you take for the second dose?
8. The pharmacokinetics of nicotinamide was reported in humans (Petley et al. 1995). Nicotinamide, a derivative of the B vitamin niacin, is under investigation as a prevention

for insulin-dependent diabetes mellitus. The drug was given orally to eight male subjects with a mean body weight of 75 kg. Two formulations were used, a standard powdered form and a sustained release form. The following data summarize the data obtained for the low dose studies (standard = 2.5 mg kg^{-1} BW^{-1} and sustained = 6.7 mg kg^{-1} BW^{-1}). Analyze these data using an appropriate pharmacokinetic model. Explain any differences in the pharmacokinetics of the two formulations used.

Time (h)	Nicotinamide Plasma Concentration (μg mL^{-1}) Standard dose	Concentration (μg mL^{-1}) Sustained release dose
0	0	0
0.3	3.3	
0.5	2.45	1.75
0.75	1.90	
1.0	1.45	1.5
1.25	1.00	
1.5	0.80	1.1
1.75	0.55	
2.0	0.50	1.05
2.5	0.30	1.1
3.0	0.20	0.67
3.5		0.50
4.0		0.45
4.5		0.50
5.0		0.55
5.5		0.15

9. The following serum insulin levels were obtained following the intranasal administration of 150 U of insulin to 12 healthy men (Jacobs et al. 1993). Analyze these data using an appropriate pharmacokinetic model.

Time (min)	Serum Insulin Concentration (μU mL^{-1})
0	10
15	150
20	220
25	230
30	260
35	250
40	210
45	170
60	140
75	105
90	80
105	50
120	45
150	35
180	35
240	35
300	35

10. The following plasma concentrations for the antibiotic imipenem were obtained from six patients suffering multiorgan failure (Hashimoto et al. 1997). All patients were also anuric because of complete renal failure and were placed on continuous venovenous hemodialysis. The antibiotic was delivered by an intravenous infusion pump for 30 min, resulting in an initial plasma concentration of 32.37 µg mL^{-1}. Following the infusion, blood samples were taken over the next 12 h. The initial dose was 500 mg and 107.7 mg of imipenem was recovered in the dialysate fluid ("urine"). The following table summarizes the plasma concentrations.

Time (min)	Plasma Concentration (µg mL^{-1})
0	32.47
30	24
60	18.5
120	13
180	10
360	4.6
540	2
720	1.12

    Use an appropriate compartment model to analyze these data.

11. Starting with Equation 7.49, derive Equation 7.50 using Laplace transforms.

12. Estimate the SC permeability for testosterone and estradiol. The molecular weight of testosterone and estradiol is 288 and 272, respectively. The respective values of their octanol/water partition coefficients are 2070 and 7000. The reported values of the SC permeability for testosterone is 2.2×10^{-3} cm h^{-1} and for estradiol the value is 3.2×10^{-3} cm h^{-1} (Joshi and Raje 2002).

13. Consider the delivery of a drug from a patch applied to the skin. Assume that the patch has a surface area for drug transport of 30 cm^2 and a thickness of 0.6 mm, that the concentration of the drug in the patch at any time after it has been applied is given by $C_{drug}(t)$, and that the SC permeability of the drug is given by P_{SC}, which for this particular drug has a value of 3.66×10^{-7} cm sec^{-1}. If the initial concentration of the drug in the patch is $C_{drug}(0) = 25$ mg mL^{-1}, how long will it take to reduce the concentration of the drug in the patch to ½ $C_{drug}(0)$, i.e., 12.5 mg mL^{-1}?

14. Alora is a transdermal system designed to deliver estradiol continuously and consistently over a 3–4 day period when placed on the skin. Alora provides systemic estrogen replacement therapy in postmenopausal women. For a 36 cm^2 patch, the release rate of the drug is 0.1 mg of estradiol per day. Clinical trials have provided the following pharmacokinetic data. In one study, following application of the patch, the peak concentration of estradiol was found to be 144 pg mL^{-1} (10^{12} pg = 1 g) and the half-life of estradiol was found to be 1 h. Over an 84 h dosing interval, the plasma clearance for estradiol from the Alora patch was found to be 61 L h^{-1}. What is the steady state plasma concentration for estradiol in picograms per milliliter? What is the apparent volume of distribution in liters?

15. Deponit is a nitroglycerin transdermal delivery system for the prevention of angina pectoris. The in vivo release rate of the 16 cm^2 patch is 0.2 mg h^{-1}. The reported volume of distribution is 3 L kg^{-1} BW^{-1} and the observed clearance rate is 1 L kg^{-1} BW^{-1} min^{-1}. What is the steady state nitroglycerine concentration (nanograms per milliliter) in a 70 kg human? What is the elimination rate constant in 1 min^{-1}? What is the half-life of nitroglycerine in minutes?

16. A pharmacokinetic study of a particular drug taken orally as a single dose provided an AUC$^{0 \to \infty}$ of 1085.5 ng h mL^{-1}, a C_{max} equal to 98.5 ng mL^{-1}, and a τ_{max} of 2.1 h. From these data, determine the values of $V_{apparent}$ (liter), k_{te} (per minute), and k_a (per minute). The dose was 80 mg and the entire drug is available, i.e., $f = 1$.

17. Design a skin patch to deliver nicotine. The goal is to have a patch that delivers nicotine for 24 h, then a new patch is added and so on. Nicotine is somewhat water soluble and has a molecular weight of 162.23 and an octanol–water partition coefficient (log $K_{o/w}$) of 1.2. The dose of nicotine to be absorbed into the body is 21 mg per 24 h and the initial concentration of nicotine within the patch is 25 mg mL^{-1}. Estimate the surface area of the patch. Assuming that the total nicotine content of the patch is 52.5 mg, how much of this will be absorbed? Also, estimate the steady state plasma concentration of nicotine in nanograms per milliliter and then estimate how thick the patch would be. The volume of distribution for nicotine is 2–3 L kg^{-1} of body weight and the average plasma clearance is 1.2 L min^{-1}.

18. Krewson et al. (1995) presented the steady state data in the following table for the distribution of nerve growth factor (NGF) in the vicinity of a thin cylindrical controlled drug release device implanted into the brain of a rat. The polymeric disks containing radiolabeled I^{125}-NGF were 2 mm in diameter and 0.8 mm in thickness. NGF has a molecular weight of 28,000 gmol^{-1}. Assuming that the NGF is eliminated from the brain tissue by a first order process (i.e., the rate of elimination is proportional to the concentration of NGF, i.e., $k_{apparent} C_{NGF}$), develop a steady state reaction–diffusion model to analyze these data. $k_{apparent}$ is the apparent first order elimination rate constant for NGF and accounts for such processes as metabolism, cellular internalization, or uptake by the brain's systemic circulation. The boundary conditions assuming the origin of the Cartesian coordinate system to be the midline of the polymeric disk are

$$C_{NGF} = C_0 \text{ at } x = a \text{ and } C_{NGF} = 0 \, x = \infty,$$

where a is the half-thickness of the polymeric disk. Estimate the value of the Thiele modulus ($a\sqrt{k_{apparent} / D_T}$) that provides the best estimate of the data. D_T represents the diffusivity of NGF in the brain tissue. After estimating a value of D_T for NGF, what is the value of $k_{apparent}$? Using Equation 7.28, calculate the half-life of NGF in the brain tissue. How does this value of the NGF half-life compare to the reported half-life of NGF in brain tissue of about 1 h?

Distance from Polymer/ Tissue Interface (mm)	NGF Concentration (μg mL^{-1})
0	37.0
0.1	31.5
0.2	20.0
0.3	15.0
0.4	10.5
0.5	8.5
0.6	6.5
0.8	3.5
1.0	2.2

19. The following table summarizes plasma nicotine levels in a 27-year-old man chewing one piece of gum containing 4 mg of nicotine (McNabb et al. 1982). Nicotine gum was developed to help people quit smoking by providing the same amount of plasma nicotine as obtained from cigarettes. Nicotine is absorbed from the gum over a 30-min period after an initial short period of chewing the gum to release the nicotine. Nicotine is then readily absorbed within the buccal cavity. From the data provided in this table, explore how well a single compartment model with first order absorption and elimination fit these data. Can a better fit to these data be obtained by assuming that the nicotine is released at a constant rate during the initial 30-min period? In your analyses, determine the pharmacokinetic parameters that describe the absorption and elimination of nicotine.

Time (min)	Plasma Nicotine Concentration (ng mL^{-1})
0	0
2	2.5
3	4
6	4
8	5
10	6
12	7
13	8.5
16	10
18	11
20	12
24	13
28	12.5
32	12
35	11
41	10
48	9
52	9
65	7.5

20. A transdermal delivery system for a drug gives an in vivo release rate of 40 mg h^{-1}. If the apparent distribution volume for this drug is 60 L and the drug half-life is 2.5 h, estimate the steady state drug concentration in milligrams per liter.

21. Following an IV injection of 500 mg of a drug, the following data were obtained for the plasma concentrations of the drug as a function of time. Estimate the apparent distribution volume, the plasma clearance, the half-life, and the value of the total elimination rate constant k_{te}.

Time (min)	Plasma Concentration (μg mL^{-1})
0	100
30	85
60	73
120	58
180	52
360	26
480	16
600	8
720	5
1000	1.6

22. Design an osmotic pump to deliver 150 μL per 90 days of a drug solution. Assume this solution has a density of 1 g cm^{-3} and a viscosity of 1 cP. The flow moderator from which the solution exits the pump consists of a tube that is 50 mm in length with a diameter of 0.05 cm. An excess of sodium chloride is placed on the side opposite the drug compartment and these compartments are separated by a frictionless piston of thickness 5 mm. At 37°C, a saturated solution of NaCl is formed and the concentration of this solution is 36.6 g NaCl per 100 g of water. The density of the NaCl solution that is formed is 1.186 g

cm^{-3} and there is always enough solid NaCl in this compartment to maintain the saturated state. Assume that the pump has a cylindrical shape made of an alloy of steel and that an osmotic membrane is placed as an end cap on the compartment containing the NaCl. This membrane has a hydraulic conductance of $L_p = 6.29 \times 10^{-5}$ µL day^{-1} cm^{-2} mmHg^{-1} and completely retains the NaCl solution. From this information, specify the diameter of the osmotic pump and provide a reasonable estimate of its length. To better acquaint yourself with the calculations that are involved, use the Bernoulli equation approach outlined in Section 7.8.1 to find your answers.

8 Extracorporeal Devices

In this chapter, we will discuss and analyze several examples of extracorporeal devices. These devices lie outside (*extra*) the body (*corporeal*) and are usually connected to the patient by an *arteriovenous shunt*.* In some respects, they may be thought of as artificial organs. Their function is based on the use of physical and chemical processes to replace the function of a failed organ or to remove an unwanted constituent from the blood. Before entering these devices, the patient's blood is infused with anticlotting drugs, such as *heparin*, to prevent clotting. A variety of ancillary equipment may also be present to complete the system, including items such as pumps, flow monitors, bubble and blood detectors, as well as pressure, temperature, and concentration control systems. It is important to note that these devices do not generally contain any living cells. Devices containing living cells are called *bioartificial organs*, and these will be discussed in Chapter 10.

8.1 APPLICATIONS

A variety of extracorporeal devices has been developed. Perhaps, the best known are *blood oxygenators*, which are used in such procedures as open heart surgery, and *hemodialysis*, which replaces the function of failed kidneys. Other examples include *aquapheresis*, for removing excess water from the body; *hemoperfusion*, wherein a bed of activated carbon particles is used for cleansing blood of toxic materials; *plasmapheresis*,† to separate erythrocytes from plasma as a first step in the subsequent processing of the plasma; *immobilized enzyme reactors*, to rid the body of a particular substance or to replace lost liver function; and *affinity columns* to remove materials such as antibodies that attack the body's own cells as in *autoimmune diseases*.

The primary functional unit of extracorporeal devices is typically provided as a sterile disposable cartridge. However, in some cases, considering mounting healthcare costs, the functional unit can be reused many times provided it is cleaned and sterilized between applications. Reused functional units are not shared between patients. Certainly, there is no end to the possibilities for extracorporeal devices. However, whatever the functions of the devices are, there will be some commonality among them in terms of their construction, use of membranes, and fluid contacting schemes.

8.2 CONTACTING SCHEMES

Most devices such as blood oxygenators and dialyzers are typically based on the use of polymeric membranes to create the surface area needed to provide mass transfer between the bloodstream and another exchange fluid stream. The membrane is physically retained within the device by a support structure that also creates the flow channels. As illustrated in Figure 8.1 for hemodialyzers, the membranes are typically arranged as stacks of flat sheets, coils of membrane sheets, or bundles of hollow fiber membranes. Hollow fiber membranes are finding increased use because of their low cost, ease of manufacture, and consistent quality.

---

* A shunt is a means of diverting flow, in this case blood from an artery through a device and then back into the body via a vein.

† See Problem 11 in Chapter 5 for a discussion and analysis of a membrane plasmapheresis system.

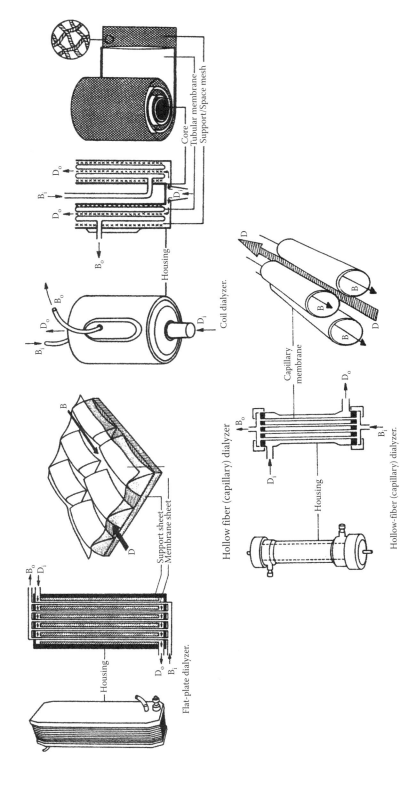

FIGURE 8.1   Membrane configurations. (From Skalak, R. and Chien, S., *Handbook of Bioengineering*, Chapter 39, McGraw-Hill, 1987. With permission.)

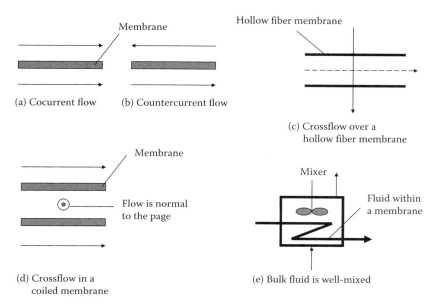

FIGURE 8.2   Fluid contacting patterns.

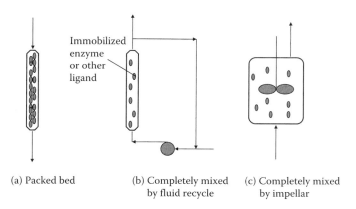

FIGURE 8.3   Immobilized enzyme and affinity reactor systems.

A variety of contacting patterns between the blood and the exchange fluid are also possible, as shown in Figure 8.2. Immobilized enzyme reactors and affinity columns have the active material, i.e., an *enzyme* or other *ligand*, firmly attached to a support particle. Support materials for immobilized enzyme reactors and affinity columns are typically made from a wide variety of polymeric materials such as alginates, agar, carrageenin, and polyacrylamide, or inorganic materials such as glass, silicas, aluminas, and activated carbon. The particles containing the immobilized material may be arranged as a packed bed or the particles may be suspended in a well-mixed device as shown in Figure 8.3. Complete mixing of the fluid phase and suspended particles can be achieved either by a mechanical impeller or by fluid recirculation.

8.3   MEMBRANE SOLUTE TRANSPORT

In many extracorporeal devices, a solute must be transported through the blood, perhaps across a membrane, and then through an exchange fluid. These transport processes are illustrated in Figure 8.4 for the case of simple diffusion. Concentrations of the solute at the fluid–membrane

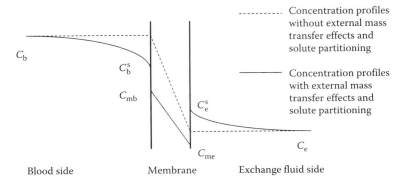

FIGURE 8.4   Concentration profiles for membrane diffusion.

interface are assumed to be at equilibrium. The *partition coefficient* (K) is used to describe the solute equilibrium at the surface of the membrane. As shown in Figure 8.4, the concentration just inside the membrane is related to the concentration adjacent to the outside of the membrane surface by the relationships, $C_{mb} = K C_b^s$ and $C_{me} = K C_e^s$.

An overall *mass transfer coefficient* (K_0) can be defined in terms of the overall mass transfer driving force based on the bulk fluid concentrations ($C_b - C_e$) as shown in Figure 8.4. The overall mass transfer coefficient can be written in terms of the individual film mass transfer coefficients on the blood (k_b) and exchange fluid (k_e) sides along with the membrane permeability (P_m) and the partition coefficient (K). Recall that the transport flux is $N_S = K_0(C_b - C_e) = k_b(C_b - C_b^s) = P_m(C_{mb} - C_{me}) = k_e(C_e^s - C_e)$, which can be rearranged and solved for the overall mass transfer coefficient (K_0) as

$$\frac{1}{K_0} = \frac{1}{k_b} + \frac{1}{KP_m} + \frac{1}{k_e}. \tag{8.1}$$

This equation states that the total mass transfer resistance, $1/K_0$, is simply the sum of the individual mass transfer resistances $((1/k_b) + (1/KP_m) + (1/k_e))$. The smallest value of k_b, KP_m, or k_e for a given solute is said to be the *controlling resistance*.

Recall that we discussed in much detail the membrane permeability in Chapter 5. The membrane permeability, although best measured experimentally (Dionne et al. 1996; Baker et al. 1997), is defined by Equation 5.90. Also as discussed in Chapter 5, the film mass transfer coefficients depend on the physical properties of the fluid and the nature of the flow. They also depend on position due to boundary layer growth, so we usually use length-averaged values of these coefficients in Equation 8.1.

8.4   ESTIMATING THE MASS TRANSFER COEFFICIENTS

Techniques for estimating film mass transfer coefficients for a variety of flow conditions were discussed in Section 5.4.8 in Chapter 5. For a laminar flow of fluids in a tube, one may use Equations 5.52 or 5.53 to estimate the film mass transfer coefficient. For other flow situations, Table 5.2 provides a summary of useful mass transfer coefficient correlations.

Blood flow through extracorporeal devices is typically laminar and, in some cases, may be fully developed. The Sherwood number (Sh) for blood for these conditions, i.e., $k_b h/D_e$ is equal to 3.66. In the Sherwood number, h is the flow channel thickness in a flat plate membrane arrangement or the tube diameter in a cylindrical hollow fiber. The effective diffusivity of the solute in blood is given by D_e and may be quite different from the solute's bulk diffusivity because of the presence of the red blood cells. This is discussed in the next section. Recall from Chapter 5 that the asymptotic limit on the *Sh* number for a fully developed laminar flow is valid when $(z/h) > 0.05\,Re\,Sc$ (concentration field) or $(z/h) > 0.05\,Re$ (velocity field). Here, z is defined as the length of the flow path, h is the characteristic dimension of the flow channel, Re is the Reynolds number, and Sc is the Schmidt number. For solutes in liquids, the Sc is on the order of 1000, so the velocity field will develop much faster than the concentration field.

The exchange fluid will be in either laminar or turbulent flow. Generally, one operates on the exchange fluid side in such a manner that its contribution to the overall mass transfer resistance is negligible. Therefore, $k_e \gg k_b$ and KP_m. If estimates of k_e are needed, then one can use Equations 5.52 or 5.53, if the exchange fluid flow is laminar. For turbulent flow of the exchange fluid, one can consult Table 5.2 or these additional references: Cussler (1984), Bird et al. (2002), and Thomas (1992).

8.5 ESTIMATING THE SOLUTE DIFFUSIVITY IN BLOOD

Transporting a solute through blood can be complicated by the presence of the red blood cells. Estimation of the effective solute diffusivity (D_e) needs to account for the effects of the red blood cells on the solute's bulk diffusivity. We need an estimate of solute diffusivity in order to calculate the blood side value of the Sherwood and Schmidt numbers, as discussed in the previous section. The factors influencing mass transfer within blood are summarized in Table 8.1 (Colton and Lowrie 1981).

The rheological behavior of blood is not a very important factor in describing solute transport, since at the relatively high shear rates encountered in extracorporeal devices blood behaves as a Newtonian fluid. Mass transport theories for Newtonian fluids are well developed and we can make use of these for the special case of blood. The rotation and translation of red blood cells can influence mass transport to some degree; however, these effects can usually be neglected unless sufficient data are available to warrant their inclusion.

A complete description of solute transport in blood requires that one knows the solute diffusivity and its red blood cell permeability, its equilibrium distribution between the plasma and the fluid within the red blood cell, and the kinetics of any chemical reactions it may undergo with other solutes, typically proteins, that may be present. Clearly, this is a complex problem. The diffusion and reaction of oxygen with hemoglobin is a special case that is important in the design of blood oxygenators and is discussed in Section 8.7. However, for most other solutes, we can treat the blood as a nonreactive fluid and assume that either of the following two cases applies.

TABLE 8.1
Factors That Affect Mass Transport in Blood

Suspension Properties
  Volume fraction of red blood cells
  Volume fraction of proteins
  Red blood cell solute permeability
Solute Behavior
  Protein binding kinetics and equilibrium
  Reactions with other solutes
Flow-dependent Behavior
  Rheological properties
  Red blood cell dynamics
  Rouleaux formation
  Migration from the wall
  Rotation and translation

First, if the red blood cell permeability for the solute of interest is extremely large, then the solute will diffuse through the blood as if it were a pseudo-homogeneous fluid and the solute diffusivity will not be affected by the presence of the red blood cells. For this case, one only needs to know the solute diffusivity in plasma, which can be obtained by estimating the aqueous diffusivity by using Figure 5.2, followed by a correction for the difference in viscosity between an aqueous solution and plasma using Equation 5.4. Solute diffusion in plasma amounts to about a 40% reduction in comparison to the aqueous solute diffusivity.

For the second case, the red blood cell may be somewhat permeable to the solute. The solute must therefore diffuse around or through the red blood cells, increasing the solute diffusion path and decreasing the solute effective diffusivity. We can use relationships such as those presented in Chapter 5, i.e., Equations 5.84 and 5.85, to describe this situation. As an example, Colton and Lowrie (1981) investigated the diffusion of urea in both stagnant and flowing blood. They presented data for the ratio of the solute permeability in blood to that in plasma alone as a function of the hematocrit. These results were then used to develop an expression for the effective solute diffusivity in nonreactive blood. The effective diffusion coefficient through blood is defined by Equation 8.2, which is similar to Fick's law (Equation 5.2):

$$N_{\text{effective}} = D_{\text{e}} \frac{dC_{\text{effective}}}{dx}. \tag{8.2}$$

The effective concentration ($C_{\text{effective}}$) represents the volume fraction weighted sum of the solute concentration in the red blood cell phase (C_{RBC}) and the continuous plasma phase (C_{plasma}) and is given by

$$C_{\text{effective}} = HC_{\text{RBC}} + (1-H)C_{\text{plasma}}. \tag{8.3}$$

where H represents the blood hematocrit. Diffusion of the solute through just the continuous or plasma phase ($H = 0$) is given by

$$N = D_{\text{plasma}} \frac{dC_{\text{plasma}}}{dx}. \tag{8.4}$$

Substituting Equation 8.3 into Equation 8.2 for $C_{\text{effective}}$ and dividing the result by Equation 8.4 provides a relationship for the effective solute diffusivity of blood in comparison to its value in plasma alone:

$$\frac{N_{\text{effective}}}{N} = \frac{D_{\text{e}}}{D_{\text{plasma}}} [HK_{\text{RBC}} + (1-H)]. \tag{8.5}$$

In this equation, $K_{\text{RBC}} = C_{\text{RBC}}/C_{\text{plasma}}$ represents the equilibrium partition coefficient for the solute between the red blood cells and the plasma. Average values of K_{RBC} for some typical solutes are summarized in Table 8.2.

Colton and Lowrie (1981) have shown that Equation 8.5 is also equal to the following expression (actually, Equation 5.84 with $\phi = H$) assuming that the red blood cell is impermeable to the solute ($D_{\text{cell}} = 0$):

$$\frac{D_{\text{effective}}}{D_{\text{plasma}}} [HK_{\text{RBC}} + (1-H)] = \frac{2(1-H)}{2+H}. \tag{8.6}$$

TABLE 8.2
Values of K_{RBC} for Some Typical Solutes

Solute	K_{RBC}
Urea	0.86
Creatinine	0.73
Uric acid	0.54
Glucose	0.95

Source: Data from Colton, C.K. and Lowrie, E.G., *The Kidney*, W.B. Saunders, Philadelphia, 1981.

Equation 8.6 can now be used to estimate the effective diffusivity of a solute in blood. This equation suggests that for a solute such as urea, its effective diffusivity through blood is about 53% of its value in plasma or about 33% of its aqueous diffusivity.

With this background on extracorporeal devices, the following discussion will focus on a description and analysis of five representative devices: hemodialysis, aquapheresis, blood oxygenators, enzyme reactors, and affinity columns.

8.6 HEMODIALYSIS

8.6.1 BACKGROUND

The basic functional unit of the kidney is the nephron (shown earlier in Figure 7.3). Each kidney contains about 1 million nephrons. Only about one-third of these nephrons are needed to maintain normal levels of waste products in the blood. If about 90% of the nephrons lose their function, then the symptoms of *uremia* will develop in the patient. Uremia results when waste products, normally removed from the blood by the kidneys, start to accumulate in the blood and other fluid spaces. For example, water generated by normal metabolic processes is no longer removed and accumulates in the body, leading to *edema*, and in the absence of additional electrolytes, almost half of this water enters the cells rather than the extracellular fluid spaces due to osmotic effects. The normal metabolic processes also produce more acid than base, and this acid is normally removed by the kidneys. Therefore, in kidney failure, there is a decrease in the pH of the body's fluids, called *acidosis*, which can result in *uremic coma*. The end products of protein metabolism include such nitrogenous substances as urea, uric acid, and creatinine. These materials must be removed to ensure continued protein metabolism in the body. The accumulation of urea and creatinine in the blood, although not life threatening by themselves, is an important marker of the degree of renal failure. They are also used to measure the effectiveness of dialysis for the treatment of kidney failure. The kidneys also produce the hormone *erythropoietin*, which is responsible for regulating the production of red blood cells in the bone marrow. In kidney failure, this hormone is diminished, leading to a condition called *anemia* and a lowered hematocrit. If kidney failure is left untreated, death can occur within a few days to several weeks.

Hemodialysis (HD) was developed in the 1940s (Kolff 1947) to treat patients with degenerative kidney failure or end-stage renal disease (ESRD). HD has the ability to replace many functions of the failed kidneys. For example, HD removes the toxic waste products from the body, maintains the correct balance of electrolytes, and removes excess fluid from the body. HD can keep patients alive for several years and, for many patients, allows them to survive long enough to receive a kidney transplant. Nearly 500,000 patients in the United States are currently being kept alive by HD.

Figure 8.5 illustrates the basic operation of HD. Heparinized blood flows through a membrane device known as the *hemodialyzer*. The dialysate or exchange fluid flows on the membrane side

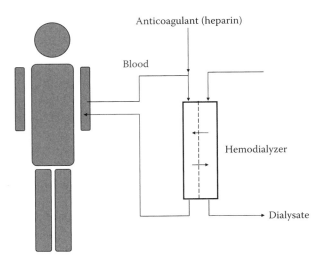

FIGURE 8.5 Hemodialysis.

opposite the blood. Solutes are exchanged by diffusion between the blood and the dialysate fluid. A variety of membrane configurations are possible and some of the typical ones are shown in Figure 8.1. A variety of flow patterns for the blood and dialysate has been used and these are also summarized in Figure 8.2. The membranes were originally made from cellulose and cellulose derivatives, such as cellulose acetate; however, these have been replaced for the most part by synthetic polymers, such as polyacrylonitrile, polycarbonate, polyvinylpyrolidone, polyarylethersulfone, and polyamide (Zelman 1987; Galletti et al. 1995; Clark and Gao 2002). The membrane surface area is on the order of 1–2 m². Blood flow rates are in the range of several hundred milliliters per minute, and the dialysate flow rate is about twice that of the blood.

8.6.2 DIALYSATE COMPOSITION

Table 8.3 compares the species present in normal and uremic plasma with those of a typical dialysate fluid. The composition of the dialysate is based on the need to restore the uremic plasma to the normal state. Note in particular the high levels of urea and creatinine in the uremic plasma. For the most part, movement of the species in Table 8.3 between the uremic plasma and the dialysate is by diffusion. The higher level of HCO_3^- in the dialysate is used to decrease the acidity of the plasma through its buffering action. Because of the reduced concentrations, the osmolarity of the dialysate is about 265 mOsM in comparison to about 300 mOsM or so for the plasma. Hence, water will tend to leave the dialysate and enter the plasma by osmosis. Since the patient needs to have excess water removed during dialysis, it is then necessary to maintain the dialysate pressure below atmospheric pressure in order to develop a transmembrane pressure gradient that is sufficient to overcome osmosis and remove water from the patient. Removal of waste products from the plasma must also be carefully controlled in order to avoid the *disequilibrium syndrome*. If waste products are removed from the blood too fast by dialysis, then the osmolarity of the plasma becomes less than that of the cerebrospinal fluid, resulting in a flow of water from the plasma and into the spaces occupied by the brain and spinal cord. This increases the local pressure in these areas and can lead to serious side effects.

8.6.3 ROLE OF ULTRAFILTRATION

The rate at which water is removed (ultrafiltration) from the plasma by the pressure gradient across the dialysis membrane can be estimated using Equation 5.102. These equations can

TABLE 8.3
Composition of Normal and Uremic Plasma and Dialysate

Species Electrolytes (mEq L^{-1})	Normal Plasma	Uremic Plasma	Dialysate
Na$^+$	142	142	133
K$^+$	5	7	1
Ca^{2+}	3	2	3
Mg^{2+}	1.5	1.5	1.5
Cl$^-$	107	107	105
HCO$_3^-$	27	14	35.7
Lactate$^-$	1.2	1.2	1.2
HPO$_4^-$	3	9	0
Urate$^-$	0.3	2	0
SO$_4^-$	0.5	3	0
Nonelectrolytes (mg dL^{-1})			
Glucose	100	100	125
Urea	26	200	0
Creatinine	1	6	0

Source: Data from Guyton, A.C., *Textbook of Medical Physiology*, W.B. Saunders, Philadelphia, 1991.

Note: Equivalents (Eq) are the amounts of substances that have the same combining capacity in chemical reactions.

be simplified by assuming only the plasma proteins are impermeable. Hence, Equation 8.7 is obtained:

$$Q = SL_P (\Delta P_{mean} - \pi_{oncotic\ plasma}), \tag{8.7}$$

where $\pi_{oncotic\ plasma}$ represents the oncotic pressure of the plasma proteins, about 28 mmHg, and ΔP_{mean} is the average transmembrane pressure difference, which is given by

$$\Delta P_{mean} = \left[\frac{P_{B,in} + P_{B,out}}{2} - \frac{P_{D,in} + P_{D,out}}{2} \right]. \tag{8.8}$$

Note that to avoid problems with the calculations it is best to simply use absolute pressures in Equation 8.8, recognizing that since the dialysate is subatmospheric, its reported gauge pressure will be negative. The pressure drop on the blood side is typically on the order of 20 mmHg and that on the dialysate side is about 50 mmHg. The mean transmembrane pressure drop is on the order of a few hundred mmHg. The hydraulic conductance, L_P, is typically about 3 mL h^{-1} m^{-2} mmHg^{-1}. High flux membranes can have values of hydraulic conductance as high as 20 mL h^{-1} m^{-2} mmHg^{-1}; however, protein deposition on the plasma side of the membrane can reduce this in a linear manner by about 6% per hour (Zelman 1987). Example 8.1 illustrates the calculation of the ultrafiltration rate in a typical membrane dialyzer.

Example 8.1

Using the membrane properties listed in the previous discussion, calculate the ultrafiltration rate of water in a hemodialyzer with a membrane surface area of 1 m^2. How much water is removed after

6 h of dialysis? Assume blood enters the device at 120 mmHg (gauge) and leaves at 100 mmHg (gauge). The dialysate fluid enters at –150 mmHg (gauge) and leaves at –200 mmHg (gauge).

Solution

$$Q = 1\,\text{m}^2 \times \frac{3\,\text{mL}}{\text{hm}^2\,\text{mmHg}} \times \left[\frac{(120+760)+(100+760)}{2} - \frac{(760-150)+(760-200)}{2} - 28 \right]\text{mmHg}$$

$$= 771\,\text{mL h}^{-1} = 13\,\text{mL min}^{-1}$$

After 6 h of dialysis, about 10 pounds of water will have been removed.

8.6.4 CLEARANCE AND DIALYSANCE

Solute transfer in the dialyzer can be analyzed with the help of the model in Figure 8.6, which shows the cocurrent flow of the blood and dialysate. The Q's represent the volumetric flow rates (milliliters per minute) of the blood (b) and the dialysate (d) and the C's represent the species concentration (usually in milligrams per deciliter). The mass transfer rate of species i across the dialysis membrane is given by

$$M_i = Q_b^{in} C_{b_i}^{in} - Q_b^{out} C_{b_i}^{out} = Q_d^{out} C_{d_i}^{out} - Q_d^{in} C_{d_i}^{in}. \tag{8.9}$$

The *clearance of the dialyzer* for a particular solute is defined as the volumetric flow rate of blood (milliliters per minute) entering the dialyzer that is completely cleared of the solute. The dialyzer clearance (CL_D) for solute i is then given by

$$CL_D = \frac{M_i}{C_{b_i}^{in}}. \tag{8.10}$$

The clearances for such solutes as urea, creatinine, and uric acid are on the order of 100 mL min⁻¹. Higher molecular weight solutes show a reduced clearance.

The term *dialysance* is also used to describe the solute removal characteristics of a dialyzer. Dialysance (D_B) is defined by Equation 8.11. We see that the mass transfer rate of species i (M_i) is now divided by the maximum solute concentration difference, i.e., ($C_{b_i}^{in} - C_{d_i}^{in}$), between the blood and the dialysate:

$$D_B = \frac{M_i}{C_{b_i}^{in} - C_{d_i}^{in}}. \tag{8.11}$$

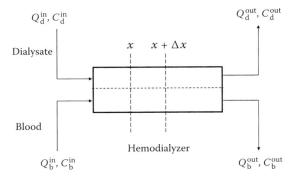

FIGURE 8.6 Solute mass balance model for a cocurrent hemodialyzer.

For a single pass dialyzer, $C_{d_i}^{in} = 0$, and the dialysance is then the same as the clearance.

Because of water removal, the inlet and outlet flow rates of the blood and dialysate streams are not equal. This water removal is by ultrafiltration across the HD membrane and is given by Equation 8.12 in terms of the flow rates of the blood and dialysate streams:

$$Q = Q_b^{in} - Q_b^{out} = Q_d^{out} - Q_d^{in}. \tag{8.12}$$

If ultrafiltration is important, then we can use this relationship in combination with Equations 8.9 and 8.10 to derive an expression for the clearance that allows for examination of the importance of ultrafiltration on the clearance of a given solute. Explicitly including the ultrafiltration flow given by Q, the solute clearance for a particular solute is expressed by

$$CL_D = Q_b^{in}\left(\frac{C_b^{in} - C_b^{out}}{C_b^{in}}\right) + Q\left(\frac{C_b^{out}}{C_b^{in}}\right). \tag{8.13}$$

Since, in general, the value of $C_b^{out} \ll C_b^{in}$, the contribution of the ultrafiltration flow to the solute clearance is less than the value of Q. As shown in Example 8.1, Q is on the order of 10 mL min^{-1}, and earlier it was stated that the clearance of such solutes as urea is on the order of 100 mL min^{-1}. Therefore, we can conclude that the effect of ultrafiltration on the clearance of small molecular weight solutes is negligible. However, for higher molecular weight solutes, the effect of ultrafiltration may be significant.

8.6.5 SOLUTE TRANSFER

We can now develop relationships between solute clearance and the mass transfer characteristics of the membrane dialyzer. In order to perform this analysis, we first must choose one of the contacting patterns shown in Figure 8.2, to describe the flow of blood and dialysate in the dialyzer. Here, we will use the cocurrent pattern, as shown in Figure 8.6. We will also assume that the overall mass transfer coefficient given by Equation 8.1 is constant. The flow rates of the blood and dialysate are also assumed to be constant, that is ultrafiltration is ignored.

A shell mass balance for the blood and dialysate can then be written for a given solute as

$$Q_b C_b|_x - Q_b C_b|_{x+\Delta x} = K_0 W \Delta x (C_b - C_d)$$
$$Q_d C_d|_x - Q_d C_d|_{x+\Delta x} = -K_0 W \Delta x (C_b - C_d). \tag{8.14}$$

The quantity, $W\Delta x$, represents the available mass transfer area within the shell volume ΔV. Hence, W is the membrane surface area per unit length of the dialyzer. If L is the length of the dialyzer, then the quantity $A = L \times W$ is the total membrane area within the dialyzer. Dividing by Δx, and taking the limit as Δx approaches zero, results in the following two differential equations that provide the position dependence of the solute concentration in the dialyzer blood and dialysate:

$$Q_b \frac{dC_b}{dx} = -K_0 W (C_b - C_d)$$
$$Q_d \frac{dC_d}{dx} = K_0 W (C_b - C_d). \tag{8.15}$$

Equation 8.15 can be integrated analytically, provided we first can relate C_d and C_b. This can be done by adding the above two equations, multiplying by dx, and then integrating between the entrance and any arbitrary value of x. The following result is then obtained for the value of C_d in terms of C_b. Note that $z \equiv Q_b/Q_d$:

$$C_d(x) = C_d^{in} - z(C_b(x) - C_b^{in}).$$ (8.16)

Substituting this equation for C_d in Equation 8.15, we obtain the following differential equation that describes the solute concentration on the blood side. We see that the only dependent variable is the blood concentration:

$$Q_b \frac{dC_b}{dx} = -K_0 W[(1+z)C_b - (C_d^{in} + zC_b^{in})].$$ (8.17)

Equation 8.17 can now be integrated analytically and rearranged to give the following result. This equation is known as the *performance equation* for a cocurrent dialyzer.

$$E = \frac{D_B}{Q_b} = \frac{C_b^{in} - C_b^{out}}{C_b^{in} - C_d^{in}} = \frac{1 - \exp[-N_T(1+z)]}{(1+z)}.$$ (8.18)

Two additional dimensionless parameters have been defined in the derivation of this equation, the *extraction ratio* (E), where $0 \leq E \leq 1$; and the *number of transfer units*, $N_T = K_0 A/Q_b$. The number of transfer units provides a measure of the mass transfer effectiveness, i.e., the amount of solute transported across the membrane vs. the amount of solute in the blood that enters the device. The concentration of a given solute in the blood exiting the dialyzer is then given by $C_b^{out} = C_b^{in}(1 - E) + EC_d^{in}$.

If the membrane area is infinite for a given value of z (i.e., for a given value of the dialysate and blood flow rates), then we achieve the maximum possible extraction ratio, i.e., $E_{maximum} = 1/(1+z)$. This corresponds to the best possible performance in a dialyzer for a given set of blood and dialysate flow rates, and is independent of the solute concentrations. This also implies that the solute concentration in the exiting blood and dialysate is equal or in equilibrium, i.e., $C_b^{out} = C_d^{out} = C_{equilibrium}$. For these conditions, the maximum achievable dialysance for any combination of blood and dialysate flows is also given by $D_{B_{maximum}} = z/(z+1) = (Q_b Q_d)/(Q_b + Q_d)$.

If $z = 0$, i.e., the dialysate flow rate is significantly higher than that of the blood, and if the membrane area was also infinite, then we obtain $E = 1$. For these conditions, the solute concentration in the blood exiting the dialyzer achieves its lowest possible value, i.e., the value of the solute concentration in the entering dialysate fluid, C_d^{in}.

Performance equations for other fluid contacting patterns, such as countercurrent, a well-mixed dialysate, and cross flow, are summarized in Table 8.4. Example 8.2 illustrates the use of these performance equations.

Example 8.2

Estimate the clearance of urea in a 1 m² cocurrent and countercurrent hollow fiber dialyzer. The dialyzer hollow fibers are 25 cm in length and they have an inside diameter of 250 μm. The wall thickness of the fibers is about 40 μm. The void volume (ε) in the dialyzer is 50%. Blood flows within the hollow fibers and the dialysate flows in a single pass on the shell side of the device parallel to the fibers. The blood flow rate is 200 mL min⁻¹ and the dialysate flow rate is 500 mL min⁻¹. The membrane permeability, KP_m, is estimated for urea to be about 10^{-3} cm sec⁻¹.

Solution

For the blood flowing through the fibers, the characteristic dimension in the *Sh* number is the internal diameter of the fiber. For flow outside and parallel to the fibers, the characteristic dimension

---

TABLE 8.4
Performance Equations for Kidney Dialyzers

Contacting Pattern	Performance Equation
Cocurrent	$E = \dfrac{1 - \exp\left[-N_T(1+z)\right]}{1+z}$
Countercurrent	$E = \dfrac{\exp\left[N_T(1-z)\right] - 1}{\exp\left[N_T(1-z)\right] - z}$
Well-mixed dialysate	$E = \dfrac{1 - \exp(-N_T)}{1 + z\left[1 - \exp(-N_T)\right]}$
Cross flow	$E = \dfrac{1}{N_T}\sum_{n=0}^{\infty}\left[S_n(N_T)S_n(N_T z)\right]$ where $S_n(x) = 1 - \exp(-x)\sum_{m=0}^{n}\left[\dfrac{x^m}{m!}\right]$

Source: From Colton, C.K. and Lowrie, E.G., *The Kidney*, W.B. Saunders, Philadelphia, 1981.

---

in the dialysate *Sh* number is the hydraulic diameter, D_H, defined by Equation 5.54 as (Yang and Cussler 1986):

$$D_H = \frac{4(\text{cross-sectional area})}{\text{wetted perimeter}}.$$

Assuming the fibers are arranged on a square pitch, the following relationship can be developed to express the relationship between the diameter of the hollow fiber (*d*), the dialyzer void fraction (ε), and the hydraulic diameter (D_H):

$$D_H = d\left(\frac{\varepsilon}{1-\varepsilon}\right) \text{ square pitch.}$$

The apparent velocity of the dialysate through the void space between the fibers is equal to the dialysate volumetric flow rate (Q_d) divided by the cross-sectional area of the void space. The cross-sectional area of the void space is equal to the void fraction (ε) times the cross-sectional area of the dialyzer itself. A little algebra provides the following result for the apparent dialysate velocity, where N_{fiber} is the number of hollow fibers in the dialyzer:

$$V_d = \frac{4(1-\varepsilon)Q_d}{N_{fiber}\varepsilon\pi d^2}.$$

The number of fibers in the dialyzer can be found from the given membrane area and the dimensions of a single hollow fiber, hence $N_{fiber} = A_{membrane}/(\pi\, d\, L)$. After performing the calculations using the performance equations given in Table 8.4 for the cocurrent and countercurrent cases, the predicted urea dialyzer clearance is about 120 mL min^{-1} for the cocurrent dialyzer and about 133 mL min^{-1} for the countercurrent case. These urea clearance values are typical of what is observed in actual practice. In comparison, the urea clearance of the two normal kidneys is about 70 mL min^{-1}. In general, the countercurrent dialyzer will provide the higher clearance, everything else being equal. This is because the countercurrent flow pattern maintains a relatively constant mass transfer driving force along the length of the dialyzer. The cocurrent pattern provides a larger driving force at the entrance of the dialyzer, but this difference rapidly decreases along the length of the dialyzer, resulting in less overall mass transfer of the solute.

8.6.6 SINGLE COMPARTMENT MODEL OF UREA DIALYSIS

Urea distributes throughout the total body water for a total distribution volume of about 40 L. The distribution volume (in liters) can also be estimated to be 58% of body weight (in kilograms) (Galletti et al. 1995). We can treat the body as a single well-mixed compartment and use the dialyzer performance equations to predict how the amount of urea within the body decreases with time during dialysis. Neglecting the effects of ultrafiltration flow, the clearance of urea from the body during dialysis is similar to the intravenous (IV) bolus injection examined in Chapter 7 (Equation 7.27), where the only elimination pathway for urea is the dialyzer. Therefore, k_{te} equals the dialyzer clearance divided by the distribution volume for urea:

$$C_{urea} = C_{urea}^0 e^{-(CL_D/V_{apparent})t}. \tag{8.19}$$

Using the urea dialyzer clearance value for a countercurrent unit obtained from Example 8.2, we can show that after 4 h of dialysis, the urea concentration in the body has been reduced by about 55%. It should be pointed out that during this dialysis period, the amount of urea generated is assumed negligible in comparison to the amount removed by dialysis. However, after dialysis, urea continues to accumulate at the rate of about 10 g each day (G_{urea}). If we assume that the kidney has no residual clearance for urea, then the urea concentration will increase linearly with time after dialysis, according to the following equation:

$$C_{urea}(t) = C_{urea}\left(\begin{smallmatrix}\text{end of}\\\text{dialysis}\end{smallmatrix}\right) + \left(\frac{G_{urea}}{V_{apparent}}\right)t. \tag{8.20}$$

This equation can be used to show that dialysis will be required about 3–4 times per week. For some solutes, the removal from the dialyzer is so fast that the rate of solute transfer from the extravascular spaces into the plasma will be the rate limiting step. For these cases, a two-compartment model may be needed to describe the kinetics of solute removal. The following references provide additional information on the artificial kidney (Ramachandran and Mashelkar 1980; Colton and Lowrie 1981; Armer and Hanley 1986, Abbas and Tyagi 1987; Zelman 1987; Lee and Chang 1988; Lee et al. 1989; Shaoting et al. 1990; Capello et al. 1994; Galletti et al. 1995).

8.6.7 PERITONEAL DIALYSIS

An alternative approach for treating kidney failure is called *continuous ambulatory peritoneal dialysis* (CAPD) (Lysaght and Moran 1995; Lysaght and Farrell 1989). Unlike the HD technique just described, CAPD has the advantage of the patient not being severely restricted by the regimen of multiple treatments per week at a dialysis center. Rather, the patient is responsible for performing a relatively simple maintenance process that can take place at home or even at work. Use of CAPD is growing rapidly. CAPD was used in about 4% of the total dialysis population in 1979 and now it is used in about 15% of the dialysis population.

CAPD is based on the addition to the peritoneal cavity of a sterile hypertonic solution of glucose and electrolytes. The peritoneal cavity is a closed space formed by the peritoneum, a membrane-like tissue that lines the abdominal cavity and covers the internal organs, such as the liver and the intestines. The peritoneum has excellent mass transfer characteristics for a process like CAPD. The peritoneum has the appearance of a fairly transparent sheath that is smooth and quite strong. The surface area (S) of the peritoneum is about 1.75 m². Its thickness ranges from 200 to over 1000 μm. The surface of the peritoneum presented to the CAPD dialysate solution consists of a single layer of mesothelial cells that is densely covered with microvilli or tiny hair-like projections. Beneath this layer of cells is the interstitium, which has the characteristics of a gel-like material.

Within the interstitium, there is a rich capillary network providing a total blood flow on the order of 50 mL min^{-1}.

The CAPD solution is added and later removed from the peritoneal cavity through an in-dwelling catheter. This process can also be automated such that all fluid exchanges are carried out by a pumping unit, even while the patient is asleep. This automated process is called *automated peritoneal dialysis* (APD).

The intraperitoneal CAPD fluid partially equilibrates across the peritoneal membrane with waste products in the plasma. Excess water from the patient's body is also removed by ultrafiltration across the peritoneal membrane as a result of osmotic gradients. Typically, four 2-L exchanges are made each day with an additional 2 L of water removed as a result of ultrafiltration. The average mass removal rate of a particular solute is given by

$$\dot{m}_{\text{CAPD}} = \frac{V_{\text{CAPD}} C_{\text{CAPD}}}{t_{\text{CAPD}}}. \tag{8.21}$$

In this equation, V_{CAPD} represents the volume of CAPD solution and C_{CAPD} is the final concentration of the solute in this solution. The time that the solution is in the peritoneal cavity is given by t_{CAPD}. Recall that the clearance of a solute is equal to its mass removal rate divided by the solute concentration in the blood. Therefore, solute clearance during CAPD is given by

$$CL_{\text{CAPD}} = \frac{\dot{m}_{\text{CAPD}}}{C_b} = \frac{V_{\text{CAPD}} C_{\text{CAPD}}}{t_{\text{CAPD}} C_b}. \tag{8.22}$$

Since the CAPD solution is usually completely equilibrated with urea in the plasma, the urea clearance is then simply $V_{\text{CAPD}}/t_{\text{CAPD}}$. About 10 L of solution is used each day, which equates to a continuous urea clearance of about 7 mL min^{-1}.

The mass transfer characteristics of CAPD can be described using relationships we have already developed. For example, the ultrafiltration rate is given by Equation 5.102a:

$$Q = S L_p \left[\Delta P - RT \sum_i \sigma_i \left(C_{\text{blood}} - C_{\text{CAPD}} \right)_i \right]. \tag{8.23}$$

CAPD usually begins with a solution containing about 4% glucose. The initial ultrafiltration rate for this solution is on the order of 20 mL min^{-1} and decreases rapidly with time as the glucose, initially in the peritoneal cavity dialysate fluid, is removed by the blood. Since ΔP across the peritoneal membrane is small during CAPD, and the osmotic pressure of the initial glucose solution is about 4300 mmHg, then the hydraulic conductance is on the order of 0.15 mL h^{-1}m^{-2} mmHg^{-1}.

The solute transport rate for species i is given by Equation 5.102b, recognizing that initially significant quantities of solute may be transported by ultrafiltration represented by the first term as

$$\dot{m}_{\text{CAPD}} = C_i (1 - \sigma_i) Q + P_{m_i} S (C_{\text{blood}} - C_{\text{CAPD}})_i. \tag{8.24}$$

Typical values of the reflection coefficient and the permeability-surface area product for several solutes are summarized in Table 8.5.

Mathematical models describing CAPD can also be developed to describe the time-course of the treatment process. The patient is considered to be represented by a single well-mixed compartment of apparent distribution volume (V_{apparent}) and solute concentration (C_{body}), which is assumed to

TABLE 8.5
Transport Properties of the Peritoneal Membrane

Solute	Molecular Weight	Reflection Coefficient	$P_m S$ (cm^3 min^{-1})
Urea	60	0.26	21
Creatinine	113	0.35	10
Uric acid	158	0.37	10
Vitamin B-12	1355		5
Inulin	5200	0.5	4
β_2 microglobulin	12,000		0.8
Albumin	69,000	0.99	

Source: Data from Lysaght, M.J. and Moran, J., *The Biomedical Engineering Handbook*, CRC Press, Boca Raton, FL, 1995.

be the same as the solute concentration in the blood (C_{blood}). The CAPD solution in the peritoneal cavity is in a much smaller variable volume (about 2 L at the start), represented by V_{CAPD} and solute concentration C_{CAPD}. The peritoneal membrane separates these two compartments. The peritoneal lymphatic system also removes fluid from the peritoneal cavity (Q_{lymph}) at a rate of about 0.1 to as high as 10 mL min^{-1}. During the initial phases of CAPD, the volume of fluid in the peritoneal cavity will increase due to osmotic flow of water into the cavity; however, as time progresses and glucose is diluted and removed from the peritoneal cavity, the volume of fluid in the peritoneal cavity will decrease as a result of the lymphatic flow.

Models for describing CAPD are reviewed by Lysaght and Farrell (1989). These models range from relatively simple analytical solutions to much more complex models, including selective transport of solutes, ultrafiltration, and the effects of lymphatic flow. These models all require numerical solutions. We will subsequently discuss two relatively simple analytical models for CAPD.

The simplest analytical model neglects ultrafiltration and lymphatic flow and assumes that only the peritoneal dialysate concentration changes with time. This diffusion-only model, with the assumption that V_{CAPD} is constant, limits use of this approach to the constant volume phase (isovolemic) that occurs about an hour or so after the exchange begins. For this diffusion-only CAPD model, the solute balance equation only needs to be written for the fluid in the peritoneal cavity:

$$\frac{d(V_{CAPD} C_{CAPD})}{dt} = V_{CAPD} \frac{dC_{CAPD}}{dt} = P_m S(C_{body} - C_{CAPD}). \tag{8.25}$$

This equation is easily integrated to obtain the following equations. In these equations, the initial concentration of the solute in the CAPD solution is given by C_{CAPD}^0:

$$P_m S = \frac{V_{CAPD}}{t} \ln\left(\frac{C_{body} - C_{CAPD}^0}{C_{body} - C_{CAPD}(t)}\right), \tag{8.26}$$

$$C_{CAPD}(t) = C_{body} - (C_{body} - C_{CAPD}^0)\exp\left(-\frac{P_m S t}{V_{CAPD}}\right). \tag{8.27}$$

Equation 8.26 allows for determination of the value of $P_m S$, given information on the solute concentrations at a particular time. Equation 8.27 allows for prediction of the solute concentration in the

CAPD dialysate solution as a function of time, assuming the transport properties of the peritoneal membrane are known.

A slightly more complex model than the one shown earlier can be developed that includes the effect of ultrafiltration (Garred et al. 1983). However, the solute concentration in the body is still assumed to be constant, the peritoneal membrane is not selective, i.e., the reflection coefficients are zero, and lymphatic flow is ignored. The solute balance for the CAPD solution for these conditions can be written as

$$\frac{d(V_{CAPD}C_{CAPD})}{dt} = P_m S(C_{body} - C_{CAPD}) + C_{body}\frac{dV_{CAPD}}{dt}. \tag{8.28}$$

The second term on the right-hand side, $(C_{body}(dV_{CAPD}/dt))$, represents the ultrafiltration flow that carries with it the solute in the body. Integration of this equation still requires knowledge of how V_{CAPD} changes with time. This could be obtained from Equation 8.23; however, an analytical solution would then be difficult to obtain. If one assumes that the initial and final CAPD volumes are known, then this equation can be integrated using an average value of V_{CAPD} to give the following two equations, where $\bar{V}_{CAPD}$ is the average volume defined as the mean of the initial and final CAPD volumes:

$$P_m S = \frac{\bar{V}_{CAPD}}{t}\ln\left[\frac{V_{CAPD}^0(C_{body} - C_{CAPD}^0)}{V_{CAPD}(t)(C_{body} - C_{CAPD}(t))}\right], \tag{8.29}$$

$$C_{CAPD}(t) = C_{body} - \frac{V_{CAPD}^0}{V_{CAPD}(t)}(C_{body} - C_{CAPD}^0)\exp\left(-\frac{P_m St}{\bar{V}_{CAPD}}\right). \tag{8.30}$$

Once again, Equation 8.29 may be used to estimate the mass transport properties of the peritoneal membrane from given values of the solute concentrations and volumes at a particular time. Equation 8.30 allows determination of the time-course of the solute concentration in the dialysate solution.

8.6.8 Aquapheresis

Congestive heart failure (CHF) is a very complex chronic disease in which the heart loses its ability to provide a sufficient flow of blood to meet the demands of the body. CHF is usually a result of myocardial infarction (i.e., heart attack), heart disease such as cardiomyopathy, high blood pressure, and problems with the valves in the heart. Symptoms of CHF include shortness of breath (*dypsnea*), venous swelling or congestion, coughing, swelling of the lower extremities, and difficulty exercising. CHF severely affects the quality of life, since there are frequent life-threatening episodes where plasma fluid volume increases significantly and fluid accumulates (congestion) in the lungs, peripheral tissues, and abdominal organs, such as the liver.

Because of the excess fluid in the body (fluid overload), treatment of CHF involves removing the excess sodium and water to achieve a proper fluid balance in the body. This can be achieved by treating the patient with a low salt diet, restricting fluid intake, and using diuretic drugs, which increase urine production. However, in many cases, these treatment methods periodically stop working and the patient requires hospitalization in order to make adjustments to their drug regimen so that the excess water can be removed. However, some of these patients with frequent hospitalization have also developed a resistance to diuretics and other means are used to remove the excess water. One approach for these cases is to use aquapheresis.

Aquapheresis involves the use of hollow fiber membrane cartridges similar in some respects to those that are used in HD. However, in aquapheresis the goal is to only remove the excess salt and water. In this case, heparanized blood from the patient is pumped through the hollow fiber membrane cartridge and then returned to the patient. Water and salt are filtered across the membrane by ultrafiltration, and this fluid is collected in a bag and disposed. The flow rate of the water is controlled by the transmembrane pressure differences and by the properties of the hollow fiber membrane. Equations 8.7 and 8.8 can be used to predict the filtration flow of the water, as shown in Example 8.1. In aquapheresis, up to 500 mL of water can be removed per hour and the average rate is about 250 mL h^{-1}. The total treatment time for aquapheresis is about 24 h.

8.7 BLOOD OXYGENATORS

8.7.1 BACKGROUND

Surgery on the heart to repair internal defects, replace valves, improve blood flow to the heart muscle by coronary artery bypass grafting, as well as heart transplants, requires that the heart be stopped or arrested (*cardioplegia*). Cardioplegia can be achieved by infusing a cold (4°C) cardioplegic solution into the coronary circulation. Once the heart is arrested, blood is no longer pumped throughout the body, and the blood is no longer oxygenated by the lungs. Blood flow to the chambers of the heart is also stopped, providing a dry and bloodless field for surgery. Over 700,000 open heart surgeries are performed in the United States every year. Special devices called *heart-lung machines* or *extracorporeal blood pump-oxygenators* (Richardson 1987; Makarewicz et al. 1993; Galletti and Colton 1995) have been developed and used for over 50 years (Gibbon 1954) to replace the gas exchange function of the lungs and the pumping action of the heart during open heart surgical procedures.

Blood flow to the blood pump-oxygenator is usually collected in a reservoir by a roller or peristaltic pump from the large systemic veins, such as the vena cava, or from the right atrium of the heart. This blood is then pumped through the oxygenator and returned to the body via the ascending portion of the aorta. Blood flow rates for a particular oxygenator are typically in the range of 0.5–7 L min^{-1}. A mixture of oxygen and carbon dioxide also passes through the blood oxygenator at flow rates between 5 and 10 L min^{-1}.

In order to meet the metabolic requirements of the body, the oxygenator will need to be able to deliver around 250 mL min^{-1} of oxygen and remove about 200 mL min^{-1} of carbon dioxide from the blood. These rates are based on body temperature and atmospheric pressure (BTP, 37°C and 1 atm). In addition, there are also blood suction devices to minimize the loss of blood and to keep the surgical field free of blood. The suctioned blood is collected and filtered before being pumped into the venous reservoir, where it is then sent to the oxygenator. Patients are also given heparin to prevent blood clotting in the pump-oxygenator flow circuit. In addition, all of the blood-contacting surfaces of the flow circuit are coated with proprietary heparin formulations to minimize the risk of blood clot formation and to minimize the amount of systemic heparin that is given to the patient.

In addition to blood pump-oxygenators for heart surgery, much interest has also focused on the development of intravascular lung assist devices (ILADs) (Vaslef et al. 1989; Jurmann et al. 1992; Makarewicz et al. 1993; Fukui et al. 1994; Hewitt et al. 1998; Guzman et al. 2005). ILADs are being used for the treatment of patients who suffer from adult respiratory distress syndrome (ARDS). ILADs consist of a bundle of hollow fiber membranes that are mounted on a catheter and then inserted into the vena cava. Pure oxygen is supplied within the hollow fibers by an external flow control system. Blood flows externally along the outer surfaces of the hollow fibers, and oxygen and carbon dioxide are exchanged.

8.7.2 OPERATING CHARACTERISTICS OF BLOOD OXYGENATORS

There are several important differences between the operational characteristics of the lungs and blood oxygenators. These are summarized in Table 8.6. The blood flow through the blood oxygenator

TABLE 8.6
Operational Characteristics of the Lungs and Blood Pump-Oxygenators

Characteristic	Lungs	Blood Oxygenator
Blood flow rate	5 L min^{-1}	0.5–7 L min^{-1}
Pressure head	12 mmHg	0–200 mmHg
Blood volume	1 L	1–4 L
Blood film thickness	5–10 μm	100–300 μm
Length of blood flow channel	100 μm	2–30 cm
Blood contact time	0.7 sec	3–30 sec
Surface area for mass transfer	140 m^2	1–10 m^2
Gas flow rates	7 L min^{-1}	2–10 L min^{-1}
pO_2 and pCO_2 blood in	40 and 45 mmHg	40 and 45 mmHg
pO_2 and pCO_2 blood out	95 and 40 mmHg	100–300 and 30–40 mmHg
Gas pO_2 and pCO_2 in	149 and 0.3 mmHg	250–713 and 0–20 mmHg
Gas pO_2 and pCO_2 out	120 and 27 mmHg	150–675 and 10–30 mmHg
pO_2 gradient	40–50 mmHg	650 mmHg
pCO_2 gradient	3–5 mmHg	30–50 mmHg

Sources: Data from Cooney, D.O., *Biomedical Engineering Principles*, Marcel Dekker, New York,
1976; Galletti, P.M. and Colton, C.K., *The Biomedical Engineering Handbook*, CRC
Press, Boca Raton, FL, 1995.

is usually the same as to the lungs, since the blood oxygenator must provide the same level of blood oxygenation; that is increasing the pO_2 (oxygen partial pressure) of the venous blood from about 40 mmHg to the arterial pO_2 of 95 mmHg. The capillaries in the lungs present the blood to the alveolar gas as a very thin film of blood with diffusional distances on the order of a few. This provides rapid gas transport in a very short time, typically a few tenths of a second. The total surface area available for oxygen transport in the lungs is about 140 m^2. In a blood oxygenator, the blood mass transfer film thickness is governed by the number and thickness of the blood flow channels, typically providing oxygen diffusional distances between the blood and the oxygen that are an order of magnitude larger than that found in the lungs. Also, the blood contact time in an oxygenator is on the order of tens of seconds.

In the lungs, the pO_2 of the alveolar air is about 104 mmHg. The average oxygen partial pressure driving force in the lungs is therefore on the order of 40–50 mmHg. Carbon dioxide has a pCO_2 of about 40 mmHg in the alveolar space and its level in the blood varies from 40 mmHg in arterial blood to 45 mmHg in venous blood. Therefore, the driving force for carbon dioxide transport is only a few mmHg and considerably smaller than that for oxygen. However, this smaller driving force is offset by the much higher permeability of carbon dioxide through the respiratory membranes, which is about 20 times that for oxygen.

One significant advantage of the blood oxygenator as compared to the lungs is the driving force for oxygen and carbon dioxide transport. In a blood oxygenator, humidified oxygen at atmospheric pressure can be used, resulting in a pO_2 of about 713 mmHg [760 – 47 mmHg (water saturation pressure at 37°C) = 713 mmHg]. Also, little or no carbon dioxide is present in this gas, so the driving force for carbon dioxide transport is considerably higher. This results in a considerable reduction in the mass transfer surface area of the blood oxygenator. Therefore, although the transport of oxygen is not as efficient within a blood oxygenator, these high mass transfer driving forces allow the resting metabolic needs of the patient to be met using blood oxygenators with mass transfer surface areas in the range from 1 to 10 m^2.

8.7.3 Types of Oxygenators

The design goal of a blood oxygenator is to present as large a surface area as possible between the blood and the oxygen-carrying gas stream. Several approaches have been developed to accomplish this and they are summarized in Figure 8.7. *Bubble oxygenators* were the earliest systems developed for open heart surgery. Bubble oxygenators achieve the required surface area for gas exchange by the production within the blood of numerous small bubbles of gas. These gas bubbles are in direct contact with the blood and are typically several millimeters in diameter. Of particular concern in the operation of bubble oxygenators is the removal of foam and gas bubbles prior to the return of the blood to the patient. This is achieved through the use of filters and defoaming sponges as the blood leaves the oxygenator. The design of bubble oxygenators is complicated by the complex nature of bubble motion. For example, bubbles exist in many different sizes, and they tend to break up and coalesce during their passage in the oxygenator.

To avoid the problem with antifoaming and bubble removal, *film oxygenators* have also been developed. In these systems, a film of blood is created on a surface that may be stationary or rotating. The blood film is then exposed directly to the gas. To increase the mass transfer rates, stationary films often employ irregular surfaces to enhance internal mixing of the blood film. Rotating discs whose bottom surface is in contact with a pool of blood create very thin films that are continuously renewed, resulting in efficient exposure of the blood to the gas stream.

Membrane oxygenators have, for the most part, now replaced the earlier bubble and film oxygenators. In these systems, the membrane physically separates the blood and gas streams. Membrane systems offer the advantage of minimizing the trauma to the blood that is created in direct contact systems, such as bubble and film oxygenators. The membrane and the associated fluid boundary layers, however, do present an additional mass transfer resistance. However, recent advances in membrane technology have resulted in membranes with significant oxygen and carbon dioxide permeabilities. The gas permeabilities for various membrane materials are shown in Table 8.7.

Typical units of the thickness normalized membrane gas permeability $(\hat{P}_m)$ are milliliters (STP) micron per minute per cubic meter per atmosphere, where the STP (standard temperature and pressure) conditions are for the pure gas at 0°C and 1 atm. The volumetric transport rate of gas (milliliter (STP) per minute) is then given by

$$N_V = \frac{\hat{P}_m \times (\text{membrane surface area}) \times \Delta(\text{gas partial pressure})}{\text{membrane thickness}}. \tag{8.31}$$

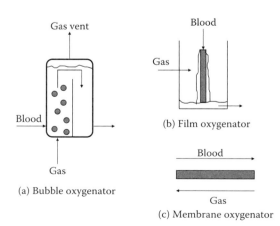

(a) Bubble oxygenator

(b) Film oxygenator

(c) Membrane oxygenator

FIGURE 8.7   Types of blood oxygenators.

TABLE 8.7

Oxygen and Carbon Dioxide Permeability of Selected Membrane Materials

Material	O_2 Permeability (mL (STP) µm min⁻¹ m⁻² atm⁻¹)	CO_2 Permeability (mL (STP) µm min⁻¹ m⁻² atm⁻¹)
Air	1.27×10^9	$1.02 - 10^9$
Polydimethysiloxane (silicone)	27,900	140,000
Water	3810	68,600
Polystyrene	1397	6985
Polyisoprene (natural rubber)	1270	7620
Polybutadiene	1016	7112
Cellulose (cellophane)	635	11,430
Polyethylene	305	1524
Polytetrafluoroethylene (teflon)	203	610
Polyamide (nylon)	2.54	10.2
Polyvinylidene chloride (saran)	0.25	1.52

Source: Data from Galletti, P.M. and Colton, C.K., *The Biomedical Engineering Handbook*, CRC Press, Boca Raton, FL, 1995.

The most common membrane material used in early versions of membrane devices was silicone with wall thicknesses on the order of 50–200 µm. The oxygen permeability for a 130-µm-thick silicone membrane is about 215 mL (STP) min⁻¹ m⁻² atm⁻¹ and that for carbon dioxide is about 1100 mL (STP) min⁻¹ m⁻² atm⁻¹ (Cooney 1976; Gray 1981, 1984).

Recent hollow fiber systems employ hydrophobic microporous polypropylene membranes. Because these membranes are hydrophobic and have very small pores, there is sufficient surface tension to prevent plasma infiltration. Hence, these pores are gas filled, resulting in significantly higher transport rates of oxygen and carbon dioxide than if the pores were filled with liquid. This can be seen by comparing, in Table 8.7, the oxygen permeability through a column of air with that through a column of water of the same thickness. Therefore, these microporous hydrophobic membranes have very high gas permeabilities and, in most cases, the resistance to transport of oxygen and carbon dioxide across these membranes is negligible in comparison to the resistance offered by the flowing blood.

The hollow fiber membranes used in blood oxygenators typically have external diameters in the range of 250–400 µm with wall thicknesses in the range of 20–50 µm. The porosity of these membranes is in the range of 40–60%, and the pores are typically on the order of 0.1 µm in diameter. These membrane-based systems come in a variety of configurations, e.g., flat plate, coil, and hollow fiber arrangements, as shown in Figure 8.1. Blood and gas stream contacting schemes can include cocurrent, countercurrent, and cross flows. Cross flow systems typically have the gas flowing through the lumen of the hollow fiber with blood flowing at right angles across the outer surface of the fibers. Cross flow of the blood results in significant improvement in performance as measured by gas exchange rates (Catapano et al. 1992; Vaslef et al. 1994; Federspiel and Henchir 2004; Wickramasinghe et al. 2002a, 2002b, 2005; Wickramasinghe and Han 2005; Nagase et al. 2005).

8.7.4 Analysis of a Membrane Oxygenator, Oxygen Transfer

The following mathematical model for a blood oxygenator is useful for illustrating the key concepts involved in the design of a blood oxygenator and serves as the foundation for other contacting schemes between the blood and the gas phases. Mathematical models such as the one shown below are also important for the evaluation of preliminary design calculations and for exploring the effects

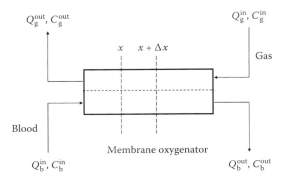

FIGURE 8.8 Solute mass balance model for a countercurrent membrane blood oxygenator.

of operating conditions on device performance. The model is developed in a manner similar to that used earlier for HD; however, now we must make use of techniques developed in Chapter 6 to account for the binding of oxygen with hemoglobin.

Figure 8.8 shows the mass balance model, where now we assume countercurrent flow of the blood and gas streams. We also assume that the total flow rates of these streams are unchanged; hence, Q_g and Q_b are constant; C_b and C_g are the bulk concentrations of oxygen on the blood (b) and gas (g) sides of the membrane; and C_b' represents the concentration of oxygenated hemoglobin. The corresponding oxygen concentration values at the membrane surface on the blood and gas sides are C_{bm} and C_{gm}.

A shell balance on oxygen may be written for the blood and gas sides of the membrane of length Δx as

$$Q_b(C_b + C_b')|_x - Q_b(C_b + C_b')|_{x+\Delta x} = k_b W_b \Delta x (C_b - C_{bm}),$$
$$Q_g C_g|_{x+\Delta x} - Q_g C_g|_x = k_g W_g \Delta x (C_g - C_{gm}),$$

(8.32)

where k_b and k_g are the blood and gas side mass transfer coefficients, and W is the membrane area per unit length in the x direction with the subscript denoting either the blood (b) or gas (g) sides. This allows application of this discussion to cylindrical fibers, where the area normal to the direction of radial transport changes with radial position. For planar membranes, we simply have that $W_b = W_g = W$.

After dividing by Δx, and taking the limit as $\Delta x \rightarrow 0$, we obtain

$$Q_b \frac{d(C_b + C_b')}{dx} = -k_b W_b (C_b - C_{bm}),$$
$$Q_g \frac{dC_g}{dx} = k_g W_g (C_g - C_{gm}).$$

(8.33)

Once again, we can make use of the fact that $m = dC_b'/dC_b$. Recall that m is related to the slope of the oxygen dissociation curve. A value for m may be estimated by Equations 6.39 or 6.40. Assuming that m is constant and evaluated at some suitable combination of the venous and arterial pO$_2$ values, the blood side equation simplifies to the following result:

$$Q_b(1 + m)\frac{dC_b}{dx} = -k_b W_b (C_b - C_{bm}).$$

(8.34)

It is now convenient to convert from the oxygen concentration to the oxygen partial pressure in our description of the mass transfer process. Recall for blood that the partial pressure of oxygen is related to the dissolved oxygen concentration by Henry's law, i.e., $pO_{2b} = HC_b$. The value of H for blood is 0.74 mmHg μM^{-1}. On the gas side, $pO_{2g} = RTC_g$, using the ideal gas law. Our equations for the blood and gas side may now be written as

$$Q_b(1+m)\frac{dpO_{2b}}{dx} = -k_b W_b(pO_{2b} - pO_{2bm}),$$

$$Q_g\frac{dpO_{2g}}{dx} = k_g W_g(pO_{2g} - pO_{2gm}). \tag{8.35}$$

These equations may be rewritten in terms of the overall partial pressure driving force, i.e., $pO_{2g} - pO_{2b}$, through definition of the overall mass transfer coefficient, K_0, defined by

$$\frac{1}{K_0} = \frac{H}{k_b} + \frac{W_b}{\rho^{STP} P_m \overline{W_L}} + \frac{RTW_b}{k_g W_g}, \tag{8.36}$$

where ρ^{STP} represents the density of the gas at STP conditions (0°C and 1 atm). The units on K_0 are typically moles per square centimeter per mmHg per second. K_0 is also now based on the membrane area on the blood side, i.e., W_b, where the log mean area of the membrane is given by $\overline{W_L} \equiv (W_g - W_b)/(\ln(W_g/W_b))$. For blood oxygenators, since the solubility of the gases in blood is very low, and as discussed earlier for microporous hydrophobic hollow fiber membranes, the blood side resistance is usually the controlling resistance, this means that Equation 8.36 simplifies to $K_0 = k_b/H$.

In terms of K_0, the blood and gas side equations become

$$\frac{Q_b(1+m)}{H}\frac{dpO_{2b}}{dx} = K_0 W_b(pO_{2g} - pO_{2b})$$

$$\frac{Q_g}{RT}\frac{dpO_{2g}}{dx} = K_0 W_b(pO_{2g} - pO_{2b}). \tag{8.37}$$

We can now solve these equations to provide an analytical solution to describe the performance of the oxygenator. First, we subtract the gas side equation from the blood side equation. This allows us to obtain a relationship between pO_{2g} and pO_{2b}:

$$\frac{Q_b(1+m)}{H}\frac{dpO_{2b}}{dx} - \frac{Q_g}{RT}\frac{dpO_{2g}}{dx} = 0. \tag{8.38}$$

This equation may now be integrated from the entrance of the bloodstream at $x = 0$ to any arbitrary value of x to give

$$pO_{2g}(x) = \left(\frac{Q_b}{Q_g}\right)\left(\frac{RT}{H}\right)(1+m)[pO_{2b}(x) - pO_{2b}^{in}] + pO_{2g}^{out}. \tag{8.39}$$

Now we can substitute this equation for pO_{2g} on the right hand side of Equation 8.37 for the blood to give

$$
\frac{dpO_{2b}}{dx} = \left(\frac{K_0 W_b H}{Q_b(1+m)} \right) \left\{ \left[\left(\frac{Q_b}{Q_g} \right)\left(\frac{RT}{H} \right)(1+m) - 1 \right] pO_{2b} \right.
$$
$$
\left. + \left[pO_{2g}^{out} - \left(\frac{Q_b}{Q_g} \right)\left(\frac{RT}{H} \right)(1+m)pO_{2b}^{in} \right] \right\}.
$$

(8.40)

This equation can then be integrated to give the following result for the blood side membrane area (A_{oxygen}) required for a given change in blood oxygenation:

$$
A_{oxygen} = \frac{\alpha}{\beta} \ln \left(\frac{\beta pO_{2b}^{out} + \gamma}{\beta pO_{2b}^{in} + \gamma} \right),
$$

(8.41)

where α, β, and γ are given by

$$
\alpha = \frac{Q_b(1+m)}{K_0 H},
$$

$$
\beta = \left(\frac{Q_b}{Q_g} \right)\left(\frac{RT}{H} \right)(1+m) - 1,
$$

(8.42)

$$
\gamma = pO_{2g}^{out} - \left(\frac{Q_b}{Q_g} \right)\left(\frac{RT}{H} \right)(1+m)pO_{2b}^{in}.
$$

The value of α will typically have units of square centimeters, β is dimensionless, and γ will be in mmHg.

The calculation of the blood side mass transfer coefficient (k_b) can be found from correlations that typically have the following general form, i.e., $Sh = a\,Re^b\,Sc^{1/3}$. The constants a and b depend on the geometry and the flow pattern that is used in a particular hollow fiber blood oxygenator. For blood in external cross flow over a bed of hollow fibers, Federspiel and Henchir (2004) averaged the values for a and b from a variety of correlations in the literature and recommend a value of $a = 0.524$ and a value of $b = 0.523$. In their study, Yang and Cussler (1986) found for a cross flow over hollow fibers that $a = 0.90$ and $b = 0.40$ when the void fraction in the hollow fiber cartridge was 0.93. When the void fraction in the fiber bundle was 0.30, the values of a and b were found to equal 1.38 and 0.34, respectively. For blood flowing inside the hollow fibers, the mass transfer coefficients for tube flow found in Table 5.2 can be used. Since the gas solubility in blood is low, the mass transfer coefficient for gases is much larger than that on the blood side and for practical purposes can be ignored.

The Reynolds number in these mass transfer correlations is defined as follows, i.e., $Re = \rho V D_H / \mu$, where D_H is the hydraulic diameter. For blood flowing within a cylindrical hollow fiber, this is the same as the inside fiber diameter (d). However, if the blood is flowing on the outside of the fibers, this will require an understanding of the fiber arrangement in order to calculate the wetted perimeter of the flow path and the cross-sectional area normal to the flow path, as required by Equation 4.78. The average velocity (V) is the volumetric flow rate of the blood divided by the cross-sectional area normal to the flow.

Recall that the Schmidt number is defined as follows, i.e., $Sc = (\mu/\rho D_e) = \nu/D_e$. The diffusivity in the Sc number is an effective diffusivity that accounts for binding of oxygen with the hemoglobin in the red blood cells. To understand this phenomenon, consider the unsteady diffusion of oxygen across a stagnant layer of blood. Only the oxygen dissolved in the plasma can diffuse, since the hemoglobin resides within the red blood cells that are not moving because the blood is stagnant. If a shell balance on oxygen is performed over a thin slice of blood from x to $x + \Delta x$, we can then write

$$S \Delta x \left[\frac{\partial C}{\partial t} + \frac{\partial C'}{\partial t} \right] = -S D \frac{\partial C}{\partial x}\bigg|_x - \left(-S D \frac{\partial C}{\partial x} \right)\bigg|_{x + \Delta x}. \tag{8.43}$$

In this equation, D is the diffusivity of oxygen in plasma and it is assumed that the red blood cell membrane is freely permeable to oxygen and offers no mass transfer resistance. The dissolved oxygen concentration is given by C, and C' represents the concentration of oxygen bound to hemoglobin, assuming that the dissolved and bound oxygen are in equilibrium. After dividing by Δx and taking the limit as $\Delta x \to 0$, Equation 8.43 becomes

$$\left[\frac{\partial C}{\partial t} + \frac{\partial C'}{\partial t} \right] = \left[\frac{\partial C}{\partial t} + \frac{\partial C'}{\partial C} \frac{\partial C}{\partial t} \right] = \left[1 + \frac{\partial C'}{\partial C} \right] \frac{\partial C}{\partial t} = D \frac{\partial^2 C}{\partial x^2}. \tag{8.44}$$

Note that in Equation 8.44, $\partial C'/\partial C$ is the slope of the oxygen–hemoglobin dissociation curve, which was previously given by Equations 6.39 and 6.40. Letting $m = \partial C'/\partial C$, Equation 8.44 can then be written as

$$\frac{\partial C}{\partial t} = \frac{D}{(1 + m)} \frac{\partial^2 C}{\partial x^2} = D_e \frac{\partial^2 C}{\partial x^2}. \tag{8.45}$$

Hence, we find from this analysis that the effective diffusivity of oxygen (D_e) in blood is given by $D/(1 + m)$ or in terms of Equations 6.39 and 6.40, we can write the effective diffusivity as

$$D_e = \frac{D}{1 + HC'_{SAT} \dfrac{dY}{dpO_2}} = \frac{D}{1 + n P_{50}^n HC'_{SAT} \dfrac{pO_2^{n-1}}{(P_{50}^n + pO_2^n)^2}}. \tag{8.46}$$

The Sherwood number (Sh) in the mass transfer coefficient correlation for k_b is defined as $k_b D_H/D$, where D is the diffusivity of oxygen in the plasma and D_H is the hydraulic diameter.

8.7.5 ANALYSIS OF A MEMBRANE OXYGENATOR, CARBON DIOXIDE TRANSFER

We also need to develop a similar set of equations for the transport of carbon dioxide in the oxygenator. Carbon dioxide can exist in blood in a variety of forms such as a dissolved gas and in combinations with water, hemoglobin, and other proteins. Like oxygen, the total amount of carbon dioxide in the blood depends on its partial pressure, i.e., pCO_2. The carbon dioxide dissociation curve shown in Figure 8.9 provides a relationship between the total amount of carbon dioxide in the blood and its pCO_2. The ordinate expresses the volume of carbon dioxide gas as a percentage of the blood volume. The volume percentages are based on body conditions (BTP) of $37°C$ and 1 atm of pure gas. For example, there is about 50 mL of carbon dioxide at BTP in each 100 mL of blood, i.e., 50 vol%, at a pCO_2 of about 42 mmHg. The change in the carbon dioxide concentration for normal blood is very narrow. For example, venous blood has a pCO_2 of about 45 mmHg and a corresponding concentration of 52 vol%. Arterial blood has a pCO_2 of about 40 mmHg and a corresponding concentration of 48 vol%.

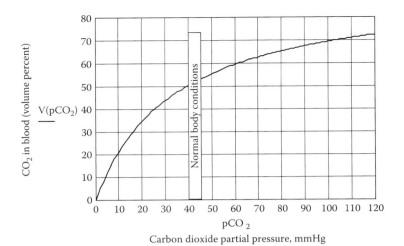

FIGURE 8.9   Carbon dioxide blood solubility curve.

Example 8.3

Show that the observed Henry's constant for carbon dioxide in blood for the physiological range of pCO_2 is about 0.0022 mmHg μM^{-1}.

Solution

First we define the Henry's constant (H) by the following relationship: $pCO_2 = HC_{CO_2}$, where C_{CO_2} is the total amount of carbon dioxide in the blood. Using the values obtained from Figure 8.9 at pCO_2s of 40 and 45 mmHg, we can calculate the corresponding values of H, i.e., H (40 mmHg) = (40 mmHg/48vol%) = 0.833 (mmHg/vol%) and H (45 mmHg) = (45 mmHg/52vol%) = 0.865 (mmHg/vol%). The average value of H in the physiological range of 40–45 mmHg is therefore 0.85 mmHg/vol%. We can then do some unit conversions on this value of H to get this result into mmHg per μM as

$$H = 0.85\frac{mmHg}{vol\%} \times \frac{1\,vol\%}{\dfrac{1\,mLCO_2\,(BTP)}{100\,mL\,(blood)} \times \dfrac{1000\,mL\,(blood)}{1\,L\,(blood)}} = 0.085\frac{mmHgL\,(blood)}{1\,mLCO_2\,(BTP)}.$$

Next we need to find how many moles of carbon dioxide there are in 1 mL of gas at BTP. From the ideal gas law, i.e., $PV = nRT$, we can then write that

$$1\,mLCO_2\,(BTP) = 0.001\,L = n_{CO_2} \times 0.082\frac{atmL}{molK} \times \frac{310\,K}{1\,atm}.$$

Solving this equation, we find, $n_{CO_2} = 3.93 \times 10^{-5}\,molCO_2$.

Therefore, we can now find the value of H in units of mmHg and μM as

$$H = 0.085\frac{mmHgL\,(blood)}{1\,mLCO_2\,(BTP)} \times \frac{1\,mLCO_2\,(BTP)}{3.93 \times 10^{-5}\,molCO_2} \times \frac{1\,mol}{LM} \times \frac{1\,M}{10^6\,\mu M},$$

$$H = 0.0022\frac{mmHg}{\mu M}.$$

Figure 8.8 can also be used to write shell balances on the blood and gas sides for carbon dioxide. A set of equations similar to those given by Equation 8.37 is obtained where m is now equal to zero. Henry's law is also assumed to describe the relationship between the carbon dioxide concentration in the blood and its pCO_2. The overall mass transfer coefficient, K_0, is still given by Equation 8.36, recognizing that the physical properties of carbon dioxide are to be used:

$$\frac{Q_b}{H}\frac{dpCO_{2b}}{dx} = K_0 W_b(pCO_{2g} - pCO_{2b}),$$

$$\frac{Q_g}{RT}\frac{dpCO_{2g}}{dx} = K_0 W_g(pCO_{2g} - pCO_{2b}).$$

(8.47)

As was done for oxygen, the gas side equation can be subtracted from the blood side equation and the result integrated to provide the following expression relating pCO_{2g} to pCO_{2b}:

$$pCO_{2g}(x) = \left(\frac{Q_b}{Q_g}\right)\left(\frac{RT}{H}\right)[pCO_{2b}(x) - pCO_{2b}^{in}] + pCO_{2g}^{out}. \quad (8.48)$$

This equation may be used in the blood equation (8.47) to obtain the following differential equation for pCO_{2b}:

$$\frac{dpCO_{2b}}{dx} = \left(\frac{K_0 W_b H}{Q_b}\right)\left\{\left[\left(\frac{Q_b}{Q_g}\left(\frac{RT}{H}\right)\right) - 1\right]pCO_{2b}\right.$$

$$\left. + \left[pCO_{2g}^{out} - \left(\frac{Q_b}{Q_g}\right)\left(\frac{RT}{H}\right)pCO_{2b}^{in}\right]\right\}. \quad (8.49)$$

When this equation is integrated over the length of the oxygenator, the following result is obtained for the blood side membrane surface area needed for the required carbon dioxide removal:

$$A_{carbon\,dioxide} = \frac{\alpha'}{\beta'}\ln\left[\frac{\beta' pCO_{2b}^{out} + \gamma'}{\beta' pCO_{2b}^{in} + \gamma'}\right]. \quad (8.50)$$

The constants α', β', and γ' are given by the next set of equations and have the same typical units as described earlier for oxygen:

$$\alpha' = \frac{Q_b}{K_0 H},$$

$$\beta' = \frac{Q_b}{Q_g}\left(\frac{RT}{H}\right) - 1,$$

(8.51)

$$\gamma' = pCO_{2g}^{out} - \left(\frac{Q_b}{Q_g}\right)\left(\frac{RT}{H}\right)pCO_{2b}^{in}.$$

8.7.6   EXAMPLE CALCULATIONS FOR MEMBRANE OXYGENATORS

Example 8.4 illustrates the calculation of the area of a countercurrent hollow fiber membrane blood oxygenator using the previous equations.

Example 8.4

Determine the hollow fiber membrane surface area for a blood oxygenator operating under the following conditions. Assume the fibers are made from microporous polypropylene. The gas membrane permeabilities are then based on diffusion through a stagnant layer of gas trapped within the pores of the membrane. The length of each fiber is 50 cm with a wall thickness of 50 μm. The inside diameter of a fiber is 400 μm. The blood flow rate through the lumen of the fibers in the device is 5,000 mL min^{-1} and the gas flow rate on the outside of the fibers is 10,000 mL min^{-1}, both at 37°C and 1 atm. The pO$_2$ of the entering blood is 40 mmHg and the amount of oxygen transported into the blood as it passes through the oxygenator must equal 250 mL min^{-1} (BTP). The pCO$_2$ of the entering blood is 45 mmHg and the amount of carbon dioxide removed from the blood must equal 200 mL min^{-1} (BTP). The entering gas is saturated with water (pH$_2$O = 47 mmHg at BTP) and contains a small amount of carbon dioxide in order to decrease the driving force for carbon dioxide transport so that a proper *respiratory exchange ratio (R)* can be achieved and for the calculated surface areas for oxygen and carbon dioxide to be the same. The respiratory exchange ratio is defined as the ratio of carbon dioxide output to oxygen uptake and should be equal to 0.8. Estimate the surface area required to deliver 250 mL min^{-1} of oxygen to the blood and remove 200 mL min^{-1} of carbon dioxide under these conditions. The gas side mass transfer resistance can be considered negligible because of the low solubility of oxygen and carbon dioxide in blood. As discussed earlier, the permeability of these polypropylene hollow fiber membranes is also very high, resulting in negligible mass transfer resistance. Hence, the bulk of the mass transfer resistance is a result of the boundary layer formed within the blood.

Solution

The solution algorithm is based on first identifying all of the relevant dimensions and physical properties. There are also three unknowns that serve as iteration variables. The first iteration variable is the membrane area required for oxygen and carbon dioxide transport. The second iteration variable is the average value of the blood pO$_2$ as it flows through the hollow fiber. This value is used to calculate, by Equation 6.40, the value of m, which is the slope of the oxygen–hemoglobin dissociation curve. The average pO$_2$ is chosen to make the oxygen mass transfer on the gas and blood sides balance. Hence, the amount of oxygen transferred to the blood must equal the amount of oxygen lost from the flowing gas. The third iteration variable is the pCO$_2$ in the incoming gas. This value is adjusted to make the oxygen and carbon dioxide membrane surface areas, calculated by Equations 8.41 and 8.50, result in the same value. In addition, an overall mass balance on oxygen and carbon dioxide is performed such that their respective transport rates are 250 and 200 mL min^{-1} at body temperature and pressure (BTP). This calculation gives a pO$_2$ of the blood exiting the hollow fibers of 90 mmHg and a pCO$_2$ of 41.55 mmHg. Once the three iteration variables are assumed, the calculation approach is then to calculate the value of m by Equation 6.40. From this value of m, the effective diffusivity of oxygen in blood is found from Equation 8.46. Next, the number of hollow fibers based on the assumed membrane area is determined, i.e., $N_{fibers} = A_{membrane}/(\pi d_{outside} L)$, where $A_{membrane}$ is the assumed membrane area, $d_{outside}$ is the external fiber diameter, and L is the fiber length. Knowing the number of hollow fibers, the average velocity of blood in a given hollow fiber can be found, i.e., $V_{blood} = 4Q_{blood}/(N_{fibers} \pi d^2)$, where Q_{blood} is the total volumetric flow rate of the blood and d is the internal diameter of a hollow fiber. In this case, the velocity was found in the converged solution to be 23.2 cm sec^{-1}. Next, the values of Re and Sc are found and from these the blood side mass transfer coefficient can be found from Equation 5.53. The pO$_2$ of oxygen in the exiting gas can then be found from Equation 8.39, which is found to be 664.8 mmHg. Next, the surface area of the membrane for oxygen transport is found from Equations 8.41 and 8.42. This process is then repeated for carbon dioxide and ends with the calculation of the membrane area required for carbon dioxide transport, using Equations 8.50 and 8.51. For carbon dioxide, it is found that the exiting pCO$_2$ in the gas is equal to 44.26 mmHg. Following this, the amount of oxygen and carbon dioxide transport is calculated from an overall mass balance for the gas and for the blood. The surface areas for oxygen and carbon dioxide transport must also come out to be the same for each respective gas. In addition, the oxygen and carbon dioxide mass balances must be equal between the gas and blood phases. If the solution has not converged in this manner, then adjustments need to be made to the iteration variables until convergence is achieved. In this case, the membrane area required for oxygen and carbon dioxide transport was found to equal 2.25 m². The partial pressure of carbon dioxide in the feed gas was found to be 29.1 mmHg and the average pO_2 used for calculating the value of m was found to be 59.5 mmHg. The respiratory exchange ratio also equals 0.80.

Hollow fiber membrane blood oxygenators in use today typically have surface areas in the range from about 1 to 3 m². In Example 8.4, it was assumed that blood flowed through the lumen of the hollow fibers in laminar flow countercurrent to the gas. Since the major resistance to oxygen transport resides in the blood, increasing the blood side oxygen mass transfer coefficient can have a significant effect on the size of the oxygenator. Hence, the surface area calculated in Example 8.4 is on the high side of what can now be achieved in actual practice. Having the blood flow across the outside of the hollow fibers can significantly increase the mass transfer rate of oxygen into the blood (Catapano et al. 1992; Vaslef et al. 1994; Wickramasinghe et al. 2005; Federspiel and Henchir 2004; Nagase et al. 2005). This is a result of repetitive mixing of the blood as it crosses successive hollow fibers, effectively decreasing the boundary layer resistance to oxygen transport.

Generally, it is recommended that experiments be performed to measure the oxygen transport rates for a given oxygenator design. Mathematical models, such as the one developed above or in the papers by Vaslef et al. (1994) and Wickramasinghe et al. (2005), can then be used to correlate the results, and provide accurate measurements of the blood side mass transfer coefficient. A calibrated model can then be used to explore operations under a wide range of operating conditions and assist in the optimal development of the device. For example, Vaslef et al. (1994) and Wickramasinghe et al. (2005) were able to assess the performance of cross flow hollow fiber oxygenators using water and other blood analogues. Subsequent tests using bovine blood provided excellent comparisons between their measurements and model predictions of the oxygen transport rates.

8.8 IMMOBILIZED ENZYME REACTORS

8.8.1 BACKGROUND

Another example of extracorporeal devices is the use of immobilized enzyme reactors to chemically change a species found in the blood. An enzyme is a protein that acts as a biochemical catalyst and offers great specificity in terms of the types of chemical species or substrates it acts on. Enzymes are usually named after the substrate whose reaction is catalyzed. For example, the enzyme that hydrolyzes the substrate urea is called *urease*. Note that the suffix *-ase* is usually added to a portion of the name of the substrate on which the enzyme acts.

Figure 8.3 shows several reactor arrangements that are possible when an enzyme is attached to or immobilized within a solid support. Enzyme immobilization offers several advantages. First, immobilization keeps the enzyme out of the bulk solution, which in the case of blood returning to the body, could be harmful or result in an allergic reaction to the foreign protein of the enzyme. Secondly, immobilization offers the potential to reuse an enzyme, which may in fact be quite expensive. Finally, in many cases, the enzyme is stabilized (less labile) when immobilized, retaining its activity for longer periods of time.

8.8.2 EXAMPLES OF MEDICAL APPLICATION OF IMMOBILIZED ENZYMES

One example where immobilized enzyme reactors have been proposed is in the treatment of neonatal jaundice (Lavin et al. 1985; Sung et al. 1986). Newborns tend to have higher levels of the greenish-yellow pigment *bilirubin* than those found in adults. Bilirubin is a natural product derived from red blood cells after they have lived out their lifespan of about 120 days. It is formed from the heme portion (the four pyrole rings) of the hemoglobin molecule after removal of the iron. The bilirubin binds to plasma albumin for transport to the liver where it is finally excreted from the body in the bile fluids.

The fetus's bilirubin readily crosses the placenta and is removed by the mother's liver. However, in the period after birth, the infant's liver is not fully functional for the first week, resulting in increased levels of plasma bilirubin. In some cases, the infant's bilirubin levels are sufficiently high, resulting in a *jaundiced* (yellow) appearance to the skin. High levels of plasma bilirubin can be toxic to a variety of tissues and, in these cases, jaundiced infants are commonly treated by *phototherapy* or blood transfusions. In phototherapy, the infant is placed under a blue light that converts the bilirubin to a less toxic byproduct. Phototherapy through the skin is not capable of controlling cases of severe jaundice. However, blood transfusions can replace the infant's blood with adult blood, effectively diluting the infant's plasma bilirubin levels. However, blood transfusions pose their own risk, particularly infectious diseases such as hepatitis and HIV.

An alternative approach to the treatment of neonatal jaundice is the use of a bilirubin-specific enzyme for removal of bilirubin from the infant's blood (Lavin et al. 1985; Sung et al. 1986). The enzyme, bilirubin oxidase, catalyzes the oxidation of bilirubin according to the following reaction stoichiometry:

$$\text{bilirubin} + \tfrac{1}{2}O_2 \rightarrow \text{biliverdin} + H_2O. \tag{8.52}$$

Calculations indicate that the amount of oxygen needed to convert all of the bilirubin found in the blood is about 100 times less than the actual oxygen content of blood. Therefore, no external supply of oxygen is needed within the enzyme reactor to carry out this reaction. Biliverdin itself is much less toxic than bilirubin and, in fact, is further oxidized by bilirubin oxidase to other less toxic substances. Experiments using a water jacketed reactor (much like that in Figure 8.3a) containing bilirubin oxidase covalently attached to agarose beads showed that plasma bilirubin levels in rats decreased by 50% after 30 min of treatment. The rat's blood was recirculated through the 6 mL reactor volume at a flow rate of 1 mL min^{-1}. Clearly, these results indicate that an immobilized bilirubin oxidase reactor could be an approach for the treatment of neonatal jaundice. It also shows the feasibility of using immobilized enzyme reactors for the specific removal of a harmful substance present in the blood.

Cells found in the liver and other organs carry out a wide variety of life-sustaining enzymatic reactions. There is considerable interest in using these cells or their enzymes to treat liver failure and other enzyme deficiency diseases. The discussion here focuses on the use of just the key enzymes. The use of immobilized cells as bioartificial organs, perhaps using liver cells (hepatocytes), is discussed in Chapter 10. However, it is important to recognize that some of the techniques used in this chapter to design immobilized enzyme reactors are directly applicable to systems that employ immobilized cells.

Another example of an immobilized enzyme reactor that we will look at in considerably more detail is that for the removal of heparin (Bernstein et al. 1987a, 1987b; Ameer et al. 1999a, 1999b). Recall that heparin is used as an anticoagulant in extracorporeal treatments such as HD and blood oxygenators. Heparin is a large, negatively charged, conjugated polysaccharide molecule that is produced by many types of cells in the body. By itself, heparin has little anticoagulant activity at the typical concentrations found in blood. However, in some regions of the body, such as the liver and lungs, it is produced in greater amounts. Therefore, it has an important role in preventing blood clots in the slow-moving venous blood flow entering the capillaries of the lungs and liver. By combining with *antithrombin III*, it increases by several orders of magnitude the ability of antithrombin to remove *thrombin*. Thrombin is an enzyme that converts the plasma protein fibrinogen into fibrin, leading to the fibrous mesh-like structure characteristic of a blood clot. Therefore, this synergistic combination of heparin with antithrombin III results in a powerful anticoagulant.

Over 20 million extracorporeal procedures using heparin are performed each year and in about 15% of these, complications due to heparin arise. Certainly, removal of heparin from the blood before it is returned to the body could significantly improve the safety of these procedures. The enzyme heparinase has the ability to degrade heparin into less harmful byproducts and one could envision an immobilized heparinase reactor for the removal of heparin from the blood returning to the patient's body.

8.8.3 ENZYME REACTION KINETICS

Our goal now is to develop relationships that can be used to describe the rates or kinetics of chemical reactions that are catalyzed by enzymes. Description of the enzyme reaction kinetics will allow us to develop mathematical models for immobilized enzyme reactors that can be used to analyze experimental data, provide information for scaleup of our devices, and explore the effects of operating conditions on reactor performance. The heparinase reactor described previously will be used as a model reaction system for these discussions.

The successful development of an immobilized enzyme reactor requires knowledge of the enzyme kinetics, an understanding of the effects on the observed reaction rate of reactant or substrate diffusion, and a design equation for the specific reactor that is used (Fogler 2005). Each of these aspects is discussed in greater detail in the following discussion.

First, we need to define the kinetics or the rate of the enzyme reaction and how it depends on the concentration of the substrate (reactant). This usually entails defining the free enzyme kinetics (enzyme in solution and not on a solid support) and the kinetics after the enzyme has been immobilized on the support material. In some cases, immobilization has no effect on the intrinsic activity of the enzyme, i.e., relative to its rate in solution. However, in most cases, immobilization of the enzyme can significantly alter the kinetics of the conversion process.

Generally, the kinetics of enzyme reactions are described by the Michaelis–Menten equation. The Michaelis–Menten equation can be derived by assuming that the conversion of substrate to product occurs in two steps. In the first step, the substrate (S) combines with the enzyme (E) to form an enzyme–substrate complex ($E*S$):

$$E + S \leftrightarrow E * S. \tag{8.53}$$

In the second step, the enzyme–substrate complex ($E*S$) is converted into product (P) and free enzyme (E), which is then available to recombine with the substrate:

$$E * S \rightarrow E + P. \tag{8.54}$$

The rate-controlling step in the previous process is assumed to be the conversion of the enzyme–substrate complex to product. Accordingly, it is therefore assumed that the reaction forming the enzyme–substrate complex, i.e., Equation 8.53, is at equilibrium. The rate of the enzyme reaction is then given by

$$r_S = \frac{dS}{dt} = -\frac{dP}{dt} = k_{cat} E * S, \tag{8.55}$$

where k_{cat} is a first order rate constant that relates the reaction rate (r_S) to the concentration of the enzyme–substrate complex, i.e., $E * S$. S and P are the respective concentrations of the substrate and the product.

From Equation 8.53, we can define the enzyme–substrate dissociation constant (reciprocal of the equilibrium constant) as $K_m = SE/E * S$ and then $E * S = SE/K_m$, with the result that Equation 8.55 becomes

$$r_S = \frac{dS}{dt} = -\frac{dP}{dt} = k_{cat} \frac{SE}{K_m}.$$
(8.56)

Letting E_0 represent the total concentration of the enzyme, then E_0 must equal the sum of the free enzyme concentration, i.e., E, and the amount of enzyme bound to substrate, i.e., $E * S$. Hence, we can write that $E_0 = E + E * S$. Since $E * S = SE/K_m$, we then have that $E = (K_m E_0)/(K_m + S)$. Substituting this result into Equation 8.56 then gives us an expression for the rate of the enzyme reaction, as follows:

$$-r_S = \frac{k_{cat} E_0 S}{K_m + S} = \frac{V_{max} S}{K_m + S},$$
(8.57)

where the reaction rate, r_S, has units of moles/((reaction volume)time); S represents the substrate or reactant concentration in units of moles per volume; V_{max} represents the maximum reaction rate for a given total enzyme concentration E_0 in units/reaction volume, where $V_{max} = k_{cat} E_0$; and k_{cat} is the reaction rate constant in moles/units time. Enzyme activity is commonly expressed in terms of *units* (U), and for heparinase, a unit of activity is defined as the amount of enzyme required to degrade 1 mg of heparin per hour. The actual amount or mass of enzyme needed is dependent on the enzyme purity and is usually reported as so many units of activity per milligram of enzyme. K_m is the Michaelis constant and is equal to the substrate concentration at which the reaction rate is equal to one-half the maximum rate (V_{max}). It is important to note that at high substrate concentrations, the reaction rate saturates at V_{max} because in total there are not enough free enzyme molecules available for the substrate reaction. The reaction rate is then independent of the substrate concentration and is therefore said to be zero order. At low substrate concentrations ($S \ll K_m$), the reaction rate is linearly proportional to the substrate concentration and is therefore said to be first order.

Example 8.5

The following table of data was obtained (Bernstein et al. 1987b) for the kinetics of heparin degradation by heparinase in solution at 37°C and a pH of 7.4. From these data, determine the values of K_m and k_{cat}.

Heparin Degradation Kinetics

Enzyme Loading (U mL^{-1})	Heparin Concentration (mg mL^{-1})	Reaction Rate (mg mL^{-1} h^{-1})	Normalized Reaction Rate (mg U^{-1} h^{-1})
24	1	20.83	0.868
	0.5	18.18	0.758
	0.25	13.70	0.571
	0.1	11.11	0.463
	0.05	7.58	0.316
43	1	38.46	0.894
	0.5	32.26	0.750
	0.25	30.30	0.705
	0.1	23.81	0.554
	0.05	16.67	0.388

Heparin Degradation Kinetics (Continued)

Enzyme Loading (U mL⁻¹)	Heparin Concentration (mg mL⁻¹)	Reaction Rate (mg mL⁻¹ h⁻¹)	Normalized Reaction Rate (mg U⁻¹ h⁻¹)
57	1	50	0.877
	0.5	40	0.702
	0.25	38.46	0.675
	0.1	26.32	0.462
67	1	55.56	0.829
	0.5	52.63	0.786
	0.25	50	0.746
	0.1	35.71	0.533

Source: Data from Bernstein, H., Yang, V. C., Langer, R., *Biotechnol. Bioeng.*, 30, 239, 1987b.

Solution

Note that for this type of analysis, it is convenient to divide the enzyme reaction rate by the corresponding enzyme concentration, thereby expressing the rate on a per unit amount of enzyme that is present in the reactor. One unit of heparinase (1 U) is defined as the amount of enzyme required to degrade 1 mg heparin h⁻¹. Hence, we can rewrite Equation 8.57 as

$$R = \frac{r_S}{E_0} = \frac{k_{cat} S}{K_m + S},$$

and this equation can then rearranged to give

$$\frac{1}{R} = \left(\frac{K_m}{k_{cat}} \right) \frac{1}{S} + \frac{1}{k_{cat}}.$$

This equation shows that if we plot the previous data as $1/R$ vs. $1/S$, the plot should be linear with an intercept equal to $1/k_{cat}$ and a slope equal to K_m/k_{cat}. This is called the *Lineweaver–Burke method* for analyzing enzyme reaction data. Using this method on the data in the table for heparin degradation by heparinase, we find that $K_m = 0.078$ mg mL⁻¹ and $k_{cat} = 0.891$ mg U⁻¹ h⁻¹. As shown in Figure 8.10, we find that the Michaelis–Menten model provides excellent representation of the heparin degradation kinetics.

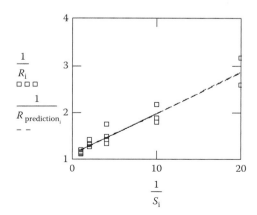

FIGURE 8.10 Lineweaver–Burke plot for Example 8.5.

8.8.4 REACTION AND DIFFUSION IN IMMOBILIZED ENZYME SYSTEMS

When an enzyme is immobilized within a supporting structure, the substrate has to diffuse into the material in order to come into contact with the immobilized enzyme. Because the substrate diffuses and then reacts within the support structure, a concentration gradient is established within the support structure and the reaction rate depends on location. A reaction–diffusion model is therefore needed to describe how the substrate diffuses through the porous support particle and reacts through the action of the immobilized enzyme. We also need to approximate the support geometry and here we will assume it is spherical. Extensions to other support geometries are straightforward once a given geometry has been examined in detail. In some cases, we may even need to know how the enzyme is distributed throughout the support particle (Bernstein et al. 1987a). In the absence of such information, we simply assume that the enzyme is uniformly distributed.

Diffusion through the support material also requires that we know the substrate diffusivity (D) in the bulk or free solution (see Figure 5.2). We also need to know how the solute diffusivity is affected by the properties of the support itself, specifically the porosity, pore size, and tortuosity. For substrates that have molecular dimensions that are comparable to the pore size, we will also need to include the effects of steric exclusion and hindered diffusion, as discussed in Section 5.6. These factors combine to define the effective diffusivity (D_e) of the substrate in the support particle, as given by

$$D_e = \frac{\varepsilon D}{\tau}\omega_r. \tag{8.58}$$

In this equation, ε represents the porosity of the support, τ is the tortuosity of the pores, and ω_r represents the reduction of the substrate diffusivity due to the proximity of the pore wall (i.e., the bracketed term in Equation 5.61). In this discussion, the substrate is already in the pore so there is no steric exclusion. We will see that steric exclusion shows up in the surface boundary condition of the differential equation that describes the reaction–diffusion phenomena occurring in the porous support.

Figure 8.11 illustrates a spherical shell of thickness Δr lying within the support particle containing a uniformly distributed and immobilized enzyme. The particle itself is porous and we can write the following steady state shell balance on the substrate:

$$4\pi r^2 D_e \left.\frac{dS}{dr}\right|_{r+\Delta r} - 4\pi r^2 D_e \left.\frac{dS}{dr}\right|_{r} = 4\pi r^2 \Delta r \left(\frac{V_{max} S}{K_m + S}\right). \tag{8.59}$$

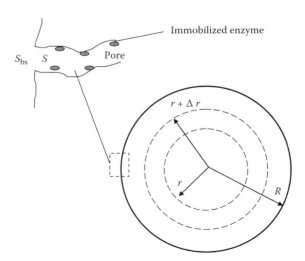

FIGURE 8.11   Shell balance on a spherical immobilized enzyme particle.

This equation simply states that the amount of substrate reacted within the shell volume is equal to the difference in the amount of substrate entering and leaving the shell volume by diffusion. After dividing by $4\pi\Delta r$, and taking the limit as Δr approaches zero, we obtain the reaction–diffusion equation for the substrate in the particle:

$$D_e \frac{d^2S}{dr^2} + \frac{2D_e}{r}\frac{dS}{dr} = \frac{V_{max}S}{K_m + S}. \tag{8.60}$$

The boundary conditions are

$$BC1:\ r = 0,\ \frac{dS}{dr} = 0, \tag{8.61}$$
$$BC2:\ r = R,\ S = KS_{bs}.$$

Once again, K represents the partition coefficient, which may include effects other than simple steric exclusion, e.g., electrostatic attraction or repulsion due to the presence of surface charges on the substrate and the particle. It is defined as the ratio of the substrate concentration on the pore side of the support surface (S) to that of the surface on the bulk side (S_{bs}), as shown in Figure 8.11. The substrate concentration at the surface of the support (S_{bs}) may also differ from the bulk substrate concentration (S_b) because of external mass transfer effects. This is discussed later.

Solution of Equations 8.60 and 8.61 is facilitated through the definition of the following dimensionless variables: $s' = S/(K\,S_{bs})$ and $r' = r/R$. On substitution of this dimensionless substrate concentration and radial dimension, we obtain

$$\frac{d^2s'}{dr'^2} + \frac{2}{r'}\frac{ds'}{dr'} = 9\phi^2\,\frac{s'}{1+\beta s'}. \tag{8.62}$$

The dimensionless boundary conditions are given by

$$BC1:\ r' = 0,\ \frac{ds'}{dr} = 0, \tag{8.63}$$
$$BC2:\ r' = 1,\ s' = 1.$$

Two additional parameters, ϕ and β, also result:

$$\phi = \frac{R}{3}\left[\frac{V_{max}}{K_m D_e}\right]^{1/2}, \tag{8.64}$$
$$\beta = \frac{S_{bs}K}{K_m}.$$

The quantity ϕ is especially important in reaction–diffusion problems and is known as the *Thiele modulus*. The square of this quantity represents the ratio of the substrate reaction rate to its diffusion rate. Its magnitude allows one to determine whether the overall reaction rate is *reaction limited* (small ϕ, minimal intraparticle substrate concentration gradient) or *diffusion limited* (large ϕ, significant intraparticle substrate concentration gradient). The factor $R/3$ is the ratio of the support volume to its external surface area. For nonspherical geometries, one can replace $R/3$ with this ratio, i.e., V/A. For diffusion in a planar material, $R/3$ is equal to L, the thickness. β is a dimensionless Michaelis constant. Large values of β indicate a zero order reaction, whereas $\beta \to 0$ indicates a first order reaction.

8.8.5 SOLVING THE IMMOBILIZED ENZYME REACTION–DIFFUSION MODEL

The solution to Equation 8.62 for the dimensionless substrate concentration profile in the immobilized enzyme support is usually defined in terms of the effectiveness factor represented by the symbol η. Because of substrate reaction and diffusional effects, the substrate concentration will generally decrease in the radial direction as one enters the support. This radial decrease in substrate concentration in the support will result in a decrease in the volume averaged reaction rate for the support in comparison to the reaction rate that would be possible if the substrate concentration was uniform throughout the support and equal to its surface concentration at the pore mouth. The effectiveness factor is used to describe this decrease in observed reaction rate and is defined by

$$\eta = \frac{\text{Observed particle rate}}{\begin{array}{c}\text{Rate obtained with no}\\ \text{concentration gradient}\\ \text{within the particle}\end{array}} = \frac{4\pi R^2 D_e \left.\dfrac{dS}{dr}\right|_{r=R}}{\dfrac{4}{3}\pi R^3 \left(\dfrac{V_{\max} S|_R}{K_m + S|_R}\right)} = \frac{\left.\dfrac{ds'}{dr'}\right|_{r'=1}}{\left(\dfrac{3\phi^2}{1+\beta}\right)}. \tag{8.65}$$

The effectiveness factor (η) varies from 0 (*diffusion limited*) to 1 (*reaction limited*). Determination of the effectiveness factor using the nonlinear Michaelis–Menten rate equation requires a numerical solution of Equations 8.62 and 8.63 for a given value of β. It is also important to point out that η, which depends on S_{bs}, will generally depend on position within the reactor due to consumption of the substrate as it flows through the reactor. Hence, the solution for η will be dependent on the local value of S_{bs}, which is found by the solution of the reactor design equation for the type of reactor being used. For the special case of a well-mixed reactor or a continuous stirred tank reactor, the substrate concentration is uniform throughout the reactor volume, so there will only be one value of the effectiveness factor, which will depend on S_{bs}.

Example 8.6

Determine the effectiveness factor for the case where the Thiele modulus and the dimensionless Michaelis constant are both unity. Assume that the particle is spherical and the kinetics are described by the Michaelis–Menten model.

Solution

For this case, Equation 8.62 must be solved numerically. This can be accomplished by first letting $y = ds'/dr'$ and $(dy/dr') = d^2s'/dr'^2$. Substituting these equations into Equations 8.62 and 8.63 reduces the original second order differential equation into two first order differential equations:

$$\frac{dy}{dr'} = -\frac{2}{r'}y + \frac{9\phi^2 s'}{1+\beta s'} \quad \text{and} \quad \frac{ds'}{dr'} = y,$$

$$\text{BC1'}: \ r' = 0, y = 0,$$

$$\text{BC2'}: \ r' = 1, s' = 1.$$

Although there are many mathematical software packages that can easily solve this problem, a relatively simple algorithm to implement is based on *Euler's method*. First, we divide the integration interval from $r' = 0$ to $r' = 1$ into N equally sized segments, i.e., $\Delta r' = 1/N$. Then, at the ith point, we have that $r_i' = i\,\Delta r'$ and we can approximate the derivatives as follows: $(dy/dr') \approx (y_{i+1} - y_i)/\Delta r'$ and

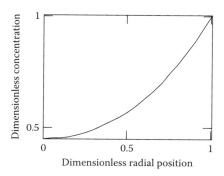

FIGURE 8.12   Dimensionless concentration profile, Example 8.6.

$(ds'/dr') \approx (s'_{i+1} - s'_i)/\Delta r'$. After substituting these relationships into the previous differential equations and solving for the values of y_{i+1} and s'_{i+1}, we obtain $y_{i+1} = y_i + ((9\phi^2 s'_i/1 + \beta s'_i) - (2/i\Delta r'))\Delta r'$ and $s'_{i+1} = s_i + y_i \Delta r'_i$.

When i is equal to zero, we are at the center of a spherical pellet and we use the known value of s' and y' from the boundary conditions. The above two equations can then be solved for the values of y_1 and s'_1. These values are then inserted into the above two equations and the values of y and s' are then calculated at the next position; this process is repeated until we reach the surface of the spherical particle. However, one complication needs to be pointed out about this particular problem, which is also known as a *two-point boundary value problem*. Note that we do not know explicitly what the value of s' is at the center of the particle, we only know that $ds'/dr'|_{r'=0} = 0$ and that $s' = 1$ at $r' = 1$. Hence, to start the solution using the algorithm described previously, we need to make a guess at what s' is at $r' = 0$. Then, we implement the algorithm described previously and march through the solution to $r' = 1$. Then, the resulting value of s' calculated at $r' = 1$ is compared to the required value of $s' = 1$ at $r' = 1$. If the calculated value of s' is greater than one, then the value of s' assumed at $r' = 0$ is too high and we must decrease the value assumed at $r' = 0$. If s' is less than one, then the value of s' assumed at $r' = 0$ is too low and we must increase the value assumed at $r' = 0$. We then repeat this process until convergence is obtained. After implementing the algorithm described earlier to solve this problem, we find that the value of η for these conditions is 0.86. Figure 8.12 shows the calculated substrate concentration profile within the spherical immobilized enzyme particle.

8.8.6 SPECIAL CASE OF A FIRST ORDER REACTION

If $K_m > K S_{bs}$, then the Michaelis–Menten kinetic model becomes that of a first order reaction. For a first order reaction, Equations 8.62 and 8.63 may be solved analytically, resulting in the following relationship between η and ϕ. This relationship between the effectiveness factor and the Thiele modulus for a first order reaction is also shown in Figure 8.13. Note that this solution is independent of the particle surface concentration and hence, the position in the reactor:

$$\eta = \frac{1}{\phi}\left(\frac{1}{\tanh 3\phi} - \frac{1}{3\phi}\right). \tag{8.66}$$

For large values of ϕ, the value of η approaches $1/\phi$.

Example 8.7

For the conditions in Example 8.6, calculate the effectiveness factor, assuming that the enzyme reaction described by Michaelis–Menten kinetics can be approximated as a first order reaction.

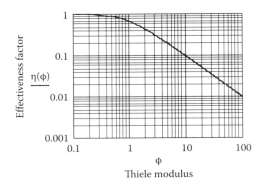

FIGURE 8.13   Effectiveness factor for a first order reaction.

Solution

The Thiele modulus is equal to unity. From Figure 8.13, we see that the effectiveness factor is about 0.70 or so. From Equation 8.66, we can calculate the value of the effectiveness factor to be 0.67.

8.8.7   OBSERVED REACTION RATE

With the effectiveness factor known, we can write the observed reaction rate for the immobilized enzyme as follows:

$$r_S = \eta \frac{V_{max} K S_{bs}}{K_m + K S_{bs}}, \tag{8.67}$$

where S_{bs} is the substrate concentration at the support surface and η accounts for the reduction in the reaction rate due to the internal substrate concentration gradient. It is important to note that the previous expression for the reaction rate is not explicitly dependent on the substrate concentration within the immobilized enzyme particle.

8.8.8   EXTERNAL MASS TRANSFER RESISTANCE

The flow of blood or other fluids around the immobilized enzyme support results in the formation of a thin boundary layer of fluid along the surface of the enzyme particle. This boundary layer provides an additional resistance to mass transfer that must be accounted for. An external mass transfer coefficient (k_m) is defined to account for this flow-induced external mass transfer resistance.

Conservation of mass requires, for a given particle, that the following steady state relationship holds between the external mass transfer rate and the reaction rate within the support:

$$4\pi R^2 k_m (S_b - S_{bs}) = \frac{4}{3}\pi R^3 \eta \left(\frac{V_{max} K S_{bs}}{K_m + K S_{bs}} \right). \tag{8.68}$$

This equation may be solved for the substrate concentration on the surface of the particle (S_{bs}) in terms of the bulk substrate concentration (S_b). Recall that it is this surface concentration that is used in the calculation of the effectiveness factor, as discussed earlier. From Table 5.3, the following equation describes the external mass transfer coefficient in packed beds:

$$\frac{k_m}{V_0} = 1.17 \left(\frac{\rho D_{particle} V_0}{\mu} \right)^{-0.42} \left(\frac{\mu}{\rho D} \right)^{-0.67}. \tag{8.69}$$

This equation is also equivalent to $Sh = 1.17\,Re^{0.58}\,Sc^{0.33}$, where $Sh = k_m\,D_{particle}/D$, $Re = \rho D_{particle}V_0/\mu$, $Sc = \mu/(\rho D)$, and $D_{particle}$ is the particle diameter. Mass transfer correlations for other situations may be found in Cussler (1984) or Bird et al. (2002). In Equation 8.69, V_0 is the superficial velocity of the fluid, i.e., the volumetric flow rate of the fluid divided by the cross-sectional area of the vessel that contains the support particles. Generally, one will want to operate the reactor in such a manner that the external mass transfer resistance is negligible, hence in many cases, $S_{bs} = S_b$. This can be accomplished by high bulk flow rates in the case of a plug flow reactor or intense mixing in a well-mixed reactor.

8.8.9 Reactor Design Equations

Finally, we need a reactor design equation that, in concert with the earlier information on the enzyme kinetics and diffusional effects, provides a relationship as to how the substrate concentration varies within the reaction volume. One usually assumes that the reaction mixture either flows through the reactor in plug flow (uniform velocity distribution in the direction normal to the flow), leading to a change in substrate concentration in the direction of flow, or that the reaction volume is well mixed, resulting in a uniform substrate concentration throughout the reaction volume.

8.8.9.1 Packed Bed Reactor

Figure 8.14 shows a packed bed immobilized enzyme reactor. The entering flow rate of the fluid is Q_b, which in biomedical applications can be, for example, blood or plasma. The substrate concentration entering the reactor is represented by S_b^{in}. A steady state shell balance can also be performed on a section of the reactor volume ($A_{xs}\,\Delta x$), where A_{xs} is the cross-sectional area of the reactor. It is assumed that the fluid flows through the reactor in plug flow, hence the velocity does not change with radial position:

$$Q_b S_b\big|_x - Q_b S_b\big|_{x+\Delta x} = (1 - \varepsilon_R)\Delta x A_{xs}\eta\left(\frac{V_{max}K S_{bs}}{K_m + K S_{bs}}\right). \tag{8.70}$$

In this equation, ε_R represents the void volume in the reactor, i.e., the volume of the reactor not occupied by the immobilized enzyme particles. Dividing by Δx and taking the limit as $\Delta x \to 0$ results in the following differential equation:

$$Q_b\frac{dS_b}{dx} = -(1 - \varepsilon_R)A_{xs}\eta\left(\frac{V_{max}K S_{bs}}{K_m + K S_{bs}}\right), \tag{8.71}$$

$$\text{BC1: } x = 0, S_b = S_b^{in}.$$

This equation can be solved numerically along with Equation 8.68, which provides the relationship between S_b and S_{bs} due to the effects of external mass transfer resistance. Since β depends on S_{bs}

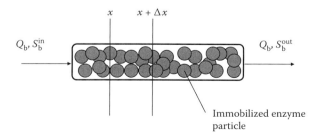

FIGURE 8.14 Shell balance for a packed bed immobilized enzyme reactor.

(see Equation 8.64), which depends on x, we also need to evaluate the value of η as a function of x. For the special case where one can assume that η is approximately constant and external mass transfer effects are negligible, i.e., $S_{bs} = S_b$, Equation 8.71 can be integrated analytically for a reactor of length L to give the following result:

$$K_m \ln\left(\frac{S_b^{out}}{S_b^{in}}\right) + K(S_b^{out} - S_b^{in}) = -\frac{(1-\varepsilon_R) A_{xs} L K V_{max} \eta}{Q_b}. \tag{8.72}$$

Oftentimes, the operation of a chemical reactor is defined in terms of the residence time (τ), i.e., the time the reacting fluid spends in the reactor. For a packed bed plug flow reactor, the residence time defined in terms of the reactor void volume is given by

$$\tau = \frac{A_{xs} L \varepsilon_R}{Q_b}. \tag{8.73}$$

In addition, we can define the conversion of the substrate, i.e., $X_S \equiv 1 - (S_b^{out}/S_b^{in})$. With these additional definitions, Equation 8.72 can be written as

$$K_m \ln(1 - X_S) - K S_b^{in}(1 - X_S) = -\frac{(1-\varepsilon_R) K V_{max} \eta}{\varepsilon_R} \tau. \tag{8.74}$$

For given values of the enzyme kinetics and size of the reactor or residence time, Equations 8.72 and 8.74 can be solved for the exiting substrate concentration and the substrate conversion.

8.8.9.2 Well-Mixed Reactor

The other type of commonly used immobilized enzyme reactor is the well-mixed reactor, shown in Figure 8.15. Perfect mixing results in a uniform substrate concentration throughout the reactor volume. The exiting concentration of the substrate is also the same as that within the reactor, hence the concentration everywhere in the bulk fluid of the reactor is S_b^{out}. A steady state substrate mass balance provides Equation 8.75, which can be solved for the exiting substrate concentration:

$$Q_b S_b^{in} = (1-\varepsilon_R) V_{reactor} \eta\left(\frac{V_{max} K S_{bs}^{out}}{K_m + K S_{bs}^{out}}\right) + Q_b S_b^{out}. \tag{8.75}$$

Equation 8.75 can be solved, along with Equation 8.68, and a value of the effectiveness factor evaluated at S_{bs}^{out}, to determine the exiting substrate concentration S_b^{out}.

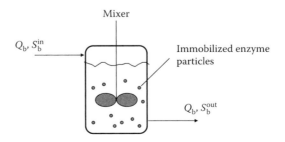

FIGURE 8.15 A well-mixed immobilized enzyme reactor.

The substrate conversion is also defined by $X_S \equiv 1 - (S_b^{out}/S_b^{in})$. In the absence of external mass transfer effects, i.e., $S_{bs}^{out} = S_b^{out}$, Equation 8.75 simplifies to

$$S_b^{in} = \frac{(1 - \varepsilon_R)}{\varepsilon_R} \, \eta \left(\frac{V_{max} K S_b^{out}}{K_m + K S_b^{out}} \right) \tau + S_b^{out}. \tag{8.76}$$

Example 8.8

The following experimental data were obtained for an immobilized heparinase reactor (Bernstein et al. 1987b) for two different inlet heparin concentrations and different enzyme loadings. Heparinase was immobilized on a cross-linked agarose support. The flow pattern in the reactor was well mixed.

Experimental Data for a Heparin Reactor

Total Support Volume (mL)	Enzyme Loading (U mL^{-1}) in the Support	Substrate Conversion (X_s) S_b^{in} = 0.2 or 0.5 mg mL^{-1}
80	100	0.58 or 0.54
85	150	0.66 or 0.62
96	120	0.69 or 0.55
90	130	0.63 or 0.58

Source: Data from Bernstein, H., Yang, V. C., Langer, R., *Biotechnol. Bioeng.*, 30, 239, 1987b.

The diffusivity of heparin is 1.2×10^{-6} cm^2 sec^{-1}. The porosity of the agarose support was 0.92 and the tortuosity was unity. The radius of a heparin molecule is 1.5 nm and the pores in the agarose support have a radius of 35 nm. The partition coefficient, K, was found from experiments to be equal to 0.36. The agarose support had a radius of 0.0112 cm. The flow rate through the reactor was 120 mL min^{-1}. The external mass transfer coefficient was estimated to be 0.154 cm min^{-1}. Using the kinetic parameters obtained in Example 8.5, use the mathematical model for a well-mixed immobilized enzyme reactor to predict the substrate conversions. Compare the results to the experimental conversions in the previous table.

Solution

The effective diffusivity of heparin within the enzyme particle is calculated by Equation 8.58 to be $D_e = 1.01 \times 10^{-6}$ cm^2 sec^{-1}. The Thiele modulus given by Equation 8.64 depends on the enzyme loading, since $V_{max} = k_{cat} E_0$. The following table summarizes the calculated Thiele modulus for each of the four enzyme loadings.

Enzyme Loading (U mL^{-1} of particle)	Thiele Modulus (ϕ)
100	2.098
150	2.569
120	2.298
130	2.392

At this point, we need to calculate the effectiveness factor, which also depends on $\beta = S_{bs} K/K_m$, and the unknown particle surface concentration, i.e., S_{bs}, which is also related to the unknown bulk or exiting substrate concentration, i.e., S_b^{out}. Since we need S_{bs} to start this process and it is unknown, we first assume that the reaction is first order, in order to start the calculation. We can then calculate the effectiveness factor, i.e., η, using Equation 8.66 or we can obtain these values from Figure 8.13. From Figure 8.13, we see that these effectiveness factors, assuming a first order reaction, are on the order of $\eta \sim 0.45$. Next, we solve Equation 8.75 for the exiting bulk substrate concentration, i.e., S_b^{out}, and substitute that result into Equation 8.68, thereby providing the following relationship between the inlet substrate concentration, i.e., S_b^{in}, and the substrate surface concentration, i.e., S_{bs}^{out}:

$$(S_b^{in} - S_{bs}^{out}) - \left[\frac{\eta k_{cat} E_0 K S_{bs}^{out}}{K_m + K S_{bs}^{out}}\right]\left[\frac{V_{reactor}(1-\varepsilon_R)}{Q_b} + \frac{R}{3 k_m}\right] = 0.$$

This is a nonlinear algebraic equation that can then be solved using Newton's method (see Example 2.10) for the given conditions to find the value of the substrate concentration at the surface of the immobilized enzyme particle. Once the value of S_{bs}^{out} has been found, then the bulk exiting substrate concentration, i.e., S_b^{out}, can be found from Equation 8.75. Recall that the first time through this process the effectiveness factors were found assuming the reaction is first order. Once new values of S_b^{out} have been found, the effectiveness values can be updated using the method discussed in Example 8.6 for Michaelis–Menten kinetics. This process is then repeated until convergence is obtained. The following table summarizes the converged values of the effectiveness factors for the two inlet substrate concentrations and the four enzyme loadings.

Calculated Effectiveness Factors

Enzyme Loading (U mL^{-1} particle^{-1})	$S_b^{in} = 0.2$ mg mL^{-1}	$S_b^{in} = 0.5$ mg mL^{-1}
100	0.52	0.62
150	0.43	0.51
120	0.47	0.56
130	0.46	0.54

The following table summarizes the resulting values obtained for the substrate surface and exiting bulk concentrations for the two inlet substrate concentrations and the four enzyme loadings.

Calculated Surface and Bulk Substrate Concentrations

Enzyme Loading (U mL^{-1} particle^{-1})	$S_b^{in} = 0.2$ mg mL^{-1}		$S_b^{in} = 0.5$ mg mL^{-1}	
	Surface	Bulk	Surface	Bulk
100	0.07	0.074	0.197	0.208
150	0.056	0.061	0.154	0.166
120	0.057	0.061	0.156	0.166
130	0.057	0.061	0.159	0.169

The following table compares the predicted to the experimental substrate conversions for the two inlet substrate concentrations and the four enzyme loadings. Figure 8.16 also shows a parity plot of the predicted conversions vs. the experimental conversions. We see that the model's predicted conversions compare rather well with the experimental conversions. The predictions, however, overestimate the substrate conversion and this is most likely due to a slight decrease in the reaction rate as a result of immobilizing the enzyme.

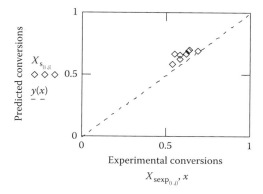

FIGURE 8.16 A comparison of experimental conversions (X_{sexp}) to the predicted heparin conversions (X_s).

Comparison of Predicted and Experimental Substrate Conversions

Enzyme Loading (U mL⁻¹ particle⁻¹)	$S_b^{in} = 0.2$ mg mL⁻¹		$S_b^{in} = 0.5$ mg mL⁻¹	
	Predicted	Experimental	Predicted	Experimental
100	0.63	0.58	0.59	0.54
150	0.70	0.64	0.67	0.62
120	0.70	0.69	0.67	0.55
130	0.69	0.63	0.66	0.58

8.9 AFFINITY ADSORPTION

Adsorption is the process where a substance in a solution literally sticks to the surface of a solid surface. Adsorption is not the same as absorption. In absorption the substance goes into the interior of another substance, which can be a solid or another liquid. There are two types of adsorption—physical and chemical. In physical adsorption, the substance binds to the solid surface through relatively weak physical interactions between the substance and the solid. The adsorbed substance retains its chemical structure. In chemical adsorption, chemical bonds are actually formed between the adsorbed substance and the solid material.

Adsorption processes can be very selective in terms of the interaction between the solid surface and the substances found in the solution. For example, in affinity adsorption, another substance, known as a *ligand*, is attached to the surface of a solid material by chemical means. The ligand has a very specific binding site for a substance that is found in the solution, as shown in Figure 8.17. Examples of various ligands and the substances they act on that are used in biomedical applications are summarized in Table 8.8.

An interesting example of the use of affinity adsorption is described in Karoor et al. (2003). As mentioned in their paper, there is great interest in using animal organs as a means of addressing the shortage of human donor organs. This use of animal organs is also known as *xenotransplantation*, and pigs are receiving the most attention because of their adequate supply and the size of their organs is comparable to those in humans.

However, transplantation of animal organs into humans leads to a severe antibody-mediated reaction to the foreign tissue, the first phase of which is also known as *hyperacute rejection*.* For

---

* A brief discussion on the immune system can be found in Section 10.2.

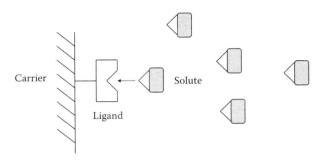

FIGURE 8.17   Ligand and solute binding.

TABLE 8.8
Examples of Ligands and Ligand Binding Substances

Ligand	Ligand Binding Substances
Enzyme inhibitor	Enzyme
Antigen	Antibodies
Complementary nucleotide sequence	Nucleic acids
Cellular receptors	Hormones
Carrier protein	Vitamin

---

this discussion, antibodies are proteins produced by certain cells of the immune system that bind with high specificity to molecules known as *antigens*. An antigen is just a chemical entity found in the foreign tissue that the immune system's antibodies recognize and then bind to.

These antibodies, also known as *preformed natural antibodies*, found in the recipient of an xeno-transplant, bind to the endothelial cells of the xenograft. These antibodies that bind to the endothelial cells then activate the complement system that destroys the transplanted organ within several hours of the transplant. The hyperacute rejection process for the most part involves a particular type of antibody, known as *IgM*. A few percent of the IgM antibody population consists of xenoreactive IgM antibodies that bind with high specificity to α-galactosyl structures (sugar-like molecules) that are expressed by nonprimate mammalian and New World monkey cells. The selective removal of these α-galactosyl-specific IgM antibodies prior to xenotransplantation could eliminate the hyper-acute rejection of the transplanted organs.

Figure 8.18 illustrates an affinity adsorption system for removing α-galactosyl reactive IgM anti-bodies (Karoor et al. 2003). Blood either from the body, or from a reservoir in the case of in vitro experiments, flows through a microfiltration hollow fiber membrane cartridge. As the blood flows through the inside of the hollow fibers, plasma is filtered across the hollow fiber membrane as a result of the transmembrane pressure difference. Recall that this filtration of blood is also known as plasmapheresis. The exiting blood and the plasma filtrate are then returned to the body or the reservoir. The pores of the hollow fiber membranes have diameters in the range from 0.2 to 0.5 μm. The diameters of these pores are many times larger than the IgM molecule. The ligand, which in this case is α-galactosyl, was immobilized on all accessible surfaces of the hollow fiber microfiltration membrane, i.e., the inner, outer, and internal surfaces. As the blood and filtered plasma flow through the hollow fiber membranes, the reactive IgM antibodies bind with the α-galactosyl ligand that are bound to the membrane surfaces, as shown in Figure 8.17.

We can use a simple, well-mixed single compartment model to describe how the concentration of reactive IgM changes with time in the body or the reservoir (Karoor et al. 2003). Letting $V_{reservoir}$ denote the reservoir volume or the IgM distribution volume in the body, and $C_{reservoir}$ denote the IgM

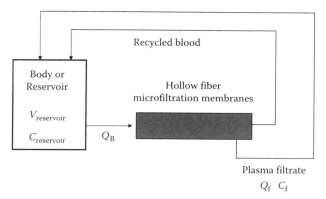

FIGURE 8.18   Affinity adsorption system.

concentration, we can write the following mass balance on IgM, which expresses the fact that the rate of change of IgM in the body or reservoir must equal the rate at which IgM is being removed from the filtrate flow:

$$V_{\text{reservoir}} \frac{dC_{\text{reservoir}}}{dt} = -Q_f (C_{\text{reservoir}} - C_f). \tag{8.77}$$

For the case where all of the IgM is bound to the α-galactosyl ligand as it passes through the hollow fiber membranes, then $C_f = 0$ and Equation 8.77 can be integrated to give the relative concentration of IgM (given by χ) in the reservoir as a function of time:

$$\chi = \frac{C_{\text{reservoir}}}{C^0_{\text{reservoir}}} = e^{-(Q_f t / V_{\text{reservoir}})}, \tag{8.78}$$

where $C^0_{\text{reservoir}}$ is the initial concentration of IgM. The filtration flow or plasmapheresis, i.e., Q_f, can be estimated from Equation 8.79 (Zydney and Colton 1986, also see Problem 5.11 for its derivation):

$$\frac{Q_f}{Q_{b\ \text{in}}} = 1 - \exp\left[-0.90\beta \ln \frac{C_w}{C_b(0)}\right], \tag{8.79}$$

where $\beta = \frac{2}{3}(a^2 L / R^3)^{2/3}$, $Q_{b\ \text{in}}$ is the inlet blood flow rate, $C_b(0)$ is the bulk red blood cell volume fraction at the inlet (0.40), C_w is the red blood cell volume fraction at the membrane surface (0.95), a is the red blood cell radius (4 μm), L is the length of the hollow fibers, and R is the inside radius of the hollow fibers.

The characteristic time for the IgM adsorption process is defined as $\tau_S \equiv V_{\text{reservoir}}/Q_f$; therefore, the number of reservoir volumes that have been filtered (i.e., RF) in a period of time (t) through the affinity adsorption membrane system is given by

$$RF = \text{reservoir volumes filtered} = \frac{Q_f t}{V_{\text{reservoir}}}. \tag{8.80}$$

Results of the studies by Karoor et al. (2003) show that greater than 90% removal of the IgM antibodies, i.e., $(C_{\text{reservoir}}/C^0_{\text{reservoir}}) < 0.10$, can be obtained after three volumes of the plasma have been filtered through the affinity adsorption system.

In designing an affinity adsorption system, one must be certain that the amount of ligand or solute to be removed, i.e., $V_{reservoir} \, C^0_{reservoir}$, is less than the maximum adsorbed solute concentration that is possible for the membrane system, i.e., $V_m Q_m$, where V_m is the membrane volume and Q_m is the solute-binding capacity of the membrane. For a human scale device to remove xenoreactive IgM antibodies, the reservoir volume would be the volume of plasma in the body, which is 3000 mL. The IgM concentration is about 0.033 mg mL^{-1}, so the total amount of IgM to be removed is on the order of 100 mg. IgM binding capacities for the systems studied by Karoor et al. (2003) had total binding capacities on the order of several hundred milligrams of IgM.

Example 8.9

Karoor et al. (2003) obtained the data shown in the following table for the relative concentration of IgM as a function of the reservoir volumes filtered. Six hundred nylon hollow fiber membranes were used, having an inner and outer diameter of 330 and 550 μm, respectively. The active length of the hollow fibers was 10.2 cm and the total luminal surface area was 718 cm². The filtration flow rate (Q_f) was 50 mL min^{-1} and the reservoir volume was 1200 mL. The initial IgM concentration was 0.019 mg mL^{-1}. Compare the model developed previously for an affinity adsorption system to the results obtained by Karoor et al. (2003). Also, calculate the predicted fractional filtrate yield, i.e., $Q_f/Q_{b\,in}$.

Reservoir Volumes Filtered, $Q_f \, t/V_{reservoir}$	Relative Concentration, $C_{reservoir}/C^0_{reservoir}$
0.025	0.89
0.13	0.68
0.25	0.58
0.55	0.41
0.83	0.30
1.1	0.25
1.38	0.22
1.63	0.21
1.9	0.20
2.20	0.19
2.5	0.16

Solution

From Equation 8.79, we calculate that the fractional filtrate yield, i.e., $Q_f/Q_{b\,in}$, is equal to 0.23. From Equation 8.78, we can calculate the relative concentration of the IgM antibody in the reservoir. Figure 8.19 shows a comparison between the affinity adsorption model developed earlier and the results from the previous table. The comparison between the model and the data is quite good.

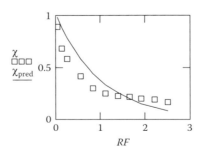

FIGURE 8.19   Relative reservoir concentration as a function of reservoir volumes filtered.

PROBLEMS

1. Estimate the clearance and fractional percent removal of urea from a flat plate dialyzer for both cocurrent and countercurrent flow under the following conditions:

   Urea diffusivity $= 2 \times 10^{-5}$ cm^2 sec^{-1}

   Four flat plate membranes each with the following dimensions:

   Dialyzer length $= 85$ cm

   Dialyzer width $= 15$ cm

   Blood flow rate $= 200$ mL min^{-1} (total)

   Dialyzer flow rate $= 500$ mL min^{-1} (total)

   Mass transfer resistances:

   Membrane $= 25$ min cm^{-1}

   Blood side $= 8$ min cm^{-1}

   Dialysate side $= 35$ min cm^{-1}

2. Estimate the urea flux [$(NH_2)_2CO$] (grams per square meter per hour) through a cellulosic membrane that is 0.001 inch thick. The blood at the membrane surface has a urea concentration of 1 mg mL^{-1}. Assume that the dialysis fluid on the other side of the membrane surface is free of urea. The void volume of the cellulosic membrane is 0.40 and the membrane tortuosity is 2.0. The pores in the membrane have a diameter such that they exclude all molecules larger than a molecular weight of 20,000.

3. The following data for the antibiotic imipenem were obtained from six patients suffering multiorgan failure (Hashimoto et al. 1997; also see Problem 10 in Chapter 7). All patients were anuric because of complete renal failure and were on continuous venovenous HD. The antibiotic was first delivered by an intravenous infusion pump for 30 min, resulting in an initial plasma concentration of 32.37 μg mL^{-1}. Following the infusion, blood samples were taken over the next 12 h. In addition to the blood samples, drug concentrations in the blood exiting the HD unit were also taken. The dialysate entering the hemodialyzer was free of drug and the concentration of the drug in the dialysate leaving the dialyzer was also measured. The blood flow rate through the hemodialyzer was set at 60 mL min^{-1} and the dialysate flow was set at 20 mL min^{-1}. The hemodialyzer had a total surface area of 0.5 m^2. Using these data, develop a model that describes the drug removal process by the patient's body and the hemodialyzer. Carefully state all of your assumptions. Determine key parameters, such as the total body clearance, the dialyzer clearance, and the overall mass transfer coefficient for the dialyzer. Compare your model to the actual data. The following table summarizes the drug concentrations in the blood entering the dialyzer, the blood leaving the dialyzer, and the exiting dialysate.

Time (min)	Blood Drug Concentration Entering Dialyzer (μg mL⁻¹)	Blood Drug Concentration Leaving Dialyzer (μg mL⁻¹)	Drug Concentration in Dialysate Leaving Dialyzer (μg mL⁻¹)
0	32.47	23	21
30	24	20.5	14
60	18.5	16	12
120	13	11	8.8
180	10	8	6.4
360	4.6	3.8	2
540	2	1.6	1.4
720	1.12	0.9	0.6

4. At the start of CAPD using a 2.5 wt% glucose solution, the filtration rate of water from the plasma was determined to be about 6 mL min^{-1}. Estimate the hydraulic conductance (L_p in milliliters per hour per square meter per mmHg) of the peritoneal membrane.

5. The following data were presented by Lysaght and Farrell (1989) for urea (MW = 60) and inulin (MW = 5200) transport during CAPD. The initial glucose concentration in the dialysate solution was equal to 1.5 wt%. Assuming a constant peritoneal fluid volume of 2.4 L, estimate the permeability of the peritoneal membrane for each solute. What thickness of a stagnant film of water is required to give an equivalent permeability for urea and inulin? Explain the difference in the observed peritoneal membrane permeabilities based on your understanding of solute diffusion through hydrogels.

Time (min)	Ratio of Dialysate to Plasma Concentration for Urea	Ratio of Dialysate to Plasma Concentration for Inulin
0	0.03	0
50	0.33	0.085
100	0.57	0.16
150	0.69	0.21
200	0.77	0.27
300	0.91	0.36
400	0.96	0.42

6. The following data were presented by Lysaght and Farrell (1989) for urea (MW = 60) transport during CAPD. The urea concentration in blood was 100 mg (100 mL)$^{-1}$. From these data, estimate the urea permeability of the peritoneal membrane.

Time (min)	Peritoneal Volume (mL)	Dialysate Urea Concentration (mg (100 mL)$^{-1}$)
0	1983	10
15	2304	30
30	2543	42
45	2938	55
60	3146	62

7. Analyze the performance of an oxygenator that uses an ultrathin (25 μm) membrane made of polyalkylsulfone cast on microporous polypropylene. The oxygen permeability of this membrane is reported (Gray 1981) to be 1100 mL (STP) min^{-1} m^{-2} atm^{-1} and that for carbon dioxide is reported to be 4600 mL (STP) min^{-1} m^{-2} atm^{-1}. What membrane area is need to deliver 250 mL min^{-1} of oxygen and remove 200 mL min^{-1} of carbon dioxide? Carefully state your assumptions.

8. Consider an oxygenator where the gas flows through the lumen of the hollow fibers and the blood flows on the outside across the hollow fibers. Assume that the gas flow is sufficiently high and that a reasonable assumption is that the gas concentrations are constant along the length of the hollow fiber. Develop a mathematical model that describes the performance of this oxygenator. Carefully state your assumptions. What membrane area is need to deliver 250 mL min^{-1} of oxygen and remove 200 mL min^{-1} of carbon dioxide? The external mass transfer coefficient for blood flowing across a bank of hollow fibers is given by the following expression (Yang and Cussler 1986):

$$Sh = 1.38 \, Re^{0.34} \, Sc^{0.33}.$$

The characteristic diameter for this equation is the external diameter of the hollow fibers. The velocity of the blood (V_b) between the fibers may be calculated by

$$V_b = \frac{Q_b}{\varepsilon_{unit} A_{unit}}.$$

Here, Q_b is the volumetric flow rate of blood, ε_{unit} is the void volume in the hollow fiber unit, and A_{unit} is the total cross-sectional area of the unit that is perpendicular to the blood flow. Show that the blood side equation in Equation 8.37 still applies with pO_{2g} now a constant.

9. Yang and Cussler (1986) considered the design of a human gill using hollow fibers. To support a man, they assumed that the human gill would have to provide up to 2500 mL min^{-1} of oxygen at 1 atm and 37°C for moderate levels of activity. Develop your design for the human gill and determine the hollow fiber membrane area that is required. How big is your human gill? What other issues need to be considered for the design of the human gill?

10. Calculate the effectiveness factor for an enzyme reaction that follows Michaelis–Menten kinetics. Assume the Thiele modulus is 4 and $\beta = 2$. Compare your result to the value obtained assuming the reaction is first order.

11. The following kinetic data (Sung et al. 1986) were obtained for the enzyme bilirubin oxidase at a pH of 7.4 and a temperature of 37°C. The substrate was bilirubin. The enzyme loading in the reactor was equivalent to 0.15 U L^{-1}, where a "Unit" of enzyme activity is defined as the amount of enzyme that catalyzes the conversion of 1 μmol min^{-1} of bilirubin. Assuming the data follow the Michaelis–Menten rate law, determine the values of V_{max}, K_m, and k_{cat}.

Bilirubin Concentration (μm)	Reaction Rate (μM bilirubin min^{-1})
1	0.015
5	0.055
10	0.089
15	0.130
25	0.185
40	0.230
50	0.244

12. Suppose the bilirubin oxidase is immobilized within an agarose gel (Sung et al. 1986). The enzyme loading in the gel equals 10 U mL^{-1} of wet gel. If external and internal mass transfer effects are negligible, calculate the conversion of bilirubin in a well-mixed reactor containing a total gel volume of 15 mL. The void fraction in the reactor is 70%. Assume that the flow rate through the reactor is equal to 1.2 mL min^{-1} and that the entering serum bilirubin concentration is 200 μm. The value of k_{cat} for these conditions is equal to 4.67×10^{-4} μmol min^{-1} U^{-1}. Assume that the value of K_m is unaffected by immobilization or other substances found in serum.

13. Fukui et al. (1994) measured in vitro oxygen transfer rates using an ILAD. The membrane oxygenator consisted of microporous polypropylene hollow fibers that were placed within a flexible polyvinyl chloride (PVC) tube. The PVC tube was 30 cm in length with a 20 mm inside diameter. The total surface area of the hollow fibers was 0.3 m^2. The blood flows along a straight path outside the fibers and the gas flows within the hollow fibers. For blood flow rates of 1, 2, and 3 L min^{-1}, the oxygen transfer rates (BTP) were 60, 120, and 140 mL min^{-1}, respectively. Estimate the number and the diameter of hollow fibers needed to reproduce their results.

14. A series of experiments was performed using various sizes of particles containing an immobilized enzyme in order to determine the importance of diffusion within the

particles. The substrate concentration is much less than the value of K_m, so the reaction may be assumed to be first order. In addition, the bulk solution was well mixed and the concentration of the substrate at the surface of the particles is 2×10^{-4} mol cm^{-3}. The partition coefficient was also found to be $K = 1$. The following table summarizes the data obtained:

Diameter of the Particle (cm)	Observed Reaction Rate(mol h^{-1} cm^{-3})
0.050	0.22
0.010	0.98
0.005	1.60
0.001	2.40
0.0005	2.40

From the data in the table, determine the values of the first order rate constant (k) and the effective diffusivity of the substrate within the particle (D_e). To assess the validity of the assumptions you make to find k and D_e, also calculate the effectiveness factor (η), the Thiele modulus (ϕ), and the predicted reaction rate for each particle size. Make a graph that compares the predicted reaction rate to the observed reaction rate for each particle size. How does it look?

15. A biotech company is designing an immobilized enzyme reactor for the removal of a toxin from blood. The device will be connected to the patient by an arteriovenous shunt. Blood will flow over a packed bed of particles (100 μm in diameter) containing the immobilized particles. The following data have been obtained for the toxin enzyme kinetics:

$K = 1$
$V_{max} = 1 \times 10^{-9}$ mol cm^{-3} sec^{-1}
$K_m = 1 \times 10^{-9}$ mol cm^{-3}
$D_e = 5 \times 10^{-7}$ cm^2 sec^{-1}

The maximum concentration of the toxin entering the reactor is estimated to be 1×10^{-11} mol cm^{-3}. Estimate the blood residence time within the reactor to obtain a 60% per pass conversion of the toxin in the blood.

16. Design a packed bed heparinase reactor to achieve a 70% conversion of heparin. Assume an enzyme loading of 120 U mL^{-1} of immobilized enzyme and a feed substrate concentration of 0.2 mg mL^{-1}. The flow rate of plasma to your reactor is 120 mL min^{-1}. Use the heparinase kinetics from Example 8.5, neglect external mass transfer effects, and base the effectiveness factor on that of a first order reaction. Other information you may need can be found in Example 8.8. Compare the enzyme gel volume for the same case solved using a well-mixed reactor in Example 8.8.

17. Derive Equation 8.1.

18. Derive the performance equation shown in Table 8.4 for a countercurrent hemodialyzer.

19. Design an aquapheresis cartridge that can remove up to 500 mL h^{-1} of fluid from the body. Specify the membrane properties, such as total membrane area, nominal molecular weight cutoff, membrane thickness, and membrane porosity that you recommend. How many hollow fibers are required and what are their dimensions in terms of length and diameter? What are the operating pressures for your system and the effective pressure drop across the cartridge?

9 Tissue Engineering

9.1 INTRODUCTION

Over eight million surgical procedures are performed each year in the United States for the treatment of lost tissue or organ function as a result of disease or injury (Holder et al. 1997). Organ transplantation has also become a routine and successful treatment method for the replacement of a failed organ. Organs routinely transplanted include the kidneys, heart, lungs, liver, intestines, and pancreas. Part of this success is due to better therapies to prevent rejection episodes, as well as improved techniques for the procurement and storage of whole organs prior to transplantation. Organ transplantation has saved and improved the quality of many lives; however, its widespread application is severely limited by a shortage of whole organ donors. In the United States, only a few thousand organ donors are available each year. Yet tens of thousands are on waiting lists for an organ transplant, and most of them will die waiting for an organ. Certainly, extracorporeal devices, such as the hemodialyzer, the bioartificial liver (to be discussed in Chapter 10), the artificial heart (Jarvik 1981; Rosenberg 1995; Spotnitz 1987), and left ventricular assist devices (Griffith et al. 2001) can be used until an organ is available. Thus, these devices serve as a *bridge* to a transplant. The continued development and improvement of these devices will help alleviate the pain and suffering while awaiting an organ transplant. However, these devices cannot be used indefinitely and by themselves are still not a perfect solution.

Many diseases do not involve an entire organ, but only certain tissues or cell types that have lost their function. In these cases, it may be simpler to restore the lost function of the tissue or cells than to replace an entire organ. A good example is the treatment of insulin-dependent diabetes mellitus. The islets of Langerhans are a specialized cluster of cells scattered throughout the pancreas. Their mass represents only about 1–2% of the pancreas. They secrete a variety of hormones and one of these, insulin, is responsible for the regulation of blood glucose levels. Loss of insulin secretion through an autoimmune disease process (Atkinson and MacLaren 1990; Notkins 1979) results in diabetes. It is estimated that perhaps only about 500,000 to 1 million islets (Warnock et al. 1988; Friedman 1989; Smith et al. 1991) will be needed to treat a patient with diabetes. With an islet averaging about 150 μm in diameter, this translates to an islet volume of 1–2 cm^3, a volume considerably less than that of the whole pancreas. Transplanting just the islets, therefore, results in a far simpler surgical procedure.

9.2 CELL TRANSPLANTATION

The selective transplantation of just those cells necessary to replace the lost function of a tissue or an organ is receiving considerable attention (Cima et al. 1991a, 1991b; Langer and Vacanti 1993; Brown et al. 2000; Ma and Choi 2001; Palsson and Bhatia 2004; van Blitterswijk et al. 2008). Selective cell transplantation is being developed for the treatment of a variety of diseases and medical problems, some of which are outlined in Table 9.1. The cited references are also a good resource for additional information concerning specific applications of cell transplantation.

Selective cell transplantation forms the basis of the relatively new fields of *tissue engineering* and *bioartificial organs*. Tissue engineering is defined as a rapidly developing multidisciplinary field that utilizes life science and engineering principles to construct biological substitutes containing viable and functioning cells for the restoration, maintenance, or improvement of tissue function. The transplanted cells can be harvested from a variety of sources. For example, cells can be

TABLE 9.1
Opportunities for Cell Transplantation and Tissue Engineering

Liver diabetes	Dental plastic/reconstructive surgery
Parkinson's disease	Nerve regeneration
Kidney	Site-specific delivery of growth factors
Chronic pain	Cornea
Esophagus	Skin
Ureter	Intestine
Bladder	Trachea
Cartilage	Bone
Muscle regeneration	Blood
Artificial blood vessels	Destruction of tumors
Gene therapy	Parathyroid
Cell production	In vitro hematopoiesis
Alzheimer's disease	Neurological diseases
Heart valves	Lymphocytes
Spine	Intervertebral discs

obtained from a human donor (*allograft*), animals (*xenograft*), the patient themselves (*autograft*), or from a genetically engineered cell line (Morgan et al. 1994; Chang 1997; Zalzman et al. 2003). There is also much interest in the use of both adult and embryonic stem cells. Adult stem cells are found within the body. Embryonic stem cells, however, are not found in the body, but are isolated from what is known as the *inner cell mass* of the *blastocyst*, which later forms the embryo. The isolation of the inner cell mass from a developing embryo for tissue engineering applications and for the treatment of disease, results in the death of the embryo, and in the case of a human embryo, this has raised ethical issues. For example, is the development from conception to birth the *development to a human being* or the *development of a human being* (van Blitterswijk et al. 2008)? This has created a moral dilemma in the sense of whether it is morally justifiable to use embryonic stem cells from human embryos to treat human diseases.

Stem cells have two unique properties that most other terminally differentiated cells do not have. First, stem cells are able to replicate themselves, which is known as *pluripotency*. Secondly, they have the potential to differentiate, under the right conditions, into other cell types that are found in the body. For example, adult bone marrow has many types of stem cells that can be coaxed to differentiate into about every type of cell found in the body. Hence, stem cells have the potential to be a rich source of cellular material for tissue engineering applications.

A particular type of stem cell found both in embryos and in adults is the *mesenchymal stem cell*. A mesenchymal stem cell can differentiate to form cartilage, bone, tendons, ligaments, marrow stroma, and connective tissue (Caplan 1991). Mesenchymal stem cells have received considerable attention because of their potential to provide a cell source for a variety of tissue-engineered constructs that can be used in such orthopedic applications as bone and cartilage repair. In many cases, the patient's own mesenchymal stem cells can be harvested from his or her bone marrow and used for these applications.

Figure 9.1 illustrates the basic steps of the tissue engineering process. This process consists of several steps and includes developing a source for the required cells, expanding these cells through in vitro tissue culture, seeding the cells into a scaffold material with the appropriate extracellular matrix materials, and providing the resulting constructs with any required mechanical stimulus or molecular signaling molecules to ensure proper differentiation of the resulting tissue. Once these steps have been completed, the tissue-engineered construct can be implanted into a patient.

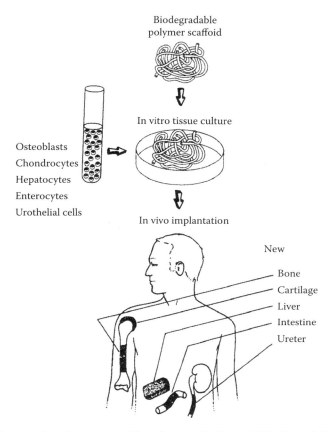

FIGURE 9.1 The tissue engineering process. (From Langer, R., Vacanti, J.P., Vacanti, C.A., Atala, A., Freed, L.E., and Vunjak-Novakovic, G., *Tiss. Eng.*, 1, 151–61, 1995. With permission.)

There is considerable overlap between the two emerging fields of bioartificial organs and tissue engineering. In some cases, they are synonymous. For our purposes, we will categorize the field of bioartificial organs as a subset of the larger field of tissue engineering. The major distinction between the two is that in a bioartificial organ, the goal is to use a cellular transplant to replace a lost organ function, e.g., the use of the islets of Langerhans or genetically engineered beta cells for the treatment of diabetes or *hepatocytes* for the treatment of liver failure. Bioartificial organs also isolate the cells from the host's immune system through the use of an *immunoisolation membrane* (Colton 1995). This membrane is permeable to small molecules, such as required nutrients and the therapeutic agent released by the cells, but impermeable to the larger molecules (antibodies and complement) and cells of the host's immune system. The immunoisolation concept is illustrated in Figure 9.2. Bioartificial organs therefore do not require the use of immunosuppressive drugs or *immunomodification* (Lanza and Chick 1994b) of the cells prior to transplantation. Because of this immunoprotection, there is the potential to use cells and tissues from animals (xenografts) or genetically engineered cell lines, which vastly increases the availability of donor tissue. Several examples of bioartificial organs will be examined in greater detail in Chapter 10.

Isolated cells or cellular aggregates can also be simply infused as is into several locations within the body. Typical sites include the liver via the portal vein, within fat, the spleen, under the kidney capsule, and the omentum. Isolated cells have the potential to grow and form larger tissue structures when implanted near existing mature tissue. However, a potential problem is their loose association, which provides no guide for attachment, restructuring, and formation of larger cellular aggregates. Additionally, cells cannot survive if they do not have an adequate supply of blood to

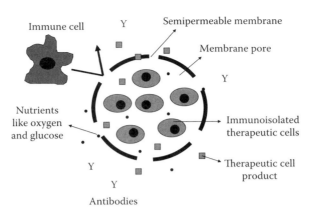

FIGURE 9.2 Concept of immunoisolation.

provide oxygen, nutrients, and a means for waste product removal. Furthermore, any therapeutic product released by the transplanted cells needs to have access to the host's vasculature in order for it to be effective. This requires that each transplanted cell be within 100 μm or so of a capillary. Initially, the existing capillary supply in the tissue surrounding the cellular transplant may not be optimal for the sustained growth of the transplanted cells. This severely limits the size of the cellular aggregates that can form, unless a means is provided for the formation of a vascular capillary bed (angiogenesis) within the transplanted cells. The goal of tissue engineering is therefore to develop methods and techniques than can enhance the success of cellular transplants.

9.3 EXTRACELLULAR MATRIX

Most of the effort in tissue engineering is now focusing on the use of polymeric support structures or scaffolds for guiding the growth and organization of the transplanted cells (Cima et al. 1991a, 1991b; Goldstein et al. 1999; Oerther et al. 1999; Petersen et al. 2002; Leach et al. 2003; Radisic et al. 2003; Pratt et al. 2004; Palsson and Bhatia 2004; van Blitterswijk et al. 2008). These scaffolds can be made from synthetic or naturally occurring polymeric materials. The materials used as scaffolds are either biodegradable or nonbiodegradable and can be formed into a variety of shapes with mechanical properties similar to the tissue they are replacing. To better understand how polymeric support structures can enhance the growth of transplanted cells, it is necessary to first examine how cells organize themselves within natural tissue structures.

Tissues are not made up entirely of cells, but also consist of an extracellular gel-like fluid containing a variety of macromolecules collectively referred to as the *extracellular matrix* (ECM) (Alberts et al. 1989; Rubin and Farber 1994; Long 1995; Naughton et al. 1995; Anderson 1994; Hubbell 1997; Parsons-Wingerter and Sage 1997; Dee et al. 2002). The major components of the ECM are glycosaminoglycans (GAGs), proteoglycans, collagens, elastic fibers, structural glycoproteins, and the basement membrane. ECM macromolecules are secreted locally for the most part by specialized cells called *fibroblasts*. Several of the most important types of macromolecules found within the ECM are summarized in Table 9.2. The materials that comprise the ECM form a unique composition for each type of tissue. These ECM materials are in intimate contact with the cells and hold them together, forming an organized three-dimensional, cross-linked, mesh-like structure. The ECM gives the tissue its mechanical strength and serves as a pipeline for intercellular signaling. Thus, the ECM plays an active role in organizing the tissue, since it not only serves to provide for the three-dimensional organization of cells and adjacent layers of tissue, but also provides a mechanism for intercellular communication and controls such cellular processes as proliferation, cell migration, attachment, differentiation, and repair. Most mammalian cells are anchorage dependent

---

TABLE 9.2
Major Macromolecules of the Extracellular Matrix

Glycosaminoglycans (GAGs): forms the ECM gel
Collagen: provides strength and organization to the ECM
Elastin: provides resilience or elasticity
Fibronectin: adhesion of fibroblasts and other cells
Laminin: adhesion of cells to the basal lamina

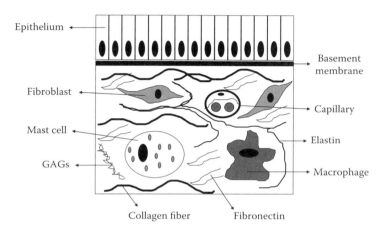

FIGURE 9.3 Connective tissue.

and possess cell-surface receptors for a variety of these ECM macromolecules. Hence, some of these ECM macromolecules possess so-called *attachment factors* that are responsible for the development, growth, and metabolic functions of cells. The ECM itself has also been used, because of these properties, as a scaffold for tissue engineering applications.

This tissue organizing principle of the ECM is shown in Figure 9.3 for the case of *epithelial cells* that are supported by a layer of *connective tissue*. Connective tissues (also called the *stroma*) provide the framework for the formation and organization of most of the larger structures found within the body. The specific functional cells that define a tissue, e.g., the hepatocytes found in the liver or the islets of Langerhans in the pancreas, are called *parenchymal cells*.

9.3.1 GLYCOSAMINOGLYCANS

GAGs form from long linear polymers consisting of a repeating disaccharide unit. Molecular weights range from several thousand to over a million. GAGs tend to have a large number of negative charges because of the presence of carboxyl and sulfate groups. They are highly extended, forming random coil-like structures. Because GAGs are also hydrophilic, they form in combination with the interstitial fluid, the hydrated gel-like material throughout which is found the other macromolecules of the ECM. In addition, GAGs bind with growth factors and intercellular messenger molecules known as *cytokines*.

There are four types of GAGs, depending on what types of sugars they are made from. They are *hyaluronic acid*, *chondroitin sulfate* and *dermatan sulfate*, *heparin sulfate* and *heparin*, and *keratin sulfate*. With the exception of hyaluronic acid, they also form what are called *proteoglycans* through covalent bonds with proteins.

9.3.2 COLLAGENS

Collagens are fibrous proteins made from three polypeptide chains that form the tough, triple-stranded helical structure of the collagen molecule. The tensile strength of tissues is provided by collagen. Collagens are also the most prevalent protein material found within the ECM and they have several unique functions. Although more than 20 types of collagen molecules have been identified, there are 4 that are most important. These are the *fibrillar collagens* (type I, type II, and type III) and type IV collagen. The first three types of collagen molecules assemble themselves into much larger structures called *collagen fibrils* that aggregate further to form collagen fibers. Type I collagen makes up about 90% of the collagen found in the body. It is found mostly in the skin, tendons, ligaments, various internal organs, and bone. Type II collagen is primarily found in cartilage and type III collagen is found in blood vessels as well as the skin. Type IV collagen organizes itself to form sheets within the basal lamina of the basement membrane. Fibroblasts have the ability to organize the collagen fibrils they secrete, forming sheets or rope-like structures, therefore they can affect the spatial organization of the matrix they produce.

9.3.3 ELASTIN

Elastin is a hydrophobic glycoprotein molecule that, through crosslinks with other elastin molecules, can form a network of sheets and filaments with the unique property of being elastic, allowing for recoil after periods of stretch. The elasticity is a result of the random coil-like structure of the elastin molecule. The elastic nature of elastin is important in blood vessels, skin, the lungs, and the uterus. Inelastic collagen fibrils are interwoven with elastin to limit and control the degree of stretching.

9.3.4 FIBRONECTIN

Fibronectin is the principal adhesive glycoprotein found within the ECM. Next to collagen, it is the second most prevalent macromolecule found in the ECM. Fibronectin binds to other ECM molecules, such as collagen and cell-surface receptors. Therefore, it has a principal role in the attachment of cells to the ECM. Fibronectin is a dimer made up of two subunit chains that are bound together at one end by a pair of disulfide bonds. Along the length of the chains are a series of functional domains that can bind to specific types of molecules (e.g., collagen or heparin) or cell-surface receptors. The cell-binding domain has a specific tripeptide sequence consisting of the amino acids: arginine (R), glycine (G), and aspartic acid (D), often referred to as the *RGD sequence*. Peptides containing this sequence will inhibit the attachment of cells to fibronectin through their competition for the RGD binding site on the cell surface.

In addition to fibronectin, there are several other ECM adhesion proteins that express the RGD sequence. These include *vitronectin*, found primarily in blood cells. *Thrombospondin* is secreted by a variety of cells involved in the development of the ECM. It serves to bind together other components of the ECM, like fibronectin. *Von Willebrand factor* is made by megakaryocytes (platelet-generating cells found in the bone marrow). It is stored in circulating platelets. Von Willebrand factor is released by platelets as a result of injury to blood vessels and this factor then binds to collagen. The platelets then bind to the von Willebrand factor that is also bound to collagen. Fibrinogen is another clotting protein found in blood. During blood clot formation, fibrinogen is converted to fibrin that forms a mesh-like structure that traps red blood cells and platelets.

9.3.5 BASEMENT MEMBRANE

The *basal lamina* is a continuous mat-like structure of ECM materials that separates specific cells, such as epithelial, endothelial, or muscle, from the underlying layer of connective tissue. The basal lamina consists of two distinct layers, the *lamina rara** directly beneath the basal membrane of the specific cells above, and the *lamina densa*[†] just below the lamina rara. Below the two layers of the basal lamina is the collagen-containing *lamina reticularis* that connects the basal lamina to connective tissue that lies below it. These three layers together constitute what is known as the *basement membrane* (see Figure 9.3). The basal lamina consists primarily of type IV collagen, proteoglycans such as those formed from heparin sulfate, and the glycoprotein, *laminin*.

Laminin is an extremely large protein with a molecular weight of 850,000, made from three long polypeptide chains that form the shape of a cross. It also contains a number of functional domains throughout its structure with sites for binding to type IV collagen, heparin sulfate, and laminin cell-surface receptors. Laminin has a major role in the formation and maintenance of blood vessels.

9.4 CELLULAR INTERACTIONS

There are three types of interactions involving cells. These interactions may be classified as cell–ECM, cell–cell, and cell–growth factor (Long 1995). Figure 9.4 illustrates the various interactions of cells with their surrounding environment. Cells interact with other cells, the ECM, and growth factors through cell-surface receptor proteins that are an integral part of the cell membrane. Binding of cell-surface receptors to surrounding ECM components, such as collagen, fibronectin, and laminin, enables the cell to link up to the surrounding matrix. Binding of cell-surface receptors between cells allow them to organize into larger structures, such as tissues and organs. Cell-surface receptor interactions with growth factors provide control over a variety of cellular functions through control of gene expression.

There are four types of cell adhesion receptors (Hubbell 1997). Three of these receptors are mainly involved in cell–cell interactions and the fourth is involved in both cell–cell and cell–ECM interactions. The adhesion receptors used in cell–cell interactions include *cadherins*, *selectins*, and *cell adhesion molecules* (CAMs). The receptors involved in cell–ECM interactions belong to a general family of adhesion receptors known as *integrins* (Horwitz 1997).

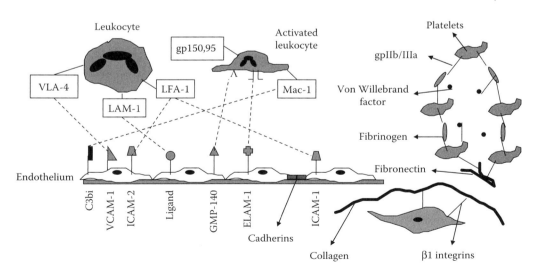

FIGURE 9.4 Cellular interactions.

---

* Under the electron microscope this layer is translucent to electrons.
[†] Under the electron microscope this layer is somewhat opaque to electrons.

9.4.1 CADHERINS

Cadherins take part in what is known as *homophilic binding*, which is the binding of a cadherin molecule on one cell with an identical cadherin on another cell of the same type. They do not bind with ECM components. Cadherin molecule binding is strongly dependent on the presence of extracellular Ca^{2+}.

9.4.2 SELECTINS

Selectins are a family of CAMs that are found on the surfaces of endothelial cells, leukocytes, and platelets. They exhibit heterophilic binding between the blood cells and the endothelial cells that line the blood vessels. They are important in the localization of leukocytes to sites of inflammation and tissue injury. The three most important selectins are GMP-140, endothelial leukocyte adherence molecule-1 (ELAM-1), and leukocyte adhesion molecule-1 (LAM-1). Like cadherins, their activity is dependent on the presence of extracellular calcium.

9.4.3 CELL ADHESION MOLECULES

CAMs are members of the immunoglobulin gene superfamily. These receptors bind independently of extracellular calcium levels. CAMs are involved in recognition of antigens (foreign agents) and cell–cell interactions. Antigen recognition involves both the T-lymphocyte receptor and two special antigen-presenting molecules called major histocompatibility class I and II (MHC I and MHC II). These specialized CAMs will be discussed further in Chapter 10 in the section on immunology. Two other CAMs are important factors in localizing leukocytes to regions of tissue injury. For example, intercellular adhesion molecule-1 (ICAM-1) is found on the surfaces of endothelial cells and leukocytes and binds to the integrin called *leukocyte function antigen*-1 (LFA-1), which is found on macrophages and neutrophils. ICAM-2, found on endothelial cells, also binds to LFA-1 and helps to localize neutrophils to sites of injured tissue. The vascular cell adhesion molecule (VCAM-1) is expressed by endothelial cells as a response to injury or inflammation and it promotes adherence of leukocytes to endothelial cells.

9.4.4 INTEGRINS

Integrins are a family of adhesion molecules that are involved in cell–cell interactions as well as binding with molecules found in the ECM. Integrin molecules are heterodimers consisting of two protein subunits, the α chain and the β chain. There are at least 15 variants of the α subunit and 8 variants of the β subunit. Overall, there are at least 20 different heterodimer $\alpha\beta$ integrins. The β subunits provide the functional aspects of the dimeric integrin molecule, hence integrins are classified according to their β chain type. Subclasses having the β_1, β_2, and β_3 chains are the most common. The β_1 and β_3 subclasses are primarily involved in interactions between the cell and the ECM. The β_2 subclass involves mostly cell–cell interactions with leukocytes.

β_1 integrins are also referred to as *very late after* (VLA) antigens (Margiotta et al. 1994). VLA antigens are numbered according to the number of the alpha subunit that forms the heterodimer with the β_1 subunit. These subclasses bind with a variety of ECM proteins, such as collagen, fibronectin, and laminin. The primary recognition site for β_1 integrins is the RGD sequence described earlier. Therefore, this group of integrins is important for the overall organization and function of a variety of cell types. VLA-4 is also a receptor found on leukocytes for VCAM-1 expressed by endothelial cells.

Cell–cell interactions, mainly leukocyte to leukocyte, involve mostly the β_2 group of integrins. Hence, this subclass is known as the "leukocyte" integrins. The common β_2 subunit is also known as CD18 (cluster designation), a protein with a molecular weight of 95,000. The most important alpha subunits, in combination with β_2, form the leukocyte integrins known as LFA-1, *macrophage antigen-1* (Mac-1), and *gp150,95* (a glycoprotein with an alpha subunit of molecular weight 150,000 along with the β_2 subunit protein with a molecular weight of 95,000). These alpha subunits are also known, respectively, by the cluster designations CD11a, CD11b, and CD11c. The expression of these integrins on leukocytes is particularly important for defense against bacterial infections.

β_3 integrins comprise the platelet glycoprotein receptor, known as gpIIb/IIIa, and the vitronectin receptor. Both of these receptors also recognize the RGD sequence of ECM matrix proteins, such as vitronectin, fibronectin, thrombospondin, fibrinogen, and von Willebrand factor. Aggregation of platelets is enhanced by the release of von Willebrand factor. gpIIb/IIIa is calcium dependent and found on the surface of activated platelets. This integrin is important for platelet aggregation and adherence of platelets to the subendothelium. This occurs through binding of this receptor to fibrinogen, fibronectin, vitronectin, and von Willebrand factor.

9.4.5 CYTOKINES AND GROWTH FACTORS

A variety of cells also secrete soluble proteins known as cytokines. Cytokines serve as intercellular chemical messengers. Many of these proteins are required for the normal development, growth, and proliferation of cells. They are also involved in such processes as inflammation, wound healing, and the varied responses of the immune system. Many are *mitogens*, that is they stimulate the proliferation of specific types of cells. They are often then referred to as *growth factors*.

Growth factors bind to specific receptors on the surface of their target cell and have the ability to induce or direct the action of specific genes in the targeted cell. A variety of growth factors have been identified and some of the more important of these are summarized in Table 9.3. All of these growth factors play a key role during the wound healing process and many are involved in the growth of new blood vessels. Growth factors can also be used to stimulate and accelerate the development of blood vessels in polymeric scaffolds used in tissue engineering applications.

Platelet-derived growth factor (PDGF) is a mitogen for such cells as smooth muscle, fibroblasts, and neuroglial cells. PDGF is stored in platelets and is released after platelet aggregation at the site

TABLE 9.3
Growth Factors

Growth Factor	Property
Platelet-derived growth factor (PDGF)	Mitogen for smooth muscle cells, fibroblasts, and neuroglial cells, chemotactic signal for immune system cells
Epidermal growth factor	Mitogen for a variety of cells types, accelerates wound healing
Fibroblast growth factor (aFGF, bFGF) and vascular endothelial cell growth factor (VEGF)	Stimulates growth of blood vessels, mitogen for endothelial cells, fibroblasts, smooth muscle cells
Tumor necrosis factor (TNF)	Stimulates growth of blood vessels
Bone morphogenetic protein (BMP)	Induces the formation of bone and cartilage
Insulin-like growth factor (INF-I)	Stimulates growth of blood vessels
Transforming growth factor (TGF-β)	Controls cellular response to other growth factors
Nerve growth factor (NGF)	Stimulates axon growth
Stromal cell-derived growth factor (SDF-1)	Stimulates growth of blood vessels
Keratinocyte-derived growth factor (KGF)	Involved in wound healing
Epithelial cell growth factor (EGF)	Important in wound healing
Hepatocyte growth factor (HGF)	Important in organ regeneration and wound healing
Interleukin-2 (IL-2)	Stimulates proliferation of T lymphocytes
Interleukin-6 (IL-6)	Stimulates proliferation and differentiation of B lymphocytes

of injury to blood vessels. Damaged endothelial and smooth muscle cells also secrete PDGF. PDGF is also a chemotactic signal for such inflammatory cells as macrophages and neutrophils. Epidermal growth factor (EGF) stimulates the growth of a variety of cell types. EGF accelerates the healing of many wounds and stimulates fibroblasts to deposit collagen.

Fibroblast growth factor (FGF) comes in two forms, acidic (aFGF) and basic (bFGF). bFGF is an order of magnitude more potent than aFGF. FGF and vascular endothelial cell growth factor (VEGF) stimulate the growth of capillaries, a process called *angiogenesis* (Folkman 1985; Folkman and Klagsbrun 1987). FGF also promotes the growth of fibroblasts, endothelial cells, and smooth muscle cells. These processes affected by FGF also accelerate wound healing.

Another angiogenic growth factor is tumor necrosis factor-α (TNF-α), which is secreted by activated macrophages during the initial stages of inflammation. Insulin-like growth factor-I (IGF-I) not only plays a major role in growth, but also stimulates angiogenesis. Transforming growth factor-β (TGF-β) can either promote or inhibit the growth of cells by controlling their response to other growth factors. Several other important growth factors include nerve growth factor (NGF), which enhances the function of neurons and stimulates axon growth, interleukin-2 (IL-2), which stimulates the proliferation of T lymphocytes, and interleukin-6 (IL-6), which stimulates the proliferation of B lymphocytes and their differentiation to antibody-secreting plasma cells. There are also several hematopoietic* growth factors that are specific to the development and differentiation of blood cells.

Other growth factors of interest to tissue engineering include stromal cell-derived growth factor (SDF-1), epithelial cell growth factor (EGF), and keratinocyte growth factor (KGF). These three growth factors are important in wound healing and angiogenesis. Hepatocyte growth factor (HGF) plays an important role in organ regeneration and wound healing, and bone morphogenetic protein (BMP) stimulates the formation of bone and cartilage.

9.5 POLYMERIC SUPPORT STRUCTURES

With this background on how cells organize themselves within the body, we can now address the properties required for an artificial polymeric support structure for transplanted cells. Table 9.4 summarizes some of the characteristics that are important to consider in selecting materials to serve as support structures for transplanted cells. A variety of biological and synthetic polymers are used as biomaterials and many of these can be used to provide the support structure for transplanted cells. Table 9.5 summarizes some of the materials that are being used or being considered for use as support structures in tissue engineering.

The biomaterials listed in Table 9.5 are grouped as to whether or not they are biodegradable. Long-term implantation of nonbiodegradable biomaterials poses the risk of localized infection, chronic inflammation, and development of fibrous encapsulation of the implant. This can compromise the proper functioning of transplanted cells if they were to reside within the porous structure of these materials. Nonbiodegradable materials, such as PTFE, PU, and Dacron, are preferred for those applications where their structural integrity is more important than the requirement for them to serve as a scaffold for transplanted cells, e.g., as vascular grafts or structural components of artificial organs. These materials can also be made into a variety of three-dimensional configurations. Other materials, such as agarose gel, are nonbiodegradable biological materials that have found wide use in cell encapsulation and have recently been proposed as scaffolds in tissue engineering (Park et al. 2010).

Biodegradable materials are now attracting significant interest for tissue engineering applications. Biodegradable materials allow the transplanted cells sufficient time to organize the desired three-dimensional structure and develop their own blood supply. The biodegradable polymers are easily hydrolyzed by the body's fluids and slowly disappear without leaving behind any foreign

---

* Hematopoiesis is the process of blood cell development and differentiation.

TABLE 9.4
Desirable Properties of Support Structures for Transplanted Cells

Biocompatible

Nonimmunogenic

Negligible toxicity

    Locally and systemically

    The polymer and any of its degradation products

Chemically and mechanically stable

Processable into a variety of shapes

    Hollow tubes, sheets, arbitrary 3-D shapes

    Open foam or sponges

    Woven or nonwoven mesh-like structures

    High porosity

    High internal surface area to volume ratio

    Controllable pore size

Promotes cell attachment

Promotes angiogenesis

Favorable interaction/mobilization of host cells

Ability to release active compounds such as growth factors

Ability of the cells and ECM materials to interact with the support structure

Ability to obtain the desired cellular response

residues. Examples of biologically derived biodegradable polymers include collagen, GAGs, chitosan, and small intestinal submucosa (SIS). Collagen is one of the ECM materials and can be prepared as fibers (Cavallaro et al. 1994) or as a gel. Collagen is being used in such applications as tissue repair and artificial skin (Morgan and Yarmush 1997). SIS has many interesting properties and is widely used as a bioscaffold and for the repair of tissues (Le Visage et al. 2006). SIS is harvested from the small intestine of pigs and consists of a rather loose scaffold of ECM materials consisting of oriented collagen fibers, GAGs, and attached growth factors, such as bFGF, VEGF, and TGF-β.

PLA, PGA, and PLGA are naturally occurring hydroxy acids. They are biodegradable and have been approved by the Food and Drug Administration (FDA) for use in sutures, controlled drug release, and surgical support fabrics (Cima et al. 1991a, 1991b). A great deal of effort is focusing on these materials for tissue engineering applications (Holder et al. 1997; Mooney et al. 1994, 1995a, 1995b; Kaufmann et al. 1997; Blitterswijk et al. 2008). They degrade by hydrolysis and form natural by-products. Their absorption rate can also be controlled from months to years (Agrawal et al. 1995; Mooney et al. 1994, 1995a, 1995b). The degradation rate of these materials is dependent on their initial molecular weight, the surface area that is exposed, the degree of crystallinity, and for the case of co-polymer blends, the ratio of the hydroxy acid monomers used (Pachence and Kohn 1997).

The resulting structure formed from a biomaterial must be mechanically strong enough to support the growth of the transplanted cells and chemically compatible with the intended duration of use. For example, cells secrete a variety of enzymes that may degrade the polymer used. For a biodegradable material, this cell-induced degradation rate must be considered during formulation of the biomaterial to achieve the desired degradation rate. In the case of nonbiodegradable materials, they must withstand cellular attack and not form by-products that are inflammatory or immunogenic and therefore capable of compromising the function of the implant. The material used to form the support structure will also generally be in direct contact with the host's immune system and connective tissue such as fibroblasts. For any biomaterial, activation of the host's immune system,

TABLE 9.5

Materials for Tissue Engineering and Other Biomedical Applications

Nonbiodegradable Polymers	Applications
Alginates	Food additives, cell encapsulation
Agarose	Cell culture and cell encapsulation, scaffold
Cellulose	Bone tissue engineering
Polydimethylsiloxane or silicone (PDMS)	Breast, penile, testicular prostheses, catheters, drug delivery, heart valves, membrane oxygenators, shunts, tubing, orthopedics
Silk fibroin	Stem cell applications for cartilage and bone
Ceramics	Bone repair
Polyurethanes (PEU)	Artificial hearts and ventricular assist devices, catheters, intraaortic balloons, wound dressings
Polytetrafluoroethylene (PTFE)	Heart valves, vascular grafts, reconstruction, shunts, membrane oxygenators, catheters, sutures, coatings
Polyethylene (PE)	Artificial hips, catheters, shunts, syringes, tubing
Polysulfone (PS)	Heart valves, penile prostheses, artificial heart
Polycarbonate	Hard contact lenses
Poly(methyl methacrylate) (PMMA)	Bone cement, fracture fixation, intraocular lenses, dentures, plasmapheresis membranes
Poly(2-hydroxyethylmethacrylate) (PHEMA)	Controlled drug release, contact lenses, catheters, coatings, artificial organs
Polyacrylonitrile (PAN)	Hemodialysis membranes
Polyamides (nylon)	Hemodialysis membranes, sutures
Polyethylene terephthalate (Dacron)	Vascular grafts, tissue patches, shunts
Polypropylene (PP)	Valve structures, plasmapheresis membranes, sutures
Polyvinyl chloride (PVC)	Tubing, plasmapheresis membranes, blood bags
Poly(ethylene-co-vinyl acetate)	Drug delivery devices
Polystyrene (PS)	Tissue culture flasks
Poly(vinyl pyrrolidone) (PVP)	Blood plasma extender
Polyvinyl alcohol (PVA)	Dental, tissue repair, scaffolds for bioartificial organs

Biodegradable Polymers	Applications
Poly(L-lactic acid), poly(glycolic acid), and poly(lactide-co-glycolide) (PLA, PGA, and PLGA)	Controlled release drug delivery devices, sutures, scaffolds for cell transplantation and tissue engineering
Collagen	Artificial skin, hemostasis, tissue regeneration scaffold
GAGs (hyaluronan)	Tissue repair, viscoelastic, wound care
Small intestinal submucosa (SIS)	Bioscaffolds and tissue repair
Chitosan	Scaffolds for tissue engineering, inhibitors of blood coagulation, cell encapsulation, membrane barriers, contact lens materials
Polyhydroxyalkanoates (PHA)	Controlled drug release, sutures, artificial skin
Poly(ε-caprolactone) (PCL)	Implantable contraceptive devices, controlled drug release, surgical staples, scaffolds for cell transplantation and tissue engineering
Polyphosphoesters (PPE)	Cartilage and bone, nerve and organ regeneration
Polyphosphazenes (PPA)	Bone and nerve regeneration
Polyanhydrides (PA)	Bone
Starch	Thermoplastic starches (TPS) for use as scaffolds

Source: From Friedman, D.W., Orland, P.J., and Greco, R.S., *Implantation Biology*, CRC Press, Boca Raton, FL, 1994; Marchant, R.E. and Wang, I-W., *Implantation Biology*, CRC Press, Boca Raton, FL, 1994; Saltzman, W.M., *Principles of Tissue Engineering,* R.G. Landes Co., Boulder, CO, 1997; Pachence, J.M. and Kohn, J., *Principles of Tissue Engineering,* R.G. Landes Co., Boulder, CO, 1997; Van Blitterswijk, C., *Tissue Engineering*, Elsevier, London, 2008.

as well as a fibrotic connective tissue response, will tend to wall off and isolate the transplanted support structure, blocking its intended function. The host's reaction and formation of a fibrotic capsule around the implant is a function of not only the materials used, but also depends on the shape and microstructure of the implant.

Unlike in vitro tissue culture of cells wherein the cells are usually grown in only two dimensions, the transplanted support structure containing the cells will be three dimensional. Furthermore, the resulting three-dimensional polymer scaffold must, in general, have high porosity, significant internal surface area, and a controlled pore size distribution. All of these properties must not compromise the structural strength of the material.

Polymer scaffolds used in tissue engineering applications must be processable into whatever shape is required for the implant. A variety of techniques may be used to process polymer scaffolds. These include such methods as fiber bonding, solvent casting, particulate salt leaching, membrane lamination, melt molding, polymer/ceramic fiber composite foams, phase separation, in situ polymerization (Lu and Mikos 1996; Thomson et al. 1997; Ma and Choi 2001), and electrospinning (Teo and Ramakrishna 2006; Nam et al. 2007, 2008, 2009). Electrospinning is receiving considerable attention because of its ability to easily produce porous structures that are highly interconnected and similar to those of the ECM. The pore sizes of these scaffolds can be controlled by adjusting the electrospinning operating conditions, such as the voltage, the distance between the spinneret and the grounded collection plate, and the solution viscosity. In addition, macroscopic porosity can be controlled by combining electrospinning with the salt-leaching technique.

Other methods for creating well-defined three-dimensional scaffolds make use of recent advances in nanofabrication, stereolithography or rapid prototyping (RP) techniques, and solid free-form fabrication (SFF). SSF builds a specific polymeric scaffold layer by layer under the control of a computer-aided design (CAD) process (Lee et al. 2010). Recently, a technique known as precision extrusion deposition (PED) has been used to make polymeric scaffolds with a controlled three-dimensional architecture consisting of interconnected pores with a uniform pore size distribution (Yildirim et al. 2008). PED also builds the scaffold through layer by layer extrusion of the polymer based on a predefined geometry controlled by a CAD program. By giving the PED scaffolds made from polycaprolactone (PCL) an oxygen-based plasma treatment, it was found that cell adhesion was enhanced and cell proliferation increased. PCL has been shown to be particularly suited for bone tissue engineering applications because of its mechanical properties, good biocompatibility, and ability to biodegrade (Yildirim et al. 2010).

Figure 9.5 shows microphotographs of a biodegradable open sponge-like scaffold and a mesh-like structure. Both of these scaffolds exhibit high porosity (>80%) and large pores. The poly(DTE

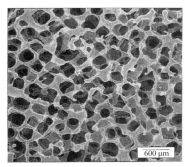
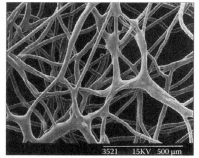

Poly(desaminotryosyl-tyrosine ethyl   Bonded poly(glycolic acid) nonwoven
ester carbonate) polymer scaffold   fiber mesh

FIGURE 9.5 Scanning electron micrographs of polymeric scaffolds for tissue engineering. (Reproduced from James, K. and Kohn, J., *MRS Bull.*, 21: 22–26, 1996; Lu, L. and Mikos, A.G., *MRS Bull.*, 21, 28–32, 1996. With permission.)

carbonate) shown on the left in Figure 9.5 was prepared by a salt-leaching technique. The resulting structure has pores in the range of 200–500 μm and is being used for bone regeneration. The PGA nonwoven mesh on the right has fibers 10–50 μm in diameter and is being considered for use in organ regeneration. These materials were shown to interact favorably with the surrounding tissue, allowing for ingrowth of tissue and the formation of a vascular network throughout the support structure. The high porosity of these structures is also important for nutrient and product transport. To avoid the presence of nonvascular regions and to optimize implant size, it is important to be able to control the pore size distribution. A high internal surface area to volume ratio will also provide numerous sites for cell adhesion to the polymer support structure. A high internal surface area also allows for surface treatment of the polymer structure with ECM materials and/or growth factors in order to enhance interaction of the support with the transplanted cells and promote their proliferation and differentiation. The ECM can also be designed using macromolecular analogues with specific properties that enhance the scaffold's use in tissue engineering applications (Welsh and Tirrell 2000). In addition, synthetic hydrogels have been proposed that have *biomimetic properties** in the sense that these materials have the ability to communicate with cells through the presentation of bound adhesion and growth factors and they can be remodeled through the action of cellular proteases (Pratt et al. 2004).

In the best of situations, the implanted support structure becomes vascularized by the host during the period of time that the transplanted cells are increasing in number. However, the metabolic demands of the transplanted cells in many cases cannot be met by the vasculature that is also developing at the same time within the implant. It may be better to first implant the support structure, allowing it to become vascularized, and then to transplant or seed the cells at a later point in time (Takeda et al. 1995). This gives the transplanted cells the opportunity to start in a well-vascularized region, thereby eliminating any transport limitations due to the metabolic requirements of the transplanted cells.

Prevascularization of the implant can also be enhanced by first seeding the implant with the host's endothelial cells (Holder et al. 1997). Endothelial cell loaded support matrices are believed to enhance vascularization by any or all of the following mechanisms. The endothelial cells can form new capillaries, they may provide chemical signals (ECM and growth factors) for growth of blood vessels, and they may merge with the host's own vascular ingrowth. Immobilization of growth factors on the support material, or within controlled release microspheres, can also be used to facilitate vascularization and improve survival of the transplanted cells (Thompson et al. 1988, 1989; Mooney et al. 1996).

9.6   BIOCOMPATIBILITY AND INITIAL RESPONSE TO AN IMPLANT

The success of a particular scaffold that is used in a tissue engineering application will depend on its biocompatibility. Biocompatibility is the ability of the biomaterial to perform its intended function with little or no harm being done to the surrounding tissues or in a larger sense to the host. A biomaterial should not release anything that is toxic, not promote a response by the immune system (nonimmunogenic), if in contact with blood it should not cause clots to form (nonthrombogenic), and a biomaterial should not cause cancer (noncarcinogenic). How the body responds to a particular biomaterial will also depend on how the biomaterial was processed. For example, the biomaterial's composition and the presence of contaminants and impurities, its degree of crystallinity, porosity, surface properties, degradation kinetics, sterility, and wearability, all play a role in determining the biocompatibility of a given biomaterial. Biocompatibility, however, is not just intrinsic to the biomaterial being used, but also depends to a significant degree on the patient. Hence, it is expected that biocompatibility will vary to some degree among patients as a result of their general health, age,

---

* Meaning to imitate.

sex, lifestyle, and the sensitivity of their tissues and immune system to the particular biomaterial being used.

The first response after implantation of a tissue engineering scaffold or support structure will be the adsorption of a variety of proteins and macromolecules on the surfaces of the scaffold material. These molecules will initiate a host response that will include both localized inflammatory and immune responses as well as processes that promote repair and regeneration of any damaged tissues. In addition, there will also be infiltration of the support by a variety of cell types. All of these processes will determine whether or not the scaffold is successfully incorporated into the host. This process is very similar to what occurs during wound healing and involves three somewhat overlapping phases referred to as *inflammation*, *proliferation*, and *maturation* (Hammar 1993; Arnold and West 1991).

The *inflammation period* lasts for several days and involves the arrival of *platelets* and *neutrophils*. Activation of the clotting process and release of growth factors (Alberts et al. 1989) (PDGF, EGF, TGF) leads to the formation of a collagen-free fibrin network that serves as a scaffold for the inflammatory cells. The neutrophils release factors that attract *monocytes* and also ingest foreign material by phagocytosis. The monocytes also enter the site and are transformed into *macrophages* that, when fused together, become *foreign body giant cells*. They continue to clean up the site, removing dead tissue and bacteria, and walling off large debris.

During the *proliferative phase*, the macrophages release a variety of factors (PDGF, TGF, EGF, FGF) that activate the migration and proliferation of fibroblasts and endothelial cells. The fibroblasts release a variety of ECM materials and begin to form a collagen network. The generally low oxygen levels within the site will also stimulate the movement and growth of capillary sprouts from the surrounding vasculature. These capillary sprouts are formed from endothelial cells and invade the site, forming a vascular bed. Over a period of weeks, the entire site becomes vascularized and this marks the end of the proliferative phase.

The *maturation phase* involves final remodeling of the site, resulting in contraction of the wound and organization of the collagen matrix. This is also a critical time for tissue engineering applications, since the reduced oxygen demand of the cells involved in wound healing can result in regression of the vascular supply. Seeding of the transplanted cells at this time and their need for oxygen may promote continued growth of the vascular supply.

9.7 TISSUE INGROWTH IN POROUS POLYMERIC STRUCTURES

The rate of tissue ingrowth can also be an important factor in the development of tissue-engineered structures. In many cases, this will require that the scaffold be seeded with cells and grown in vitro within a specially designed bioreactor to ensure that the cells have populated the entire scaffold. Otherwise, the slow rate of ingrowth of vascularized tissue from the host will severely limit the thickness of most tissue-engineered structures (Holder et al. 1997). Much effort is therefore being focused on optimizing the rate of tissue ingrowth into support materials so that the thickness of these devices can be increased from millimeters to centimeters (Galban and Locke 1997). For example, Mikos et al. (1993), in a series of now classic experiments, prepared highly porous polymer disks, 13.5 mm in diameter and about 5 mm thick. The polymers used were made either from PLA, PGA, or PLGA. In forming the polymer disks, NaCl particles on the order of several hundred microns were incorporated in the polymer solution at 80–90 wt%, to form a composite membrane. By leaching out the salt particles, a highly porous foam-like support structure was formed. These structures typically had porosities in the range of 64–90% with mean pore sizes ranging from as low as 36 to 179 μm. Porosity and pore size was controllable by adjusting, respectively, the relative amount and particle size of the salt used. The devices were then implanted in the *mesentery* of male rats.

Figure 9.6 illustrates some typical results for a prewetted PLA implant (83% porosity and 166 μm mean pore size) at two sites in the mesentery. Tissue ingrowth was determined from

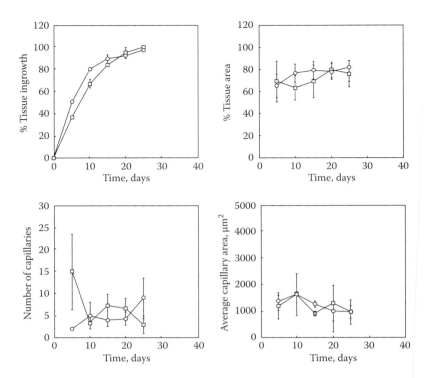

FIGURE 9.6 Normalized tissue ingrowth, percent tissue area, number of capillaries, and capillary area for prewetted PLLA devices. (From Mikos, A.G., Sarakinos, G., Lyman, M.D., Ingber, D.E., Vacanti, J.P., and Langer. R., *Biotechnol. Bioeng.*, 42, 716–23, 1993. With permission.)

histological sections of implants removed at 5, 15, and 25 days. For these results, it appears that 25 days is sufficient to fill the void space of the support structure with vascularized tissue. However, the regions occupied by tissue were not completely filled. The percent tissue area provides a measure of the device total volume fraction occupied by tissue. At day 25, this volume fraction of the tissue is 79%. Since the device total porosity (ε_p, recall it is 83%) is the sum of the tissue volume fraction (ε_T) and the remaining void volume (ε_V), we conclude that the remaining void fraction is 4% at day 25.

Also determined from the histological sections of the implant were the number of capillaries and their average cross-sectional surface area within a 1×1 mm^2 field of view. The number of capillaries (N_c) after 25 days was found to be about 5 per field, whereas the average capillary cross-sectional area ($A_{capillaries}$) was about 1000 μm^2.

Example 9.1

Using the above data obtained by Mikos et al. (1993), calculate the capillary radius, the volume fraction of the capillaries, and the Krogh tissue cylinder radius for the tissue formed in their scaffolds.

Solution

From the measured average cross-sectional area of the capillaries and the number of capillaries in that field, we have that

$$1000 \ \mu\text{m}^2 = 5 \text{ capillaries} \times \pi \, r_C^2.$$

Solving this equation for the capillary radius, we find $r_C = 8$ μm. The volume fraction of capillaries within the tissue (v) is then given by

$$v = \frac{A_{capillaries}}{\varepsilon_T \times 1 \text{ mm}^2} = \frac{1000 \text{ μm}^2}{0.79 \times 1 \text{ mm}^2 \times \dfrac{1 \text{ m}^2}{(1000 \text{ mm})^2} \times \dfrac{(10^6 \text{ μm})^2}{1 \text{ m}^2}} \times 100.$$

This gives a capillary tissue volume fraction of 0.13%. Assuming the idealized Krogh tissue cylinder model shown in Figure 5.16, we can calculate the Krogh tissue cylinder radius using Equation 5.120, i.e., from $v = (r_C/r_T)^2$. With a capillary radius (r_C) determined above of 8 μm, we obtain, as shown by the following calculation, a Krogh tissue cylinder radius of about 220 μm:

$$r_T = \frac{r_C}{\sqrt{v}} = \frac{8 \text{ μm}}{\sqrt{0.0013}} = 220 \text{ μm}.$$

In a subsequent study, Wake et al. (1994) focused on disks with the same geometry as the earlier Mikos et al. (1993) study, forming an amorphous PLLA device with pore sizes ranging from 91 to 500 μm. Figure 9.7 illustrates tissue ingrowth for various implantation times out to 35 days. Here we see the strong effect that mean pore size has on the rate of tissue ingrowth. A 500 μm pore size provides 100% ingrowth after only 5 days, compared to a requirement of 25 days for the 179 μm mean pore size. For the 91 μm pore size device after 25 days, tissue ingrowth is only about 90%. Once again, however, the void space available for cell transplant after tissue ingrowth was quite small.

Similar studies by Holder et al. (1997) showed significantly higher numbers of new capillaries during the first 3 weeks for implants that were initially seeded with endothelial cells. They also reported cellular ingrowth rates into their matrices of about 0.5–1 mm per week. These rates of ingrowth are comparable to the values obtained by Mikos et al. (1993) and Sarver et al. (1995).

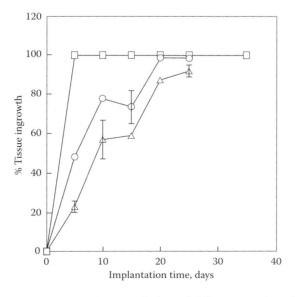

FIGURE 9.7   Normalized tissue ingrowth in PLLA devices of different pore sizes: 500 μm, square; 179 μm, circle; 91 μm, triangle. (From Wake, M.C., Patrick, Jr., C.W., and Mikos, A.G., *Cell Transplant.*, 3, 339–43, 1994. With permission.)

9.8 MEASURING BLOOD FLOW WITHIN SCAFFOLDS USED FOR TISSUE ENGINEERING

Another way to monitor the ingrowth of tissue and the development of a vascular supply within the scaffold structure is through the use of the radioactive or fluorescent microsphere technique (Selman et al. 1984; Sarver et al. 1995). This technique is illustrated for the rat in Figure 9.8.

In this technique, a syringe pump, which we will see later serves as a reference flow rate, is connected to the left femoral artery and blood is removed at a constant rate. After about 1 min, 15 μm diameter radioactive or fluorescent microspheres are injected into the left ventricle of the heart via a cannula in the right common carotid artery. The microspheres are then allowed to distribute throughout the animal's body for 1 min. Because of their small size, the microspheres become trapped within the capillaries in the body. They are also being removed from the body by the syringe pump at a known flow rate. The animal is then euthanized, and the major organs and tissue engineering scaffold being studied are removed from the animal.

The radioactivity or fluorescent levels in samples from the organs, the scaffold, and the syringe pump are determined. The blood flow rate to each of these samples is then proportional to their level of radioactivity or fluorescence, recognizing that the amount of radioactivity or fluorescence in the syringe pump sample is used to calibrate the flow rate.

Sarver et al. (1995) used this technique to determine the blood flow rate for various times after implantation within two layers of a polyester support matrix that were placed within a thin disc-shaped device that was only open on one side. The results are shown in Figure 9.9. The layer adjacent to the mesentery clearly shows the progressive development of the capillary bed as reflected by the increase in the support matrix blood flow. After about 14 days, a plateau is reached with little further improvement in the blood flow rate, indicating that perhaps 14 days after implantation is an appropriate time for cell transplantation into a support matrix of this thickness. The innermost layer

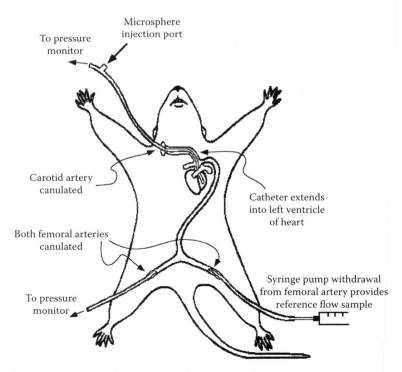

FIGURE 9.8 Experimental setup for measuring blood flow using the radioactive microsphere technique. (Reprinted from Sarver, J.G., PhD dissertation, The University of Toledo, 1994.)

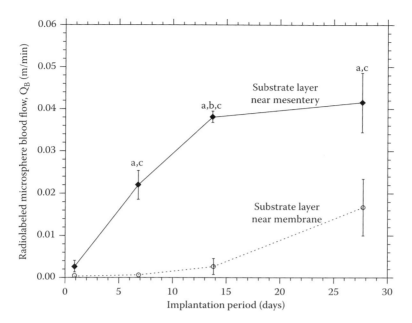

FIGURE 9.9 Blood flow measured in a polymer matrix for each implantation period. (From Sarver, J.G., et al., *Cell Transplant.*, 4, 201–17, 1995. With permission.)

shows a delay in blood flow rate development, which is expected, recognizing the time needed for the tissue front to reach this layer.

Example 9.2

From the data shown in Figure 9.9 for the substrate layer adjacent to the mesentery, calculate the tissue blood perfusion rate in milliliters per minute (100 g of tissue)$^{-1}$.

Solution

Focusing on the layer near the mesentery, we see that the maximum blood flow rate is about 0.042 mL min^{-1} after 28 days of implantation. This layer of support material has a volume of $(\pi/4) \times 1$ cm$^2 \times 0.15$ cm $= 0.118$ cm^3, of which tissue occupies 84%, giving a tissue volume of about 0.1 cm^3. The blood perfusion rate for the tissue in this layer adjacent to the mesentery is

$$0.042\frac{\text{mL}}{\text{min}} \times \frac{1}{0.1\text{ cm}^3} = 0.42\frac{\text{mL}}{\text{cm}^3\text{ min}} = 42\frac{\text{mL}}{\text{min}\,100\text{ g of tissue}}.$$

This value of the tissue blood perfusion rate within the support structure compares favorably with values obtained for other tissues within the body, as summarized in Table 6.2.

9.9 CELL TRANSPLANTATION INTO POLYMERIC SUPPORT STRUCTURES

The previous discussion shows that porous polymeric scaffolds are capable of developing within their confines a neovascularized tissue region. Furthermore, we find that this new tissue region has blood flow rates comparable to those found in other tissues. We can now ask the following questions related to cell transplantation into these vascularized support structures. When is the best time to transplant cells into the scaffold? Is it possible to maintain the viability of transplanted cells and can they continue to perform their normal function? To answer these questions, consider the classic results reported by Thompson et al. (1988, 1989). They induced the formation of *organoid* neovascular structures in rats using polytetrafluoroethylene (PTFE) fibers coated with collagen and the angiogenesis initiator, aFGF. The fibers were about 20 µm in diameter and a bundle or cotton ball-like mass of these fibers was implanted within the abdominal cavity adjacent to the liver.

To demonstrate the efficacy of a cellular transplant using the vascularized PTFE support structure, they used as a host for the implant the Gunn rat. *Homozygous* Gunn rats lack the liver enzyme needed for the conjugation of bilirubin. Recall that bilirubin is a product obtained from hemoglobin at the end of a red blood cell's lifespan. The conjugated form of bilirubin is readily passed from the liver via the bile and then removed from the body in the feces. Since the Gunn rat cannot conjugate bilirubin, its plasma bilirubin levels are increased significantly from the normal level of <1 mg dL^{-1}. The Wistar rat is genetically identical to the Gunn rat except that it has the ability to conjugate bilirubin. In their initial set of experiments, they obtained hepatocytes from the livers of the syngeneic Wistar rats. These hepatocytes were then seeded onto collagen-coated PTFE fibers that did not contain the angiogenesis promoter aFGF. This structure was then implanted adjacent to the liver of the Gunn rat. Because the Gunn and Wistar rats are genetically the same, there is no need for immunosuppressive agents or immunoisolation of the transplanted cells. For the first 10 days after implantation, the plasma bilirubin levels remained unchanged, as shown in Figure 9.10a. By the twentieth day, the plasma bilirubin levels had decreased by about 50%. Shortly thereafter, the plasma bilirubin levels returned to their original high levels and remained there for the remaining days of the experiment.

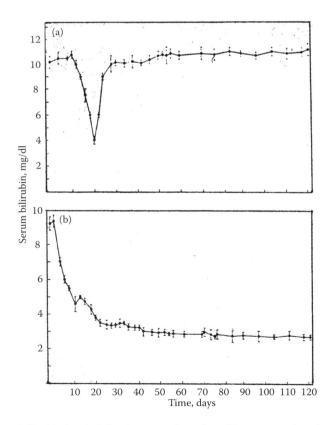

FIGURE 9.10   Plasma bilirubin levels following transplantation of hepatocytes into Gunn rats. (a) Collagen-coated PTFE support structure with hepatocytes added at time of implantation; (b) collagen-coated PTFE support structure with aFGF. Hepatocytes added 28 days after implantation. (From Thompson, J.A., Haudenschild, C.C., Anderson, K.D., DiPietro, J.M., Anderson, W.F., and Maciag, T., *Proc. Natl. Acad. Sci. USA*, 86, 7928–32, 1989. With permission.)

The lack of long-term function of the hepatocytes in these preliminary studies can be attributed to several effects. First, the angiogenesis factor, aFGF, was not used in these first experiments. This resulted in less vascularization of the support structure. Furthermore, the cells were seeded into the support structure at the time of implant. Lack of a good vascular supply at the time of cell seeding is clearly not optimal for cell growth and the transport of nutrients and bilirubin. This accounts for the delayed response with regard to any effect on the plasma bilirubin levels. For the first 10 days, the implant was still in the process of tissue ingrowth and the development of a blood supply that is needed for efficient mass transport. It also allowed for the accumulation of bile acids that led to the death of the cells after about 20 days. Clearly, these results show that seeding the cells at the time the support structure is implanted is not the best approach.

In their second set of experiments, the collagen-coated PTFE fibers containing adsorbed aFGF were first implanted. After 28 days, a suspension of Wistar rat hepatocytes was seeded into the network of the now vascularized fibers. As shown in Figure 9.10b, we see that the plasma bilirubin levels began to decrease within 1 day. After about 10 days, the plasma bilirubin levels had decreased by about 50% and at the end of the 120-day experiment this reduction was >60%.

These results strongly suggest that polymeric scaffold structures, through their ability to form their own vascular supply, can be used to sustain the long-term function of transplanted cells. Additionally, it appears from these results that it is better to first vascularize the support structure for several weeks before seeding the implant with the transplanted cells. Furthermore, these results demonstrate that transplanted cells carrying a normal gene are able to restore the function lost by the host's own genetically compromised cells. If the host's genetically deficient cells can be removed, the appropriate gene inserted, and their numbers expanded in tissue culture prior to their transplantation, then this approach becomes a powerful treatment method for a variety of diseases. The only limitation that needs to be addressed is the problem of immunorejection of the transplanted cells if they are not genetically identical to those of the host. Chapter 10 addresses how immunoisolation of the transplanted cells can be incorporated to create bioartificial organs.

9.10   BIOREACTOR DESIGN FOR TISSUE ENGINEERING

Critical to the success of tissue engineering for many applications is the in vitro culturing of cells or tissue within the three-dimensional scaffold. In vitro culturing will allow the cells to establish the desired tissue within the scaffold prior to implantation into the patient. This in vitro culture is facilitated through the use of specially designed bioreactors that provide precise control over the environmental conditions that are conducive to the establishment of a viable tissue-engineered construct that is ready at the appropriate time for transplantation into the patient (Martin and Vermette 2005; Mazzei et al. 2010). The use of these custom-designed bioreactors improves the level of control as well as facilitating the reproducibility and scalability of the processes used to manufacture a tissue-engineered construct. Hence, a bioreactor will provide much better control over the culture conditions than what is achievable in standard tissue culturing techniques. A bioreactor therefore provides an opportunity for the seeded cells to differentiate and populate the construct and create an ECM.

A bioreactor can be designed and customized for the unique three-dimensional shape of a tissue-engineered construct and will also provide control of such process variables as temperature, pressure, pH, dissolved oxygen levels, nutrient supply, and waste product removal. In addition, a bioreactor can be designed to provide for mechanical stimulation or conditioning of the construct, which in many cases is crucial for the proper differentiation of the cells (Carver and Heath 1999). The use of bioreactors also improves control over the manufacturing process, reducing production costs while improving overall product quality. This ensures that the tissue-engineered construct meets product specifications and that these products comply with good manufacturing practices (GMP) and meet the criteria set by regulators such as the FDA.

The design of a bioreactor for a specific tissue engineering application involves the consideration of several key factors. First, we need to address how the cells are seeded or loaded into the

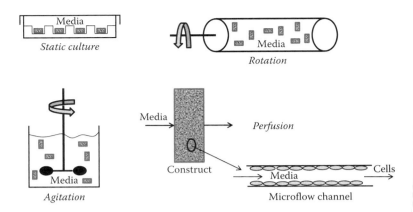

FIGURE 9.11   Bioreactor configurations.

polymeric scaffold. Cell seeding can be as simple as pipetting a solution containing the cells onto the surface of the construct, allowing capillary action to draw this solution of cells into the scaffold structure. A more efficient method of seeding the cells is to pump or perfuse a solution of cells through the pores of the construct. Perfusion seeding of the cells will give a more uniform initial distribution of cells in the construct.

Once the cells have been distributed inside the scaffold, the bioreactor design must be capable of keeping these cells alive. Without proper distribution* of key nutrients and oxygen throughout the scaffold, there will be regions near the interior where the cells cannot survive, resulting in a nonuniform cell density throughout the scaffold. The deposition of ECM materials like GAG has also been shown to be dependent on the local oxygen concentration (Obradovic et al. 2000). Thus, oxygen transport is often the limiting nutrient, not only for keeping the cells alive, but also for ensuring that the cells proliferate and establish an ECM. In addition, waste products cannot be allowed to accumulate within the construct, since these products can harm the cells.

Figure 9.11 illustrates the four basic approaches that can be used for the design of a bioreactor. Static culture techniques simply rely on diffusion to distribute key nutrients and oxygen to the cells and this approach is similar in concept to a multiwell plate. Static culture has significant internal and external diffusion limitations and, in general, cannot be used as a large-scale tissue engineering bioreactor. Another approach is to eliminate external diffusion limitations by inducing fluid motion over the surface of the construct. This can be accomplished by agitation, which causes a flow of media across the surfaces of the construct. A rotating bioreactor can also achieve a flow of media across the surfaces of the construct and eliminates any damage to the constructs by the agitator. Although agitation or rotation can eliminate external diffusion limitations, there can still be significant internal diffusional barriers[†] that can affect the viability and the properties of the tissue in the interior regions of the construct.

Hence, bioreactors for tissue engineering applications may need to provide some means of perfusing the inner most regions of the construct with the nutrient media. Perfusion of the construct with the culture medium results in the convective transport of these nutrients into the inner most regions of the construct, which is much more efficient than what can be obtained by diffusion alone. Perfusing the construct with the culture medium will therefore improve the transport of oxygen and key nutrients (Zhao and Ma 2005; Kitigawa et al. 2005; Lee et al. 2010).

---

* The transport of oxygen and key nutrients in a bioreactor can be described using the principles outlined in Chapters 5 and 6. Reactor design principles are discussed in Sections 8.8.4 through 8.8.9.

† See Sections 8.8.4 through 8.8.7 for a discussion on internal diffusion and reaction phenomena in porous materials. In this case the enzyme kinetics will be replaced by the cellular kinetics for a particular substrate of interest.

As shown in Figure 9.11, perfusion bioreactors provide for a convective flow of the nutrient media over the surface of a very thin layer of cells.* This involves designing the porous structure of a scaffold with nutrient channels to ensure that the culture media is uniformly distributed to all regions of the construct (Mehta and Linderman 2006). These microflow channels can be designed into the porous construct using microfabrication techniques (Lee et al. 2010; Park et al. 2010). This will ensure that key nutrients as well as oxygen can reach those cells located in the most interior regions of the construct. In addition, the perfusion medium can also be loaded with oxygen carriers, such as perfluorocarbons (PFCs) (Radisic et al. 2004, 2005). As discussed in Chapter 6 (Section 6.12), PFC solutions overcome the low solubility of oxygen in water with the result that significantly more oxygen can be transported.

In many applications, cells that are loaded into the construct must be exposed to the proper biomechanical environment in order to differentiate properly (Gemmiti and Guldberg 2009; Mathes et al. 2010; Shaikh et al. 2010). This is known as mechanical or physical conditioning of cells as they grow and differentiate in the in vitro environment of the bioreactor. For example, it is well known that cells can respond to a variety of forces caused by fluid flow, pressure, and dynamic loading. Hence, cells can respond to forces related to shear, strain, pressure, and mechanical loads. In many cases, it is therefore necessary to incorporate these physiological forces into the operation of the bioreactor so that cells differentiate into the desired tissue. Examples where mechanical conditioning has been shown to be important include tissue engineering of blood vessels, heart valves, tendons and ligaments, cartilage, bone, oral tissues, and intervertebral disks.

Example 9.3

A construct for a tissue engineering application consists of a polymeric scaffold that incorporates cylindrical flow channels so that culture media can be perfused through the construct during in vitro culture. Each flow channel in the construct is 300 μm in diameter and the length of each flow channel is 25 mm. Surrounding and concentric to each flow channel is a cylindrical volume of tissue whose thickness is 50 μm. The tissue in the construct consumes oxygen at a rate of 10 μM sec^{-1} based on the total construct volume in the region that contains the tissue. The pO$_2$ of the culture media entering the construct is 150 mmHg. Estimate the flow rate of the media per flow channel (cubic centimeter per hour) so that the pO$_2$ in the tissue region is greater than 20 mmHg. Also calculate the pO$_2$ of the media leaving the flow channel.

Solution

The cylindrical flow channels within the construct are similar in concept to the Krogh tissue cylinder discussed in Section 5.10.2. In this case, oxygen is just a dissolved solute in the culture media, so we do not have to take into account the binding of oxygen with hemoglobin as was done in Sections 6.10 and 6.11. We can therefore use Equation 5.115 to describe the oxygen concentration in the tissue region that surrounds each flow channel in the construct:

$$\bar{C}(r,z) = C_0 - \frac{R_0}{V\,r_c^2}\left[r_T^2 - (r_c + t_m)^2\right]z - \frac{R_0}{2r_c\,K_0}\left[r_T^2 - (r_c + t_m)^2\right]$$

$$+ \frac{R_0}{4D_T}\left[r^2 - (r_c + t_m)^2\right] - \frac{R_0\,r_T^2}{2D_T}\ln\left(\frac{r}{r_c + t_m}\right).$$

The critical oxygen level in the tissue is specified as 20 mmHg and this will occur at the position defined by $L = 25$ mm and $r_T = ((300\ \mu m/2) + 50\ \mu m) = 200\ \mu m$. Also, the flow channel is in intimate contact with the surrounding tissue, so t_m and P_m are both equal to zero. The overall mass transfer coefficient (K_0) is then equal to the mass transfer coefficient (k_m) for the media in the

---

* See Section 6.9 for a discussion of oxygen transport in a microchannel perfusion bioreactor.

flow channel. Using Henry's law ($pO_2 = HC$) to replace the oxygen concentration with pO_2, the previous equation becomes

$$pO_2(r_T, L) = pO_2^{in} - \frac{R_0 H}{V r_c^2}[r_T^2 - r_c^2]L - \frac{R_0 H}{2 r_c k_m}[r_T^2 - r_c^2]$$

$$+ \frac{R_0 H}{4 D_T}[r_T^2 - r_c^2] - \frac{R_0 H r_T^2}{2 D_T}\ln\left(\frac{r_T}{r_c}\right).$$

This equation can then be solved for the average velocity of the culture media in each flow channel, recognizing that k_m depends on the unknown value of V. The analysis of this problem shows that the flow is laminar and that the concentration profile in the media flowing through the flow channel is not fully developed, so Equation 5.52 is used to calculate the mass transfer coefficient. When Equation 5.52 is used in the above equation, the required average velocity of the media flowing in the flow channel is found to be 0.15 cm sec^{-1}, which gives a volumetric flow rate of media per flow channel of 0.39 cm^3 h^{-1}. From Equation 5.113, the exiting pO_2 of the media can be found as

$$C(z) = C_0 - \frac{R_0}{V r_c^2}\left[r_T^2 - (r_c + t_m)^2\right]z,$$

$$pO_2^{out} = pO_2^{in} - \frac{H R_0 L}{V r_c^2}(r_T^2 - r_c^2) = 150\ \text{mmHg} - \frac{0.74\dfrac{\text{mmHg}}{\mu M} \times 10\dfrac{\mu M}{\text{sec}} \times 2.5\ \text{cm}}{0.15\dfrac{\text{cm}}{\text{sec}} \times 0.015^2\ \text{cm}^2}(0.02^2 - 0.015^2)\ \text{cm}^2,$$

$$pO_2^{out} = 54.1\ \text{mmHg}.$$

PROBLEMS

1. Consider a bioreactor for growing a cornea. Within this bioreactor, glucose diffuses from a well-stirred bulk liquid solution across a thin hydrogel layer. The glucose then diffuses and reacts within a multilayer of epithelium that lies on top of the hydrogel layer (Perez et al. 1995). The epithelium consists of N layers of cells, where each cell is h_c in thickness. The cells consume glucose at a first order volumetric rate given by $k_1 C$, where C is the local concentration of glucose and k_1 is the first order rate constant. The surface of the cells exposed to the gas phase above is a no flux boundary for glucose, and C will therefore attain its minimum value at that interface (C_{min}). Calculate the fractional drop (C_b/C_{min}) in the glucose concentration from the bulk liquid medium (C_b) to the outermost epithelial surface (C_{min}). Assume that there is no mass transfer resistance between the bulk solution and the surface of the hydrogel. Some additional information needed to solve this problem is shown in the following table.

Parameter	Value	Description
h_c	15 μm	Thickness of single cell layer
k_1	1.5×10^{-3} sec^{-1}	Glucose rate constant
C_b	3.2 mg mL^{-1}	Bulk glucose concentration
D_G	3×10^{-6} cm^2 sec^{-1}	Diffusivity of glucose in the hydrogel
K_p	2	Glucose hydrogel partition coefficient, bulk-gel or gel-epithelium
L	0.05 cm	Hydrogel thickness
N	5	Layers of epithelial cells
D_e	6×10^{-6} cm^2 sec^{-1}	Glucose diffusivity in epithelium

2. Review the research literature on tissue engineering and write a short paper on an interesting design for a bioreactor. Explain how the proposed bioreactor addresses any transport limitations and provides for mechanical conditioning of the cells in the construct.

3. From Table 9.1, select a subject for tissue engineering. Prepare a presentation for your class that summarizes recent advancements in the area you selected. Topics covered in your presentation should include a description of the clinical need, the methodology used, results that have been obtained (in vitro, in animals, and the status of any human clinical trials), the potential clinical impact, potential market impact, safety issues, and additional development needs.

4. Design a tissue-engineered system for the delivery of human growth hormone (hGH) to the systemic circulation. hGH has a molecular weight of 22,000 gmol^{-1}. Assume that hGH has an apparent distribution volume in the body of 30 L. Autologous cells were transfected with a recombinant gene for hGH. The production rate of hGH from these cells is about 2500 ng/10^6 cells 24 h. The plasma concentration of hGH should be maintained at about 10 ng mL^{-1}. Assume that hGH has a half-life in the body of about 2.3 h. Describe your system for delivering hGH and carefully state all assumptions. How many cells will be required for the delivery of hGH?

5. Write a paper that addresses some of the ethical issues associated with tissue engineering.

6. Redo Example 9.3 assuming that the construct is perfused with culture media containing an insoluble PFC material at a volume fraction of 0.60. The PFC material forms an emulsion that has an enhanced solubility for oxygen. The Henry's constant for the PFC material is equal to 0.04 mmHg μM^{-1}, where μM refers to the volume of PFC oxygen-carrying material only. The Henry's constant for dissolved oxygen in the aqueous portion of artificial blood is 0.85 mmHg μM^{-1}.

7. Radisic et al. (2006) measured the oxygen gradients in cultured constructs containing cardiac tissue. The constructs were made from collagen sponges that were thin disks nominally 1.8 mm in radius and 1.8 mm in thickness. In addition to measuring the oxygen gradient within these constructs, they also measured cell viability, which was found to decrease exponentially with depth (z), according to $\delta(z) = ae^{-bz}$, where $\delta(z)$ is the live cell density at position z. Assuming that oxygen concentration only depends on the construct radius and depth, show that the oxygen concentration within the construct is given by

$$D_{\text{oxygen}}\left(\frac{1}{r}\frac{\partial}{\partial r}\left(r\frac{\partial C}{\partial r}\right) + \frac{\partial^2 C}{\partial z^2}\right) = \frac{V_{\text{max}}\,\delta(z)\,C}{K_{\text{oxygen}} + C} = \frac{V_{\text{max}}\,a\,e^{-bz}C}{K_{\text{oxygen}} + C}.$$

The boundary conditions for this equation are based on known measurements of the oxygen concentration at the top, side, and bottom surfaces of the construct and are given by

$$\text{BC1: } C(r,0) = C_{\text{top}},$$

$$\text{BC2: } C(r,L) = C_{\text{bottom}},$$

$$\text{BC3: } C(R,z) = C_{\text{top}},$$

$$\text{BC4: } \left.\frac{\partial C}{\partial r}\right|_{r=0,z} = 0.$$

BC4 expresses the fact that the oxygen concentration is symmetric about the centerline of the construct along the z axis. In the previous equation, D_{oxygen} is the diffusivity of oxygen within the construct, C is the concentration of oxygen, and it was assumed by these researchers that the oxygen consumption can be described by the Michaelis–Menten equation. Hence, V_{max} is the maximum oxygen consumption rate per cell. K_{oxygen} is the Michaelis constant and is the concentration of oxygen where the oxygen consumption rate is equal to one half of V_{max}. The following table summarizes the values of key parameters that were reported in their study.

Parameter	Value
R and L (construct radius and thickness)	1.8 mm
V_{max} (maximum O_2 consumption rate)	1.5 nmol min^{-1} (10^6 cells)$^{-1}$
K_{oxygen} (Michaelis constant)	6.875 μM
a (pre-exponential factor for cell distribution)	1.7053×10^8 cells cm^{-3}
b (live cell density exponential decay constant)	0.0042 μm^{-1}
D_{oxygen} (O_2 diffusivity in the tissue)	2.0×10^{-5} cm^2 sec^{-1}
C_{top} (O_2 concentration at top surface)	175.6 μM
C_{bottom} (O_2 concentration at bottom surface)	22.4 μM

Using the above mathematical model and the parameters provided in the table, solve the differential equation for the oxygen concentration as a function of r and z in the cardiac tissue construct. Compare your results for $C(0,z)$ to the following data that were presented in this paper.

Depth (μm)	Oxygen Concentration (μM)
0	175.6
100	149
200	140
300	130
400	120
500	110
600	105
700	100
800	95
900	85
1000	75
1100	65
1200	60
1300	55
1400	50
1500	45
1600	40
1700	35
1800	22.4

10 Bioartificial Organs

10.1 BACKGROUND

Tissue engineering is a very promising approach for the treatment of a variety of medical conditions. Treatment of a disease or medical problem using this approach requires the availability of the appropriate cells and the creation of an artificial support structure to contain them. The cells may be obtained from a variety of sources, for example, from an expanded population of the host's own cells (perhaps genetically modified), from other compatible human donors, from animal sources, or even genetically engineered cell lines. The cells to be transplanted are then seeded into a polymeric scaffold or construct. The construct can then be placed within a bioreactor and then implanted in the patient at the appropriate time, as discussed in Chapter 9. However, with the exception of *autologous cells*, one of the major obstacles that must be overcome is rejection of the transplanted cells by the host's immune system. The transplanted cells will be destroyed quickly by the host's immune system unless they are immunologically similar to the host's own cells (Benjamin and Leskowitz 1991).

Immunosuppressive drugs can be used to suppress the host's immune system and prolong the function of transplanted cells that are a relatively close match to the host. However, immunosuppressive drugs have potent side effects, and in the case, for example, of transplanting the islet of Langerhans to treat insulin-dependent diabetes, these drugs may provide a situation where the cure is worse than the disease. Genetically engineered cell lines, in addition to their possibly being rejected by the host's immune system, also pose additional risks that need to be considered. For example, direct implantation of an immortal (usually of tumor origin) cell line can lead to the unchecked growth of the implanted cells. There is also the risk of these cells moving and proliferating at a site other than the desired site, and they can change to a potentially hazardous form with a loss of their original therapeutic function. The full potential of tissue engineering will therefore be limited, unless techniques can be developed to either restrict the host's immune response, or somehow modify the transplanted cells to make them more acceptable to the host's immune system, i.e., *immunomodification* (Lanza and Chick 1994b).

10.2 SOME IMMUNOLOGY

The immune response to foreign materials (*antigens*) such as transplanted cells consists primarily of a *cell-mediated component* and a *humoral component* (Benjamin and Leskowitz 1991; Gray 2001). The major cellular components are *B lymphocytes* and *T lymphocytes*. B lymphocytes or B cells form in the *bone marrow*, and when properly activated by antigens, form proteins called *antibodies* that comprise the active agents of the humoral component of the immune system. T lymphocytes originate in the *thymus* and come in two basic types: *CD4+* (*helper*) *T cells* and *CD8+* (*killer or cytotoxic*) *T cells*.

The immune system is activated by its intimate contact with foreign molecules called *antigens* (*anti*body *gen*erating). Antigens must possess *foreignness*, meaning that the antigen is unlike anything the immune system has seen before. Immature B and T lymphocytes that react against *self-antigens* (the body's own proteins for example) are eliminated during their maturation phase in the bone marrow or the thymus. This is the so-called *clonal deletion theory*. Thus, the mature B and T cells are *self-tolerant* and do not normally react against the body's own tissues.

Antigens generally have molecular weights > 6000 g mol^{-1} and possess some degree of molecular complexity. In fact, there can be many sites along the surface of such large molecules that are *immunogenic*.* These sites are referred to as *epitopes*. Fortunately, many synthetic polymers, although of high molecular weight, do not possess a sufficient amount of molecular complexity to provoke an immune response. Lower molecular weight materials, although not immunogenic by themselves, can associate for example with larger carrier molecules, like a protein, to give an immune response directed at the lower molecular weight compound. In this case, the lower molecular weight compound is referred to as a *hapten*.

10.2.1 B LYMPHOCYTES

Each B cell has a unique antibody receptor on its surface that only recognizes a specific antigen. The B cell inventory within an individual is capable of producing as many as 100 million distinct antibodies. This allows the immune system to respond to almost any known or unknown antigen. If a specific B cell comes into contact with its corresponding antigen, it becomes activated through a process that we shall later see also involves the CD4$^+$ or helper T cells. The activated B cell then begins to reproduce, rapidly increasing the number of B cells that are specific for a given antigen. This expanded collection of lymphocytes with a given specificity is referred to as a *clone of lymphocytes*. This selection by the antigen of a specific reactive clone of lymphocytes from a large pool of existing lymphocytes, each with their own unique antigen specificity, is called the *clonal selection theory*. These B cells then differentiate to form *plasma cells* that begin to secrete antibodies with the same antigen specificity.

10.2.2 ANTIBODIES

Antibodies are proteins and are also called *gamma globulins* or *immunoglobulins*. They comprise about 20% of the total amount of protein found in plasma. There are five classes of antibodies, IgA, IgD, IgE, IgG, and IgM, where Ig stands for immunoglobulin and the letter designates the class. The basic structure of an antibody molecule consists of two light polypeptide chains and two heavy polypeptide chains held together by disulfide bonds, as shown in Figure 10.1.

Conceptually, the antibody molecule is *Y shaped*, with each light chain paired in the upper branches of the *Y* with the heavy chains that form the *Y* structure. The molecule consists of three

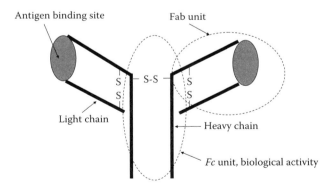

FIGURE 10.1 Structure of an antibody.

---

* An immunogen is a molecule that can induce a specific response of the immune system. An antigen refers to the ability of a given molecule to react with the products of an immune response, e.g., the binding of an antigen to an antibody. In this discussion, we will assume that antigen and immunogen are synonymous.

fragments. Two of these fragments are identical and reside at the top of the *Y* and each binds with antigen. These fragments are referred to as *Fab*, for *fragment antigen binding*. The base of the *Y* is called the *Fc fragment*, for *fragment crystallizable*. The *Fc* fragment does not bind with antigen, but is responsible for the biological activity of the antibody after it binds with antigen. The *Fc* fragment is referred to as a *constant region* in terms of its amino acid sequence, since it is the same for all antibodies within a given class. Since the *Fab* fragments bind to antigen, their structure is highly *variable* in order to provide the multitudinous shapes or conformations required for antigen specificity and recognition. The specific antigen-binding characteristics of the *Fab* fragments are determined by their unique sequence of amino acids. Although an activated B cell (or plasma cell) makes antibodies with only a single antigen specificity, it can switch to make a different class of antibody while still retaining the same antigen specificity.

IgG is the major immunoglobulin in the body and is found in all fluid spaces. It consists of a single *Y*-shaped molecule with a molecular weight of about 150,000 g mol^{-1}. IgG has several important biological properties that are also found in some of the other antibody classes. Because of its ability to bind two antigens per IgG molecule, IgG can cause the clumping or *agglutination* of particulate antigens, such as those of invading microorganisms. These large antibody–antigen complexes are then readily phagocytized by phagocytic cells, such as *neutrophils* and *macrophages*. In the case of smaller soluble antigens, this crosslinking of antigen and IgG leads to much larger complexes that become insoluble and *precipitate* out of solution and are then phagocytized.

IgG is also capable of binding to a variety of epitopes found on the surfaces of invading cells. The cell then becomes covered with IgG molecules that have their *Fc* fragments sticking out. Many phagocytic cells have surface receptors for the IgG *Fc* fragments, allowing the phagocyte to bind to, engulf, and then destroy the invading organism. IgG is therefore an *opsonizing* antibody (*opsonin* is a Greek word that means to prepare for eating), since it facilitates the subsequent "eating" of the invading microorganism or cell by phagocytic cells.

The *Fc* receptor on an IgG molecule bound to a cellular antigen also binds to specific receptors on so-called *natural killer lymphocytes* (*NK cells*). The NK cells will then be attracted to and destroy the invading cell by the release of toxic substances called *perforins* (Ojcius et al. 1998). Perforins form channels in the cell membrane of the target cell, causing them to leak and resulting in death of the cell. This process is called *antibody-dependent cell-mediated cytotoxicity* (ADCC).

IgG is also capable of neutralizing many toxins by binding to their active site. Viruses are also neutralized in a similar manner. In this case, the IgG molecule binds and blocks sites on the virus's surface coat that is used by the virus for attachment to the cells it normally invades. IgG when bound to antigen is also capable of activating the *complement system*. The complement system consists of a group of protein enzyme precursors that when activated undergo an amplifying cascade of reactions that generate cell *membrane attack complexes* (MACs) that literally punch holes through the cell membrane. This results in lysis and death of the targeted cell.

IgA is found primarily in the fluid secretions of the body, e.g., in saliva, tears, mucous, and in the gastrointestinal fluids. It is found in a dimeric form, that is, it consists of two *Y*-shaped antibodies linked at the base of the *Y*. The molecular weight of this antibody is about 400,000 g mol^{-1}. Its primary role is as a defense against local infections, where it stops the invading micro organism from penetrating the body's epithelial surfaces that line the respiratory and gastrointestinal tract. It does not activate the complement system.

IgM forms a planar star-like pentamer and consists of five antibody molecules joined together at the base of the *Fc* region. Its molecular weight is about 900,000 g mol^{-1} and it is found mainly in the intravascular spaces. It has only five antigen-binding sites because its restricted pentameric structure does not allow the *Fab* fragments to open fully. Therefore, large antigens bound to one *Fab* will block the adjacent *Fab*. IgM is a very efficient agglutinating antibody since its large structure provides for binding between epitopes that are widely spaced on the antigen. IgM is also an excellent activator of the complement system. IgM is also found on the surface of mature B lymphocytes, where it serves as the specific antigen receptor.

IgD is also present primarily on the cell surface of B lymphocytes and is believed to be involved in the maturation of these cells. It exists as a single antibody molecule and has a molecular weight of about 150,000 g mol^{-1}. Not much is really known about this antibody.

Finally, *IgE* also exists as a single antibody molecule and has a molecular weight of about 200,000 g mol^{-1}. The *Fc* region of the IgE molecule binds with high affinity to specific receptors found on *mast cells* and *basophils*. Mast cells are found outside the capillaries within the connective tissue region (see Figure 9.3). They are involved in the process of inflammation and are responsible for the secretion of *heparin* into the blood, as well as *histamine*, *bradykinin*, and *serotonin*. Basophils are a type of white blood cell or *granulocyte* that mediates the inflammation process through the release of these same substances. When antigen binds to two adjacent cell surface-bound IgE molecules, the cell (mast or basophil) becomes activated and releases a host of potent biologically active compounds, such as histamine. These agents are responsible for the dilation and increased permeability of the blood vessels. In normal situations, these changes facilitate the movement of the immune system components, such as white blood cells, antibodies, and complement, into localized sites of inflammation and infection. However, in people with allergies, a particular antigen, in this case called an *allergen*, stimulates the IgE on the surface of mast cells and basophils, leading to the unwanted effects of an allergic reaction.

10.2.3   T Lymphocytes

T lymphocytes are the other major cell type that forms the basis of the cell-mediated immune response. T cells are characterized by the presence of an antigen-specific *T cell receptor* or *TcR*. The T cell receptor consists of an antigen-recognizing molecule called *Ti* in close association with another polypeptide complex called *CD3*. T cells, unlike B cells, are not activated by free antigen. T cell activation requires that antigen be *presented* by other cells, such as macrophages or other B cells. Cells that present antigen to the T cells are collectively called *accessory cells* or *antigen-presenting cells* (APCs).

The APCs ingest and breakdown the polypeptide antigen into much smaller fragments. These fragments then become associated with special molecules called the *major histocompatibility complex* (MHC). There are two major types of MHC molecules, called *MHC class I* and *MHC class II*. Recall that these molecules belong to the class of cell adhesion molecules (CAMs). The MHC class I molecule is expressed by almost every nucleated cell found in the body. The MHC class II molecule is only found in specialized APCs, such as B cells and macrophages. The MHC molecule forms an antigen-binding area in the shape of a cleft or pocket. This pocket can accept polypeptide antigens consisting of up to 20 amino acids. The complex of small antigen and MHC is then transported to the surface of the cell where the antigen is presented for recognition by the T cell receptor. Thus, the T cell only recognizes antigen in combination with the MHC molecule.

T cell antigen recognition is shown in Figure 10.2. On the left side of this figure, we see a foreign material such as a dead virus, a cancer cell, a microorganism, or a large polypeptide, being ingested and broken down by an APC. The antigen, in combination with an MHC class II molecule, is then transported to the cell surface for antigen presentation. On the right side of this figure, we see a live virus infecting the cell. As the virus takes over the operation of the cell to increase the number of viruses, many of the viral peptides that are produced are transported to the cell surface by MHC class I molecules for antigen presentation. The presence of antigen and MHC will then be recognized by the T cell.

However, in either case, the T cell receptor by itself has a low affinity for antigen bound to MHC. To facilitate the recognition and binding of the T cell receptor with the antigen/MHC complex, two important *accessory molecules* called *CD4* and *CD8* also serve to distinguish between the two principal types of T cells.

T cells that have the *CD4 molecule* are known as *CD4$^+$* or *helper T cells*, whereas those that have the *CD8 molecule* are known as *CD8$^+$* or *cytotoxic T cells*. CD4$^+$ T cells only recognize

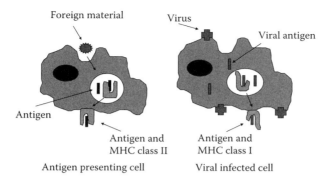

FIGURE 10.2   Presentation of antigen by MHC class I and II molecules.

antigen that is bound to the MHC class II molecule on the surface of an APC (see the left side of Figure 10.2). CD8⁺ T cells only recognize antigen that is bound to the MHC class I molecule on the surface of all other nucleated cells found in the body, as shown on the right side of Figure 10.2. This is called *MHC restriction*, where the response of CD4⁺ T cells is restricted to only that antigen bound to MHC class II molecules, whereas the response of CD8⁺ T cells is restricted to only that antigen bound to MHC class I molecules.

10.2.4   Interaction between APCs, B Cells, and T Cells

With this background, we can now examine the cooperative relationship that exists between APCs, B cells, and T cells during the immune response. First, let us consider the interaction between a resting CD4⁺ T cell, an APC, and a B cell. This is illustrated in Figure 10.3. The first step involves ingestion by an APC of an antigen that, in this example, contains both a B cell epitope and a T cell epitope. The APC could be a macrophage that engulfs the antigen, or a B cell that internalizes the antibody–antigen complex. The T cell epitope is then expressed on the surface of the APC in combination with an MHC class II molecule. A CD4⁺ T cell that has a Ti-CD3 receptor that is specific for this particular antigen, then binds with the antigen/MHC class II complex. Note that the CD4 molecule stabilizes the binding of the antigen/MHC complex with the T cell receptor.

This binding of the CD4⁺ T cell and the antigen/MHC class II complex constitutes the first signal for activation of the CD4⁺ T cell. This is then followed by the release of a second signal by the APC of a soluble substance called *interleukin-1* (IL-1), which is also essential for T cell activation. IL-1 is a small protein with a molecular weight of 15,000 g mol⁻¹ that belongs to a class of substances called *lymphokines*. Lymphokines are cellular messengers that have an effect on other lymphocytes and are a subcategory of a broader class of intercellular messengers called *cytokines*. The CD4⁺ T cell then becomes activated and begins to secrete its own interleukins, specifically IL-2, IL-4, and IL-5. IL-2 induces the CD4⁺ T cell to proliferate, rapidly forming a clone of CD4⁺ T cells that are reactive to the specific antigen presented by the APC. IL-4 then activates the B cells and IL-5 induces the activated B cells to proliferate in number, forming a clone of B cells.

It is important to note that binding of antigen with the B cell receptor is not sufficient for the B cell to become activated and then differentiate to antibody-secreting plasma cells. The B cell must become activated through the process shown in Figure 10.3, which also involves the CD4⁺ T cell. Additionally, the B cell need not bind with the same epitope that activated the T cell. Hence, the activated B cell may release antibodies with a different epitope specificity. Since the CD4⁺ T cell is responsible for the activation of the B cell, it is referred to as the *helper* T cell. The activated CD4⁺ T cell also secretes other lymphokines, such as γ-*interferon*, which serve to attract and activate macrophages and NK cells and inhibit viral replication.

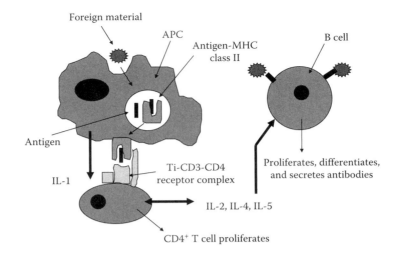

FIGURE 10.3   Interaction between APC, CD4⁺ T cell, and B cell.

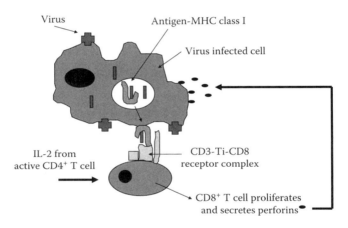

FIGURE 10.4   Activation of CD8⁺ T cells.

Figure 10.4 illustrates how the CD8⁺ T cell becomes activated. Here, we see that the CD8⁺ T cell receptor binds with the antigen/MHC class I complex presented for example, by a virally infected cell. Once again, the CD8 molecule stabilizes the interaction between the T cell receptor and the antigen/MHC complex. This binding of T cell receptor and antigen/MHC complex represents the first signal. In order for the CD8⁺ T cell to proliferate (i.e., form a clone of CD8⁺ T cells) and become activated, it also must receive a second signal that is provided by the IL-2, which is released as a result of the activation of the CD4⁺ T cell, as shown in Figure 10.3. The activated CD8⁺ T cell will then kill any cell that expresses the appropriate combination of antigen and MHC class I. The activated CD8⁺ T cell accomplishes this cellular destruction through the release of special molecules called *perforins*, which destroy the cell membranes of the target cell. Because of their role in cell death, the CD8⁺ T cells are also known as cytotoxic T cells or killer T cells.

10.2.5   Immune System and Transplanted Cells

The earlier discussion illustrates the complexity as well as the coordination that exists between the different components of the immune system. Through an elegant process, the immune system is capable of recognizing *self* from *non-self*. Our interest here is to prevent or restrict the action of the immune system toward the transplanted cells. The immune system response to transplanted cells involves a combination of effects resulting from antibodies, complement, macrophages, B cells, and T cells.

Antibodies recognize the foreign antigens presented by the transplanted cells and induce the destruction of the transplanted cells through activation of the complement system and NK cells by the process of ADCC. Rejection of the transplanted cells also occurs by a T cell response against the MHC molecules that are expressed by the transplanted cells. The transplanted cells contain foreign MHC class I and class II molecules. The foreign MHC class II molecules are present on *passenger leukocytes* and macrophages, which are present in the transplanted tissue. The foreign MHC class II molecules are sufficiently similar to the host's own combination of antigen/MHC class II complex to trigger the activation of the CD4$^+$ T cells. This is also true for the foreign MHC class I molecules, leading to the activation of the host's CD8$^+$ T cells. Activation of these T cells and the antibody-producing B cells then leads to the destruction of the transplanted cells.

10.3 IMMUNOISOLATION

A promising method for restricting the host's immune response is to immunoisolate the transplanted cells (Lanza et al. 1995; Zielinski et al. 1997; Gray 2001). This concept was shown earlier in Figure 9.2. Figure 10.5 presents a more detailed view of *immunoisolation*, showing the possible pathways for rejection of the transplanted cells (Colton 1995). Immunoisolation can be accomplished through the use of a specially designed polymeric membrane that prevents passage of the major components of the immune system, which are the immune cells, antibodies, and complement.

The immune response occurs as a result of antigens shed by the transplanted cells. These antigens may be products secreted by the functioning cells or materials released by the death of the transplanted cells. These antigens will cross the immunoisolation membrane and when recognized and presented by the host's immune system lead to the cellular and humoral immune responses discussed earlier.

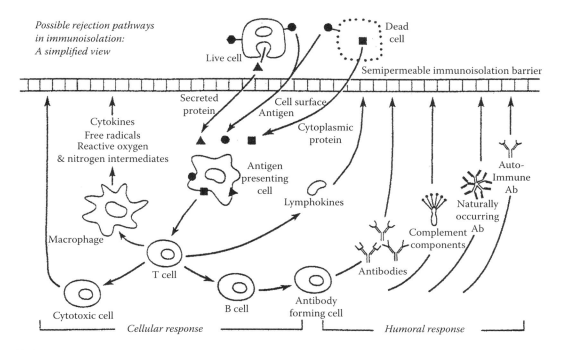

FIGURE 10.5 Pathways for rejection of immunoisolated cells. (From Colton, C.K., *Cell Transplant.*, 4, 415–36, 1995. With permission.)

To restrict the passage of cells such as lymphocytes and macrophages, the pores or spaces within the immunoisolation membrane must be no larger than a micron or so. Recent evidence indicates that blocking the entry of the cellular component of the immune system is sufficient to prevent the rejection of allografts (Colton 1995). This is of special interest for applications involving allogeneic cells for gene therapy (Chang et al. 1993; Liu et al. 1993; Hughes et al. 1994; Al-Hendy et al. 1995; Chang 1997; Zalzman et al. 2003).

However, openings that restrict the entry of immune cells will still allow the passage of antibodies and complement, which must be blocked for the successful use of xenogeneic cells in a bioartificial organ. Therefore, the solute permeability characteristics of the immunoisolation membrane are critical for achieving successful immunoisolation of xenogeneic transplanted cells. This requires that the membrane has significant permeability to essential small molecular weight solutes such as oxygen, glucose, growth factors, and molecular carriers such as albumin and transferrin.* The membrane must also be permeable to waste products and to the therapeutic product, for example in the case of diabetes, this would be insulin. On the other hand, the membrane must have negligible permeability to the humoral components of the immune system, i.e., antibodies, complement, and various cytokines and lymphokines.

As we discussed earlier, antibodies have molecular weights ranging from 150,000 to 900,000 g mol^{-1}. Naturally occurring antibodies of the IgM class exist in the host and are reactive against the MHC of xenografts. In addition, antibodies produced from autoimmune diseases such as diabetes would be expected to bind to antigens on xenograft *islets of Langerhans*.

In the absence of immune cells, antibodies binding to antigens on the surface of the transplanted cells are usually not sufficient to damage the cells. However, if both antibodies and complement are present, then the transplanted cells can be destroyed by complement activation and formation of the MAC (Iwata et al. 1996).

The complement system is activated by two different routes called the *classical pathway* and the *alternative pathway*. The classical pathway requires the formation of antigen–antibody complexes for its initiation, whereas the alternative pathway does not. The alternative pathway becomes activated in response to recognition of large polysaccharide molecules present on the surface of the cell membranes of invading microorganisms. As far as the immunoisolation of transplanted cells is concerned, the classical pathway is of most interest for this discussion.

Activation of the classical pathway involves nine protein components, known as C1 through C9. C1 is the first component to become activated and requires the binding with antigen of either two or more adjacent IgG antibodies or a single IgM antibody. C1 consists of three subunits called C1q, C1r, and C1s. Their molecular weights are, respectively, 400,000, 95,000, and 85,000 g mol^{-1}. The C1q subunit is the first subunit activated and it binds with the *Fc* region of the bound antibodies. Its activation then activates the C1r subunit, which then activates the C1s subunit, which activates C4 and so on.

As we discussed earlier, in the absence of immune cells, binding of antibody with antigen is not sufficient to damage the transplanted cells. Destruction of the antibody-laden cell will require the presence and activation of complement. Since the C1q subunit must bind with either IgM or several IgG to start the cascade of complement reactions, the immunoisolation membrane must be capable of preventing the passage of the C1q molecule.

C1q has an interesting molecular shape, as shown in Figure 10.6. The critical dimension of C1q is the span of the six projections from the cylindrical base, which is about 30 nm. Because of the presence of proteins coating the walls of the pores in the immunoisolation membrane, the maximum pore size allowable for blocking C1q is on the order of 50 nm (Colton and Avgoustiniatos 1991; Colton 1995; Zielinski et al. 1997).

Several other issues with regard to immunoisolation also need to be considered. The immunoisolation membrane may be capable of preventing the passage of immune cells and complement C1q. However, activation of the immune cells by antigens released by the transplanted cells will also

---

* Albumin carries fatty acids and transferrin carries iron.

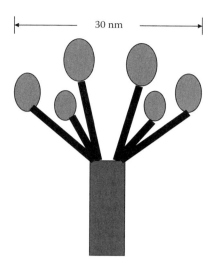

FIGURE 10.6 Complement C1q molecule.

result in the release of such lymphokines as IL-1 and other cytotoxic agents, such as nitric oxide, peroxides, and free radicals. These substances can be toxic to the transplanted cells. Lymphokines have small molecular weights, typically around 20,000 g mol^{-1}, and because of their small size, may not be retained by the immunoisolation membrane. Recent studies, however, have shown that immunoisolated cells were not affected by the presence of IL-1 (Zekorn et al. 1990). Substances such as IL-1, and other highly reactive species, may be consumed by other reactions before they can penetrate the immunoisolation membrane to any significant distance (Colton 1995). In addition, many bioartificial organs operate at very low tissue densities in order to provide for effective oxygenation of the transplanted cells. The reduced tissue density results in a lower concentration of shed antigens and hence a reduced concentration of humoral agents.

The shed antigens could also cause the host to undergo a life-threatening *anaphylactic reaction** or to suffer from *antibody-mediated hypersensitivity reactions.*[†] Lanza et al. (1994) examined these issues by transplanting immunoprotected canine and porcine islets into rats. The islets were encased within an acrylic hollow fiber membrane with a *nominal molecular weight cutoff* (NMWCO) of about 80,000 g mol^{-1}. Their studies showed that the immunoprotected islet xenografts caused the host to generate antibodies to antigens given off by the transplanted islets. However, there was no evidence of any other pathological effects of these antibody–antigen immune complexes. Subsequent studies using immunoprotected human islets in human patients with diabetes (Scharp et al. 1994; Shiroki et al. 1995) have shown that the immunoisolation membrane can protect against not only the allogeneic immune response, but also against the autoimmune components responsible for the development of insulin-dependent diabetes mellitus (IDDM).

10.4 PERMEABILITY OF IMMUNOISOLATION MEMBRANES

Figure 10.7 presents permeability data obtained on a hollow fiber immunoisolation membrane for a variety of solutes of different molecular weight (Dionne et al. 1996). This particular membrane was made from a copolymer of acrylonitrile and vinyl chloride. The permselective membrane consists of a thin skin on either side of a much thicker spongy wall region that provides overall structural

---

* A result of a significant release of inflammatory mediators such as histamine causing severe hypotension and bronchiolar constriction: can cause death due to circulatory and respiratory failure.

[†] An inappropriate or exaggerated response of the immune system to an antigen. Binding of antibodies to the antigen starts the process.

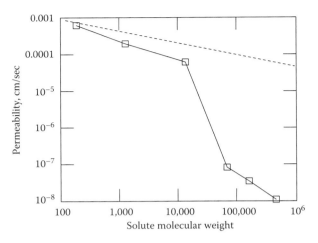

FIGURE 10.7   Immunoisolation membrane permeability. (Data from Dionne, K.E., et al., *Biomaterials*, 17, 257–66, 1996.)

strength. The thickness of the spongy wall was about 100 μm and these membranes were similar to those that were successfully used in preliminary tests to immunoisolate human islets in patients with diabetes (Scharp et al. 1994). These membranes had a reported NMWCO of 65,000 g mol^{-1}.

Notice how the permeability of this membrane at first decreases gradually as the solute molecular weight increases. As the solute molecular weight continues to increase, the permeability decreases much more rapidly. This can be explained by recalling our equation for solute permeability, given by Equation 5.90. The size of low molecular weight solutes is much smaller than the pores in the membrane. For these solutes, steric exclusion (K) and solvent drag (ω_r) effects caused by the pore wall are negligible. The decrease in permeability is therefore directly related to the decrease in solute diffusivity, which, according to Equations 5.4 and 5.5, is inversely proportional to the solute size, or the one-third power of the solute molecular weight (shown as the dashed line in Figure 10.7). As the solute size reaches a critical fraction of the membrane average pore size, steric exclusion and solvent drag become more important, and the solute permeability rapidly decreases with increasing solute molecular weight.

These data also clearly show the problem of defining the NMWCO for the membrane. No molecular weight exists above which the solute permeability drops to zero. Surprisingly, we see that even very large molecules, here at the extreme represented by a molecular weight of about 440,000 in Figure 10.7, have a measurable permeability. Therefore, a membrane's molecular weight cutoff is very subjective and dependent on the definition that one chooses to apply. For example, some membrane manufacturers define the molecular weight cutoff on the basis of 90% retention of a given solute after so many hours of dialysis. Others base it on ultrafiltration of a solution in which 90% of a particular solute is retained.

Although the exact nature of the pores or openings within the membrane is not known, descriptions such as that based on the hydrodynamic pore model developed in Chapter 5 (see Equation 5.61) provide a useful framework for understanding the transport of various solutes and also provide a means for correlating data. The data shown in Figure 10.7 can therefore be used to obtain an estimate of the average pore size in the membrane. This is shown in the following example.

Example 10.1

Using the solute permeability data shown in Figure 10.7, determine the average size of the pores in the membrane. Recall that the membrane is 100 μm thick. The following table summarizes the physical properties of the solutes that were used. Use the hindered diffusion model developed by

Bungay and Brenner (1973) to describe the effect of the pore size on the diffusion of the solute. This equation is used, rather than the Renkin model given in Equation 5.61, because the Bungay and Brenner equation is generally valid for the entire range of the ratio of solute to pore size, i.e., $0 \leq (a/r) \leq 1$. The Renkin equation is only valid for $(a/r) < 0.4$ and may not provide a correct description of the diffusion of larger molecules in comparably sized pores. In the Bungay and Brenner model, $\omega_r = 6\pi/K_t$ in the permeability equation given by Equation 5.90. K_t is given by the following equation, where a is the molecular radius and r is the pore radius:

$$K_t = \frac{9}{4}\pi^2\sqrt{2}\left(1-\frac{a}{r}\right)^{-5/2}\left[1+\sum_{n=1}^{2}z_n\left(1-\frac{a}{r}\right)^n\right]+\sum_{n=0}^{4}z_{n+3}\left(\frac{a}{r}\right)^n,$$

with $z_1 = -1.2167$; $z_2 = 1.5336$; $z_3 = -22.5083$; $z_4 = -5.6117$; $z_5 = -0.3363$; $z_6 = 1.216$; $z_7 = 1.647$.

Solute	Molecular Weight	Stokes Radius (nm)	$D_{water} \times 10^6$ cm^2 sec^{-1}, 37°C	$P_m \times 10^8$ cm sec^{-1}	$D_{water} \times 10^6$ cm^2 sec^{-1}, 37°C
Glucose	180	0.35	9.24	63,200	9.24
Vitamin B-12	1300	0.77	5.00	20,000	5.00
Cytochrome C	13,400	1.65	1.78	6160	1.78
Bovine serum albumin (BSA)	67,000	3.61	0.964	7.95	0.964
IgG	155,000	5.13	0.629	3.59	0.629
Apoferritin	440,000	5.93	0.611	1.07	0.611

Source: Data from Dionne, K.E., et al., *Biomaterials,* 17, 257–66, 1996.

Solution

Since the membrane consists of two permselective skins and a much thicker spongy wall region, the observed solute permeabilities must account for the resistances of all three regions. Therefore, the observed permeability is given by the following equation, assuming that the two skins are equivalent in their resistance to mass transfer:

$$\frac{1}{P_m} = \frac{2}{\dfrac{\varepsilon_{skin}DK\omega_r}{\tau_{skin}t_{skin}}} + \frac{1}{\dfrac{\varepsilon_{sponge}D}{\tau_{sponge}t_{sponge}}}.$$

The porosity of the sponge is assumed to be about 0.8 and the tortuosity of the sponge is assumed to be unity. The nonlinear regression analysis is based on two variables, the pore radius (r) and a parameter α. The parameter α accounts for the group $\varepsilon_{skin}/(\tau_{skin}t_{skin})$ for which the individual values of the parameters are not known. Since the permeability values vary over a considerable range, it is appropriate to define the objective function for the regression in terms of the natural logarithm of the permeabilities. Hence, we write the sum of the square of the errors as follows for the data points ranging from $i = 1$ to N:

$$SSE(r,\alpha) = \sum_{i=1}^{N}\left[\ln(P_i) - \ln(P_{calc}(i,r,\alpha))\right]^2.$$

The result of the regression analysis is shown in Figure 10.8 and provides an estimated pore diameter of 14.2 nm.

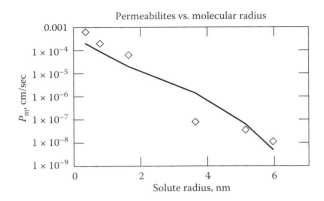

FIGURE 10.8   Measured and predicted permeabilities in an immunoisolation membrane.

10.5   MEMBRANE SHERWOOD NUMBER

The solute permeabilities given in Figure 10.7 can also be compared to the permeability the solute would have in an aqueous layer that has the same thickness as the membrane itself. Therefore, we may write Equation 10.1, which expresses the ratio of these respective quantities:

$$\frac{P_i}{P_{\text{aqueous layer}}} = \frac{P_m}{D/t_m} = \frac{P_m t_m}{D} = \frac{D_{\text{effective}}}{D} = Sh_m. \tag{10.1}$$

We see that this ratio is equivalent to defining a Sherwood number for the membrane, i.e., Sh_m, which represents the ratio of the effective diffusivity of the solute through the membrane to the aqueous solute diffusivity. This process has the effect of eliminating membrane thickness as a variable when correlating membrane permeability data. For a given solute and membrane structure, the Sh_m should be independent of the membrane thickness.

The data shown in Figure 10.7 are replotted as the membrane Sherwood number vs. solute molecular weight in Figure 10.9. We see that for low molecular weight solutes, the membrane Sherwood number is relatively constant. In this example, it is about 0.68. As the solute molecular weight is increased, and the solute size becomes comparable to that of the pores, the membrane Sherwood number begins to decrease rapidly.

10.6   EXAMPLES OF BIOARTIFICIAL ORGANS

Bioartificial organs contain living tissue or cells that are immunoisolated by polymeric membranes with the properties discussed earlier. These devices are also known as *hybrid artificial organs*, since they consist of both artificial materials and living tissue or cells. Bioartificial organs are being considered for the treatment of a variety of diseases, such as diabetes, liver failure, kidney failure, for neurological disorders such as Parkinson's or Alzheimer's disease, for the control of pain, and for delivery of a variety of therapeutic products secreted by genetically engineered cell lines.

For diabetes, the bioartificial organ contains cells that secrete, minute by minute, the appropriate amount of insulin needed to maintain blood glucose levels within a very narrow range. Bioartificial livers and kidneys attempt to replace all of the physiological functions of these complex organs. Treatment of neurological disorders involves the use of specialized genetically engineered cells that secrete neuroprotective products, such as dopamine, nerve growth factor, glial cell-line derived neurotrophic factor, or brain-derived neurotrophic factor. Immunoprotected cells that secrete pain-reducing neuroactive compounds such as β-endorphin are being considered for the treatment of

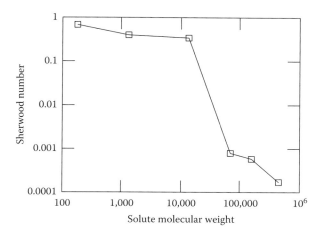

FIGURE 10.9   Immunoisolation membrane Sherwood number. (Data from Dionne, K.E., et al., *Biomaterials,*
17, 257–66, 1996.)

pain resulting from cancer. Genetically engineered cell lines can also be used to treat a variety of
other diseases. Examples include providing a specific hormone (e.g., growth hormone), replacing
missing clotting factors in hemophilia (factor VIII or IX), or for providing needed enzymes in such
diseases as adenosine deaminase deficiency that results in severe combined immunodeficiency disease (SCID). More details about these specific applications of bioartificial organs may be found in
the list of references. In the following discussion, we will look at bioartificial organ replacements for
the endocrine pancreas, the liver, and the kidney as representative systems.

10.6.1 Bioartificial Pancreas

The focus of much of the research on bioartificial organs has been on the treatment of IDDM or
type I diabetes with a *bioartificial pancreas*. This is because IDDM is a chronic disease with many
serious complications, such as blindness, kidney disease, gangrene, heart disease, and stroke, which
lead to a poorer quality of life for those afflicted with this disease. IDDM is believed to be caused
by an autoimmune process (Notkins 1979; Atkinson and MacLaren 1990) that destroys the insulin-
secreting β *cells* found within the islets of Langerhans. The islets of Langerhans are found scattered throughout the pancreas and represent about 1–2% of the pancreas mass. Both environmental
(viruses or chemicals) and hereditary factors are involved in its development. IDDM therefore represents a major health problem in the United States. Over 1 million people in the United States
have IDDM and an additional 30,000 cases are diagnosed each year in the United States alone.
The cost of healthcare for patients with IDDM is tens of billions of dollars each year. Although a
pancreas transplantation can eliminate the need for exogenous insulin, this approach is limited by
the shortage of donor organs and the complications and side effects associated with the required
immunosuppressive drugs.

Insulin is a small protein (6000 g mol^{-1}) and the key hormone involved in the regulation of the
body's blood glucose levels. In diabetes, insulin is no longer produced by the β cells found in the
islets of Langerhans. In the absence of insulin, glucose levels in the blood exceed normal values by
several times. This is a result of the body's cells not being able to metabolize glucose. Without the
ability to use glucose as an energy source, the body responds by metabolizing fats and proteins with
a corresponding increase in the body fluids of ketoacids, such as acetoacetic acid. This excess acid
results in a condition known as *ketoacidosis*, which, if left untreated, can lead to death.

The conventional treatment for IDDM involves the daily administration of exogenous insulin
to replace the insulin that is no longer produced by the patient's β cells found within the islets of

Langerhans. This results in the almost normal metabolism of carbohydrates, fats, and proteins. However, even with exogenous insulin, the patient's blood glucose levels are not controlled as well as normal. The healthy islet β cells are able to regulate the release of insulin in such a manner as to maintain blood glucose levels within a narrow range, about 80–120 mg dL^{-1}. Insulin injections are not capable of providing such close control because of the difficulty of properly timing the injections and the inherent insulin transport delays from the site of injection. The abnormally high blood glucose levels, or *hyperglycemia*, in patients treated with exogenous insulin is now believed to be the primary cause of the long-term complications of diabetes (American Diabetes Association 1993). These elevated blood glucose levels lead to protein *glycosylation* that damages the microvasculature and other tissues in the body.

Methods have recently been developed that allow for the isolation of mass quantities of the islets of Langerhans from the pancreas of humans and from large mammals such as pigs, thereby providing a potentially unlimited supply of donor islet tissue (Lanza and Chick 1994a; Ricordi et al. 1988, 1990a, 1990b; Warnock et al. 1988, 1989; Inoue et al. 1992; Ricordi 1992; Lacy 1995; Maki et al. 1996; Lakey et al. 1996, Cheng et al. 2004; Grundfest-Broniatowski et al. 2009). Use of these allogenic or xenogenic cells without immunosuppression requires that they be immunoisolated in a device such as a bioartificial organ. Transplanting the donor islets without immunoprotection is only warranted in those cases where the patient is already receiving immunosuppressive therapy, e.g., as a result of a kidney transplant to treat end-stage renal failure. In an otherwise healthy patient with diabetes, immunosuppressive drugs and their attendant problems would not be needed if the islets were immunoisolated. Immunoisolation of the islets would not only protect them from the host's normal response to foreign tissue, but also from those agent's of the immune system that caused the original autoimmune destruction of the patient's own cells. These immune system *memory cells* are still capable of responding to the antigens that resulted in the original destruction of the patient's own islets and could possibly destroy the donor islets or donor β cells.

Immunoisolation requires the use of an immunoisolatory membrane that has the following properties. First, it must be biocompatible and have a solute permeability profile that allows for efficient transport of small molecules, such as oxygen, key nutrients, and insulin. In addition, the immunoisolation membrane must exclude immune cells and immunoactive molecules. Because of the limitations on oxygen transport, the transplanted cells need to be within a few hundred microns of the capillaries, which serve as the oxygen source. The insulin-secreting cells and the immunoisolation membrane also need to be easy to assemble and sterilize before implantation in the patient. Since there is the possibility that the device can lose function over time, ease of removal from the body also needs to be considered.

The goal of the bioartificial pancreas is therefore to utilize the glucose regulating capability of healthy donor islets to provide improved blood glucose control in patients with IDDM. This approach should minimize the complications of the disease and improve the quality of the patient's life. Because of the advances in genetic engineering, there is also the possibility of using genetically engineered β cells (Efrat 1999; Zalzman et al. 2003; Tatake et al. 2007) instead of the islets of Langerhans. Genetically engineered beta cells that can respond to changing glucose levels with the appropriate insulin response will ensure a consistent source of cellular material for use in a bioartificial pancreas.

10.6.1.1 Bioartificial Pancreas Approaches

A variety of approaches have been described over the years for the bioartificial pancreas, and most have been tested extensively in laboratory animals. An excellent review of these devices can be found in the paper by Colton and Avgoustiniatos (1991) and in reviews by Mikos et al. (1994), Lanza et al. (1995), and Colton (1995). A series of books related to islet transplantation and the bioartificial pancreas have also been published (Ricordi 1992; Lanza and Chick 1994c).

These devices can be broadly classified into the following four categories: *intravascular, microencapsulation, macroencapsulation,* and *organoid.* These approaches are illustrated in Figure 10.10.

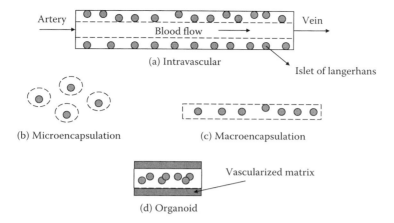

FIGURE 10.10 Approaches for a bioartificial pancreas.

These devices are, for the most part, general and are being considered for the immunoisolation of a variety of other cell types.

Of concern in any of these approaches for a bioartificial pancreas is the ability of the device to be readily fabricated and seeded with islets or cells, their stability and reliability over an extended period of time, and their cost. These issues also relate to the overall biocompatibility of the device. Specifically, device toxicity, immune system response, potential for fibrotic encapsulation, and for intravascular devices, the potential for thrombosis. Furthermore, the devices must be capable of maintaining the long-term viability of the donor tissue and have the proper mass transfer characteristics to normalize blood glucose control. This requires proper understanding and integration during the design process of the islet glucose-insulin kinetics, the mass transfer properties of the immunoisolation membrane, and the islet or cell density within the device.

10.6.1.2 Intravascular Devices

The intravascular devices (Figure 10.10a) (Chick et al. 1977; Moussy et al. 1989; Lepeintre et al. 1990; Lanza et al. 1992, 1993; Sarver and Fournier 1990; Maki et al. 1991, 1993; Petruzzo et al. 1991; Sullivan et al. 1991) involve a direct connection to the vascular system of the host via an arteriovenous shunt. Blood flows through the lumen of small polymeric hollow fibers or in the spaces between flat membrane sheets. Surrounding the blood flow path and protected by the membrane, are the islets or cells. An advantage of this type of device is that the convective flow of the blood provides the potential for better mass transfer rates. In some cases, an ultrafiltration flow can cross the membrane at the arterial end and re-enter at the venous end, much like the Starling flow found in capillaries. The close contact with a sizable flow of blood also offers the potential for good tissue oxygenation. The primary disadvantage of the intravascular approach is the surgical risk associated with device implantation, connection to the host's vascular system, potential to form blood clots, and a complicated surgery is needed for device removal at a later time.

One of the first pioneering experiments to demonstrate the feasibility of the intravascular approach for a bioartificial pancreas was performed by Chick et al. (1977). A rat with chemically induced diabetes was connected ex vivo to an intravascular device containing cells obtained from neonatal rats. The device used in these experiments consisted of a bundle of 100 hollow fibers with diameters less than 1 mm and a fiber length of 11 cm. The fibers had an NMWCO of about 50,000 g mol⁻¹.

After connection of the animal to the ex vivo intravascular device, there followed a rapid decrease in blood glucose levels, reaching normalized values of 110–130 mg dL⁻¹ (considered normal for a rat) after about 6 h. After the removal of the device, blood glucose levels rapidly increased and returned to their diabetic levels. These results indicated the feasibility and the potential of the bioartificial pancreas for the treatment of IDDM.

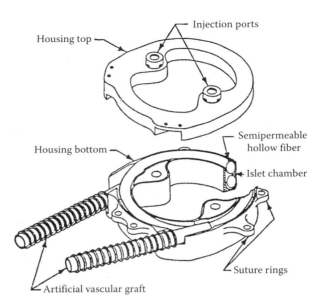

FIGURE 10.11   The biohybrid artificial pancreas. (From Maki, T., et al., *Transplantation,* 51, 43–51, 1991. With permission.)

A particularly serious problem with devices that contained numerous hollow fiber membranes was the formation of blood clots in the entrance and exit header regions of the device. These regions of the device, where the fluid stream is either expanding or contracting, tend to form secondary flows or swirls that induce the formation of blood clots.

In an effort to minimize the formation of blood clots without the use of anticoagulants such as heparin, a device with a single coiled hollow fiber membrane tube was developed by Maki et al. (1991). Their device is shown in Figure 10.11. This device consists of an annular-shaped acrylic chamber, 9 cm in diameter and 2 cm thick. The chamber contains a 30–35 cm length of a coiled hollow fiber tubule with an inner diameter of 5–6 mm and a wall thickness of 120–140 μm. The hollow fiber tubule was connected at each end to polytetrafluoroethylene (PTFE) arterial grafts that provided connection of the device to the vascular system by an arteriovenous shunt. The larger bore size of the single hollow fiber tubule eliminates the thrombosis observed in smaller bore hollow fiber tubes and in the header regions of the device described earlier. Low-dose aspirin therapy was also found to contribute to long-term patency through aspirin's ability to prevent platelet activation.

The hollow fiber membrane had an NMWCO of about 80,000 g mol^{-1}. A cavity surrounded the coiled tubular membrane and provided a 5–6 mL volume for the placement of the islets of Langerhans. This device was evaluated in pancreatectomized dogs using allogeneic canine islets and xenogeneic bovine and porcine islets. The device was implanted within the abdominal cavity and *anastomosed* to the left common iliac artery and the right common iliac vein.

In many cases, this device was able to demonstrate improved blood glucose control for varying periods of time using both allo- and xenogeneic islets (Lanza et al. 1992). Figure 10.12 illustrates an example of long-term function in a dog implanted with a single device containing about 42,000 canine islet equivalents.* Prior to device implantation, this particular dog had a fasting blood glucose level of 230 mg dL^{-1} and required about 16 U day^{-1} of insulin.† After implantation, no exogenous insulin was provided for the 280-day period of the experiment. During the first 140 days, the

---

* The equivalent islet number (EIN) is defined as the number of islets 150 μm in diameter that are equivalent in volume to a given sample of islets.
† Insulin dose is based on U(nits) of insulin where 1 U of insulin = 40 μg of insulin.

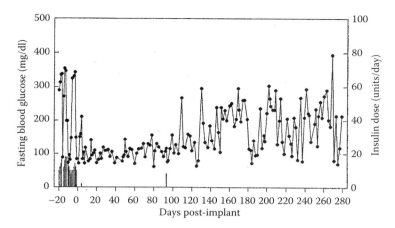

FIGURE 10.12 Fasting glucose and insulin requirements after implantation of device shown in Figure 10.11. (From Lanza, R.P., Sullivan, S.J., and Chick, W.L., *Diabetes,* 41, 1503–10, 1992d. With permission.)

fasting blood glucose was nearly normal and averaged 116 mg dL^{-1}, which compares favorably to the value of 91 mg dL^{-1} observed in normal dogs. After 140 days, glucose control began to deteriorate over the rest of the experiment.

Results obtained with xenogeneic implants containing porcine islets in this device were also very encouraging (Maki et al. 1996). Prior to receiving porcine islets, one particular dog had a fasting blood glucose level of 479 mg dL^{-1} and required 39 U day^{-1} of insulin. After this dog received 216,000 EIN of porcine islets, fasting blood glucose averaged 185 mg dL^{-1} and exogenous insulin averaged about 10 U day^{-1}. This device failed after 271 days of operation. Analysis of the porcine islets after device removal did not show any evidence of immune cell infiltration. These tests therefore presented clear evidence that xenogeneic porcine islets could be protected from the host's immune system using immunoisolation membranes. The only major complication in the use of these devices in dogs was vascular thrombosis. However, this clotting problem may be unique to dogs because of their known hypercoagulability.

10.6.1.3 Microencapsulation

The microencapsulation approach (see Figure 10.10b) involves the placement of one or several islets within small polymeric capsules (Altman et al. 1986; Gharapetian et al. 1986; Lum et al. 1991; Soon-Shiong et al. 1992a, 1992b, 1993; Sugamori and Sefton 1989; Sun et al. 1977, 1996; Lanza et al. 1995; Tun et al. 1996). The microcapsules are then injected within the peritoneal cavity. The diameter of the microcapsule is typically in the range of 300–600 μm. Immunoprotective surface coatings of the islets have also been described (Pathak et al. 1992).

Microencapsulation provides diffusion distances on the order of 100–200 μm and very high surface areas per volume of islet tissue. Accordingly, the small size of a microencapsule provides good diffusion characteristics for nutrients and oxygen, which improves islet or cell viability. This also provides for a good glucose/insulin response that offers the potential for normalization of blood glucose levels.

The microcapsules usually consist of the islet or cellular aggregates immersed within a hydrogel material with another eggshell-like layer that provides the immunoisolation characteristics and mechanical strength. A wide variety of polymer chemistries have been described for the hydrogel and the immunoprotective layer and some of these are described in the following references (Douglas and Sefton 1990; Gharapetian et al. 1986, 1987; Matthew et al. 1993; Sefton et al. 1987; Sun et al. 1996; Lanza et al. 1995). Of particular concern in selecting the microcapsule chemistry is the formation of a fibrotic capsule around the encapsulated islet. Fibrotic capsule formation can severely limit the diffusion of nutrients and oxygen, resulting in loss of islet or cellular function. The success of the microencapsulation approach is therefore strongly dependent on choosing membrane materials that minimize this fibrotic reaction.

A particularly promising microencapsulation approach was described by Soon-Shiong et al. (1994) and demonstrated in large animals and several human patients. They used an alginate-poly-L-lysine encapsulation system. Alginates are natural polymers composed of the polysaccharides, mannuronic acid and guluronic acid. Soon-Shiong et al. (1994) showed that high mannuronic acid residues in the alginate are responsible for the fibrotic response. Mannuronic acid was shown to induce the lymphokine IL-1 and tumor necrosis factor (TNF), which are known to promote the proliferation of fibroblasts and lead to fibrotic capsule formation. By reducing the alginate's mannuronic acid content, and increasing the guluronic content (>64%), they were able to minimize the fibrotic response. The higher guluronic acid content also provided another benefit. It was found that alginates with higher guluronic contents were mechanically stronger.

These modified alginate microcapsules were evaluated in a series of nine spontaneously diabetic dogs. Three dogs received free unencapsulated islets and six dogs received encapsulated islets. The donor islets were obtained from other dogs and each recipient received their quantity of islets by intraperitoneal injection. The islets were provided at an average dose of about 20,000 EIN kg⁻¹ of body weight.

Figure 10.13 summarizes the plasma glucose levels prior to and following the injection of the islets. Exogenous insulin was stopped 4 days prior to islet injection and the plasma glucose levels at that time averaged 312 mg dL⁻¹. The first day after receiving the islets, the blood glucose levels were reduced to an average of 116 mg dL⁻¹ in those animals that received encapsulated islets, and averaged 120 mg dL⁻¹ in those receiving free islets. Rejection of the unprotected free islets occurred rapidly with hyperglycemia returning in about 6 days. The animals receiving encapsulated islets exhibited normoglycemia for periods of time ranging from 63 to 172 days, for a median period of 105 days. Failure of the encapsulated islets was attributed to membrane failure as a result of the water soluble nature of the alginate system.

Sun et al. (1996) developed a microencapsulation system for islets using alginate-polylysine-alginate microcapsules. Most microcapsules contained only a single islet and had diameters in

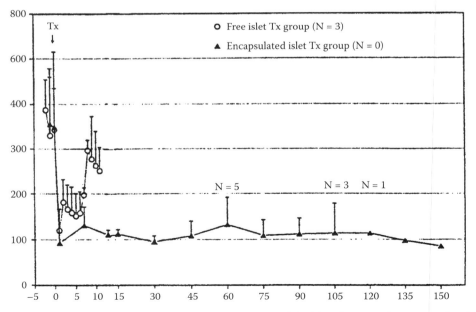

FIGURE 10.13 Fasting glucose levels following treatment with either free or microencapsulated islets. (From Soon-Shiong, P., et al., *Transplantation,* 54, 769–74, 1992a. With permission.)

the range of 250–350 µm. Preclinical studies were done using microencapsulated porcine islets implanted into spontaneously diabetic cynomologus monkeys. Seven monkeys receiving an average of 16,100 EIN islets kg^{-1} body weight were insulin independent for an average of 289 days. The shortest duration of control was 120 days and the longest was 803 days. Prior to receiving the islets, fasting blood glucose levels averaged 353 mg dL^{-1}. Following the transplant, fasting blood glucose levels averaged 112 mg dL^{-1}. During the experiments, blood samples from the animals were tested repeatedly for the presence of antiporcine islet antibodies. No antibodies were detected, confirming the presence of effective immunoisolation of the porcine islets. Diabetic control animals also received unencapsulated islets. However, euglycemia was only maintained for about 9 days, after which the islets were destroyed by the host's immune system with a return to pretransplant diabetic levels of the blood glucose.

These results using the microencapsulation approach are very promising. However, there are still some technical difficulties with this approach that must be overcome. The strength and integrity of the microcapsules need to be improved to provide long-term functioning and viability of the islets. Aggregation of the beads is also a problem as this affects their mass transfer characteristics and impacts glucose control and islet survival. The total volume for the smaller microcapsules of Sun et al. (1996) is modest. For example, using their monkey results, and assuming a 70 kg patient, a total of 1.1 million EIN would be required (16,100 EIN kg^{-1} body $weight^{-1}$). With one islet per 300 µm diameter microcapsule, this amounts to a total bead volume of only 16 mL. Loss of islet function with time can be compensated for by injection of fresh microcapsules. Retrieval of the small microcapsules could be difficult if this should be necessary after their introduction.

10.6.1.4 Macroencapsulation

Macroencapsulation (see Figure 10.10c) involves placing numerous islets within the immunoisolation membrane structure. Typical approaches involve the use of hollow fibers that encase the islets (Altman et al. 1986; Lanza et al. 1991, 1992a, 1992b, 1992c; Lacy et al. 1991), a much larger flat bag-like structure (Gu et al. 1994; Inoue et al. 1992; Hayashi et al. 1996), thin polymeric hydrogel sheets (Storrs et al. 2001), or a thin scaffold laminated to a thin immunoisolation membrane (Grundfest-Broniatowski et al. 2009). Their larger size and smaller number also make retrieval less of a problem in comparison to microcapsules. The membranes used for macroencapsulation must not only possess the needed permeability requirements, but because of their larger size, must also provide sufficient mechanical strength to maintain their integrity. This is a difficult objective, since the membrane wall must be thin for diffusional purposes and this may compromise overall strength. In many cases, it has been found that thin-wall hollow fiber membranes are prone to breaking (Lanza et al. 1995). Rupture of the membrane wall will result in rejection of the islet tissue and loss of function. As is the case for microcapsules, membrane chemistry is an important factor in minimizing the formation of fibrous tissue around the implant.

Most of the early attention given to the macroencapsulation technique has focused on the use of hollow fibers. A tradeoff exists on the selection of the fiber diameter. A large diameter fiber will result in a shorter overall length, but can lead to diffusional limitations. Lack of oxygen transport and the accumulation of waste products can therefore result in a central core of necrotic tissue. On the other hand, a small fiber diameter, while improving the transport characteristics, can result in an incredibly long fiber length. This longer length increases the probability of breakage and makes implantation of the fibers more difficult.

An example of the use of hollow fibers as a vehicle for the macroencapsulation of the islets of Langerhans has been reported (Lanza et al. 1991, 1992a, 1992b, 1992c, 1995). This particular system has been studied in some detail in both rats and dogs, using a variety of sources for the islets. The islets were enclosed within semipermeable hollow fiber membranes that were 2–3 cm in length and had an internal diameter of 1.8–4.8 mm. The membrane wall thickness ranged from 69 to 105 µm and the membrane had an NMWCO of 50,000–80,000 g mol^{-1}. Depending on the fiber diameter, approximately 9–50 fibers were implanted within the peritoneal cavity of each rat.

When these fibers containing bovine islets (20,000 EIN) were implanted in the peritoneal cavity of rats whose diabetes was chemically induced using the drug *streptozotocin*, the diabetic state was reversed within 24 h after implantation. In some cases, normoglycemia was maintained for 1 year. However, the wider bore membranes (4.5–4.8 mm ID) that were retrieved after several months were found to contain a necrotic core as a result of oxygen transport limitations. Viable islets were only found to exist within a 0.5–1 mm layer along the membrane wall. Some of the late failure of the implants was also attributable to membrane breakage as a result of the thin membrane wall.

Similar results were obtained using canine islets contained within the same hollow fibers and implanted within the peritoneal cavity of diabetic BB/Wor rats (Lanza et al. 1992b). This type of rat develops autoimmune diabetes with a pathology similar to that found in humans. The canine islet xenografts provided normoglycemia in these rats for more than 8 months. This is an important result since it shows that the immunoisolation membrane protected the islet xenografts from the host's normal graft rejection response, as well as from the autoimmune disease process that produced the original diabetes. Protection of the islets from the autoimmune disease process is important in order to successfully treat human patients with diabetes.

The hollow fiber macroencapsulation of islets was also tested in pancreatectomized dogs. Figure 10.14 (Lanza et al. 1992) summarizes results obtained for two dogs that were insulin-free for at least 70 days after receiving islet allografts contained within these fibers. Each dog received between 155 and 248 fibers containing a total of about 300,000 EIN. This equates to a total fiber

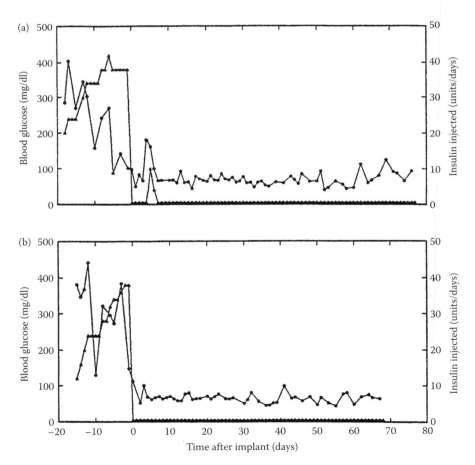

FIGURE 10.14 Fasting glucose levels in two dogs receiving hollow fiber implants containing canine islets. (From Lanza, R.P., Sullivan, S.J., and Chick, W.L., *Diabetes*, 41, 1503–10, 1992. With permission.)

length of about 500 cm. The dogs had an average weight of about 17 kg, providing an islet loading of nearly 20,000 EIN kg^{-1} body weight. For a 70 kg human, the total number of islets would therefore be about 1.4 million EIN, or a total fiber length on the order of 2100 cm (70 ft).

Storrs et al. (2001) have described a macroencapsulation approach known as the *islet sheet*. In their approach, a mixture of highly purified alginate with a reinforcing polymeric mesh is sandwiched between acellular alginate layers. The resulting flat sheet construct is then allowed to gel. The ultrathin acellular alginate layers are highly crosslinked and form the immunoisolation barrier. The thickness of the immunoisolation barrier is about 50–75 μm. The islet sheets are typically 250 μm thick with an active width of 3 cm and a length of 7 cm. The thinness of the islet sheets provides for excellent transport of oxygen and allows for an islet volume fraction of up to 40%. Because of the high tissue density, it is estimated that 10 sheets containing 400,000 islet equivalents would be needed to treat a human with diabetes. In these early studies on the islet sheet, a pancreatectomized beagle dog that received six islet sheets was able to maintain normal fasting glucose levels for 84 days.

10.6.1.5 Organoid

The organoid approach (see Figure 10.10d) (Sarver et al. 1995; Hill et al. 1994) represents the union of the tissue engineering and macroencapsulation concepts. The bioartificial organoid consists of a thin circular or rectangular compartment for the tissue that is enclosed by an immunoisolation membrane. Adjacent or integral to the membrane is a thin layer of a porous scaffold material that becomes vascularized by the host after implantation. The use of tissue engineering allows control over the environment adjacent to the immunoisolation membrane. Therefore, the fibrotic response can be minimized, and through the growth of a vascular bed, intimate contact with the host's vasculature can be obtained. This has the potential to improve the viability and long-term functioning of the transplanted tissue. Additionally, these devices, unlike the microcapsule or macroencapsulation approaches, have ports that allow the device to be seeded with the islet tissue once the porous material has become sufficiently vascularized. The ports also allow for islet tissue to be replaced on a regular basis by a relatively minor procedure.

10.6.2 Number of Islets Needed

One issue confronting all of the devices discussed thus far concerns the mass of tissue required to achieve normoglycemia in a human patient with diabetes. The previously mentioned studies in rats and dogs have used islet loadings on the order of 20,000 EIN kg^{-1} body weight. This is considerably higher than the benchmark study reported by Warnock et al. (1988). They found that the threshold islet loading for treating pancreatectomized dogs corresponded to $\approx$ 4.3 μL of islet tissue per kilogram. This equates to about 2500 EIN kg^{-1}. It is important to point out, however, that these were unencapsulated autologous islets that were implanted within the liver or the spleen. These islets had no artificial mass transfer resistance to overcome and no host immune response that would lead to rejection. This result most likely defines the lower limit on islet loading. Certainly, improvements in the mass transfer characteristics of the devices discussed so far could lead to islet loadings more on the order of 5000 to 10,000 EIN kg^{-1}. Overall device size is dependent on the amount of tissue needed, as well as on the device mass transfer characteristics for nutrients, oxygen, waste products, and the therapeutic agent itself. Therefore, the mass transfer characteristics are a critical factor in defining the islet tissue loading within the device itself.

The results presented so far demonstrate the potential of the bioartificial pancreas as a method for treating diabetes. All of the approaches, however, suffer from the problem of poor long-term islet survival. Islets have the potential to function for many years, as evidenced by the fact that, unless they become diseased, they exist for as long as the lifespan of the animal they are found in. Therefore, islets should be capable of surviving within a bioartificial pancreas for many years.

Islet survival is a function of many variables, including the amount of trauma they experience during their isolation and purification; the device mass transfer characteristics for nutrients, oxygen, and waste products; the integrity of the membrane immunoisolation system; and the host's response to the implant. This latter factor determines the nature of the physiological environment in the vicinity of the implant. Certainly, the formation of a region of under-vascularized fibrotic tissue will not provide the conditions needed for the long-term functioning and survival of the islets. Because of questions concerning the long-term viability and functioning of the islet tissue, it may be necessary to periodically recharge the patients with islets. A bioartificial pancreas that has the ability to be reseeded would certainly offer an advantage.

10.6.3 ISLET INSULIN RELEASE MODEL

The design of a bioartificial pancreas requires an understanding of the insulin release rate from an islet or insulin-producing cells and its dependence on plasma glucose levels. Islets that have been isolated from a pancreas or genetically engineered cells are usually challenged in culture with a step change in the glucose concentration. The glucose challenge is used to assess islet or cell viability, the insulin release rate, and glucose responsiveness. When challenged in this manner, islets or insulin-secreting cells release insulin in a biphasic fashion (Grodsky 1972).

Figure 10.15 (Nomura et al. 1984) illustrates the insulin response of rat islets to a rapid ramp increase in the glucose concentration from an initial value of 80 mg dL^{-1} to a final value of 200 mg dL^{-1}. Initially, the islets respond to the rate of change of the plasma glucose concentration, resulting in a maximum of the first phase insulin release rate within a few minutes of the glucose challenge. Following this, the second phase islet insulin release rate is proportional to the difference between the plasma glucose concentration and its fasting value of about 80 mg dL^{-1}. We see that this second phase release rate is relatively constant once the glucose level has achieved a steady state value.

A mathematical description of the islet insulin release rate during a glucose challenge can be obtained through application of control theory. Nomura et al. (1984) have shown that the dynamics of the glucose-induced secretion of insulin can be expressed as the sum of the proportional response to the glucose concentration, and a derivative response to the rate of change in the glucose concentration, each with a first order lag time, respectively given by T_1 and T_2. The Laplace transform of the islet insulin release rate can therefore be expressed as

$$r_{\text{islet}}(s) = \left[\frac{K_{\text{p}}}{1 + T_1 s} + \frac{T_{\text{d}} s}{1 + T_2 s} \right] C_{\text{G}}(s). \tag{10.2}$$

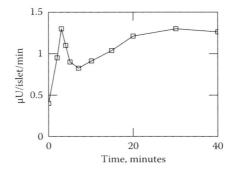

FIGURE 10.15 Rat islet insulin secretion rate. (Data from Nomura, N., Schihiri, M., Kawamori, R., Yamasaki, Y., Iwama, N., and Abe, H., *Comput. Biomed. Res.*, 17, 570–79, 1984.)

This equation may be inverted (in the Laplace transform sense) to give Equation 10.3 for the islet insulin release rate in terms of the glucose concentration and its rate of change:

$$r_{\text{islet}}(t) = \int_{-\infty}^{t} C_p C_G(z) e^{-(t-z)/T_1} dz + \int_{-\infty}^{t} C_d \frac{dC_G(z)}{dz} e^{-(t-z)/T_2} dz. \tag{10.3}$$

The parameters in this equation, i.e., $C_p(= K_p/T_1)$, $C_d(= T_d/T_2)$, T_1, and T_2, may be obtained by performing a nonlinear regression analysis of islet or cellular insulin release rate data, such as that shown in Figure 10.16. This can be accomplished by first representing the ramp glucose profile as

$$C_G(t) = C_G^0, \text{ for } t < 0,$$

$$C_G(t) = \left(\frac{C_G^{SS} - C_G^0}{t_0} \right) t + C_G^0, \text{ for } 0 \le t \le t_0, \tag{10.4}$$

$$C_G(t) = C_G^{SS}, \text{ for } t \ge t_0.$$

These equations may be substituted into Equation 10.3 to obtain algebraic expressions for the islet insulin release rate for a ramp change in the glucose concentration:

$$r_{\text{islet}}(t) = C_p T_1 C_G^0 e^{-t/T_1} + C_p T_1^2 \left(\frac{C_G^{SS} - C_G^0}{t_0} \right) \left[\frac{t}{T_1} - (1 - e^{-t/T_1}) \right]$$

$$+ C_p C_G^0 T_1 (1 - e^{-t/T_1}) + C_d T_2 \left(\frac{C_G^{SS} - C_G^0}{t_0} \right) (1 - e^{-t/T_2}), \text{ for } 0 \le t \le t_0, \tag{10.5}$$

$$r_{\text{islet}}(t) = C_p C_G^0 T_1 e^{-t/T_1} + C_p T_1^2 \left(\frac{C_G^{SS} - C_G^0}{t_0} \right) \left[e^{-(t-t_0)/T_1} \left(\frac{t_0}{T_1} - 1 \right) + e^{-t/T_1} \right]$$

$$+ C_p C_G^0 T_1 [e^{-(t-t_0)/T_1} - e^{-t/T_1}] + C_p C_G^{SS} T_1 (1 - e^{-(t-t_0)/T_1})$$

$$+ C_d \left(\frac{C_G^{SS} - C_G^0}{t_0} \right) T_2 [e^{-(t-t_0)/T_2} - e^{-t/T_2}], \text{ for } t > t_0. \tag{10.6}$$

Example 10.2 illustrates how the parameters in the Nomura et al. (1984) islet insulin release model may be determined for rat islets through the use of the previous equations.

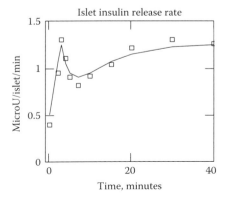

FIGURE 10.16 Comparison of measured and predicted insulin release rate.

Example 10.2

Determine the constants in the Nomura et al. (1984) insulin release model for the rat islet perfusion data shown in Figure 10.15.

Solution

A nonlinear regression analysis finds that the values of $T_1 = 10.11$ min, $T_2 = 1.88$ min, $C_p = 6.26 \times 10^{-4}$ µU dL mg^{-1} min^{-2} islet^{-1}, and $C_d = 0.011$ µU dL mg^{-1} min^{-1} islet^{-1} best fit the data shown in Figure 10.15. Figure 10.16 shows that the Nomura model provides an excellent fit to the islet insulin release data.

Rat islets have a very potent response to a step change in glucose concentration. Islets from dogs, pigs, and human pancreas typically show significantly lower insulin secretion rates on a per islet basis.

The actual in vivo response of islets depends not only on glucose levels, but also, in a significant way, on the levels of other nutrients such as amino acids and gastrointestinal hormones.* These latter substances, in the presence of rising glucose levels, can almost double the insulin secretion rate of an islet. However, no quantitative framework is available to describe the effect of these additional secretagogues on the insulin secretion rate.

Under physiological conditions, the variation in plasma glucose levels is less pronounced than a glucose challenge test on islets and the biphasic insulin release pattern is not as evident. In fact, the insulin secretion rate under these conditions displays a strong sigmoidal dependence on the blood glucose levels (Guyton 1991; Sturis et al. 1991). The insulin secretion rate for the human pancreas saturates at a value of about 200,000 µU min^{-1} when the glucose concentration reaches 300 mg dL^{-1}. Since the human pancreas contains about 1 million islets, this saturation insulin release rate is 0.2 µU islet^{-1} min^{-1}.

Sturis et al. (1991) represented the insulin secretion rate under physiological conditions as a sigmoidal function of the plasma glucose concentration. Equation 10.7, proposed by Sturis et al. (1991), can then be used to describe the in vivo insulin response of an islet:

$$r_{islet} = \frac{0.209}{1 + \exp(-3.33 G_B + 6.6)}. \tag{10.7}$$

In this equation, the islet insulin release rate is in units of microunits per islet per minute and the glucose concentration is in milligrams per milliliter.

10.6.4 Pharmacokinetic Modeling of Glucose and Insulin Interactions

Although device testing in experimental diabetic animals provides conclusive proof of a device's efficacy and potential, mathematical models are also useful tools for investigating the myriad parameters that affect device performance. Mathematical models also provide the opportunity to perform scaleup studies. For example, they allow for critical examination of the results obtained in small laboratory animals and extrapolation of these results to understand how the device may perform in humans.

Several physiological pharmacokinetic models of glucose and insulin metabolism have been described (Guyton et al. 1978; Sorensen et al. 1982; Berger and Rodbard 1989; Sturis et al. 1991, 1995). These models can be combined with an islet glucose-insulin response model, and a device model, to assess quantitatively the level of glucose control that can be achievable (Smith et al. 1991).

---

* Gastrointestinal hormones include gastrin, secretin, cholecystokinin, and gastric inhibitory peptide.

The pharmacokinetic model for glucose and insulin metabolism proposed by Sturis et al. (1991, 1995) is particularly attractive because of its relative simplicity and its ability to represent the experimentally observed temporal oscillations of insulin and glucose levels (Kraegen et al. 1972). Secretion of insulin in humans has been found to exhibit two distinct types of periodic oscillations. A rapid oscillation with a period of 10–15 min, and a longer or ultradian oscillation, with a period of 100–150 min. The rapid oscillations are of small amplitude (insulin <1–2 μU mL^{-1} and glucose <1 mg dL^{-1}) and may be the result of an intrinsic pacemaker in the islets. The ultradian oscillations exhibit a much larger amplitude and are self-sustained when the stimulus is continuously presented, e.g., by a constant infusion of glucose or by continuous enteral glucose feeding. On the other hand, the ultradian oscillations are damped when the glucose stimulus is presented as a discrete event, e.g., by either a meal or ingestion of a fixed quantity of glucose.

Figure 10.17 presents a block diagram of the Sturis et al. (1991) model for describing the interactions between glucose and insulin. This model is based on four negative feedback loops involving interactions between glucose and insulin: (1) elevated glucose levels stimulate the secretion of insulin and the resulting increase in insulin levels decreases the endogenous production of glucose, which then has the effect of lowering glucose levels; (2) elevated glucose levels stimulate the secretion of insulin and the resulting increase in insulin levels enhances glucose utilization, which also has the effect of reducing glucose levels; (3) rising levels of glucose inhibit production of glucose; and (4) increasing levels of glucose stimulate its utilization. These four loops control the amount of glucose and insulin in the body. However, because of the nonlinear and dynamic interaction between glucose and insulin, the amount of these substances in the body is never at a stable equilibrium.

The model also includes two time delays that are important for describing the observed oscillatory dynamics. The first delay affects the suppression by insulin of glucose production or the recovery of this process when insulin levels decrease. The second delay considers the fact that the biological action of insulin correlates better with the concentration of insulin in an interstitial compartment that equilibrates slowly with the plasma insulin concentration.

The three main variables in the model are the concentration of glucose in the plasma (G_B, milligrams per milliliter), the concentration of insulin in the plasma (I_B, microunits per milliliter), and the concentration of insulin in the interstitial fluid (I_{IF}, microunits per milliliter). Three additional variables are also introduced to account for the delay between the plasma insulin level and its effect on glucose production (x_1, x_2, x_3 with time lag τ_d).

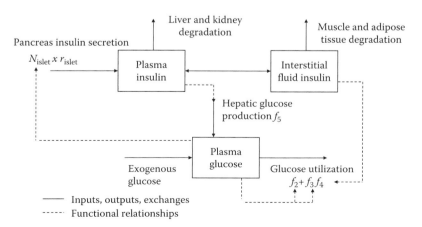

FIGURE 10.17 Model of glucose and insulin interactions in the body. (After Sturis, J., Polonsky, K.S., Mosekilde, E., and Van Cauter, E., *Am. J. Physiol.*, 260, E801–E809, 1991.)

A total of six differential equations are needed to describe the state of the system as a function of time (here in minutes). These equations are

$$V_{Bi} \frac{dI_B}{dt} = r_{islet}(G_B) \, N_{islets} - E(I_B - I_{IF}) - \frac{I_B V_{Bi}}{\tau_p},$$

$$\frac{dI_{IF}}{dt} = E\left(\frac{I_B}{V_{IF}} - \frac{I_{IF}}{V_{IF}}\right) - \frac{I_{IF}}{\tau_i},$$

$$(10.8)$$

$$V_{Bg} \frac{dG_B}{dt} = r_{Gin}(t) - f_2(G_B) - f_3(G_B) f_4(I_{IF}) + f_5(x_3),$$

$$\frac{dx_1}{dt} = 3\frac{I_B V_{Bi} - x_1}{\tau_d}, \ \frac{dx_2}{dt} = 3\frac{x_1 - x_2}{\tau_d}, \ \frac{dx_3}{dt} = 3\frac{x_2 - x_3}{\tau_d}.$$

The insulin secretion rate of an islet, r_{islet}, was given by Equation 10.7 and is multiplied by the number of islets in the pancreas (N_{islets}).

The insulin that is released distributes in the plasma space of volume given by V_{Bi}, or it enters the interstitial fluid space represented by the distribution volume V_{IF}. The insulin transport rate into the interstitial fluid is proportional to the difference in the insulin concentration in the two compartments and is described by a rate constant E. Insulin is degraded within the plasma space by a first order rate process with time constant given by τ_p. Interstitial insulin is also degraded with a time constant of τ_i and enhances glucose utilization as described by the function $f_4(I_{IF})$.

Glucose is assumed to distribute throughout a single compartment (V_{Bg}) and affects its own utilization through the functions f_2 and f_3. Glucose utilization, represented by f_2, is insulin independent, whereas f_3 is multiplied by an additional term, f_4, which is dependent on the interstitial insulin concentration. The functions f_2 (milligrams per minute) and $f_3 \times f_4$ (milligrams per minute) are given by

$$f_2 = 72[1 - \exp(-6.94 G_B)]$$

$$f_3 \times f_4 = 10 G_B \times \left[\frac{90}{1 + \exp\left(-1.772 \ln\left(I_{IF}\left(1 + \frac{V_{IF}}{E \tau_i}\right)\right) + 7.76\right)} + 4 \right].$$

$$(10.9)$$

Insulin also inhibits production of glucose by the liver (f_5) via a process that is dependent on a time delay (τ_d) between the appearance of insulin in the plasma and its inhibitory effect on glucose production. The function f_5 (milligrams per minute) is given by Equation 10.10 and is dependent on a time-delayed plasma insulin concentration represented by x_3:

$$f_5 = \frac{180}{1 + \exp\left(\frac{0.29 x_3}{V_{Bi}} - 7.5\right)}.$$

$$(10.10)$$

Glucose can enter the body in an arbitrary time-varying fashion, as given by the function $r_{G \, in}(t)$ (milligrams per minute). For example, the glucose input function for a meal or an *oral glucose tolerance test* (OGTT) can be described by Equation 10.11, which describes the absorption of glucose from the gastrointestinal tract. In this equation, D (milligram) represents the total amount of

TABLE 10.1

Parameter Values for the Sturis et al. Model of Glucose and Insulin Interactions

Parameter	Value
E (rate constant for exchange of insulin between plasma and the interstitial fluid compartment)	200 mL min^{-1}
$V_{B\,i}$ (insulin plasma distribution volume)	3000 mL
V_{IF} (insulin interstitial fluid distribution volume)	11,000 mL
$V_{B\,g}$ (glucose plasma distribution volume)	10,000 mL
τ_p (time constant for plasma insulin degradation)	6 min
τ_i (time constant for interstitial fluid insulin degradation)	100 min
τ_d (time delay between plasma insulin and glucose production)	36 min

Source: Sturis, J., Polonsky, K.S., Mosekilde, E., and Van Cauter, E., *Am. J. Physiol.*, 260, E801–E809, 1991; Sturis, J., et al., *Dynamical Disease: Mathematical Analysis of Human Illness*, AIP Press, 1995.

glucose ingested, k_a (minutes) is the absorption rate constant, and k_e (minutes) is the elimination rate constant:

$$r_{G\,in}(t) = D k_e \left(\frac{k_a}{k_a - k_e} \right) \left(e^{-k_e t} - e^{-k_a t} \right). \tag{10.11}$$

Table 10.1 summarizes the values of all the parameters used in the model of Sturis et al. (1991). For a given glucose input function, the previous equations can be solved numerically to provide the time course of the plasma glucose and insulin levels.

10.6.5 USING THE PHARMACOKINETIC MODEL TO EVALUATE THE PERFORMANCE OF A BIOARTIFICIAL PANCREAS

We can now use the Sturis et al. (1991) pharmacokinetic model outlined earlier, along with an insulin release model for an islet, to explore the level of glucose control that is possible with a bioartificial pancreas. There are two basic types of tests that can be used to measure the level of glucose control. One stringent measure of glucose control that is frequently used is called the *k-value*. The *k*-value represents the slope of the least squares fit of the logarithm of the plasma glucose concentration plotted as a function of time for the 50 min following the intravenous administration of a dose of glucose. This glucose dose is typically 0.5 g kg^{-1} body weight and is referred to as an *intravenous glucose tolerance test* (IVGTT).

Following the glucose dose for the IVGTT, plasma glucose levels rapidly rise within a few minutes to levels exceeding 300 mg dL^{-1} from the initial fasting level of around 80 mg dL^{-1}. The rate of glucose decay following an IVGTT is indicative of the glucose and insulin responsiveness of the patient's pancreas. The *k*-value is expressed in units of percent per minute. If the *k*-value following an IVGTT is less than 1% min^{-1}, then the patient is assumed to have diabetes.

Normal *k*-values range from about 2 to 3% min^{-1}. A bioartificial pancreas would need to provide a *k*-value greater than about 1.5% min^{-1} to be considered effective at controlling blood glucose levels in a patient with diabetes. This somewhat lower than normal *k*-value is acceptable because the IVGTT provides a major glucose challenge that is not normally observed following a meal.

The OGTT involves the oral administration of a beverage containing 50 g of glucose. Glucose levels greater than 140 mg dL^{-1} 2 h after administration of the glucose may be indicative of diabetes.

The simulation of blood glucose and insulin levels following an IVGTT or an OGTT using the pharmacokinetic model outlined above is straightforward. Assume that the islets are macroencapsulated

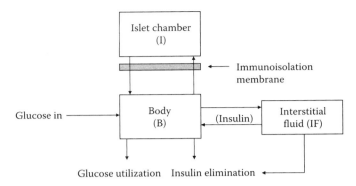

FIGURE 10.18 Compartmental model for evaluation of a bioartificial pancreas.

within a thin sheet and that the immunoisolation membrane of the bioartificial pancreas is the controlling mass transfer resistance for glucose and insulin transport. As shown in Figure 10.18, two well-mixed compartments are used to define the distribution of glucose and insulin within the device and the rest of the body. The distribution of insulin also includes an additional interstitial fluid compartment to account for the delay in its action, as discussed earlier. Unsteady mass balances for glucose and insulin can now be written for each of the compartments.

Islet or cell chamber:

$$V_I \frac{dI_I}{dt} = N_{islets}\, r_{islet} - P_{mI} S_m (I_I - I_B),$$

$$V_I \frac{dG_I}{dt} = P_{mG} S_m (G_B - G_I). \tag{10.12}$$

Body distribution volume:

$$V_{Bi} \frac{dI_B}{dt} = P_{mI} S_m (I_I - I_B) - E(I_B - I_{IF}) - \frac{I_B V_{Bi}}{\tau_p},$$

$$V_{Bg} \frac{dG_B}{dt} = r_{Gin}(t) - P_{mG} S_m (G_B - G_I) - f_2(G_B) - f_3(G_B) f_4(I_{IF}) + f_5(x_3). \tag{10.13}$$

Interstitial fluid space for insulin:

$$V_{IF} \frac{dI_{IF}}{dt} = E(I_B - I_{IF}) - \frac{V_{IF} I_{IF}}{\tau_i}. \tag{10.14}$$

Equations for delayed insulin action on glucose production:

$$\frac{dx_1}{dt} = 3\frac{V_{Bi} I_B - x_1}{\tau_d}, \quad \frac{dx_2}{dt} = 3\frac{x_1 - x_2}{\tau_d}, \quad \frac{dx_3}{dt} = 3\frac{x_2 - x_3}{\tau_d}. \tag{10.15}$$

For the special case of an IVGTT, the value of $r_{Gin}(t)$ would be set equal to zero.

The initial conditions for the solution of the previous equations for either an IVGTT or an OGTT are based on fasting levels. Fasting plasma glucose levels are typically about 80 mg dL^{-1}. Assuming negligible consumption of glucose by the tissue in the bioartificial pancreas, one may reasonably assume that the initial glucose concentration in the bioartificial pancreas is the same as that found in the rest of the body. However, since the IVGTT glucose dose is rapidly distributed throughout the body distribution volume, in comparison to the length of time required for its subsequent removal, we assume that at $t = 0 (+)$, the plasma glucose level is given by

$$G_B(t = 0+) = G_B(\text{fasting}) + \frac{500(\text{mg/kg bw}) \times (\text{bw, kg})}{V_{Bg}}. \tag{10.16}$$

Fasting plasma insulin levels are about 10 μU mL^{-1}. During the fasting period, we assume that the bioartificial pancreas is at a steady state equilibrium with the rest of the body as far as insulin is concerned. Accordingly, the islet insulin secretion rate based on the fasting glucose concentration must match the rate of insulin removal from the body. The steady state solution of the set of compartmental insulin equations provides the following relationships for the initial insulin concentration in each of the compartments:

$$I_B^0 = \frac{N_{\text{islets}} r_{\text{islet}}^0 (G_B^0) \tau_i}{V_{IF}} \times \frac{1}{\dfrac{V_{Bi} \tau_i}{V_{IF} \tau_p} - \dfrac{1}{\dfrac{V_{IF}}{\tau_i E} + 1}},$$

$$I_{IF}^0 = \frac{I_B^0}{\left(\dfrac{1}{\tau_i E} + \dfrac{1}{V_{IF}}\right) V_{IF}}, \tag{10.17}$$

$$I_I^0 = I_B^0 + \frac{N_{\text{islet}} r_{\text{islet}}^0 (G_B^0)}{P_{mI} S_m}.$$

Solution of Equations 10.12 through 10.17 provides the temporal change in the plasma glucose concentration following administration of either an OGTT or an IVGTT to a patient with a bioartificial pancreas. The previous equations can be easily modified to evaluate the glucose control for a variety of bioartificial pancreas approaches. The following example illustrates a prediction of the plasma glucose levels using a bioartificial pancreas following an OGTT.

Example 10.3

Estimate the glucose concentrations following an OGTT using islets that are macroencapsulated between two immunoisolation membrane disks. A total of 50 g of glucose is taken orally. In Equation 10.11, the glucose absorption rate constant is 0.042 min^{-1} and the glucose elimination rate constant is 0.0083 min^{-1}. Base the calculation on a 70 kg patient receiving a total of 750,000 EIN. Assume the islet insulin secretion rate is given by Equation 10.7. The half-thickness of the islet chamber is 75 μm. The void volume within the islet chamber must be at least 65% for sufficient oxygen transport. The immunoisolation membrane permeabilities for glucose and insulin are based on the membrane developed by Baker et al. (1997), i.e., 4×10^{-4} cm sec^{-1} for glucose and 8×10^{-5} cm sec^{-1} for insulin.

Solution

Figure 10.19 provides the solution for the glucose and insulin concentrations as a function of time. Note the damped ultradian oscillations of the glucose and insulin concentrations with time. The

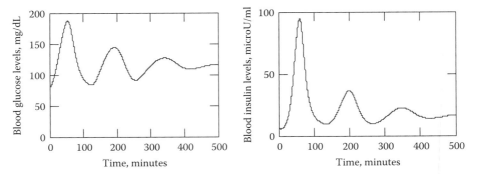

FIGURE 10.19   Predicted glucose and insulin levels in a bioartificial pancreas following an OGTT.

bioartificial pancreas provides a glucose and insulin response that is very similar to that observed in normal patients (Kraegen et al. 1972). We therefore would expect that the bioartificial pancreas has the potential to restore normoglycemia in patients with diabetes. The size of the device is 18 cm (7 in) in diameter and only several millimeters in thickness. Rather than one large unit, several smaller units could be implanted. For example, four units each 9 cm (3.5 in) in diameter would be equivalent.

10.7 BIOARTIFICIAL LIVER

The liver is a very complex organ that performs a variety of life-sustaining functions (Yarmush et al. 1992; Galletti and Jauregui 1995; Davidson et al. 2010). These functions include detoxifying the blood, regulating glucose levels, and making proteins. The human liver weighs about 1500 g and is highly vascularized. From Table 6.2, we see that the liver receives over 25% of the cardiac output. The liver blood supply comes from two major sources. These are the *hepatic artery* and the *portal vein*. The portal vein collects the venous drainage from the spleen, the pancreas, and the intestines. The portal vein input from these regions passes through the liver before entering the rest of the systemic circulation. The liver, therefore, has the ability to detoxify potentially harmful substances that are absorbed from the gastrointestinal tract.

The basic functional cellular units of the liver are called the *hepatocytes*. Each hepatocyte is about 25 μm in diameter and there are close to 250 billion of them in the human liver, accounting for 75% of the liver volume. The hepatocytes are metabolically very active and provide an incredible variety of functions. Their role in carbohydrate metabolism includes the storage of excess glucose as *glycogen* and the release of this stored form of glucose (*glycogenolysis*) when blood glucose levels are low. In addition, the liver converts other sugars, such as galactose and fructose, to glucose. If blood glucose levels are low, and the glycogen stores are also depleted, then the liver performs a process known as *gluconeogenesis* wherein glucose is synthesized from amino acids.

The liver also plays a major role in fat metabolism. The liver is principally responsible for the body's ability to derive energy from fats. The liver can also convert excess carbohydrates and proteins into fat, the liver synthesizes cholesterol and phospholipids, and it forms the lipoprotein carrier molecules. The lipoproteins are responsible for transporting cholesterol, phospholipids, and fats to other tissues throughout the body. Cholesterol and phospholipids are important components of cellular membranes. The liver has an extremely important role in protein metabolism. These functions include the deamination of amino acids as a prelude to their use as an energy source, or their conversion to carbohydrates or fats. Deamination of amino acids produces large amounts of ammonia. The liver converts ammonia into urea, which is then excreted by the kidneys. Without this function, ammonia levels in the blood rapidly increase leading to hepatic coma and death.

With the exception of the gamma globulins (immunoglobulins or antibodies), practically all of the plasma proteins are made in the liver. The most important being albumin. The liver also makes most of the substances found in the blood that are responsible for the clotting of blood. These include fibrinogen, prothrombin, and most of the other *clotting factors*. The liver also provides storage for a variety of nutrients, such as vitamins and iron, which is stored as a protein complex called *ferritin*. The hepatocytes, because of their high enzyme content, also play a major role in the detoxification and conjugation of a variety of materials, such as drugs, environmental toxins, and hormones that are produced elsewhere in the body. These byproduct materials are water soluble and are excreted in the bile or by the kidneys.

The liver has the unique capability of being able to regenerate itself following tissue damage. This is extremely beneficial considering the role it plays in detoxifying materials from the gastrointestinal tract. However, if the tissue damage rate exceeds the liver's ability to regenerate, then liver failure results, leading to a life-threatening situation. Liver failure accounts for over 60,000 deaths each year in the United States.

There are principally two types of liver failure: *cirrhosis* and *fulminant hepatic failure*. Cirrhosis of the liver accounts for over half of the deaths due to liver disease. In cirrhosis, fibrotic tissue forms in place of the damaged liver tissue, severely compromising the liver's ability to regenerate. Common causes of cirrhosis include alcoholism and chronic hepatitis. Fulminant hepatic failure is a rapidly progressing failure of the liver that can lead to death within several weeks of onset. It can be caused by chemical and viral hepatitis.

Liver failure causes a variety of life-threatening abnormalities, including the accumulation in the plasma of ammonia, bilirubin, and decreased levels of albumin and clotting factors. There is also a buildup of toxins and overactivity of the hormonal systems, which are believed to lead to a condition known as *hepatic encephalopathy* that can cause irreversible brain damage, coma, and death.

The only long-term and relatively successful treatment method for liver failure is transplantation. However, liver transplantation is severely limited by the shortage of donor organs. Thus, many patients die before a liver becomes available, or they succumb shortly after the transplant because of the complications and brain damage associated with liver failure.

10.7.1 Artificial Liver Systems

Treatment of liver failure by artificial means has been the focus of considerable research. Table 10.2 provides a summary of some of the artificial liver systems that have been evaluated (Yarmush et al. 1992). The primary goal of these artificial liver systems is to provide a means to maintain the patient in a stable state until a liver transplant is possible or in some cases, artificial liver support can provide the liver with the chance to regenerate and the patient can return to a normal life without a transplant.

The difficulty in developing an artificial liver is the multitude of functions the liver performs. Initially it was thought that the best approach is to minimize the buildup of toxins. Accordingly, efforts focused on such approaches as hemodialysis, hemoperfusion, and immobilized enzyme reactors for removal of these materials.

The *hemodialysis* systems were similar to those used in kidney dialysis, with the exception that they used membranes with a higher molecular weight cutoff to allow passage of the larger-sized toxic materials. This approach was still not effective at removing large protein-bound toxins.

TABLE 10.2
Artificial Liver Systems

Hemodialysis	Cross circulation
Hemoperfusion	Extracorporeal perfusion
Immobilized enzymes	Cross hemodialysis
Plasma exchange	Hepatocyte hemoperfusion

Hemoperfusion employed beds of activated carbon that adsorbed the toxic molecules. A particular problem with this approach is its nonspecificity, removing beneficial substances as well. *Immobilized enzyme reactors* used liver enzymes to provide for more specific removal of the toxic molecules. The major difficulty is providing a complete set of liver enzymes. These three approaches also suffered from the disadvantage that they do not restore any of the synthetic functions of the liver.

Other approaches that were used attempted to both reduce the level of toxins and restore substances normally synthesized by the liver. These methods included plasma exchange, cross circulation, extracorporeal perfusion, and cross hemodialysis. *Plasma exchange* involves replacing the patients plasma with donor plasma. Provided the exchange rate is sufficiently high, this approach can replace the lost liver function. However, the major limitation is the large amount of donor plasma needed and the increased risk to the patient of viral infections. *Cross circulation* involves connecting the patient's circulation to that of another human. In this way, the healthy liver is shared. However, the risk to the human donor limits the widespread use of this approach. *Extracorporeal perfusion* of the patient's blood through an xenogeneic liver has also been tried. Some success with this approach has been achieved using pig and baboon livers. However, the liver function degrades quickly as a result of the host's immune rejection response. *Cross hemodialysis* attempts to minimize the host's immune response to the donor liver by perfusing the donor liver with a separate supply of blood. This blood is then sent through a hemodialyzer where it is contacted across a dialysis membrane with the patient's own blood. In this way, toxins are removed and synthetic substances are added to the patient's blood. *Hemoperfusion of liver tissue* has also been tried. Liver tissue pieces are directly contacted with the patient's blood. Frozen or freeze-dried dead tissue was comparable in activity to fresh tissue, indicating that only those few layers of cells near the surface of the fresh liver pieces are adequately oxygenated. The dead tissue still contained enzymes with some residual activity.

10.7.2 BIOARTIFICIAL LIVERS

Hemoperfusion of liver tissue pieces is not capable of providing optimal conditions for the long-term function of the liver tissue. In the absence of an intact vasculature, the mass transfer resistances within the liver pieces are just too large for simple diffusion to overcome. Recently, attention has focused on the development of *bioartificial livers* that are based on the use of isolated hepatocytes (Jauregui et al. 1997; Legallais et al. 2001; Allen et al. 2001; Nose 2001; Patzer 2001; Sauer et al. 2001; Kobayashi et al. 2003; Davidson et al. 2010). The use of isolated hepatocytes in a properly designed bioartificial liver has the potential to mimic the synthetic, metabolic, biliary excretion, and detoxifying functions of the normal liver. Isolated hepatocytes are not as severely affected by mass transfer limitations and, furthermore, they can be immunoprotected. The challenge is to design a bioartificial liver so that the isolated hepatocytes can maintain their differentiated functions and continue to detoxify the blood, regulate glucose levels in the body, and synthesize proteins.

Isolated hepatocytes have been proposed for both implantable and extracorporeal systems. Implantable systems are based on the tissue engineering concepts discussed in Chapter 9 and involve the encapsulation of the hepatocytes in a manner similar to the techniques already discussed for islets (Cai et al. 1988; Cima et al. 1991a, 1991b; Johnson et al. 1994; Yang et al. 1994; Powers et al. 2002).

The hepatocytes can come from primary sources such as from human donors as well as animals like the pig and the rabbit. Alternatively, cells lines based on hepatic tumors and immortalized cells have also been developed. In addition, stem cells and progenitor cells have the potential, given the proper signals, to differentiate into hepatocytes.

The mass of hepatocytes that is needed to treat liver failure is estimated to be on the order of 10–40% of the original liver mass (Yarmush et al. 1992; Rozga et al. 1994; Davidson et al. 2010).

At the low end of this range, this amounts to 25 billion cells or a cellular volume on the order of 200 mL. Attachment to a support structure, or encapsulation of this quantity of cells, while still maintaining a reasonable device volume, is a major obstacle with implantation approaches. For example, for hepatocytes attached to microcarriers,* the estimated total volume is 500 mL. For microencapsulation within a polymeric capsule, this volume is 2500 mL, whereas macroencapsulation in hollow fibers yields a volume that is estimated to be 1300 mL.

Because of these size constraints, extracorporeal bioartificial livers are an attractive approach for providing temporary liver function until an organ is available for transplantation or the patient's liver can regenerate. The patient's blood or plasma flows through an external circuit that provides contact with the hepatocytes. The hepatocytes can be presented to the patient's blood or plasma in a number of ways. The simplest is as a suspension of cells. However, this provides no immunoprotection and, like most mammalian cells, hepatocytes do better if they are attached to something. In some cases, the suspension of cells can be placed within a hemodialyzer and separated from the blood or plasma stream by a dialysis membrane. Although this approach provides immunoprotection, the low molecular weight cutoff of the dialysis membrane severely limits the molecular sizes that can be detoxified or synthesized. Other approaches involve perfusing the patient's blood or plasma through columns packed with microcarrier-attached hepatocytes, or hepatocytes encapsulated within polymeric beads. Hepatocytes attached to microcarriers, however, provide no immunoprotection. Hollow fiber units have also been proposed. In these systems, the hepatocytes are within the shell space and attach themselves to the outer surface of the hollow fibers. Blood or plasma flows through the lumen of the hollow fibers.

One major limitation of the membrane-based approaches discussed so far involves the significant mass transfer resistance created by the presence of the membrane. Solute transport through the membrane and into the space containing the hepatocytes is by diffusion only. This not only limits the supply of essential nutrients and oxygen to the hepatocytes, but also limits the detoxification and biosynthesis rates of the hepatocytes as well. To overcome these membrane diffusional limitations, some bioartificial livers incorporate perfusion of plasma through the region that contains the hepatocytes.

10.7.3 EXAMPLES OF EXTRACORPOREAL BIOARTIFICIAL LIVERS

Figure 10.20 illustrates one approach that provides three distinct compartments (Nyberg et al. 1992, 1993a, 1993b). The first compartment consists of a gel-like collagen matrix that traps the hepatocytes within the hollow fiber lumen. The second compartment consists of the space within the lumen of the hollow fiber that results after contraction of the original collagen suspension containing the hepatocytes. The third compartment is the shell space that surrounds the hollow fibers. The hollow fibers have an NMWCO of $100,000$ g mol^{-1} and immunoprotect the hepatocytes from the patient's immune system.

In operation, the patient's blood flows through the shell space. Substances to be detoxified readily diffuse through the hollow fiber membrane and into the collagen gel that contains the hepatocytes. Similarly, substances produced by the hepatocytes diffuse across the membrane and are carried away by the blood to the patient. Of particular note in this system is the presence of the space within the lumen that is formed after the gel contracts. This space provides an additional flow path that allows simultaneous perfusion of the hepatocytes with specific nutrients, growth factors, and hormones that are needed to maintain their differentiated function and viability. ECM type materials can also be incorporated into the gel that contains the hepatocytes. In this way, the hepatocytes can be presented with an optimal culture environment while performing their liver support functions.

---

* A microcarrier is a small particle about 200 μm in diameter, sometimes coated with ECM materials, that provides for cell attachment and formation of a confluent monolayer of cells.

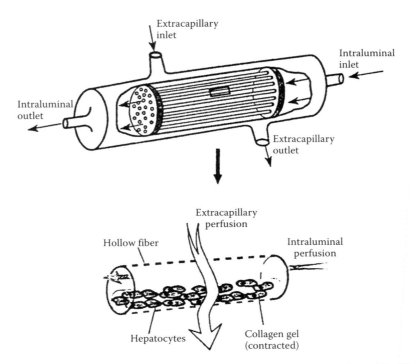

FIGURE 10.20 A three compartment hollow fiber bioartificial pancreas. (From Nyberg, S.L., Shatford, R.A., Peshwa, M.V., White, J.G., Cerra, F.B., and Hu, W-S., *Biotechnol. Bioeng.*, 41, 194–203, 1993b. With permission.)

In one particular test, this bioartificial liver was evaluated in vitro over a period of 7 days. Approximately 5×10^7 cells were loaded into the device. Hepatocyte function was evaluated on the basis of albumin production, oxygen consumption, and lidocaine clearance. It is important to point out that unlike the bioartificial pancreas, there is no single measure of differentiated hepatocyte function. However, albumin was steadily produced during the 7-day period at rates comparable to that seen in static cultures. Furthermore, oxygen consumption remained relatively constant following an initial decline over the first 2 days of culture. This initial decrease in the oxygen consumption, although possibly being related to cell death, was also considered to be a result of the increased metabolic demands on the hepatocytes as a result of the trauma during their isolation and placement in a new environment.

The cytochrome P-450 system (Hantsen 1998) is a major pathway for the biotransformation in the liver of many substances and is an important function that must be provided by a bioartificial liver. Lidocaine clearance can be used as a measure of oxidative metabolism provided by the cytochrome P-450 system. Lidocaine clearance in this bioartificial liver was relatively constant, and the presence of lidocaine metabolites demonstrated that lidocaine biotransformation was occurring. Electron micrographs after 7 days of operation showed the presence of differentiated viable hepatocytes.

Figure 10.21 illustrates an extracorporeal bioartificial liver system, known as HepatAssist, which consists of several integrated components (Rozga et al. 1993, 1994; Giorgio et al. 1993; Jauregui et al. 1997). A plasmapheresis unit is first used to form a plasma stream from arterial blood. This plasma stream feeds into a high-flow plasma recirculation loop that forms the core of the bioartificial liver support system. Within the recirculation loop, plasma first enters a column loaded with activated cellulose-coated charcoal. The activated charcoal column is used to enhance the detoxification capability of the overall system and to protect the hepatocytes from any toxic materials found in the patient's plasma. The detoxified plasma then enters a membrane oxygenator. This ensures an

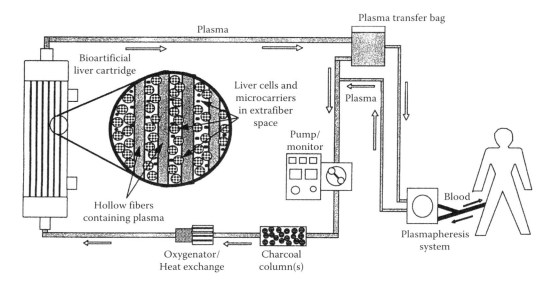

FIGURE 10.21 HepatAssist bioartificial liver. (From Jauregui, H.O., Mullon, C.J.P., and Solomon, B.A., *Principles of Tissue Engineering*, R.G. Landes., Austin, TX, 1997. With permission.)

adequate supply of oxygen to maintain the viability and function of the hepatocytes. The oxygenated plasma then flows through the hollow fiber module that contains the hepatocytes. The plasma flows through the hollow fiber lumens and the hepatocytes are in the shell space that surrounds the hollow fibers. Plasma exiting the hollow fiber cartridge is then recombined with the cellular components of the blood and returned to the patient's body.

Hepatocytes within the bioartificial liver cartridge are attached to collagen-coated microcarrier beads that occupy the shell space. About 6 billion porcine hepatocytes are used. The hollow fibers have large pore sizes (0.2 μm) that allow for significant fluid convection through the hollow fiber membrane as a result of the transfiber pressure drop. From the fiber entrance to its midpoint, there is a significant flow of plasma out of the fiber and into the shell space that contains the anchored hepatocytes. From the fiber midpoint to the fiber exit, this flow reverses itself and re-enters the fiber lumen. This is much like the Starling flow phenomena discussed in Chapter 3 for capillaries. This convective flow into the shell space passes freely between the microcarrier beads that contain the hepatocytes on their surfaces. This convective flow through the shell space provides for very efficient contacting of the plasma solutes and the hepatocytes, and far surpasses what is achievable by diffusion alone.

One limitation of the approach shown in Figure 10.21 is the lack of complete immunoisolation of the xenogeneic porcine hepatocytes. Plasma perfusion eliminates the cellular components of the immune system; however, antibodies and complement will readily flow across the membrane and into the shell space that contains the hepatocytes. However, the treatment time proposed for this device is relatively short and the patient is immunosuppressed in anticipation of a liver transplant. So, the tradeoff between a high transmembrane flow resulting in good mass transport, and the resultant permeability to the humoral components of the immune system, seems appropriate. Certainly longer-term use of this device will require improved methods of immunoisolation that do not compromise the mass transport properties of the hepatocyte bioreactor.

Figure 10.22 illustrates the extracorporeal liver assist device or ELAD (Sussman et al. 1992; Kelly and Sussman 1994). This is a hollow fiber device that uses a cloned human cell line in place of hepatocytes. The cell line is derived from an hepatoblastoma and is selected for liver-specific functions. These cells exhibit the ability to synthesize protein, urea, and glucose, and are capable of detoxifying substances via the cytochrome P-450 system. The use of a human cell line provides

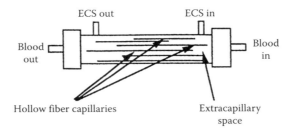

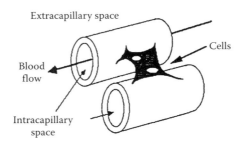

FIGURE 10.22 The extracorporeal liver assist device (ELAD). (From Sussman, N.L., et al., *Hepatology*, 16, 60–65, 1992. With permission.)

an unlimited supply of cells and minimizes the immune problems that may result from the use of xenogeneic hepatocytes. The ELAD is seeded with 2 g of cells and is placed in culture until 200 g of cells are obtained. This quantity of cells is capable of producing therapeutic levels of plasma proteins. For example, 5 g of albumin are produced each day, which is about one-half the normal daily production rate of a healthy adult human liver.

Blood flows from the patient's vein through a standard hemodialysis pump. The blood exiting the pump is mixed with heparin to prevent blood clots and then passes through the lumen of the hollow fibers of the ELAD device. A portion of the blood flowing through the hollow fibers forms a plasma ultrafiltrate that perfuses the shell space where it comes into direct contact with the cells. Ultrafiltration of the blood allows delivery of higher molecular weight solutes at rates significantly higher than that possible by diffusion alone. The ultrafiltrate then passes through a 0.45 μm filter before it is returned to the patient.

Evaluation of this device in dogs with fulminant hepatic failure provided rapid improvement in blood chemistry. There was also a rapid rise in plasma human albumin and α-fetoprotein levels following connection of the device. Factor V-dependent clotting times showed a corresponding rapid decrease. After 48 h of treatment, dogs with fulminant hepatic failure had sufficiently recovered that they could be disconnected from the device. Compared to control animals that die, the treated animals regenerated their own livers.

To overcome some of the transport limitations of the bioartificial liver devices described earlier, a novel approach has been developed that uses three separate hollow fiber membrane bundles (Sauer et al. 2001; Schmelzer et al. 2009). As shown in Figure 10.23, the hepatocytes are distributed throughout the complex three-dimensional extracapillary space formed by these three sets of hollow fiber bundles. Two sets of these hollow fiber bundles contain hydrophilic membranes with an NMWCO of about 300,000 g mol⁻¹. One end of each bundle is closed off, thereby making the plasma that enters the bioreactor in the first hollow fiber bundle flow into the extracapillary space before exiting the bioreactor through the second hollow fiber bundle. The hollow fibers in the third bundle are hydrophobic and are used to supply oxygen and remove carbon dioxide. The unique arrangement of

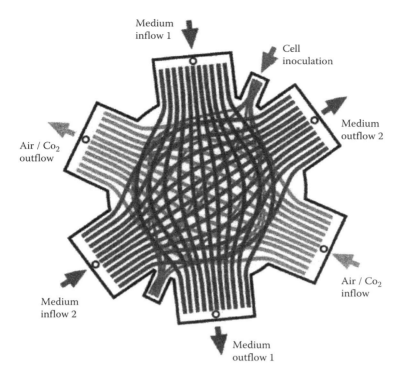

Medium
inflow 1

Cell
inoculation

Medium
outflow 2

Air / Co$_2$
outflow

Air / Co$_2$
inflow

Medium
inflow 2

Medium
outflow 1

FIGURE 10.23 A bioartificial liver comprising three bundles of hollow fiber membranes. (From Schmelzer, E., Mutig, K., Schrade, P., Bachmann, S., Gerlach, J.C., and Zeilinger, K., *Biotech. Bioeng.,* 103, 817–27, 2009. With permission.)

these three hollow fiber bundles provides a means for plasma perfusion of the extracapillary space containing the hepatocytes, while also providing for efficient transport of oxygen. This bioartificial liver containing human hepatocytes has been successfully used as a bridge to transplantation in patients suffering from liver failure.

10.8   THE BIOARTIFICIAL KIDNEY

Another potential area for the application of bioartificial organ technology is in the development of a bioartificial kidney (Ip and Aebischer 1989; Cieslinski and Humes 1994; Humes 1995, 1997; Fissell et al. 2001). Patients with end-stage renal disease can be treated for many years either by hemodialysis or by the newer technique of *continuous ambulatory peritoneal dialysis* (CAPD). However, these are not permanent solutions and these patients ultimately require a kidney transplant in order to survive. The mortality on an annualized basis for hemodialysis patients with end-stage renal disease is about 13%, whereas in age-matched patients that receive a kidney transplant, the annual mortality is 4% (Fissell et al. 2001). These data show that a normal or transplanted kidney provides physiologic functions that are not provided by hemodialysis.

The kidney provides not only a filtration and waste removal function, but also several other important functions that are important to the metabolic, immune, and endocrine systems of the body. For example, *erythropoietin* is released by specialized cells found in the kidney in response to hypoxia. Erythropoietin is a major stimulus for the production of red blood cells in the bone marrow. Uremic patients therefore suffer from *anemia*. Low blood pressure causes the release of *renin* from the juxtaglomerular cells that are found in the kidney. Renin initiates the formation of *angiotensin II*, a potent vasoconstrictor, which results in an increase in blood pressure. Angiotensin II also acts on the kidneys, decreasing the excretion of both salt and water. This expands the extracellular fluid volume with the result that the blood pressure is increased. The kidneys are also responsible for the conversion of vitamin D into a substance* that promotes the absorption of calcium from the intestine. Without this substance, the bones become severely weakened because of the loss of calcium. These other very important functions that are performed by the healthy kidney are therefore compromised as a result of kidney failure.

As shown in Figure 7.3, the functional unit of the kidney is called the *nephron*. Recall that it consists of two major components, the glomerulus and the renal tubule. The glomerulus is responsible primarily for the selective ultrafiltration of waste products from the blood. It must perform this waste removal function and, at the same time, retain essential blood components such as albumin. The glomerular filtrate that is formed then passes through the various segments of the renal tubule. The specialized segments of the renal tubule regulate the amount of urine that is formed and its final solute composition. Thus, the renal tubule cells have the ability to control both the fluid reabsorption rate and the transport rate of an individual solute. This is accomplished in such a manner as to maintain homeostasis with regard to the body's fluid volume and overall composition. Because of the chemical sensing and selective transport ability of the renal tubule, it is unlikely that this sophisticated function could ever be reproduced artificially. Accordingly, artificial kidneys or hemodialyzers will primarily function at the level of simply removing waste products by dialysis.

A bioartificial kidney has the potential of reproducing many of the homeostatic, regulatory, and endocrine functions that are performed by the healthy kidney. This would far surpass the filtration function of existing dialysis systems and allow blood purification to occur in a more physiologic fashion. A bioartificial kidney consists of two main components, an artificial glomerulus and a bioartificial renal tubule (Ip and Aebischer 1989; Fissell et al. 2001; Humes et al. 2004). The artificial glomerulus can be fabricated from polymeric membranes that have a high hydraulic conductance. The transmembrane pressure gradient can be controlled to provide the desired bulk flow rate of plasma across the membrane. This bulk flow, or ultrafiltration of the plasma, will also provide for significant convective transport of solutes across the membrane. The convective transport of all solutes will essentially be the same up to the molecular weight cutoff of the membrane. This is especially important for the larger-size solutes, where much more can be removed by convection than by simple diffusion. The artificial glomerular membranes can be configured as hollow fibers or perhaps spiral wound sheets. The major requirement being that the membrane arrangement is conducive to the production of a significant ultrafiltration flow.

Several problems still need to be addressed with regard to the fabrication of an artificial glomerulus. These include uncontrolled bleeding as a result of anticoagulation, a decreased ultrafiltration rate over time as a result of protein deposition in the membrane, and the challenge of developing a nonthrombogenic membrane. One exciting possibility for overcoming the clotting tendency of most polymeric materials is to cover the blood-contacting surfaces with a monolayer of autologous endothelial cells. Endothelial cells form the blood-contacting lining of the body's blood vessels and, in the absence of injury, are responsible for preventing the blood from clotting.

The bioartificial renal tubule would consist of viable renal epithelial cells supported within a tubular structure such as a hollow fiber membrane. These renal tubule cells can be obtained from the renal proximal tubule since it is this section of the renal tubule that reclaims the majority of the water, salt, glucose, amino acids, and other species that are filtered from the glomerular capillaries (Fissell et al. 2001).

Another possibility for creating a renal tubule is to use renal tubule stem cells (Cieslinski and Humes 1994; Humes and Cieslinski 1992). *Stem cells* are a class of highly proliferative cells that

---

* 1,25-dihydroxycholecalciferol.

give rise to cells that have a differentiated function. The resulting differentiated cells provide the ultimate physiologic structure and function found in the body's tissues and organs. Perhaps the best known class of stem cells are the *pluripotential hemopoietic stem cells* (PHSC) (Koller and Palsson 1993), which are responsible for the formation of all the differentiated cells found in blood. The differentiated cells arising from the PHSC consist of the red blood cells, white blood cells, and platelets. In many cases, the terminally differentiated state of the cell is incapable of reproducing. Therefore, the stem cell has the responsibility of producing and replacing those terminally differentiated cells that have become damaged or injured for a variety of reasons. For example, renal tubule stem cells would regenerate renal tubules following their injury. In the presence of the proper growth factors, renal tubule stem cells have the capability to form tubule-like structures in culture. The isolation and culture of human renal tubule stem cells could therefore be an important source of tissue for the formation of bioartificial renal tubules.

Figure 10.24 shows the extracorporeal hemoperfusion system described by Fissell et al. (2001; Humes et al. 2004). In experiments performed on nephrectomized dogs, venous blood is pumped through a standard hollow fiber dialyzer or hemofilter (artificial glomerulus). The ultrafiltrate that is formed on the shell side of the hemofilter is then pumped through the luminal space of a hollow fiber cartridge (artificial proximal tubule), which they called the renal assist device (RAD). The blood leaving the hemofilter flows through the extracapillary space of the RAD. Proximal tubule cells line the luminal surface of the hollow fibers in the RAD, providing for reabsorption into the blood of

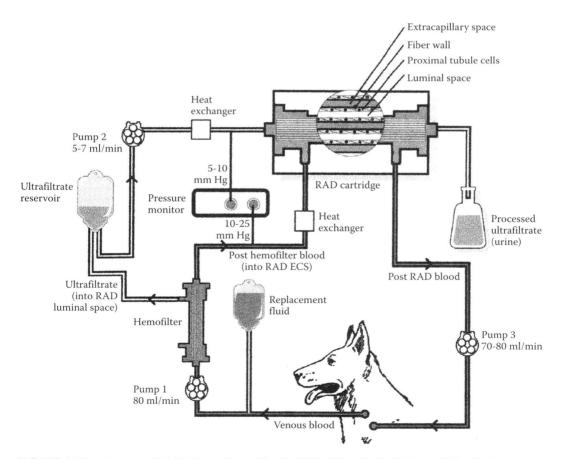

FIGURE 10.24   A bioartificial kidney. (From Fissell, W.H., Kimball, J., Mackay, S.M., Funke, A., and Humes, H.D., *Ann. NY Acad. Sci.*, 944, 284–95, 2001. With permission.)

about 50% of the hemofilter ultrafiltrate. The ultrafiltrate flow to the RAD cartridge is controlled by adjusting the pressure difference between the luminal and extracapillary sides of the RAD cartridge. The luminal effluent from the RAD cartridge is the urine and the blood leaving the extracapillary space of the RAD cartridge is returned to the body. The system shown in Figure 10.24 is currently being evaluated in clinical trials in humans with acute renal failure and multiorgan failure (Humes et al. 2004).

10.9   DESIGN CONSIDERATIONS FOR BIOARTIFICIAL ORGANS

The design of a bioartificial organ is based on the mass transfer principles outlined in Chapters 5 and 6 as well as the reactor design principles described in Sections 8.8.4 through 8.8.9. The design of a bioartificial organ must provide sufficient cell mass to affect the desired physiological response in the patient. Furthermore, the proposed design must allow for the efficient transport of key nutrients and the removal of any waste products. In most cases, the limiting nutrient is oxygen and the device design must be capable of transporting oxygen to all the cells (Davidson et al. 2010). This limitation on oxygen transport also requires intimate contact of the cells with a source of oxygen. For implantable devices, tissue engineering concepts can be used to enhance the growth of capillaries in the tissue region adjacent to the immunoisolated cells. For extracorporeal devices, either blood or plasma can flow through the device. In this case, the blood or plasma flow rates are typically in the range of 100–300 mL min^{-1}.

Example 10.4

A perfusion bioreactor is being designed to contain hepatocytes for the treatment of liver failure. The bioreactor will consist of thin rectangular microflow channels with the hepatocytes encased within a gel-like material that coats the surfaces of the walls of the microflow channel. Assume that the cell fraction of the hepatocytes in the gel-like layer is 0.35 and that the pO$_2$ of the perfusion fluid flowing through the microflow channel is 95 mmHg. The gel itself has a negligible effect on the diffusivity of oxygen and the hepatocytes consume oxygen at the rate of 35 µM sec^{-1}. What is the maximum thickness of the gel-like layer of hepatocytes? Assume that the value of H is 0.74 mmHg µM^{-1}.

Solution

The situation that is described is similar to the discussion found in Section 6.8 for oxygen transport in tissue-engineered constructs and bioartificial organs. In this case, we can use Equation 6.18 to describe the oxygen profile in the cell containing gel-like material that coats the walls of the flow channel. This equation can be solved for the maximum thickness (δ) assuming that the pO$_2$ is 0 mmHg at the interface between the gel layer and the wall of the flow channel. Equation 6.18 can then be written as

$$pO_2(\delta) = pO_2\big|_{x=0} + \left[\frac{\Gamma_{\text{metabolic}} H(1-\varepsilon)\delta^2}{2D_e}\right](-1).$$

Assuming the perfusion fluid has a negligible change in pO$_2$, and that there is no mass transfer resistance between the flowing perfusion fluid and the surface of the gel-like material, then pO$_2\big|_{x=0}$ = 95 mmHg. The void fraction (ε) of the layer of cells coating the channel walls is equal to $(1-0.35) = 0.65$. The diffusivity of oxygen in the perfusion fluid at 37°C using Equation 5.3 is 2.11×10^{-5} cm^2 sec^{-1}. This value can then be corrected for diffusion through the layer of cells using Equation 5.84, where $D_T = D_e$ and $D_0 = D$. This gives a value of $D_e = 1.17 \times 10^{-5}$ cm^2 sec^{-1}. Solving the previous equation results in a value of $\delta = 156$ µm.

PROBLEMS

1. Prepare a paper and presentation that describes a specific application of a bioartificial organ. Include a description of the clinical need, the approach used, results that have been obtained (in vitro, in animals, and the status of any clinical trials), the potential clinical impact, the potential market impact, safety issues, and additional development needs.

2. In Example 10.1, the hindered diffusion model of Bungay and Brenner (1973) was used to estimate the effective pore size of the hollow fiber immunoisolation membrane based on the measured solute permeabilities. This hindered diffusion model provides at best a semiquantitative description of the membrane's solute permeability. Discuss factors that may impact the ability of the hindered diffusion model to describe solute permeability data in immunoisolation membranes.

3. Baker et al. (1997) developed an immunoisolation membrane by incorporating a high water content polyvinyl alcohol (PVA) hydrogel into a thin disk-shaped microporous polyethersulfone (PES) filter. The PES filter had a thickness of 150 μm, a porosity of 81%, and a mean pore size of 0.2 μm. The hydrogel formed in the pores of the microporous membrane was found to contain 86% water by weight. The 0.2 μm pores of the microporous membrane should block the cellular components of the immune system and the PVA hydrogel through control of the PVA crosslinking can be tailored to control the passage of high molecular weight solutes. The following table summarizes the physical properties of the solutes that were used and their measured permeability in this composite membrane.

Solute	Molecular Weight	Solute Radius (nm)	$D \times 10^6$ (cm^2 sec^{-1}, 23°C)	$P_m \times 10^8$ (cm sec^{-1})
Glucose	180	0.36	6.38	40,000
Vitamin B12	1355	0.75	3.10	16,000
Inulin	5500	1.41	1.65	8000
Lysozyme	14,500	1.92	1.22	80
Myoglobin	16,890	1.9	1.21	70
Chymo-trypsinogen	25,000	2.24	1.04	60
Ovalbumin	45,000	2.73	0.851	25
Albumin	69,000	3.55	0.655	4
IgG	160,000	5.35	0.434	4.5
Blue dextran	2,000,000	9.26	0.251	0.5

Source: Data from Baker, A.R., et al., *Cell Transplant.,* 6, 585–95, 1997.

What is the effective pore size of the composite membrane based on the hindered diffusion model? An alternative model for solute diffusion through this composite membrane can be based on the methods outlined in Chapter 5 for gels. In this case, the solute moves through a random fibrous network consisting of the crosslinked polymer chains of the PVA hydrogel. Use Equations 5.86 and 5.87 to describe the solute permeability results.

4. It is estimated that as many as 500,000 patients with diabetes could benefit from a bioartificial pancreas. Assuming that each patient needs about 1 million islets, and that the islets need to be replaced each year, then about 1000 pancreatic islet isolations need to be performed each day to meet this demand for islets. Prepare a report that describes a process for the massive isolation of islets of Langerhans from the pancreas of large mammals, such as the dog or the pig. Be sure to describe any safety issues. The references should first be consulted for an assessment of the current state of islet isolation methods for a single pancreas.

5. Derive an expression for the steady state islet insulin secretion rate based on the Nomura et al. (1984) model.

6. Derive Equations 10.5 and 10.6 for the islet insulin release rate following a step change in the glucose concentration.

7. The following perifusion data were obtained by Lakey et al. (1995) for canine islets of Langerhans. The islets were challenged with glucose that was initially at 50 mg dL^{-1} followed by a prompt increase to 500 mg dL^{-1}. Use the Nomura et al. (1984) islet insulin release model to describe these data.

Time (min)	Islet Insulin Release Rate (μU islet^{-1} min^{-1})
0	0.10
5	0.83
8	0.75
28	0.61
50	0.68
60	0.50

8. Use the Sturis et al. (1991, 1995) pharmacokinetic model for glucose and insulin to describe an OGTT for a normal person. How do the normal person glucose and insulin profiles compare to those obtained in Example 10.3 using a bioartificial pancreas?

9. Repeat Example 10.3 assuming that the 70 kg patient receives an IVGTT. What is the k-value?

10. Repeat Example 10.3 assuming that human islets are used. Use the Nomura et al. islet model and the parameters obtained in Problem 7 for human islets.

11. Repeat Example 10.3, assuming that the islets are macroencapsulated in the hollow fibers described in Example 10.2. These hollow fiber membranes have a nominal external diameter of 950 μm. What is the entire length of fibers needed?

12. For the device described in Example 10.3, the oxygen permeability of the membrane is estimated to be 0.001 cm sec^{-1}. Based on the method outlined in Example 6.4, show that the maximum volume fraction of the islets is 0.35.

13. Repeat Example 10.3 assuming that the half-thickness of the device is 350 μm. The membrane oxygen permeability is estimated to be 0.001 cm sec^{-1}. How are the glucose and insulin profiles affected by a change in device thickness? What volume fraction of islets is allowable based on oxygen transport limitations?

References

Abbas, M. and V.P. Tyagi. 1987. Analysis of a hollow fiber artificial kidney performing simultaneous dialysis and ultrafiltration. *Chem. Eng. Sci.* 42:133–42.

Agrawal, C.M., G.G. Niederauer, and K.A. Athanasiou. 1995. Fabrication and characterization of PLA-PGA orthopedic implants. *Tiss. Eng.* 1:241–52.

Alberts, B., D. Bray, J. Lewis, M. Raff, K. Roberts, and J.D. Watson. 1989. *Molecular Biology of the Cell.* 2nd ed. New York: Garland Publishing.

Al-Hendy, A., G. Hortelano, G.S. Tannenbaum, and P.L. Chang. 1995. Correction of the growth defect in dwarf mice with nonautologous microencapsulated myoblasts – an alternate approach to somatic gene therapy. *Hum. Gene Ther.* 6:165–75.

Allen, J.W. and S.N. Bhatia. 2003. Formation of steady state oxygen gradients in vitro: Application to liver zonation. *Biotechnol. Bioeng.* 82:253–62.

Allen, J.W., T. Hassanein, and S.N. Bhatia. 2001. Advances in bioartificial liver devices. *Hepatology* 34:447–55.

Altman, J.J., D. Houlbert, P. Callard, P. McMillan, B.A. Solomon, J. Rosen, and P.M. Galletti. 1986. Longterm plasma glucose normalization in experimental diabetic rats with microencapsulated implants of benign human insulinomas. *Diabetes* 35:625–33.

Ameer, G.A., W. Harmon, R. Sasisekharan, and R. Langer. 1999a. Investigation of a whole blood fluidized bed Taylor–Couette flow device for enzymatic heparin neutralization. *Biotechnol. Bioeng.* 62:602–8.

Ameer, G.A., S. Raghavan, R. Sasisekharan, W. Harmon, C.L. Cooney, and R. Langer. 1999b. Regional heparinization via simultaneous separation and reaction in a novel Taylor–Couette flow device. *Biotechnol. Bioeng.* 63:618–24.

American Diabetes Association. 1993. Position statement. Implications of the diabetes control and complications trial. *Diabetes* 42:1555–58.

Anderson, J.L. and J.A. Quinn. 1974. Restricted transport in small pores. *Biophys. J.* 14:130–49.

Anderson, J.M. 1994. The extracellular matrix and biomaterials. In *Implantation Biology*, ed. R.S. Greco, 113–30. Boca Raton, FL: CRC Press.

Armer, T.A. and T.R. Hanley. 1986. Characterization of mass transfer in the hollow fiber artificial kidney. *Chem. Eng. Commun.* 47:49–71.

Arnold, F. and D.C. West. 1991. Angiogenesis in wound healing. *Pharmacol. Ther.* 52:407–22.

Arpaci, V.S. 1966. *Conduction Heat Transfer.* Reading, MA: Addison-Wesley.

Atkinson, M.A. and N.K. Maclaren. 1990. What causes diabetes? *Sci. Am.* July: 62–71.

Baker, A.R., R.L. Fournier, J.G. Sarver, J.L. Long, P.J. Goldblatt, J.M. Horner, and S.H. Selman. 1997. Evaluation of an immunoisolation membrane formed by incorporating a polyvinyl alcohol hydrogel within a microporous filter support. *Cell Transplant.* 6:585–95.

Barbee, J.H. and G.R. Cokelet. 1971a. Prediction of blood flow in tubes with diameters as small as 29μ. *Microvasc. Res.* 3:17–21.

———. 1971b. The Fahraeus effect. *Microvasc. Res.* 3:6–16.

Bass, L. and S. Keiding. 1988. Physiologically based models and strategic experiments in hepatic pharmacology. *Biochem. Pharmcol.* 37:1425–31.

Bawa, R., R.A. Siegel, B. Marasca, M. Karel, and R. Langer. 1985. An explanation for the controlled release of macromolecules from polymers. *J. Control. Release* 1:259–67.

Bazilevsky, A.V., K.G. Kornev, A.N. Rozhkov, and A.V. Neimark. 2003. Spontaneous absorption of viscous and viscoelastic fluids by capillaries and porous substrates. *J. Coll. Interface Sci.* 262:16–24.

Beck, R.E. and J.S. Schultz. 1970. Hindered diffusion in microporous membranes with known pore geometry. *Science* 170:1302–5.

Benjamin, E. and S. Leskowitz. 1991. *Immunology: A Short Course.* New York: Wiley-Liss.

Berger, M. and D. Rodbard. 1989. Computer simulation of plasma and glucose dynamics after subcutaneous insulin injection. *Diabetes Care* 12:725–36.

Bernstein, H., V.C. Yang, and R. Langer. 1987a. Distribution of heparinase covalently immobilized to agarose: Experimental and theoretical studies. *Biotechnol. Bioeng.* 30:196–207.

———. 1987b. Immobilized heparinase: In vitro reactor model. *Biotechnol. Bioeng.* 30:239–50.

Bird, R.B., W.E. Stewart, and E.N. Lightfoot. 2002. *Transport Phenomena.* 2nd ed. New York: John Wiley.

Braasch, D., and W. Jenett. 1968. Erythrocytenflexibilität, Hämokonzentration und Reibungswiderstand in Glascapillaren mit Durchmessern zwischen 6 bis 50 microns. *Pflugers Arch.* 302, 245–254.

Brinkman, H.C. 1947. A calculation of the viscous force exerted by a flowing fluid on a dense swarm of particles. *Appl. Sci. Res.* A1:27–34.

Brown, A.N., B-S. Kim, E. Alsberg, and D.J. Mooney. 2000. Combining chondrocytes and smooth muscle cells to engineer hybrid soft tissue constructs. *Tiss. Eng.* 6:297–305.

Bungay, P.M. and H. Brenner. 1973. The motion of a closely fitting sphere in a fluid filled tube. *Int. J. Multiph. Flow.* 1:25.

Burton, A.C. 1972. *Physiology and Biophysics of the Circulation.* 2nd ed. Chicago: Year Book Medical Publishers.

Bush, M.B., P.E. Papa Petros, and B.R. Barrett-Lennard. 1997. On the flow through the human urethra. *J. Biomech.* 30: 967–969.

Cai, Z., Z. Shi, G.M. O'Shea, and A.M. Sun. 1988. Microencapsulated hepatocytes for bioartificial liver support. *Artif. Org.* 12:388–93.

Caplan, A. 1991. Mesenchymal stem cells. *J. Orthopaedic Res.* 9:641–50.

Cappello, A., F. Grandi, C. Lamberti, and A. Santoro. 1994. Comparative evaluation of different methods to estimate urea distribution volume and generation rate. *Int. J. Artif. Org.* 17:322–30.

Carver, S.E. and C.A. Heath. 1999. Semi-continuous perfusion system for delivering intermittent physiological pressure to regenerating cartilage. *Tiss. Eng.* 5:1–11.

Castro, C.I. and J.C. Briceno. 2010. Perfluorocarbon-based oxygen carriers: Review of products and trials. *Artif. Org.* 34:622–34.

Catapano, G., A. Wodetzki, and U. Baurmeister. 1992. Blood flow outside regularly spaced hollow fibers: The future concept of membrane devices? *Int. J. Artif. Org.* 15:327–30.

Cavallaro, J.F., P.D. Kemp, and K.H. Kraus. 1994. Collagen fabrics as biomaterials. *Biotechnol. Bioeng.* 43:781–91.

Chang, P.L. 1997. Nonautologous gene therapy with implantable devices. *IEEE Eng. Med. Biol.* September/October: 145–50.

Chang, P.L., N. Shen, and A.J. Wescott. 1993. Delivery of recombinant gene products with microencapsulated cells in vivo. *Hum. Gene Ther.* 4:433–40.

Charm, S.E. and G.S. Kurland. 1974. *Blood Flow and Microcirculation.* New York: John Wiley.

Cheng, S-Y., J. Gross, and A. Sambanis. 2004. Hybrid pancreatic tissue substitute consisting of recombinant insulin-secreting cells and glucose responsive material. *Biotechnol. Bioeng.* 87:863–73.

Chick, W.L., J.J. Perna, V. Lauris, D. Low, P.M. Galletti, G. Panol, A.D. Whittemore, A.A. Like, C.K. Colton, and M.J. Lysaght. 1977. Artificial pancreas using living beta cells: Effects on glucose homeostasis in diabetic rats. *Science* 197:780–82.

Cieslinski, D.A. and H.D. Humes. 1994. Tissue engineering of a bioartificial kidney. *Biotechnol. Bioeng.* 43:678–81.

Cima, L.G., D.E. Ingber, J.P. Vacanti, and R. Langer. 1991a. Hepatocyte culture on biodegradable polymeric substrates. *Biotechnol. Bioeng.* 38:145–58.

Cima, L.G., J.P. Vacanti, C. Vacanti, D. Ingber, D. Mooney, and R. Langer. 1991b. Tissue engineering by cell transplantation using degradable polymer substrates. *J. Biomech. Eng.* 113:143–51.

Clark, W.R. and D. Gao. 2002. Properties of membranes used for hemodialysis therapy. *Semin. Dial.* 15:191–95.

Colton, C.K. 1995. Implantable biohybrid artificial organs. *Cell Transplant.* 4:415–36.

Colton, C.K. and E.S. Avgoustiniatos. 1991. Bioengineering in development of the hybrid artificial pancreas. *J. Biomech. Eng.* 113:152–70.

Colton, C.K. and E.G. Lowrie. 1981. Hemodialysis: Physical principles and technical considerations. In *The Kidney*, ed. B.M. Brenner and F.C. Rector, Jr. 2nd ed., 2425–89, Vol. II. Philadelphia: W.B. Saunders.

Cooney, D.O. 1976. *Biomedical Engineering Principles.* New York: Marcel Dekker.

Curry, F.E. and C.C. Michel. 1980. A fiber matrix model of capillary permeability. *Microvasc. Res.* 20:96–99.

Cussler, E.L. 1984. *Diffusion: Mass Transfer in Fluid Systems.* Cambridge: Cambridge University Press.

Davidson, A.J., M.J. Ellis, and J.B. Chaudhuri. 2010. A theoretical method to improve and optimize the design of bioartificial livers. *Biotechnol. Bioeng.* 106:980–88.

Dee, K.C., D.A. Puleo, and R. Bizios. 2002. *Tissue–Biomaterial Interactions.* Hoboken, NJ: John Wiley.

Deen, W.M. 1987. Hindered transport of large molecules in liquid-filled pores. *AIChE J.* 33:1409–25.

Dionne, K.E., C.K. Colton, and M.L. Yarmush. 1989. Effect of oxygen on isolated pancreatic tissue. *ASAIO Trans.* 35:739–41.

_____. 1991. A microperfusion system with environmental control for studying insulin secretion by pancreatic tissue. *Biotechnol. Prog.* 7:359–68.

Dionne, K.E., B.M. Cain, R.H. Li, W.J. Bell, E.J. Doherty, D.H. Rein, M.J. Lysaght, and F.T. Gentile. 1996. Transport characterization of membranes for immunoisolation. *Biomaterials* 17:257–66.

Douglas, J.A. and M.V. Sefton. 1990. The permeability of EUDRAGIT RL and HEMA-MMA microcapsules to glucose and inulin. *Biotechnol. Bioeng.* 36:653–64.

Durbin, R.P. 1960. Osmotic flow of water across permeable cellulose membranes. *J. Gen. Physiol.* 44:315–26.

Efrat, S. 1999. Genetically engineered pancreatic beta-cell lines for cell therapy of diabetes. *Ann. NY Acad. Sci.* 875:286–93.

Fissell, W.H., J. Kimball, S.M. Mackay, A. Funke, and H.D. Humes. 2001. The role of a bioengineered kidney in renal failure. *Ann. NY Acad. Sci.* 944:284–95.

Fogler, H.S. 2005. *Elements of Chemical Reaction Engineering.* 4th ed. New York: Prentice–Hall.

Folkman, J. 1985. Tumor angiogenesis. *Can. Res.* 43:175–203.

Folkman, J. and M. Klagsbrun. 1987. Angiogenic factors. *Science* 235:442–47.

Foster, T.H., D.F. Hartley, M.G. Nichols, and R. Hilf. 1993. Fluence rate effects in photodynamic therapy of multicell tumor spheroids. *Cancer Res.* 53:1249–1254.

Freeman, B. 1995. Osmosis. In *Encyclopedia of Applied Physics*, ed. G.L. Trigg. Vol. 13. Berlin: VCH Publishers.

Friedman, D.W., P.J. Orland, and R.S. Greco. 1994. Biomaterials: An historical perspective. In *Implantation Biology*, ed. R.S. Greco, Chapter 1. Boca Raton, FL: CRC Press.

Friedman, E.A. 1989. Toward a hybrid bioartificial pancreas. *Diabetes Care* 12: 415–19.

Fukui, Y., A. Funakubo, and T. Kawamura. 1994. Development of an intra blood circuit membrane oxygenator. *ASAIO J.* M732–34. 40: M732–734.

Fung, Y.C. 1993. *Biomechanics: Mechanical Properties of Living Tissues.* 2nd ed. New York: Springer-Verlag.

Gabriel, J.L., T.F. Miller, M.R. Wolfson, and T.H. Shaffer. 1996. Quantitative structure–activity relationships of perfluorinated hetero-hydrocarbons as a potential respiratory media. *ASAIO J.* 42:968–73.

Gaehtgens, P. 1980. Flow of blood through narrow capillaries: Rheological mechanisms determining capillary hematocrit and apparent viscosity. *Biorheology* 17:183–89.

Galban, C.J. and B.R. Locke. 1997. Analysis of cell growth in a polymer scaffold using a moving boundary approach. *Biotechnol. Bioeng.* 56:422–32.

Galletti, P.M. and C.K. Colton. 1995. Artificial lungs and blood-gas exchange devices. In *The Biomedical Engineering Handbook*, ed. J.D. Bronzino, 1879–1997. Boca Raton, FL: CRC Press.

Galletti, P.M., C.K. Colton, and M.J. Lysaght. 1995. Artificial kidney. In *The Biomedical Engineering Handbook*, ed. J.D. Bronzino, 1898–1922. Boca Raton, FL: CRC Press.

Galletti, P.M. and H.O. Jauregui. 1995. Liver support systems. In *The Biomedical Engineering Handbook*, ed. J.D. Bronzino, 1952–66. Boca Raton, FL: CRC Press.

Garred, L.J., B. Canaud, and P.C. Farrell. 1983. A simple kinetic model for assessing peritoneal mass transfer in chronic ambulatory peritoneal dialysis. *ASAIO J.* 6:131–37.

Gemmiti, C.V. and R.E. Guldberg. 2009. Shear stress magnitude and duration modulates matrix composition and tensile mechanical properties in engineered cartilaginous tissue. *Biotechnol. Bioeng.* 104:809–20.

Gharapetian, H., N.A. Davies, and A.M. Sun. 1986. Encapsulation of viable cells within polyacrylate membranes. *Biotechnol. Bioeng.* 28:1595–1600.

Gharapetian, H., M. Maleki, G.M. O'Shea, R.C. Carpenter, and A.M. Sun. 1987. Polyacrylate microcapsules for cell encapsulation: Effects of copolymer structure on membrane properties. *Biotechnol. Bioeng.* 29:775–79.

Gibaldi, M. and D. Perrier. 1982. *Pharmacokinetics.* 2nd ed. New York: Marcel Dekker.

Gibbon, J.H. 1954. Application of a mechanical heart and lung apparatus to cardiac surgery. *Minnesota Med.* 37:71.

Giorgio, T.D., A.D. Moscioni, J. Rozga, and A.A. Demetriou. 1993. Mass transfer in a hollow fiber device used as a bioartificial liver. *ASAIO J.* 39:886–92.

Goldstein, A.S., G. Zhu, G.E. Morris, R.K. Meszlenyi, and A.G. Mikos. 1999. Effect of osteoblastic culture conditions on the structure of poly(DL-lactic-co-glycolic acid) foam scaffolds. *Tiss. Eng.* 5:421–32.

Goorha, Y.K., P. Deb, T. Chatterjee, P.S. Dhot, and R.S. Prasad. 2003. Artificial blood. *Med. J. Armed Forces India* 59:45–50.

Gray, D.N. 1981. The status of olefin-SO$_2$ copolymers as biomaterials. In *Biomedical and Dental Applications of Polymers*, ed. C.G. Gebelein and F.F. Koblitz, 21–27. New York: Plenum.

———. 1984. Polymeric membranes for artificial lungs. ACS Symposium Series No. 256. In *Polymer Materials and Artificial Organs*, ed. C.G. Gebelein, 151–61. Washington, DC: American Chemical Society.

Gray, D.W.R. 2001. An overview of the immune system with specific reference to membrane encapsulation and islet transplantation. *Ann. NY Acad. Sci.* 944:226–39.

Griffith, B.P., R.L. Kormos, H.S. Borovetz, K. Litwak, J.F. Antaki, V.L. Poirier, and K.C. Butler. 2001. HeartMate II left ventricular assist system: From concept to first clinical use. *Ann. Thorac. Surg.* 71:S116–20.

Grodsky, G.M. 1972. A threshold distribution hypothesis for packet storage of insulin and its mathematical modeling. *J. Clin. Invest.* 51:2047–59.

Grundfest-Broniatowski, S.F., G. Tellioglu, K.S. Rosenthal, J. Kang, G. Erdodi, B. Yalcin, M. Cakmak, et al. 2009. A new bioartificial pancreas utilizing amphiphilic membranes for the immunoisolation of porcine islets. *ASAIO J.* 55:400–5.

Gu, Y.J., K. Inoue, S. Shinohara, R. Doi, M. Kogire, T. Aung, S. Sumi, et al. 1994. Xenotransplantation of bioartificial pancreas using a mesh-reinforced polyvinyl alcohol bag. *Cell Transplant.* 3:S19–S21.

Guyton, A.C. 1991. *Textbook of Medical Physiology.* 8th ed. Philadelphia: W.B. Saunders.

Guyton, J.R., R.O. Foster, J.S. Soeldner, M.H. Tan, C.B. Kahn, L. Koncz, and R.E. Gleason. 1978. A model of glucose insulin homeostasis in man that incorporates the heterogeneous fast pool theory of pancreatic insulin release. *Diabetes* 27:1027–42.

Guzman, A.M., R.A. Escobar, and C.H. Amon. 2005. Methodology for predicting oxygen transport on an intravenous membrane oxygenator combining computational and analytical models. *J. Biomech. Eng.* 127:1127–40.

Hachiya, H.L., P.A. Halban, and G.L. King. 1988. Intracellular pathways of insulin transport across vascular endothelial cells. *Am. J. Physiol.* 255 (*Cell Physiol.* 24): C459–64.

Hacker, M., K. Bachmann, and W. Messer. 2009. *Pharmacology: Principles and Practice.* Burlington, MA: Academic Press.

Hammar, H. 1993. Wound healing. *Int. J. Dermatol.* 32:6–15.

Hantsen, P.D. 1998. Understanding drug–drug interactions. *Sci. Med.* 5:16–25.

Hashimoto, S., M. Honda, M. Yamaguchi, M. Sekimoto, and Y. Tanaka. 1997. Pharmacokinetics of imipenem and cilastatin during continuous venovenous hemodialysis in patients who are critically ill. *ASAIO J.* 43:84–88.

Hayashi, H., K. Inoue, T. Aung, T. Tun, G. Yuanjun, W. Wenjing, S. Shinohara, et al. 1996. Application of a novel B cell line MIN6 to a mesh reinforced polyvinyl alcohol hydrogel tube and three layer agarose microcapsules: An in vitro study. *Cell Transplant.* 5:S65–S69.

Haynes, R.F. 1960. Physical basis of the dependence of blood viscosity on tube radius. *Am. J. Physiol.* 198:1193.

Hewitt, T.J., B.G. Hattler, and W.J. Federspiel. 1998. A mathematical model of gas exchange in an intravenous membrane oxygenator. *Ann. Biomed. Eng.* 26:166–78.

Heywood, H.K., D.L. Bader, and D.A. Lee. 2006. Rate of oxygen consumption by isolated articular chondrocytes is sensitive to medium glucose concentration. *J. Cell Physiol.* 206 (2): 402–10.

Hill, R.S., L.A. Martinson, S.K. Young, and R.W. Dudek. 1994. Macroporous devices for diabetes correction. In *Immunoisolation of Pancreatic Islets*, ed. R.P. Lanza and W.L. Chick, Chapter 12. Boulder, CO: R.G. Landes.

Holder, W.D., H.E. Gruber, W.D. Roland, A.L. Moore, C.R. Culberson, A.B. Loebsack, K.J.L. Burg, and D.J. Mooney. 1997. Increased vascularization and heterogeneity of vascular structures occurring in polyglycolide matrices containing aortic endothelial cells implanted in the rat. *Tiss. Eng.* 3:149–60.

Horwitz, A.E. 1997. Integrins and health. *Sci. Am.* 276:68–75.

Hubbell, J.A. 1997. Matrix effects. In *Principles of Tissue Engineering*, ed. R.P. Lanza, R. Langer, and W.L. Chick, 247–62. Austin, TX: R.G. Landes.

Hughes, M., A. Vassilakos, D.W. Andrews, G. Hortelano, J. Belmont, and P.L. Chang. 1994. Construction of a secretable adenosine deaminase for a novel approach to somatic gene therapy. *Hum. Gene Ther.* 5:1445–54.

Humes, H.D. 1995. Tissue engineering of the kidney. In *The Biomedical Engineering Handbook*, ed. J.D. Bronzino, 1807–24. Boca Raton, FL: CRC Press.

———. 1997. Application of cell and gene therapies in the tissue engineering of renal replacement devices. In *Principles of Tissue Engineering,* ed. R.P. Lanza, R. Langer, and W.L. Chick, 577–89. Boulder, CO: R.G. Landes.

Humes, H.D. and D.A. Cieslinski. 1992. Interaction between growth factors and retinoic acid in the induction of kidney tubulogenesis. *Exp. Cell Res.* 201:8–15.

Humes, H.D., W.F. Weitzel, R.H. Bartlett, F.C. Swaniker, E.P. Paganini, J.R. Luderer, and J. Sobota. 2004. Initial clinical results of the bioartificial kidney containing human cells in ICU patients with acute renal failure. *Kidney Int.* 66:1578–88.

Inoue, K., T. Fujisato, Y.J. Gu, K. Burczak, S. Sumi, M. Kogire, T. Tobe, et al. 1992. Experimental hybrid islet transplantation: Application of polyvinyl alcohol membrane for entrapment of islets. *Pancreas* 7:562–68.

Intaglietta, M. 1997. Whitaker Lecture 1996. Microcirculation, biomedical engineering, and artificial blood. *Ann. Biomed. Eng.* 25:593–603.

Intaglietta, M. and R.M. Winslow. 1995. Artificial blood. In *The Biomedical Engineering Handbook,* ed. J.D. Bronzino, 2011–24. Boca Raton, FL: CRC Press.

Ip, T.K. and P. Aebischer. 1989. Renal epithelial-cell-controlled solute transport across permeable membranes as the foundation for a bioartificial kidney. *Artif. Org.* 13:58–65.

Iwata, H., N. Morikawa, and Y. Ikada. 1996. Permeability of filters used for immunoisolation. *Tiss. Eng.* 2:289–98

Jacobs, M.A., R.H. Schreuder, K. Jap-A-Joe, J.J. Nauta, P.M. Andersen, and R.J. Heine. 1993. The pharmacodynamics and activity of intranasally administered insulin in healthy male volunteers. *Diabetes.* 42:1649–1655.

James, K. and J. Kohn. 1996. New biomaterials for tissue engineering. *MRS Bull.* 21:22–26.

Jarvik, R.K. 1981. The total artificial heart. *Sci. Am.* 244:74–80.

Jauregui, H.O., C.J.P. Mullon, and B.A. Solomon. 1997. Extracorporeal artificial liver support. In *Principles of Tissue Engineering*, ed. R.P. Lanza, R. Langer, and W.L. Chick, 463–79. Austin, TX: R.G. Landes.

Johnson, L.B., J. Aiken, D. Mooney, B.L. Schloo, L. Griffith-Cima, R. Langer, and J.P. Vacanti. 1994. The mesentery as a laminated vascular bed for hepatocyte transplantation. *Cell Transplant.* 3:273–81.

Joshi, A. and J. Raje. 2002. Sonicated transdermal drug transport. *J. Control. Release* 83:13–22.

Jurmann, M.J., S. Demertzis, H-J. Schaeffers, T. Wahlers, and A. Haverich. 1992. Intravascular oxygenation for advanced respiratory therapy. *ASAIO J.* 38:120–24.

Karoor, S., J. Molina, C.R. Buchmann, C. Colton, J.S. Logan, and L.W. Henderson. 2003. Immunoaffinity removal of xenoreactive antibodies using modified dialysis or microfiltration membranes. *Biotechnol. Bioeng.* 81:134–48.

Kaufmann, P.M., S. Heimrath, B.S. Kim, and D.J. Mooney. 1997. Highly porous polymer matrices as a three-dimensional culture system for hepatocytes. *Cell Transplant.* 6:463–68.

Kedem, O. and A. Katchalsky. 1958. Thermodynamic analysis of the permeability of biological membranes to nonelectrolytes. *Biochem. Biophys. Acta* 27: 229.

Kelly, J.H. and N.L. Sussman. 1994. The hepatix extracorporeal liver assist device in the treatment of fulminant hepatic failure. *ASAIO J.* 40:83–85.

Kitigawa, T., T. Yamaoka, R. Iwase, and A. Murakami. 2005. Three-dimensional cell seeding and growth in radial-flow perfusion bioreactor for in vitro tissue reconstruction. *Biotechnol. Bioeng.* 93:947–54.

Kobayashi, N., T. Okitsu, S. Nakaji, and N. Tanaka. 2003. Hybrid bioartificial liver: Establishing a reversibly immortalized human hepatocyte line and developing a bioartificial liver for practical use. *J. Artif. Org.* 6:236–44.

Kolff, W.J. 1947. *New Ways of Treating Uremia: The Artificial Kidney, Peritoneal Lavage, Intestinal Lavage.* London: J.A. Churchill.

Koller, M.R. and B.O. Palsson. 1993. Tissue engineering: Reconstitution of human hematopoiesis ex vivo. *Biotechnol. Bioeng.* 42:909–30.

Kornev, K.G. and A.V. Neimark. 2001. Spontaneous penetration of liquids into capillaries and porous membranes revisited. *J. Col. Int. Sci.* 235:101–13.

Koushanpour, E. 1976. *Renal Physiology.* Philadelphia: W.B. Saunders.

Kraegen, E.W., J.D. Young, E.P. George, and L. Lazarus. 1972. Oscillations in blood glucose and insulin after oral glucose. *Horm. Met. Res.* 4:409–13.

Krewson, C.E., M. Klarman, and W.M. Saltzman. 1995. Distribution of nerve growth factor following direct delivery to brain interstitium. *Brain Res.* 680:196–206.

Krogh, A. 1919. The number and distribution of capillaries in muscles with calculations of the oxygen pressure head necessary for supplying the tissue. *J. Physiol.* 52:409–15.

Lacy, P.E. 1995. Treating diabetes with transplanted cells. *Sci. Am.* 273:501–48.

Lacy, P.E., O.D. Hegre, A. Gerasimidi-Vazeou, F.T. Gentile, and K.E. Dionne. 1991. Maintenance of normoglycemia in diabetic mice by subcutaneous xenografts of encapsulated islets. *Science* 254:1782–84.

Lagerlund, T.D. and P.A. Low. 1993. Mathematical modeling of time dependent oxygen transport in rat peripheral nerve. *Comput. Biol. Med.* 23:29–47.

Lakey, J.R.T., R.V. Rajotte, G.L Warnock, and N.M. Kneteman. 1995. Human pancreas preservation prior to islet isolation. Cold ischemic tolerance. *Transplantation* 59:689–694.

Lakey, J.R.T., G.L. Warnock, Z. Ao, and R.V. Rajotte. 1996. Bulk cryopreservation of isolated islets of Langerhans. *Cell Transplant.* 5:395–404.

Langer, R. and J.P. Vacanti. 1993. Tissue engineering. *Science* 260:920–26.

Langer, R., J.P. Vacanti, C.A. Vacanti, A. Atala, L.E. Freed, and G. Vunjak-Novakovic. 1995. Tissue engineering: Biomedical applications. *Tiss. Eng.* 1:151–61.

Lanza, R.P., A.M. Beyer, and W.L. Chick. 1994. Xenogeneic humoral responses to islets transplanted in biohybrid diffusion chambers. *Transplantation* 57:1371–75.

Lanza, R.P., A.M. Beyer, J.E. Staruk, and W.L. Chick. 1993. Biohybrid artificial pancreas. *Transplantation* 56:1067–72.

Lanza, R.P., D.H. Butler, K.M. Borland, J.E. Staruk, D.L. Faustman, B.A. Solomon, T.E. Muller, et al. 1991. Xenotransplantation of canine, bovine, and porcine islets in diabetic rats without immunosuppression. *Proc. Natl. Acad. Sci. USA* 88:11100–4.

Lanza, R.P., D.H. Butler, K.M. Borland, J.M. Harvey, D.L. Faustman, B.A. Solomon, T.E. Muller, et al. 1992a. Successful xenotransplantation of a diffusion based biohybrid artificial pancreas: A study using canine, bovine, and porcine islets. *Transplant. Proc.* 24:669–71.

Lanza, R.P., K.M. Borland, P. Lodge, M. Carretta, S.J. Sullivan, T.E. Muller, B.A. Solomon, T. Maki, A.P. Monaco, and W.L. Chick. 1992c. Treatment of severely diabetic pancreatectomized dogs using a diffusion-based hybrid pancreas. *Diabetes* 41:886–89.

Lanza, R.P., K.M. Borland, J.E. Staruk, M.C. Appel, B.A. Solomon, and W.L. Chick. 1992b. Transplantation of encapsulated canine islets into spontaneously diabetic BB/Wor rats without immunosuppression. *Endocrinology* 131:637–42.

Lanza, R.P. and W.L. Chick. eds. 1994a. *Procurement of Pancreatic Islets.* Vol. 1. Boulder, CO: R.G. Landes.

_____. 1994b. *Immunoisolation of Pancreatic Islets.* Vol. 2. Boulder, CO: R.G. Landes.

_____. 1994c. *Immunomodulation of Pancreatic Islets.* Vol. 3. Boulder, CO: R.G. Landes.

Lanza, R.P., W.M. Kuhtreiber, and W.L. Chick. 1995. Encapsulation technologies. *Tiss. Eng.* 1:181–96.

Lanza, R.P., A.P. Monaco, S.J. Sullivan, and W.L. Chick. 1992d. The hybrid artificial pancreas: Diffusion and vascular devices. *Pancreatic Islet Cell Transplantation*, ed. C. Ricordi, Chapter 22. Boulder, CO: R.G. Landes.

Lanza, R.P., S.J. Sullivan, and W.L. Chick. 1992. Islet transplantation with immunoisolation. *Diabetes* 41:1503–10.

Lauffenburger, D.A. and J.J. Linderman. 1993. *Receptors*, 73–78. New York: Oxford University Press.

Lavin, A., C. Sung, A.L. Klibanov, and R. Langer. 1985. A potential treatment for neonatal jaundice. *Science* 230:543–45.

Leach, J.B., K.A. Bivens, C.W. Patrick, Jr., and C.E. Schmidt. 2003. Photocrosslinked hyaluronic acid hydrogels: Natural, biodegradable tissue engineering scaffolds. *Biotechnol. Bioeng.* 82:579–89.

Lee, C-J. and Y-L. Chang. 1988. Theoretical evaluation of the performance characteristics of a combined hemodialysis/hemoperfusion system – I. Single pass flow. *Can. J. Chem. Eng.* 66:263–70.

Lee, C-J., S-T. Hsu, and S-C. Hu. 1989. A one compartment single pore model for extracorporeal hemoperfusion. *Comput. Biol. Med.* 19:83–94.

Lee, W., V. Lee, S. Polio, P. Keegan, J.-H. Lee, K. Fischer, J.-K. Park, and S.-S. Yoo. 2010. On-demand three-dimensional freeform fabrication of multi-layered hydrogel scaffold with fluidic channels. *Biotechnol. Bioeng.* 105:1178–86.

Legallais, C., B. David, and E. Dore. 2001. Bioartificial livers (BAL): Current technological aspects and future developments. *J. Membr. Sci.* 181:81–95.

Lepeintre, J., H. Briandet, F. Moussy, D. Chicheportiche, S. Darquy, J. Rouchette, P. Imbaud, J.J. Duron, and G. Reach. 1990. Ex vivo evaluation in normal dogs of insulin released by a bioartificial pancreas containing isolated rat islets of Langerhans. *Artif. Org.* 14:20–27.

Levick, J.R. 1991. *An Introduction to Cardiovascular Physiology.* Boston, MA: Butterworths & Co.

Le Visage, C., S-H. Yang, L. Kadakia, A.N. Sieber, J.P. Kostuik, and K.W. Leong. 2006. Small intestinal submucosa as a potential bioscaffold for intervertebral disc regeneration. *Spine* 31:2423–30.

Lewis, R. 1997. New directions in research on blood substitutes. *Genetic Eng. News* 17:1.

Li, Z., T. Yipintsoi, and J.B. Bassingthwaighte. 1997. Nonlinear model for capillary-tissue oxygen transport and metabolism. *Ann. Biomed. Eng.* 25:604–19.

Lifson, N., K.G. Kramlinger, R.R. Mayrand, and E.J. Lender. 1980. Blood flow to the rabbit pancreas with special reference to the islets of Langerhans. *Gastroenterology* 79:466–73.

Liu, H., F.A. Ofosu, and P.L. Chang. 1993. Expression of human factor IX by microencapsulated recombinant fibroblasts. *Hum. Gene Ther.* 4:291–301.

Long, M.W. 1995. Tissue microenvironments. In *The Biomedical Engineering Handbook*, ed. J.D. Bronzino, 1692–1709. Boca Raton, FL: CRC Press.

Lu, L. and A.G. Mikos. 1996. The importance of new processing techniques in tissue engineering. *MRS Bull.* 21:28–32.

Lum, Z-P., I.T. Tai, M. Krestow, J. Norton, I. Vacek, and A.M. Sun. 1991. Prolonged reversal of diabetic state in NOD mice by xenografts of microencapsulated rat islets. *Diabetes* 40:1511–16.

Lysaght, M.J. and P.C. Farrell. 1989. Membrane phenomena and mass transfer kinetics in peritoneal dialysis. *J. Mem. Sci.* 44:5–33.

Lysaght, M.J. and J. Moran. 1995. Peritoneal dialysis equipment. In *The Biomedical Engineering Handbook*, ed. J.D. Bronzino, 1923–35. Boca Raton, FL: CRC Press.

Ma, P.X. and J-W. Choi. 2001. Biodegradable polymer scaffolds with well-defined interconnected spherical pore network. *Tiss. Eng.* 7:23–33.

Maki, T., J.P.A. Lodge, M. Carretta, H. Ohzato, K.M. Borland, S.J. Sullivan, and J. Staruk. 1993. Treatment of severe diabetes mellitus for more than one year using a vascularized hybrid artificial pancreas. *Transplantation* 55:713–18.

Maki, T., I. Otsu, J.J. O'Neil, K. Dunleavy, C.J.P. Mullon, B.A. Solomon, and A.P. Monaco. 1996. Treatment of diabetes by xenogeneic islets without immunosuppression. *Diabetes* 45:342–47.

Maki, T., C.S. Ubhi, H. Sanchez-Farpon, S.J. Sullivan, K. Borland, T.E. Muller, B.A. Solomon, W.L. Chick, and A.P. Monaco. 1991. Successful treatment of diabetes with the biohybrid artificial pancreas in dogs. *Transplantation* 51:43–51.

Makarewicz, A.J., L.F. Mockros, and R.W. Anderson. 1993. A pumping intravascular lung with active mixing. *ASAIO J.* 39:M466–69.

Malda, J., J. Rouwkema, D.E. Martens, E.P. le Comte, F.K. Kooy, J. Tramper, C.A. van Blitterswijk, and J. Riesle. 2004. Oxygen gradients in tissue engineered PEGT/PBT cartilaginous constructs: Measurement and modeling. *Biotechnol. Bioeng.* 86:9–18.

Mann, G.E., L.H. Smaje, and D.L. Yudilevich. 1979. Permeability of the fenestrated capillaries in the cat submandibular gland to lipid-insoluble molecules. *J. Physiol.* 297:335–54.

Marchant, R.E. and I-W. Wang. 1994. Physical and chemical aspects of biomaterials used in humans. In *Implantation Biology*, ed. R.S. Greco, 13–38. Boca Raton, FL: CRC Press.

Margiotta, M.S., L.D. Benton, and R.S. Greco. 1994. Integrins, adhesion molecules, and biomaterials. In *Implantation Biology*, ed. R.S. Greco, 149–63. Boca Raton, FL: CRC Press.

Martin, Y. and P. Vermette. 2005. Bioreactors for tissue mass culture: Design, characterization, and recent advances. *Biomaterials* 26:7481–7503.

Mathes, S.H., L. Wohlend, L. Uebersax, R. von Mentlen, D.S. Thoma, R.E. Jung, C. Gorlach, and U. Graf-Hausner. 2010. A bioreactor test system to mimic the biological and mechanical environment of oral soft tissues and to evaluate substitutes for connective tissue grafts. *Biotechnol. Bioeng.* 107:1029–39.

Matthew, H.W., S.O. Salley, W.D. Peterson, and M.D. Klein. 1993. Complex coacervate microcapsules for mammalian cell culture and artificial organ development. *Biotechnol. Prog.* 9:510–19.

Maxwell, J.C. 1873. *A Treatise on Electricity and Magnetism.* Vol. 1. Oxford: Clarendon Press.

Mazzei, D., M.A. Guzzardi, S. Giusti, and A. Ahluwalia. 2010. A low shear stress modular bioreactor for connected cell culture under high flow rates. *Biotechnol. Bioeng.* 106:127–37.

McCabe, W.L., J.C. Smith, and P. Harriott. 1985. *Unit Operations of Chemical Engineering.* 4th ed. New York: McGraw-Hill.

McNabb, M.E., R.V. Ebert, and K. McCusker. 1982. Plasma nicotine levels produced by chewing nicotine gum. *JAMA* 248:865–68.

Mehta, K. and J.J. Linderman. 2006. Model-based analysis and design of a microchannel reactor for tissue engineering. *Biotechnol. Bioeng.* 94:596–609.

Merrill, E.W., A.M. Benis, E.R. Gilliland, T.K. Sherwood, and E.W. Salzman. 1965. Pressure flow relations of human blood in hollow fibers at low flow rates. *J. Appl. Physiol.* 20:954–67.

Mikos, A.G., M.G. Papadaki, S. Kouvroukoglou, S.L. Ishaug, and R.C. Thomson. 1994. Mini-review: Islet transplantation to create a bioartificial pancreas. *Biotechnol. Bioeng.* 43:673–77.

Mikos, A.G., G. Sarakinos, M.D. Lyman, D.E. Ingber, J.P. Vacanti, and R. Langer. 1993. Pevascularization of porous biodegradable polymers. *Biotechnol. Bioeng.* 42:716–23.

Mirhashemi, S., K. Messmer, and M. Intaglietta. 1987. Tissue perfusion during normovolemic hemodilution investigated by a hydraulic model of the cardiovascular system. *Int. J. Microcirc. Clin. Exp.* 6:123–36.

Mitragotri, S. 2003. Modeling skin permeability to hydrophilic and hydrophobic solutes based on four permeation pathways. *J. Control. Release* 86:69–92.

Mooney, D.J., C. Breuer, K. McNamara, J.P. Vacanti, and R. Langer. 1995a. Fabricating tubular devices from polymers of lactic and glycolic acid for tissue engineering. *Tiss. Eng.* 1:107–17.

Mooney, D.J., P.M. Kaufmann, K. Sano, S.P. Schwendeman, K. Majahod, B. Schloo, J.P. Vacanti, and R. Langer. 1996. Localized delivery of epidermal growth factor improves the survival of transplanted hepatocytes. *Biotechnol. Bioeng.* 50:422–29.

Mooney, D.J., G. Organ, J.P. Vacanti, and R. Langer. 1994. Design and fabrication of biodegradable polymer devices to engineer tubular tissues. *Cell Transplant.* 3:203–10.

Mooney, D.J., S. Park, P.M. Kaufmann, K. Sano, K. McNamara, J.P. Vacanti, and R. Langer. 1995b. Biodegradable sponges for hepatocyte transplantation. *J. Biomed. Mat. Res.* 29:959–65.

Morgan, J.R., R.G. Tompkins, and M.L. Yarmush. 1994. Genetic engineering and therapeutics. In *Implantation Biology*, ed. R.S. Greco, 387–400. Boca Raton, FL: CRC Press.

Morgan, J.R. and M.L. Yarmush. 1997. Bioengineered skin substitutes. *Sci. Med.* 4:6–15.

Moussy, F., J, Rouchette, G. Reach, R. Cannon, and M.Y. Jaffrin. 1989. In vitro evaluation of a bioartificial pancreas under various hemodynamic conditions. *Artif. Org.* 13:109–15.

Nagase, K., F. Kohori, and K. Sakai. 2005. Oxygen transfer of a membrane oxygenator composed of crossed and parallel hollow fibers. *Biochem. Eng. J.* 24:105–13.

Nam, J., Y. Huang, S. Agarwal, and J. Lannutti. 2007. Improved cellular infiltration in electrospun fiber via engineered porosity. *Tiss. Eng.* 13:2249–57.

_____. 2008. Materials selection and residual solvent retention in biodegradable electrospun fibers. *J. Appl. Polym. Sci.* 107:1547–54.

Nam, J., B. Rath, T.J. Knobloch, J.J. Lannutti, and S. Agarwal. 2009. Novel electrospun scaffolds for the molecular analysis of chondrocytes under dynamic compression. *Tiss. Eng.* 15:513–23.

Naughton, G.K., W.R. Tolbert, and T.M. Grillot. 1995. Emerging developments in tissue engineering and cell technology. *Tiss. Eng.* 1:211–19.

Niro, R., J.P. Byers, R.L. Fournier, and K. Bachmann. 2003. Application of a convective-dispersion model to predict in vivo hepatic clearance from in vitro measurements using cryopreserved human hepatocytes. *Curr. Drug Metab.* 4:357–69.

Nomura, N., M. Schihiri, R. Kawamori, Y. Yamasaki, N. Iwama, and H. Abe. 1984. A mathematical insulin secretion model and its validation in isolated rat pancreatic islets perfusion. *Comput. Biomed. Res.* 17:570–79.

Nose, Y. 2001. Historical perspectives of hybrid hepatic assist devices. *Ann. NY Acad. Sci.* 944:18–34.

Notari, R.E. 1987. *Biopharmaceutics and Clinical Pharmacokinetics.* 4th ed. New York: Marcel Dekker.

Notkins, A.L. 1979. The causes of diabetes. *Sci. Am.* 241:62–73.

Nugent, L.J. and R.K. Jain. 1984a. Extravascular diffusion in normal and neoplastic tissues. *Cancer Res.* 44:238–44.

_____. 1984b. Pore and fiber-matrix models for diffusive transport in normal and neoplastic tissues. *Microvasc. Res.* 28:270–74.

Nyberg, S.L., H.J. Mann, R.P. Remmel, W-S. Hu, and F.B. Cerra. 1993a. Pharmacokinetic analysis verifies P450 function during in vitro and in vivo application of a bioartificial liver. *ASAIO J.* 39:M252–56.

Nyberg, S.L., J.L. Platt, K. Shirabe, W.D. Payne, W-S. Hu, and F.B. Cerra. 1992. Immunoprotection of xenocytes in a hollow fiber bioartificial liver. *ASAIO J.* 38: M463–67.

Nyberg, S.L., R.A. Shatford, M.V. Peshwa, J.G. White, F.B. Cerra, and W-S. Hu. 1993b. Evaluation of a hepatocyte entrapment hollow fiber bioreactor: A potential bioartificial liver. *Biotechnol. Bioeng.* 41:194–203.

Obradovic, B., J.H. Meldon, L.E. Freed, and G. Vunjak-Novakovic. 2000. Glycosaminoglycan deposition in engineered cartilage: Experiments and mathematical model. *AIChE J.* 46:1860–71.

Oerther, S., H. LeGall, E. Payan, F. Lapicque, N. Presle, P. Hubert, J. Dexheimer, P. Netter, and F. Lapicque. 1999. Hyaluronate-alginate gel as a novel biomaterial: Mechanical properties and formation mechanism. *Biotechnol. Bioeng.* 63:206–15.

Ogston, A.G. 1958. The spaces in a uniform random suspension of fibers. *Trans. Faraday Soc.* 54:1754–57.

Oie, S. and T.N. Tozer. 1979. Effect of altered plasma protein binding on apparent volume of distribution. *J. Pharm. Sci.* 68:1203–5.

Ojcius, D.M., C-C. Liu, and J.D. Young. 1998. Pore-forming proteins. *Sci. Med.* 5:44–53.

Owen, D.H., J.J. Peters, and D.F. Katz. 2000. Rheological properties of contraceptive gels. *Contraception* 62:321–26.

Pachence, J.M. and J. Kohn. 1997. Biodegradable polymers for tissue engineering. In *Principles of Tissue Engineering*, ed. R.P. Lanza, R. Langer, and W.L. Chick, 273–93. Austin, TX: R.G. Landes.

Palsson, B.O. and S. Bhatia. 2004. *Tissue Engineering*. Upper Saddle River, NJ: Pearson Prentice Hall.

Panchagnula, R. 1997. Transdermal delivery of drugs. *Indian J. Pharmacol.* 29:140–56.

Park, J.H., B.G. Chung, W.G. Lee, J. Kim, M.D. Brigham, J. Shim, S. Lee, et al. 2010. Microporous cell-laden hydrogels for engineered tissue constructs. *Biotechnol. Bioeng.* 106:138–48.

Parsons-Wingerter, P. and E.H. Sage. 1997. Regulation of cell behavior by extracellular proteins. In *Principles of Tissue Engineering*, ed. R.P. Lanza, R. Langer, and W.L. Chick, 111–31. Austin, TX: R.G. Landes.

Pathak, C.P., A.S. Sawhney, and J.A. Hubbell. 1992. Rapid photopolymerization of immunoprotective gels in contact with cells and tissues. *J. Am. Chem. Soc.* 114:8311–12.

Patzer, J.F. 2001. Advances in bioartificial liver assist devices. *Ann. NY Acad. Sci.* 944:320–33.

Perez, E.P., E.W. Merrill, D. Miller, and L.G. Cima. 1995. Corneal epithelial wound healing on bilayer composite hydrogels. *Tiss. Eng.* 1:263–77.

Petersen, E.F., R.G.S. Spencer, and E.W. McFarland. 2002. Microengineering neocartilage scaffolds. *Biotechnol. Bioeng.* 78:802–5.

Petley A., B. Macklin, A.G. Renwick, T.J. Wilkin. 1995. The pharmacokinetics of nicotinamide in humans and rodents. *Diabetes.* 44:152–155.

Petruzzo, P., L. Pibiri, M.A. DeGiudici, G. Basta, R. Calafiore, A. Falorni, P. Brunetti, and G. Brotzu. 1991. Xenotransplantation of microencapsulated pancreatic islets contained in a vascular prosthesis: Preliminary results. *Transplant Int.* 4:200–4.

Poling, B.E., J.M. Prausnitz, and J.P. O'Connell. 2001. *The Properties of Gases and Liquids.* 5th ed. New York: McGraw-Hill.

Potts, R.O. and R.H. Guy. 1992. Predicting skin permeability. *Pharm. Res.* 9:663–69.

Powers, M.J., K. Domansky, M.R. Kaazempur-Mofrad, A. Kalezi, A. Capitano, A. Upadhyaya, P. Kurzawski, et al. 2002. A microfabricated array bioreactor for perfused 3D liver culture. *Biotechnol. Bioeng.* 78:257–69.

Pratt, A.B., F.E. Weber, H.G. Schmoekel, R. Muller, and J.A. Hubbell. 2004. Synthetic extracellular matrices for in situ tissue engineering. *Biotechnol. Bioeng.* 86:27–36.

Prausnitz, J.M., R.N. Lichtenthaler, and E.G. de Azevedo. 1986. *Molecular Thermodynamics of Fluid-Phase Equilibria.* 2nd ed. Englewood Cliffs, NJ: Prentice-Hall.

Prausnitz, J.M. and F.H. Shair. 1961. A thermodynamic correlation of gas solubilities. *AIChE J.* 7:682.

Radisic, M., W. Deen, R. Langer, and G. Vunjak-Novakovic. 2005. Mathematical model of oxygen distribution in engineered cardiac tissue with parallel channel array perfused with culture medium containing oxygen carriers. *Am. J. Physiol. Heart Circ. Physiol.* 288:H1278–89.

Radisic, M., M. Euloth, L. Yang, R. Langer, L. Freed, and G. Vunjak-Novakovic. 2003. High density seeding of myocyte cells in cardiac tissue engineering. *Biotechnol. Bioeng.* 82:403–14.

Radisic, M., J. Malda, E. Epping, W. Geng, R. Langer, and G. Vunjak-Novakovic. 2006. Oxygen gradients correlate with cell density and cell viability in engineered cardiac tissue. *Biotechnol. Bioeng.* 93:332–42.

Radisic, M., L. Yang, J. Boublik, R.J. Cohen, R. Langer, L.E. Freed, and G. Vunjak-Novakovic. 2004. Medium perfusion enables engineering of compact and contractile cardiac tissue. *Am. J. Physiol. Heart Circ. Physiol.* 286:H507–16.

Ramachandran, P.A. and R.A. Mashelkar. 1980. Lumped parameter model for hemodialyzer with application to simulation of patient-artificial kidney system. *Med. Biol. Eng. Comput.* 18:179–88.

Reid, R.C., J.M. Praunsnitz, and T.K. Sherwood. 1977. *The Properties of Gases and Liquids.* 3rd ed. New York: McGraw-Hill.

Renkin, E.M. 1954. Filtration, diffusion, and molecular sieving through porous cellulose membranes. *J. Gen. Physiol.* 38:225.

_____. 1977. Multiple pathways of capillary permeability. *Circ. Res.* 41:735–43.

Renkin, E.M. and F.E. Curry. 1979. Transport of water and solutes across capillary endothelium. In *Membrane Transport in Biology,* ed. G. Giebisch, and D.C. Tosteson, Vol. 4, Chapter 1. New York: Springer-Verlag.

Replogle, R.L., H.J. Meiselman, and E.W. Merrill. 1967. Clinical implications of blood rheology studies. *Circulation* 36:148–60.

Richardson, P.D. 1987. Artificial lungs and oxygenation devices. In *Handbook of Bioengineering*, ed. R. Skalak and S. Chien, Chapter 27. New York: McGraw-Hill Co.

Ricordi, C. (ed.) 1992. *Pancreatic Islet Cell Transplantation.* Boulder, CO: R.G. Landes Co.

Ricordi, C., P.E. Lacy, E.H. Finke, B.J. Olack, and D.W. Scharp. 1988. Automated method for isolation of human pancreatic islets. *Diabetes* 37:413–20.

Ricordi, C., C. Socci, A. Davalli, C. Staudacher, A. Vertova, P. Baro, M. Freschi, et al. 1990b. Swine islet isolation and transplantation. *Horm. Metab. Res.* 25:26–30.

Ricordi, C., C. Socci, A. Davalli, A. Vertova, P. Baro, I. Sassi, S. Braghi, N. Giuzzi, G. Pozza, and V. DiCarlo. 1990a. Application of the automated method to islet isolation in swine. *Transplant. Proc.* 22:784–85.

Riley, M.R., F.J. Muzzio, and H.M. Buettner. 1995c. The effect of structure on diffusion and reaction in immobilized cell systems. *Chem. Eng. Sci.* 50:3357–67.

Riley, M.R., F.J. Muzzio, H.M. Buettner, and S.C. Reyes. 1994. Monte Carlo calculation of effective diffusivities in two- and three-dimensional heterogeneous materials of variable structure. *Phys. Rev. E.* 49:3500–3.

———. 1995a. Diffusion in heterogeneous media: Application to immobilized cell systems. *AIChE J.* 41:691–700.

———. 1995b. Monte Carlo simulation of diffusion and reaction in two-dimensional cell structures. *Biophys. J.* 68:1716–26.

———. 1996. A simple correlation for predicting effective diffusivities in immobilized cell systems. *Biotechnol. Bioeng.* 49:223–27.

Rosen, S.L. 1993. *Fundamental Principles of Polymeric Materials.* 2nd ed. New York: John Wiley.

Rosenberg, G. 1995. Artificial heart and circulatory assist devices. In *The Biomedical Engineering Handbook*, ed. J.D. Bronzino, 1839–46. Boca Raton, FL: CRC Press.

Rowland, M. and T.N. Tozer. 1995. *Clinical Pharmacokinetics: Concepts and Applications.* Philadelphia: Lippincott, Williams & Wilkins.

Rozga, J., E. Morsiani, E. LePage, A.D. Moscioni, T. Giorgio, and A.A. Demetriou. 1994. Isolated hepatocytes in a bioartificial liver: A single group view and experience. *Biotechnol. Bioeng.* 43:645–53.

Rozga, J., F. Williams, M-S. Ro, D.F. Neuzil, T.D. Giorgio, G. Backfisch, A.D. Moscioni, R, Hakim, and A.A. Demetriou. 1993. Development of a bioartificial liver: Properties and function of a hollow fiber module inoculated with liver cells. *Hepatology* 17:258–65.

Rubin, E. and J.L. Farber. 1994. *Pathology.* 2nd ed. Philadelphia: J.B. Lippincott.

Saltzman, W.M. 1997. Cell interactions with polymers. In *Principles of Tissue Engineering*, ed. R.P. Lanza, R. Langer, and W.L. Chick, 225–46. Boulder, CO: R.G. Landes.

Sandler, S.I. 1989. *Chemical Engineering Thermodynamics.* 2nd ed. New York: John Wiley.

Sarver, J.G. 1994. Development of a tracer technique that utilizes a physiological pharmacokinetic model to quantitatively assess the change in mass transfer rates associated with vascular growth and the application of this technique to the evaluation of a novel bioartificial organoid. PhD dissertation. The University of Toledo.

Sarver, J.G. and R.L. Fournier. 1990. Numerical investigation of a novel spiral wound membrane sandwich design for an implantable bioartificial pancreas. *Comput. Biol. Med.* 20:105–19.

Sarver, J.G., R.L. Fournier, P.J. Goldblatt, T.L. Phares, S.E. Mertz, A.R. Baker, R.J. Mellon, J.M. Horner, and S.H. Selman. 1995. Tracer technique to measure in vivo chemical transport rates within an implantable cell transplantation device. *Cell Transplant.* 4:201–17.

Sauer, I.M., N. Obermeyer, D. Kardassis, T. Theruvath, and J.C. Gerlach. 2001. Development of a hybrid liver support system. *Ann. NY Acad. Sci.* 944:308–19.

Scharp, D.W., C.J. Swanson, B.J. Olack, P.P. Latta, O.D. Hegre, E.J. Doherty, F.T. Gentile, K.S. Flavin, M.F. Ansara, and P.E. Lacy. 1994. Protection of encapsulated human islets implanted without immunosuppression in patients with type I or type II diabetes and in nondiabetic control subjects. *Diabetes* 43:1167–70.

Schlichting, H. 1979. *Boundary Layer Theory.* New York: McGraw-Hill.

Schmelzer, E., K. Mutig, P. Schrade, S. Bachmann, J.C. Gerlach, and K. Zeilinger. 2009. Effect of human patient plasma ex vivo treatment of gene expression and progenitor cell activation of primary human liver cells in multi-compartment 3D perfusion bioreactors for extra-corporeal liver support. *Biotechnol. Bioeng.* 103:817–27.

Schrezenmeir, J., J. Kirchgessner, L. Gero, L.A. Kunz, J. Beyer, and W. Mueller-Klieser. 1994. Effect of microencapsulation on oxygen distribution in islets organs. *Transplantation* 57:1308–14.

Scott, M.D., K.L. Murad, F. Koumpouras, M. Talbot, and J.W. Eaton. 1997. Chemical camouflage of antigenic determinants: Stealth erythrocytes. *Proc. Natl. Acad. Sci. USA* 94:7566–71.

Secomb, T.W., R. Hsu, M.W. Dewhirst, B. Klitzman, and J.F. Gross. 1993. Analysis of oxygen transport to tumor tissue by microvascular networks. *Int. J. Radiat. Oncol. Biol. Phys.* 25:481–89.

Sefton, M.V., R.M. Dawson, R.L. Broughton, J. Blysniuk, and M.E. Sugamori. 1987. Microencapsulation of mammalian cells in a water insoluble polyacrylate by coextrusion and interfacial precipitation. *Biotechnol. Bioeng.* 29:1135–1143.

Selman, S.L., M. Kreimer-Birnbaum, J.E. Klaunig, P.J. Goldblatt, R.W. Keck, and S.L. Britton. 1984. Blood flow in transplantable bladder tumors treated with hematoporphyrin derivative and light. *Cancer Res.* 44:1924–27.

Shah, N. and A. Mehra. 1996. Modeling of oxygen uptake in perfluorocarbon emulsions. *ASAIO J.* 42:181–89.

Shaikh, F.M., T.P. O'Brien, A. Callanan, E.G. Kavanagh, P.E. Burke, P.A. Grace, and T.M. McGloughlin. 2010. New pulsatile hydrostatic pressure bioreactor for vascular tissue-engineered constructs. *Artif. Org.* 34:153–58.

Shaoting, W., Z. Fengbao, M. Tengxiang, and G. Hanqing. 1990. Investigation on patient-artificial kidney system using compartment models. *Chem. Eng. Sci.* 45:2943–48.

Shuler, M. and F. Kargi. 2001. *Bioprocess Engineering: Basic Concepts.* 2nd ed. Upper Saddle River, NJ: Prentice Hall.

Shiroki, R., T. Mohanakumar, and D.W. Scharp. 1995. Analysis of the serological and cellular sensitization induced by encapsulated human islets transplantation in type I and type II diabetes patients. *Cell Transplant.* 4:535–38.

Smith, B., J.G. Sarver, and R.L. Fournier. 1991. A comparison of islet transplantation and subcutaneous insulin injections for the treatment of diabetes mellitus. *Comput. Biol. Med.* 21:417–27.

Soon-Shiong, P. 1994. Encapsulated islet transplantation: Pathway to human clinical trials. In *Immunoisolation of Pancreatic Islets*, ed. R.P. Lanza and W.L. Chick, Chapter 6, Vol. 3. Boulder, CO: R.G. Landes.

Soon-Shiong, P., E. Feldman, R. Nelson, R. Heintz, N. Merideth, P. Sandford, T. Zheng, and J. Komtebedde. 1992b. *Trans. Proc.* 24:2946–47.

Soon-Shiong, P., E. Feldman, R. Nelson, R. Heintz, Q. Yao, O. Smidsrod, and P. Sandford. 1993. Longterm reversal of diabetes by the injection of immunoprotected islets. *Proc. Natl. Acad. Sci. USA* 90:5843–47.

Soon-Shiong, P., E. Feldman, R. Nelson, J. Komtebedde, O. Smidsrod, G. Skjak-Braek, T. Espevik, R. Heintz, and M. Lee. 1992a. Successful reversal of spontaneous diabetes in dogs by intraperitoneal microencapsulated islets. *Transplantation* 54:769–74.

Sorenson, J.T., C.K. Colton, R.S. Hillman, and J.S. Soeldner. 1982. Use of a pharmacokinetic model of glucose homeostasis for assessment of performance requirements for improved insulin therapies. *Diabetes Care* 5:148–57.

Spotnitz, H.M. 1987. Circulatory assist devices. In *Handbook of Bioengineering*, ed. R. Skalak and S. Chien, 38.1–38.18. New York: McGraw-Hill.

Staverman, A.J. 1948. Non-equilibrium thermodynamics of membrane processes. *Trans. Faraday Soc.* 48:176–85.

Storrs, R., R. Dorian, S.R. King, J. Lakey, and H. Rilo. 2001. Preclinical development of the islet sheet. *Ann. NY Acad. Sci.* 944:252–66.

Stryer, L. 1988. *Biochemistry.* New York: W.H. Freeman and Company.

Sturis, J., C. Knudsen, N.M. O'Meara, J.S. Thomsen, E. Mosekilde, E. Van Cauter, and K.S. Polonsky. 1995. Phase-locking regions in a forced model of slow insulin and glucose oscillations. In *Dynamical Disease: Mathematical Analysis of Human Illness*, ed. J. Belair, L. Glass, U. An der Heiden, and J. Milton. New York, NY: AIP Press.

Sturis, J., K.S. Polonsky, E. Mosekilde, and E. Van Cauter. 1991. Computer model for mechanisms underlying ultradian oscillations of insulin and glucose. *Am. J. Physiol.* 260:E801–E809.

Sugamori, M.E. and M.V. Sefton. 1989. Microencapsulation of pancreatic islets in a water insoluble polyacrylate. *Trans. Am. Soc. Artif. Org.* 35:791–99.

Sullivan, S.J., T. Maki, K.M. Borland, M.D. Mahoney, B.A. Solomon, T.E. Muller, A.P. Monaco, and W.L. Chick. 1991. Biohybrid artificial pancreas: Longterm implantation studies in diabetic, pancreatectomized dogs. *Science* 252:718–21.

Sun, A.M., W. Parisius, G.M. Healy, I. Vacek, and H.G. Macmorine. 1977. The use, in diabetic rats and monkeys, of artificial capillary units containing cultured islets of Langerhans. *Diabetes* 26:1136–39.

Sun, Y., X. Ma, D. Zhou, I. Vacek, and A.M. Sun. 1996. Normalization of diabetes in spontaneously diabetic cynomologus monkeys by xenografts of microencapsulated porcine islets without immunosuppression. *J. Clin. Invest.* 98:1417–22.

Sung, C., A. Lavin, A.M. Klibanov, and R. Langer. 1986. An immobilized enzyme reactor for the detoxification of bilirubin. *Biotechnol. Bioeng.* 28:1531–39.

Sussman, N.L., M.G. Chong, T. Koussayer, D. He, T.A. Shang, H.H. Whisennand, and J.H. Kelly. 1992. Reversal of fulminant hepatic failure using an extracorporeal liver assist device. *Hepatology* 16:60–65.

Takeda, T., S. Murphy, S. Uyama, G.M. Organ, B.L. Schloo, and J.P. Vacanti. 1995. Hepatocyte transplantation in swine using prevascularized polyvinyl alcohol sponges. *Tiss. Eng.* 1:253–62.

Tatake, R.J., M.M. O'Neill, C.A. Kennedy, V.D. Reale, J.D. Runyan, K.-A.D. Monaco, K. Yu, W.R. Osborne, R.W. Barton, and R.D. Schneiderman. 2007. Glucose-regulated insulin production from genetically engineered human non-beta cells. *Life Sci.* 81:1346–54.

Teo, W.E. and S. Ramakrishna. 2006. A review on electrospinning design and nanofibre assemblies. *Nanotechnology* 17:R89–R106.

Thomas, L.C. 1992. *Heat Transfer*. New York: Prentice-Hall.

Thompson, J.A., K.D. Anderson, J.M. DiPietro, J.A. Zwiebel, M. Zametta, W.F. Anderson, and T. Maciag. 1988. Site-directed neovessel formation in vivo. *Science* 241:1349–52.

Thompson, J.A., C.C. Haudenschild, K.D. Anderson, J.M. DiPietro, W.F. Anderson, and T. Maciag. 1989. Heparin-binding growth factor 1 induces the formation of organoid neovascular structures in vivo. *Proc. Natl. Acad. Sci. USA* 86:7928–32.

Thomson, R.C., M.J. Yaszemski, and A.G. Mikos. 1997. Polymer scaffold processing. In *Principles of Tissue Engineering*, ed. R.P. Lanza, R. Langer, and W.L. Chick, 263–72. Austin, TX: R.G. Landes.

Tilles, A.W., H. Baskaaran, P. Roy, M.L. Yarmush, and M. Toner. 2001. Effects of oxygenation and flow on the viability and function of rat hepatocytes cocultured in a microchannel flat-plate bioreactor. *Biotechnol. Bioeng.* 74:379–89.

Tinoco, I., K. Sauer, J.C. Wang, and J.D. Puglisi. 2001. *Physical Chemistry: Principles and Applications in Biological Sciences*. 4th ed. New York: Prentice-Hall.

Tong, J. and J.L. Anderson. 1996. Partitioning and diffusion of proteins and linear polymers in polyacrylamide gels. *Biophys. J.* 70:1505–13.

Tozer, T.N. and M. Rowland. 2006. *Introduction to Pharmacokinetics and Pharmacodynamics*. Baltimore, MD: Lippincott, Williams & Wilkins.

Tsai, A.G., K.-E. Arfors, and M. Intaglietta. 1990. Analysis of oxygen transport to tissue during extreme hemodilution. *Adv. Exp. Med. Biol.* 277:881–87.

Tun, T., H. Hayashi, T. Aung, Y-J. Gu, R. Doi, H. Kaji, Y. Echigo, et al. 1996. A newly developed three layer agarose microcapsule for a promising biohybrid artificial pancreas: Rat to mouse xenotransplantation. *Cell Transplant.* 5:S59–S63.

Tyn, M.T., and T.W. Gusek. 1990. Prediction of diffusion coefficients of proteins. *Biotechnol. Bioeng.* 35:327–38.

Van Blitterswijk, C., P. Thomsen, J. Hubbel, and R. Cancedda. (ed.) 2008. *Tissue Engineering*. London: Elsevier.

Vaslef, S.N., L.F. Mockros, and R.W. Anderson. 1989. Development of an intravascular lung assist device. *ASAIO Trans.* 35:660–64.

Vaslef, S.N., L.F. Mockros, R.W. Anderson, and R.J. Leonard. 1994. Use of a mathematical model to predict oxygen transfer rates in hollow fiber membrane oxygenators. *ASAIO J.* 40:990–96.

Verma, R.K., D. Murali Krishna, and S. Garg. 2002. Formulation aspects in the development of osmotically controlled oral drug delivery systems. *J. Control. Release* 79:7–27.

Wake, M.C., C.W. Patrick, Jr., and A.G. Mikos. 1994. Pore morphology effects on the fibrovascular tissue growth in porous polymer substrates. *Cell Transplant.* 3:339–43.

Warnock, W.L., D. Ellis, R.V. Rajotte, I. Dawidson, S. Baekkeskov, and J. Egebjerg. 1988. Studies of the isolation and viability of human islets of Langerhans. *Transplantation* 45:957–63.

Warnock, W.L., D.K. Ellis, M. Cattral, D. Untch, N.M. Kneteman, and R.V. Rajotte. 1989. Viable purified islets of Langerhans from collagenase perfused human pancreas. *Diabetes* 38:136–39.

Warnock, W.L. and R.V. Rajotte. 1988. Critical mass of purified islets that induce normoglycemia after implantation into dogs. *Diabetes* 37:467–70.

Welling, P.G. 1986. *Pharmacokinetics: Processes and Mathematics*, ACS monograph 185. Washington, DC: American Chemical Society.

Welsh, E.R. and D.A. Tirrell. 2000. Engineering the extracellular matrix: A novel approach to polymeric biomaterials. I. Control of the physical properties of artificial protein matrices designed to support adhesion of vascular endothelial cells. *Biomacromolecules* 1:23–30.

Wickramasinghe, S.R., J.D. Garcia, and B. Han. 2002a. Mass and momentum transfer in hollow fiber blood oxygenators. *J. Membr. Sci.* 208:247–56.

Wickramasinghe, S.R. and B. Han. 2005. Designing microporous hollow fibre blood oxygenators. *Trans. IChemE, Part A.* 83:256–67.

Wickramasinghe, S.R., B. Han, J.D. Garcia, and P. Specht. 2005. Microporous membrane blood oxygenators. *AIChE J.* 51:656–70.

Wickramasinghe, S.R., C.M. Kahr, and B. Han. 2002b. Mass transfer in blood oxygenators using blood analogue fluids. *Biotechnol. Prog.* 18:867–73.

Winslow, R.M. 1997. Blood substitutes. *Sci. Med.* 4:54–63.

Wright, J.C., S.T. Leonard, C.L. Stevenson, J.C. Beck, G. Chen, R.M. Jao, P.A. Johnson, J. Leonard, and R.J. Skowronski. 2001. An in vivo/in vitro comparison with a leuprolide somotic implant for the treatment of prostate cancer. *J. Control. Release* 75:1–10.

Yang, M-C. and E.L. Cussler. 1986. Designing hollow-fiber contactors. *AIChE J.* 32:1910–16.

Yang, M.B., J.P. Vacanti, and D.E. Ingber. 1994. Hollow fibers for hepatocyte encapsulation and transplantation: Studies of survival and function in rats. *Cell Transplant.* 3:373–85.

Yarmush, M.L., J.C.Y. Dunn, and R.G. Tompkins. 1992. Assessment of artificial liver support technology. *Cell Transplant.* 1:323–41.

Yildirim, E.D., R. Besunder, S. Guceri, F. Allen, and W. Sun. 2008. Fabrication and plasma treatment of 3D polycaprolactone tissue scaffolds for enhanced cellular function. *Virtual Phys. Prototyping* 3:199–207.

Yildirim, E.D., R. Besunder, D. Pappas, F. Allen, S. Guceri, and W. Sun. 2010. Accelerated differentiation of osteoblast cells on polycaprolactone scaffolds driven by a combined effect of protein coating and plasma modification. *Biofabrication* 2:1–12.

Zalzman, M., S. Gupta, R.K. Giri, I. Berkovich, B.S. Sappal, O. Karnieli, M.A. Zern, N. Fleischer, and S. Efrat. 2003. Reversal of hyperglycemia in mice by using human expandable insulin-producing cells differentiated from fetal liver progenitor cells. *Proc. Natl. Acad. Sci. USA* 100:7253–58.

Zekorn, T., U. Siebers, R.G. Bretzel, M. Renardy, H. Planck, P. Zschocke, and K. Federlin. 1990. Protection of islets of Langerhans from interleukin-1 toxicity by artificial membranes. *Transplantation* 50:391–94.

Zelman, A. 1987. The artificial kidney. In *Handbook of Bioengineering*, ed. R. Skalak and S. Chien, Chapter 39. New York: McGraw-Hill.

Zhao, F. and T. Ma. 2005. Perfusion bioreactor system for human mesenchymal stem cell tissue engineering: Dynamic cell seeding and construct development. *Biotechnol. Bioeng.* 91:482–93.

Zhmud, B.V., F. Tiberg, and K. Hallstensson. 2000. Dynamics of capillary rise. *J. Col. Int. Sci.* 228:263–69.

Zielinski, B.A., M.B. Goddard, and M.J. Lysaght. 1997. Immunoisolation. In *Principles of Tissue Engineering*, ed. R.P. Lanza, R. Langer, and W.L. Chick, 323–32. Austin, TX: R.G. Landes.

Zydney, A.L. 1995. Therapeutic apheresis and blood fractionation. In *The Biomedical Engineering Handbook*, ed. J.D. Bronzino, 1936–51. Boca Raton, FL: CRC Press.

Zydney, A.L. and C.K. Colton. 1986. A concentration polarization model for the filtrate flux in cross-flow microfiltration of particulate suspensions. *Chem. Eng. Comm.* 47:1–21.

Index